T0250471

KEY ENGINEERING MATERIALS

Volume I : Current State of the Art on
Novel Materials

KEY ENGINEERING MATERIALS

MATERIALS

Volume I : Current State of the Art on
Novel Materials

Edited by
**Devrim Balköse, PhD, Daniel Horak, PhD,
and Ladislav Šoltés, DSc**

A. K. Haghi, PhD, and Gennady E. Zaikov, DSc
Reviewers and Advisory Board Members

Apple Academic Press

TORONTO   NEW JERSEY

Apple Academic Press Inc.	Apple Academic Press Inc.
3333 Mistwell Crescent	9 Spinnaker Way
Oakville, ON L6L 0A2	Waretown, NJ 08758
Canada	USA

©2014 by Apple Academic Press, Inc.

First issued in paperback 2021

Exclusive worldwide distribution by CRC Press, a member of Taylor & Francis Group

No claim to original U.S. Government works

ISBN 13: 978-1-77463-300-7 (pbk)
ISBN 13: 978-1-926895-73-4 (hbk)

This book contains information obtained from authentic and highly regarded sources. Reprinted material is quoted with permission and sources are indicated. Copyright for individual articles remains with the authors as indicated. A wide variety of references are listed. Reasonable efforts have been made to publish reliable data and information, but the authors, editors, and the publisher cannot assume responsibility for the validity of all materials or the consequences of their use. The authors, editors, and the publisher have attempted to trace the copyright holders of all material reproduced in this publication and apologize to copyright holders if permission to publish in this form has not been obtained. If any copyright material has not been acknowledged, please write and let us know so we may rectify in any future reprint.

Trademark Notice: Registered trademark of products or corporate names are used only for explanation and identification without intent to infringe.

Library of Congress Control Number: 2013956108

Library and Archives Canada Cataloguing in Publication

Key engineering materials.

Includes bibliographical references and index.
Contents: v. 1. Current state of the art on novel materials/edited by Devrim Balköse, PhD, Daniel Horak, PhD, and Ladislav Šoltés, DSc; A. K. Haghi, PhD, and G. E. Zaikov, DSc, reviewers and advisory board members -- v. 2. Interdisciplinary concepts and research/edited by François Kajzar, PhD, Eli M. Pearce, PhD, Nikolai A. Turovskij, PhD, and Omari Vasilii Mukbaniani, DSc; A. K. Haghi, PhD, and Gennady E. Zaikov, DSc, reviewers and advisory board members.
ISBN 978-1-926895-73-4 (v. 1 : bound)

1. Materials. I. Balköse, Devrim, editor of compilation II. Horák, D. (Daniel), editor of compilation III. Šoltés, Ladislav, editor of compilation IV. Title: Current state of the art on novel materials. V. Title: Interdisciplinary concepts and research.

| TA403.K49 2014 | 620.1'1 | C2013-908007-4 |

Apple Academic Press also publishes its books in a variety of electronic formats. Some content that appears in print may not be available in electronic format. For information about Apple Academic Press products, visit our website at **www.appleacademicpress.com** and the CRC Press website at **www.crcpress.com**

KEY ENGINEERING MATERIALS

Volume I : Current State of the Art on Novel Materials

Edited by
**Devrim Balköse, PhD, Daniel Horak, PhD,
and Ladislav Šoltés, DSc**

A. K. Haghi, PhD, and Gennady E. Zaikov, DSc
Reviewers and Advisory Board Members

Apple Academic Press

TORONTO   NEW JERSEY

Apple Academic Press Inc. | Apple Academic Press Inc.
3333 Mistwell Crescent | 9 Spinnaker Way
Oakville, ON L6L 0A2 | Waretown, NJ 08758
Canada | USA

©2014 by Apple Academic Press, Inc.

First issued in paperback 2021

Exclusive worldwide distribution by CRC Press, a member of Taylor & Francis Group

No claim to original U.S. Government works

ISBN 13: 978-1-77463-300-7 (pbk)
ISBN 13: 978-1-926895-73-4 (hbk)

This book contains information obtained from authentic and highly regarded sources. Reprinted material is quoted with permission and sources are indicated. Copyright for individual articles remains with the authors as indicated. A wide variety of references are listed. Reasonable efforts have been made to publish reliable data and information, but the authors, editors, and the publisher cannot assume responsibility for the validity of all materials or the consequences of their use. The authors, editors, and the publisher have attempted to trace the copyright holders of all material reproduced in this publication and apologize to copyright holders if permission to publish in this form has not been obtained. If any copyright material has not been acknowledged, please write and let us know so we may rectify in any future reprint.

Trademark Notice: Registered trademark of products or corporate names are used only for explanation and identification without intent to infringe.

Library of Congress Control Number: 2013956108

Library and Archives Canada Cataloguing in Publication

Key engineering materials.

Includes bibliographical references and index.
Contents: v. 1. Current state of the art on novel materials/edited by Devrim Balköse, PhD, Daniel Horak, PhD, and Ladislav Šoltés, DSc; A. K. Haghi, PhD, and G. E. Zaikov, DSc, reviewers and advisory board members -- v. 2. Interdisciplinary concepts and research/edited by François Kajzar, PhD, Eli M. Pearce, PhD, Nikolai A. Turovskij, PhD, and Omari Vasilii Mukbaniani, DSc; A. K. Haghi, PhD, and Gennady E. Zaikov, DSc, reviewers and advisory board members.
ISBN 978-1-926895-73-4 (v. 1 : bound)

1. Materials. I. Balköse, Devrim, editor of compilation II. Horák, D. (Daniel), editor of compilation III. Šoltés, Ladislav, editor of compilation IV. Title: Current state of the art on novel materials. V. Title: Interdisciplinary concepts and research.

TA403.K49 2014   620.1'1   C2013-908007-4

Apple Academic Press also publishes its books in a variety of electronic formats. Some content that appears in print may not be available in electronic format. For information about Apple Academic Press products, visit our website at **www.appleacademicpress.com** and the CRC Press website at **www.crcpress.com**

ABOUT THE EDITORS

Devrim Balköse, DSc

Devrim Balköse, DSc, graduated from the Middle East Technical University (Ankara, Turkey) Chemical Engineering Department in 1969. She received her MS and PhD degrees from Ege University, İzmir, Turkey, in 1974 and 1977 respectively. She became Associate Professor in macromolecular chemistry in 1983 and Professor in process and reactor engineering in 1990. She worked as research assistant, assistant professor, associate professor, and professor between 1970–2000 at Ege University. She was the Head of Chemical Engineering Department at İzmir Institute of Technology, İzmir, Turkey, between 2000–2009. She is now a faculty member in the same department. Her research interests are in polymer reaction engineering, polymer foams and films, adsorbent development, and moisture sorption. Her research projects are on nanosized zinc borate production, ZnO polymer composites, zinc borate lubricants, antistatic additives, and metal soaps.

Daniel Horak, PhD

Daniel Horak, PhD, graduated from the Institute of Chemical Technology in Prague, Czech Republic, where he received MSc degree in macromolecular chemistry. His PhD degree in chemistry was obtained from the Institute of Macromolecular Chemistry, Academy of Sciences of the Czech Republic, where he is employed as the Head of the Department of Polymer Particles. He was a post-doctoral fellow at the University of Ottawa in Canada with Professor Frechet in 1983–84, winner of the scholarship from the Japanese Society for Promotion of Science, at the Technological University of Nagaoka with Professor Imai in Japan in 1987–88, visiting scientist at the Cornell University in Ithaca, New York, with Professor Frechet in 1993–94. He also served as a visiting scientist at the University of Montreal in Canada in 2001. His research activity includes magnetic nano- and microspheres, polymer particles by heterogeneous polymerization techniques including emulsion, suspension, and dispersion polymerization; properties of the particles, their chemical modifications and applications in medicine, biochemistry, and biotechnology; immobilization of enzymes and antibodies; hydrogels, sorbents, and ion exchangers; and advanced separation media. He is member of the International Polymer Colloid Group, organizer and chairman of the Polymer Colloid Symposium in Prague 2008 and 2014, editorial board member of *Journal of Colloid Science and Biotechnology*, and supervisor of PhD students. He has published more than 150 original scientific papers, 7 book chapters (in *Polymeric Materials Encyclopedia* and *Strategies in Size Exclusion Chromatography*), 10 reviews, many lectures and communications at international symposia, and 7 patents. He has published in *Biomaterials, Bioconjugate Chemistry, Journal of Polymer Science,*

Polymer Chemistry Edition, Polymer, Journal of Materials Chemistry, Chemistry of Materials, among others.

Ladislav Šoltés, PhD

Ladislav Šoltés, PhD, has expertise in macromolecular and analytical chemistry. He has been employed for over 30 years at academic research institutes in Bratislava, Slovakia. His research related to the polysaccharides, which started over two decades ago, resulted in patenting a novel approach "clathrate complexes formed by hyaluronic acid derivatives and use thereof as pharmaceuticals". His current research interests are focused on the studies of hyaluronan oxidative damage and the regulation of this process. Dr. Šoltés is the only distinguished representative of Slovakia in the International Society for Hyaluronan Sciences, USA. In 2007 he was named Scientist of the Year of the Slovak Republic.

REVIEWERS AND ADVISORY BOARD MEMBERS

A. K. Haghi, PhD

A. K. Haghi, PhD, holds a BSc in urban and environmental engineering from University of North Carolina (USA); a MSc in mechanical engineering from North Carolina A&T State University (USA); a DEA in applied mechanics, acoustics and materials from Université de Technologie de Compiègne (France); and a PhD in engineering sciences from Université de Franche-Comté (France). He is the author and editor of 65 books as well as 1000 published papers in various journals and conference proceedings. Dr. Haghi has received several grants, consulted for a number of major corporations, and is a frequent speaker to national and international audiences. Since 1983, he served as a professor at several universities. He is currently Editor-in-Chief of the *International Journal of Chemoinformatics and Chemical Engineering* and *Polymers Research Journal* and on the editorial boards of many international journals. He is a member of the Canadian Research and Development Center of Sciences and Cultures (CRDCSC), Montreal, Quebec, Canada.

Gennady E. Zaikov, DSc

Gennady E. Zaikov, DSc, is Head of the Polymer Division at the N. M. Emanuel Institute of Biochemical Physics, Russian Academy of Sciences, Moscow, Russia, and Professor at Moscow State Academy of Fine Chemical Technology, Russia, as well as Professor at Kazan National Research Technological University, Kazan, Russia. He is also a prolific author, researcher, and lecturer. He has received several awards for his work, including the Russian Federation Scholarship for Outstanding Scientists. He has been a member of many professional organizations and on the editorial boards of many international science journals.

CONTENTS

LIST OF CONTRIBUTORS

Nima Ahmadi
Mechanical Engineering Department, Urmia University of Technology, Iran.

K. V. Aleksanyan
N. N. Semenov Institute of Chemical Physics of Russian Academy of Sciences, Kosygin str. 4, Moscow-119991, Russia Federation.

V. Z. Aloev
Kabardino-Balkarian State Agricultural Academy, Nal'chik-360030, Tarchokov str., 1a, Russian.

Mohamad Sadeghi Azad
Mechanical Engineering Department, Urmia University of Technology, Iran.

V. A. Babkin
Volgograd State Architect-build University.

Devrim Balköse
Department of Chemical Engineering, İzmir Institute of Technology, Gulbahce, Urla-35430 İzmir, Turkey.
E-mail: devrimbalkose@iyte.edu.tr

A. A. Berlin
N. N. Semenov's Institute of Chemical Physics, RAS, Kosygin str. 4, Moscow-119996, Russian Federation.

M. C. Bignozzi
Dipartimento di Chimica Applicata e Scienza dei Materiali, Facoltà di Ingegneria, Università di Bologna, Italy.

G. A. Bonartseva
A. N. Bach's Institute of Biochemistry, RAS, Leninskiy pr. 33, Moscow.

Jakub Czakaj
AIB Ślączka, Szpura, Dytko spółka jawna Knurów
Institute for Engineering of Polymer Materials and Dyes, Department of Elastomers and Rubber Technology in Piastow.

Hasan Demir
Osmangazi Korkut Ata Universitesi Kimya Mühendisliği Bölümü, Osmangazi, Turkey.
E-mail: demirrhasan.hd@gmail.com

Tahereh Dodel
Amirkabir University of Technology, Iran.

Peter Duchovič
Vipo a.s., Gen.Svobodu 1069/4, 95801 Partizánske, Slovakia.

Pavol Duchovic
Vipo a.s., Gen. Svobodu 1069/4, 95801 Partizánske, Slovakia.

Theresa O. Egbuchunam
Department of Chemistry, Federal University of Petroleum Resources, Effurun, Nigeria.

M. S. Mohy Eldin
Polymer Materials Research Department, Advanced Technologies and New Materials Research Institute (ATNMRI), City of Scientific Research and Technological Applications (SRTA–City), New Borg El-Arab City 21934, Alexandria, Egypt.

Kazuo Furuya
High voltage electron microscopy station, National Institute for Materials Science, 3–13 Sakura, Tsukuba-305-0003, Japan.

Mehmet Gönen
Department of Chemical Engineering, Süleyman Demirel Üniversitesi, Isparta, Turkey.

K. Z. Gumargalieva
N. N. Semenov's Institute of Chemical Physics, RAS, Kosygin str. 4, Moscow-119996, Russian Federation.

A. K. Haghi
University of Guilan, Rasht, Iran.

Mahdi Hasanzadeh
Imam Hossein University, Iran.
Department of Textile Engineering, University of Guilan, Rasht, Iran.
E-mail: hasanzadeh_mahdi@yahoo.com

A. I. Hashem
Organic Chemistry Department, Faculty of Science, Ain-Shams University, Cairo, Egypt.

Fikret İnal
Department of Chemical Engineering, İzmir Institute of Technology Gulbahce, Urla-35430 İzmir, Turkey.
E-mail: devrimbalkose@iyte.edu.tr

Akihisa Inoue
Institute for Materials Research, Tohoku University, Sendai 980-8577, Japan.
WPI Advanced Institute for Materials Research, Tohoku University, Sendai 980-8577, Japan.

A. L. Iordanskii
Semenov Institute of Chemical Physics, Russian Academy of Sciences, Moscow, Russia.
N. N. Semenov's Institute of Chemical Physics, RAS, Kosygin str. 4, Moscow-119996, Russian Federation.
A. N. Bach's Institute of Biochemistry, RAS, Leninskiy pr. 33, Moscow-119071, Russian Federation.

Peter Jurkovič
Vipo a.s., Gen.Svobodu 1069/4, 95801 Partizánske, Slovakia.

S. G. Karpova
Emanuel Institute of Biochemical Physics, Russian Academy of Sciences, ul. Kosygina 4, Moscow-119991, Russia.
E-mail: livanova@sky.chph.ras.ru

G. V. Kozlov
Institute of Applied Mechanics of Russian Academy of Sciences, Leninskii pr., 32 A, Moscow-119991, Russian Federation.
Kabardino-Balkarian State Agricultural Academy, Nal'chik-360030, Tarchokov st., 1 a, Russian Federation.

N. M. Livanova
Emanuel Institute of Biochemical Physics, Russian Academy of Sciences, ul. Kosygina 4, Moscow-11999,1 Russia.
E-mail: livanova@sky.chph.ras.ru

S. M. Lomakin
Emanuel Institute of Biochemical Physics, Russian Academy of Sciences, Moscow, Russia.
N. M. Emanuel Institute of Biochemical Physics of Russian Academy of Sciences, Kosygin str. 4, Moscow-119334, Russia.
E-mail: lomakin@sky.chph.ras.ru

Ramin Mahmoodi
Amirkabir University of Technology, Iran.

Meysam Masoodi
Department of Chemical Engineering Faculty of Engineering, Imam Hossein University, Tehran

Ján Matyšovský
Vipo a.s., Gen.Svobodu 1069/4, 95801 Partizánske, Slovakia.

Bentolhoda Hadavi Moghadam
Department of Textile Engineering, University of Guilan, Rasht, Iran.

Tahereh Moieni
Amirkabir University of Technology, Iran.

Igor Novák
Polymer Institute, Slovak Academy of Sciences, Dúbravská cesta 9, 845 41 Bratislava 45, Slovakia.

Filiz Ozmıhçı Omurlu
İzmir Institute of Technology Chemical Engineering Department 35430, Gülbahçe köyü, Urla, İzmir Turkey.
E-mail: filizozmihci@iyte.edu.tr

Yu. N. Pankova
N. N. Semenov's Institute of Chemical Physics, RAS, Kosygin str. 4, Moscow-119996, Russian Federation.

Elena L. Pekhtasheva
G. V. Plekhanov Russian Economic University, 36 Stremyannyi way, Moscow-17997, Russia.
E-mail: pekhtashevael@mail.ru

A. A. Popov
Emanuel Institute of Biochemical Physics, Russian Academy of Sciences, Moscow, Russia.

E. V. Prut
N. N. Semenov Institute of Chemical Physics of Russian Academy of Sciences, Kosygin str. 4, Moscow-119991, Russia.
Email: s.rogovina@mail.ru

Maria Rajkiewicz
Director of the Institute for Research of Composite Elastomer Materials.
Institute for Engineering of Polymer Materials and Dyes, Department of Elastomers and Rubber Technology in Piastów.

S. Z. Rogovina
N. N. Semenov Institute of Chemical Physics, Russian Academy of Sciences, Kosygin str. 4, Moscow-119996, Russian Federation.

M. M. Sabet
Polymer Materials Research Department, Advanced Technologies and New Materials Research Institute (ATNMRI), City of Scientific Research and Technological Applications (SRTA-City), New Borg El-Arab City 21934, Alexandria, Egypt.

Jan Sedliacik
Faculty of Wood Sciences and Technology, Technical University in Zvolen, Zvolen, Slovakia.

N. G. Shilkina
Emanuel Institute of Biochemical Physics, Russian Academy of Sciences, Moscow, Russia.

Satyajit Shukla
Ceramic Technology Department, Materials and Minerals Division (MMD), National Institute for Interdisciplinary Science and Technology (NIIST), Council of Scientific and Industrial Research (CSIR), Thiruvananthapuram-695019, Kerala, India.

Anamika Singh
Indian Council of Medical Research, Ansari Nagar, New Delhi.

Rajeev Singh
Indian Council of Medical Research, Ansari Nagar, New Delhi.

Marcin Ślączka
AIB Ślączka, Szpura, Dytko spółka jawna, Knurów

E. A. Soliman
Polymer Materials Research Department, Advanced Technologies and New Materials Research Institute (ATNMRI), City of Scientific Research and Technological Applications (SRTA-City), New Borg El-Arab City, 21934, Alexandria, Egypt.

Ladislav Šoltes
Institute of Experimental Pharmacology of the Slovak Academy of Sciences, Bratislava, Slovakia.

Minghui Song
High voltage electron microscopy station, National Institute for Materials Science, 3–13 Sakura, Tsukuba-305-0003, Japan.

Milena Špirková
Institute of Macromolecular Chemistry AS CR, Heyrovsky sq. 2, 162 53 Prague 6, Czech Republic.

Petr Sysel
Department of Polymers, Institute of Chemical Technology, Technicka 5, 166 28 Prague 6, Czech Republic.

T. M. Tamer
Polymer Materials Research Department, Advanced Technologies and New Materials Research Institute (ATNMRI), City of Scientific Research and Technological Applications (SRTA- City), New Borg El-Arab City 21934, Alexandria, Egypt.

Semra Ülkü
Department of Chemical Engineering, İzmir Institute of Technology Gulbahce, Urla 35430 İzmir, Turkey. E-mail: devrimbalkose@iyte.edu.tr

Khodadad Vahedi
Mechanical Engineering Department, Imam hosein University, Tehran, Iran.

Guoqiang Xie
Institute for Materials Research, Tohoku University, Sendai-980-8577, Japan.

Yu. G. Yanovskii
Institute of Applied Mechanics of Russian Academy of Sciences, Leninskii pr., 32 A, Moscow-119991, Russian Federation.

Mirkazem Yekani
Aerospace Engineering Department, Imam hosein University, Tehran, Iran. E-mail: meyeksani@yahoo.com

G. E. Zaikov
N. M. Emanuel Institute of Biochemical Physics, Russian Academy of Sciences, Kosygin str. 4, Moscow-119334, Russia Federation.
E-mail: chembio@sky.chph.ras.ru
N. M. Emanuel's Institute of Biochemical Physics, Russian Academy of Sciences, Moscow-119996, Russia Federation.

Z. M. Zhirikova
Kabardino-Balkarian State Agricultural Academy, Nal'chik-360030, Tarchokov st., 1 a, Russian Federation.

LIST OF ABBREVIATIONS

ACA	ε-Amino caproic acid
AIDS	Acquired immune deficiency syndrome
AI	Activity index
ATP	Adenosine triphosphate
AP	Ammonium perchlorate
APEP	Ammonium Perchlorate Experimental Plant
AFM	Atomic force microscopy
ASM	Atomic-power microscopy
ATR-FTIR	Attenuated total reflectance-Fourier transform infrared spectroscopy
BET	Brunauer–Emmet–Teller
BZONO	4-Benzoate-2,2,6,6-tetramethyl-1-piperidinyloxy
BSR	Butadiene-styrene rubber
BTTN	Butanetholtrinitrate
CNT	Carbon nanotubes
CMC	Cell membrane complex
CAB	Cellulose acetate butyrate
CVD	Chemical vapor deposition
CoSt2	Cobalt stearate
CVID	Common variable immune deficiency
CFD	Computational fluid dynamics
CuSt2	Copper stearate
RDX	Cyclotrimethylene-trinitramine
DD	Deacetylation degree
DNA	Deoxyribonucleic acid molecule
DBTL	Dibutyltindilaurate
DEGDN	Diethyleneglycol dinitrate
DSC	Differential scanning calorimetry
DTA	Differential thermal analysis
DOP	Dioctylphthalate
DFM	Dynamic force microscope
DMTA	Dynamic mechanical thermal analysis
EPR	Electron paramagnetic resonance
EBID	Electron-beam-induced deposition
EDS	Energy dispersive spectroscopy
EDX	Energy dispersive X-ray
EAA	Ethylene–acrylic acid
EDA	Ethylenediamine

EOE	Ethylene-octene elastomers
EPDM	Ethylene-propylene-diene elastomer
EVA	Ethylene–vinyl acetate
EXAFS	Extended X-ray absorption fine structure
FD	Formaldehyde
FdUR	5-Fluoro-2'deoxyuridine
FTIR	Fourier transform infrared spectroscopy
GDLs	Gas diffusion layers
GlcN	Glucose amine
HDPE	High density polyethylene
HIV	Human immunodeficiency virus
HCl	Hydrochloric acid
HBP	Hyperbranchedpolymers
IR	Infrared
IIR	Isobutylene-isoprene rubber
IPP	Isotactic polypropylene
KTGF	Kinetic theory granular flow model
LB	Langmuir–Blodgett multilayers
LB	Lattice Boltzmann model
LOI	Loss of ignition
LDPE	Low density polyethylene
LMC	Low-molecular compound
MgSt2	Magnesium stearate
MALDI-TOF	Matrix-assisted laser desorption/ionization time of flight
MW	Matt waste
MDF	Medium density fiberboard
MUF	Melamine urea-formaldehyde
MEF	Melamine-formaldehyde
MUPF	Melamine-urea-phenol-formaldehyde
MFI	Melt flow index
MHS	Metal hydroxide salts
MB	Methylene blue
MW	Molecular weight
NK	Natural killer
NEPE	Nitrate ester plasticized polyether
NG	Nitroglycerin
MNA	N-Methyl-p-nitroaniline
NBRs	Acrylonitrile–butadiene rubbers
NMR	Nuclear magnetic resonance
OSA	Objective-based simulated annealing
OPC	Ordinary Portland cement
OWRK	Owens–Wendt–Rabel–Kaelble method
pBQ	Para-benzoquinone
PF	Phenol-formaldehyde
PPMS	Physical property measurement system

PSA	Pixel-based simulated annealing
PVC	Poly vinyl chloride
PHB	Poly(3-hydroxybutyrate)
PEO	Poly(ethylene oxide)
PIS	Poly(imide-co-siloxane)
PPG	Poly(propylene glycol)
PVA	Poly(vinyl alcohol)
PVC	Poly(vinyl chloride)
PB	Polybutadienes
PCA	Polycaproamide
PEG	Polyethylene glycol
PET	Polyethylene terephthalate
PGA	Polyglycolides
PHB	Polyhydroxybutyrate
PIB	Polyisobutylene
PLA	Polylacticacid
PEFC	Polymer electrolyte fuel cell
PEMFC	Polymer electrolyte membrane fuel cell
PNC	Polymer-based nanocomposite
PP	Polypropylene
PSI	Polysiloxane
PUR	Polyurethane
PVC	Polyvinyl chloride
PVAc	Polyvinylacetate
KOH	Potassium hydroxide
PDA	Potato dextrose agar
PIDs	Primary immunodeficiency diseases
PSA	Prostate-specific antigen
PEMFC	Proton exchange membrane fuel cell
RNA	Ribonucleic acid
SEM	Scanning electron microscope
SPIP	Scanning probe image processor
SPM	Scanning probe microscope
STM	Scanning tunneling microscopy
SAED	Selected area electron diffraction
SCID	Severe combined immune deficiency
SENs	Single-enzyme nanoparticles
SEE	Surface energy evaluation
TC	Technical carbon
TEMPO	2,2,6,6-Tetramethyl-1-piperidinyloxy
TEOS	Tetraethylorthosilicate
TG	Thermogravimetric
TGA	Thermogravimetric analysis
TPE	Thermoplastic elastomers
TPE-V	Thermoplastic vulcaniates

TPV	Thermoplastic vulcanizates
TEM	Transmission electron microscope
TEA	Triethylamine
TEGDN	Triethyleneglycoldi nitrate
TMETN	Trimethylolethanetrinitrate
TPB	Triphenyl bismuth
TNF-α	Tumor necrosis factor alfa
UV	Ultraviolet
UF	Urea-formaldehyde
VSM	Vibrating sample magnetometer
WBC	White blood cells
XRD	X-ray diffraction
XPS	X-ray photoelectron spectroscopy
$Zn(O2CCH3)2$	Zinc acetate
$ZnCl2$	Zinc chloride
ZHC	Zinc hydroxide chloride
ZnOHCl	Zinc hydroxychloride
ZnO	Zinc oxide

PREFACE

Materials with multifield coupling properties are an important aspect of modern science and technology with applications in many industrial fields. This book is a collection of the accepted papers concerning key engineering materials. Depending on the scale at which they are analyzed and used, we can talk about modern science and technology with applications in engineering and science.

Different parts of the research presented here were partially conducted by the authors.

The book is intended for researchers, engineers, designers and students interested in the materials and their use in engineering and science.

The fundamental aims of the book are:

- To expand design horizons with a thorough, interdisciplinary knowledge of materials science;
- To cover a more complete and broad spectrum of current problems and scientific researches in the area of the design of materials and structures;
- To highlight an entire range of possibilities of the use of various chemical materials for different problems encountered in practice—it demonstrates the advisability and sense of their use;
- To focus on the importance and significance of taking into account advanced materials and further in the optimization of their properties.

— **Devrim Balköse, PhD, Daniel Horak, PhD,
and Ladislav Šoltés, DSc**

CHAPTER 1

PREPARATION AND PROPERTIES OF ANIMAL PROTEIN HYDROLYSATES FOR OPTIMAL ADHESIVE COMPOSITIONS

PETER JURKOVIČ, JÁN MATYŠOVSKÝ, PETER DUCHOVIČ, and IGOR NOVÁK

CONTENTS

1.1 INTRODUCTION

Determination of mathematical models of the kinetics of polycondensation of proteinous hydrolysates reactions with selected crosslinking agents with the regard to the content of free formaldehyde and phenol in final products. Optimization of adhesive compositions with the respect to their applicability in the wood processing industry.

Dried collagen hydrolysates were laboratory prepared at Liptospol Liptovský Mikuláš, Slovak producer of leather and leather glue, Gelima Liptovský Mikuláš, Slovak producer of food and technical gelatine and CSIC Barcelona in Spain, leather glue prepared by oxidation method from chrome tanned shavings, with the aim to compare their influence on formaldehyde emission, physical and mechanical parameters of board materials. Hydrolysates were used for trials for preparation of adhesive mixtures with the application of biopolymers and supplementary additives.

Analytic analysis of powdered collagen hydrolysate from Cr-shavings (sample from CSIC Barcelona) proved, that the presence of chromium by the method of atomic absorptive spectrophotometry was not determined at sensitivity of the method less than $0{,}0012$ ppm of Cr. Series of application trials of described biopolymers were carried out in conditions at technical university zvolen with the aim to evaluate the influence of selected mixtures on ecological, physical, and mechanical parameters on board materials – plywood. Preparation of collagen samples for experimental applications into other types of adhesives (polyvinylacetate (PVAc) and polyurethane (PUR)), possibilities of direct application of modified hydrolysate of collagen and keratin, as the input raw material for preparation of polycondensation resins.

1.2 DETERMINATION OF MATHEMATICAL MODELLING (KINETICS) OF POLYCONDENSATION REACTIONS CONTROL ALGORITHMS AND REACTOR DYNAMICS

Obtaining, processing, and interpretation of kinetics thermodynamic data related to polycondensations reaction urea-formaldehyde (UF), phenol-formaldehyde (PF), melamine urea-formaldehyde (MUF), and other similar resins modified by the addition of protein hydrolysate.

In conditions of VIPO, following works were realized for assurance of required polycondensation kinetics of UF and PF adhesives with the addition of biopolymers and their influence on physical and mechanical parameters and formaldehyde emission:

- The way of preparation of collagen and keratin hydrolysates,
- Selection of analytic parameters of biopolymers evaluation, content of inorganic salts, and viscosity,
- Determination of optimal concentration of biopolymer in adhesive mixtures,
- The way of biopolymer modification,
- Temperature and time of polycondensation, condensation time.

For application into polycondensation adhesives, there were prepared collagen hydrolysates by acid hydrolysis (HCl, H_2SO_4, HCOOH, $Al_2(SO_4)_3$, and so on), alkaline hydrolysis (NaOH, $Ca(OH)_2$, and so on), enzymatic hydrolysis (alkaline protease and tripsin), and lyotropic agents (urea, $CaCl_2$, and so on).

For application into polycondensation adhesives, there is optimal technology with the addition of proteolytic enzyme and eventually lyotropic agent—urea. Collagen hydrolysate has the value of pH neutral, minimal content of inorganic salts, and required concentration minimal 40% of the dry content matter. Measurement of condensation time of adhesive mixtures confirmed significant worsening of kinetics—rate of polycondensation (condensation time in the test tube at the temperature of 100°C was prolonged up to 100% in comparison with the standard). For improving of polycondensation kinetics, collagen hydrolysates were modified with organic acid (HCOOH, inorganic acid HCl, and H_2SO_4, $Al_2(SO_4)_3$), while the pH was gradually adjusted to the value of 1, 2, 3, 4, and 5. At the value pH 4, optimal condensation times were reached and comparable with the standard (57–65 sec, at temperature of 100°C), hydrolysate modified to the value pH less than 3 in UF mixtures caused shortening of the workability time, approximate than 15 min at pH 1 up to approximate 3 hr at pH less than 3.

Collagen hydrolysate with the dry content matter 40% must be viscous liquid at the temperature of processing, and not semi-rigid gel and for standardization of the time of hydrolysis (molecular weight) there is necessary a measurement of the viscosity.

Laboratory trials confirmed, that the addition of modified collagen hydrolysate up to 5% related to dry content of adhesive do not grows worse physical and mechanical parameters of prepared products and at the same time significantly reduces the formaldehyde emission.

Temperature of polycondensation of UF adhesive mixtures with the addition of hydrolysate was optimized in the range of 120–140°C, the temperature of 160°C during 30 and 60 min caused the destruction of the hardener with following increasing of the formaldehyde content in hardened resins.

For application of biopolymers into PF adhesives, there were prepared keratin hydrolysates by oxidation and reduction technology in alkaline medium. Concentrated hydrolysates with the dry content matter of 20–30% and pH min. 10 and 5 were evaluated at dosing up to 10% as expressly positive—convenient physical and mechanical parameters, convenient viscosity of adhesive mixtures, and sufficient storage life.

Presented possibilities of application of biopolymers describe the kinetics of polycondensation of commercially produced adhesives, (Diakol M1—UF adhesive and Fenokol A—PF adhesive), in the dependence on the way of modification, temperature, and time. The other possibility of biopolymer application at the synthesis of polycondensation adhesive is preparing at the present time.

1.3 DETERMINATION OF ADHESIVES COMPOSITIONS AND OPTIMIZATION OF PROTEIN HYDROLYSATE COMPOSITIONS

Modification of recipes for preparation of adhesive compounds with respect to the results of prior analyses and trials with aim of obtaining to best possible quality of adhesive joints in wood processing applications. Preparation of adhesive compounds and necessary mechanical and chemical testing.

For the completion of results with the sample of hydrolysate from CSIC Barcelona there was realized the series of comparative trials, with the aim to consider ecological, physical, and mechanical parameters of adhesive mixtures of the three types of collagen biopolymers:

- Acid hydrolysis—VIPO Partizánske,
- Enzymatic hydrolysis—UTB Zlín,
- Oxidation method—CSIC Barcelona

With the evaluation of the influence of collagen hydrolysate prepared by oxidation method from Cr-shavings on formaldehyde emission.

Comparative measurements of powdered samples of collagen hydrolysates from CSIC Barcelona—oxidation method, sample from Liptospol Liptovský Mikuláš (technology VIPO —acid hydrolysis) and Gelima—Weishardt Liptovský Mikuláš (standard—producer of food gelatines) were realized, while it can be stated that:

- Dry content matter of samples is almost the same and do not shows larger deviations,
- Values of pH of water solutions are comparable and we can consider them as very slightly acid or neutral,
- Temperature of gelation was in the range of 12–15°C,
- Viscosity of 2% solutions at 20°C after 24 hr was:
    a. sample from CSIC Barcelona—5.50 mPas,
    b. sample from Gelima—5.07 mPas,
    c. sample from Liptospol—4.76 mPas ,

and its comparison with standards of similar parameters from Gelima and Liptospol—viscosity at concentration of 6.67% and temperature of 60°C—29.7 mPas, the strength of gel –97 Bloom, confirmed, that hydrolysate from CSIC Barcelona has comparable parameters with collagen applied to UF and PF adhesives in previous trials.

The Cr^{6+} was not determined in samples qualitatively with diphenylcarbazide, either presence of overall Cr by the method of atomic absorptive spectrophotometry on equipment "Shimadzu AA 6601 F". Following analysis were realized at UTB Zlín.

In laboratory conditions, there were consequently prepared liquid modified forms enzymatically and with lyotropic agent from collagen powder hydrolysates. Neutral or slightly acid with the value of pH 5–6.5 was adjusted with inorganic acid to the value of pH = 4 and consequently adhesive mixtures were condensed at temperature of 140°C during 30 and 45 min. After polycondensation and conditioning, samples were ground and formaldehyde emission determined colorimetrically from water extract prepared by absorption and also extraction method. Results of emissions confirmed the decrease of formaldehyde at absorption determination (decrease about approximate 30%).

Reached results are presented in the Table 1.

Table 1   Properties of prepared collagen hydrolysates

Sample	VIPO Partizánske	UTB Zlín	CSIC Barcelona
dry content matter percent	89.3	91.9	87.3

TABLE 1 *(Continued)*

viscosity 20°C of 2% solution (after 1 hr) [mPas]	15.57	12.68	11.08
extinction 405/495nm (transparency of solution)	0.072/0.015	1.03/0.734	1.241/0.913
pH 2% solution	6.2	6.9	6.75
Cr^{6+}	negative	negative	negative

We have prepared the reference trials with the application of 3 hydrolysates, which will be applied in eco-adhesives. The CSIC Barcelona glue powder was applied as: 40% solution (substitution of 5% adhesive), Second case as fine powder (substitution of 5% adhesive)— fine powder was impossible to homogenize in adhesive and the decrease of formaldehyde was minimal, from this reason laboratory trials continued only with 40% water gels.

Results of the formaldehyde content in hardened adhesive mixture are presented in the Table 2.

TABLE 2 Formaldehyde content in hardened adhesive mixtures

Sample	pH of collagen hydrolysate	Content of fd /mg/litre/
Standard UF adhesive + hardener	–	0.131
+ Hydrolysate VIPO	6	0.054
+ Hydrolysate VIPO	4	0.047
+ Hydrolysate CSIC	6	0.057
+ Hydrolysate CSIC	4	0.047

KEYWORDS

- **Animal proteins**
- **Biopolymers**
- **Formaldehyde**
- **Polycondensation adhesives**
- **Plywood**
- **Shear strength properties**

ACKNOWLEDGMENT

This publication was prepared as part of the project "Application of Knowledge-based Methods in Designing Manufacturing Systems and Materials" co-funded by the Ministry of Education, Science, Research, and Sport of the Slovak Republic within the granted stimuli for research and development from the State Budget of the Slovak Republic pursuant to Stimuli for Research and Development Act No. 185/2009 Coll. and the amendment of Income Tax Act No. 595/2003 Coll. in the wording of subsequent regulations in the wording of Act. No. 40/2011 Coll.

REFERENCES

1. Blažej, A., et al. *Technologie kže a kožešin*, SNTL Praha (1984).
2. Matyašovský, J., Kopný, J., Meluš, P., Sedliačik, J., and Sedliačik, M. Modifikácia polykondenzačných lepidiel bielkovinami. In: *Pokroky vo výrobe a použití lepidiel v drevopriemysle*. TU Zvolen, pp. 37–42 (2001).
3. Matyašovský, J., Kopný, J., Jurkovič, P., and Sedliačik, J. Modification of polycondensation adhesives with animal proteins. In: *Annals of Warsaw Agricultural University*. Forestry and Wood Technology. No 53, SGGW Warszawa, pp. 228–231 (2003).
4. Matyašovský, J., Kopný, J., Jurkovič, P., Sedliačik, J., and Kasala, J. Modification of polycondensation adhesives with animal proteins. Part II. In: *Annals of Warsaw Agricultural University*. Forestry and Wood Technology. No 55, SGGW Warszawa, pp. 354–359 (2004).
5. Restorm – WP1, 24 months Progress Meeting Report. Ecoadhesives. Priebežná hodnotiaca správa projektu 5. rámcového programu EÚ. Freiberg, 12 p.
6. Sedliačik, J. *Lepidlá a ich aplikácia. Vedecké štúdie.* TU Zvolen, p. 51 (2002).
7. Sedliačik, J. *Procesy lepenia dreva, plastov a kovov.* TU Zvolen, p. 220 (2005) ISBN 80-228-1500-4.
8. Sedliačik, M. and Sedliačik, J. Utilisation of leather proteins for glue compositions. In: *Wood science and engineering in the third millenium*. Transilvania university of Brasov, pp. 159–160 (2004) ISBN 973-635-385-0.
9. Sedliačik, M., Sedliačik, J., Matyašovský, J., and Kopný, J. *Bielkoviny ako nadstavovadlo pre močovinové a fenolické lepidlá.* Drevo 56, No. 8, pp. 164–165 (2001).

CHAPTER 2

COLLAGEN MODIFIED HARDENER FOR MELAMINE-FORMALDEHYDE ADHESIVE FOR INCREASING WATER-RESISTANCE OF PLYWOOD

JÁN MATYŠOVSKÝ, PETER JURKOVIČ, PAVOL DUCHOVIČ, and IGOR NOVÁK

CONTENTS

2.1   INTRODUCTION

One of the very important technological operations in woodworking industry is gluing. The aim of this work was preparation of hardener for melamine-formaldehyde (MEF) adhesives suitable for gluing of plywood. Commercial hardener was modified by biopolymers (waste animal polymers). Glued joints were expected to be classified as resistant to water in class 3.

In the experiments, two types of leather collagen hydrolysates (VIPOTAR I and VIPOTAR II) were applied into MEF adhesive. Leather collagen hydrolysates were obtained from waste produced by leather industry. Glued plywood specimens were preliminary conditioned by two different ways. Plywood glued with MEF adhesive with the modified hardener showed good strength properties when evaluated according to the standard STN EN 314-1, 2. Glued joints can be graded in class 3. Great attention is paid to improvement of technology of gluing and development of new types of adhesives. The important effort is exploitation of available products that could improve effectiveness of adhesive mixtures and reduce cost in production of adhesives. For the improvement of product quality (from the point of view of hygienic criteria), searching and using of raw materials reducing release of formaldehyde from glued joints is very important. Biopolymers could be such materials (for example waste from leather or food industry).

The MEF adhesives are thermo-reactive adhesives curing at neutral or acidic pH at higher temperatures (130–140°C) usually at presence of hardeners. Laser scanning microscopy was used to investigate the distribution of adhesive in wood fibers. Cyr et al. [1] researched penetration of melamine-urea-formaldehyde (MUF) Medium density fiberboard (MDF) production. Atomic force microscopy (AFM) enabled to recreate the finest detail of fiber surface. Adhesive can penetrate into any layers of wood cell walls, uses its affinity to both water and wood polymers to penetrate through pores from surface to lumen.

The improvement the water resistance for challenge expositions, or modification of certain properties of joints can be achieved by a mixture of adhesives for example urea-formaldehyde (UF) with resorcinol, melamine or polyvinylacetate (PVAC). Problems of influence of melamine content in MUF adhesives on formaldehyde emission and cured resine structure was investigated by Tohmura et al. [2]. They used six MUF adhesives synthesized with different F/(M + U) and M/U molar ratios. The ^{13}C nuclear magnetic resonance (NMR) spectroscopy of cured MUF resins revealed that more methylol groups, dimethylene-ether, and branched methylene structures were present in the MUF resins with a higher F/(M + U) molar ratio, leading to increased bond strength and formaldehyde emission. The lower formaldehyde emission from cured MUF adhesives with a higher M/U molar ratio may be ascribed to the stronger linkages between triazine carbons of melamine than those of urea carbons.

Dukarska and Lecka [3] researched in preparation of adhesive mixture based on melamine adhesive for production of exterior plywood. Melamine-urea-phenol-form-aldehyde (MUPF) and phenol-formaldehyde (PF) resins were filled by the waste from polyurethane (PUR) foam. Usage of adhesive mixtures based on MUF adhesive was searched by Jozwiak [4]. Fillers used were potato starch and rye flour. Obtained results showed that glued plywood met the standard for bond quality grade 3 and the mixture could be used for wood gluing at various levels of wood moisture content (6–21%).

Cellulose and lignin, as the basic wood component, are able to interact with proteins. Experiments were carried out on the interaction with dried animal blood plasma and egg albumin [5]. Infrared FTIR spectroscopy was used to analyze chemical changes in cellulose and lignin during the reaction. Obtained spectra indicated on possible chemical reaction between the peptide chain and reactive groups associated with cellulose.

Shitij Chaba and Anil N. Netravali [6] presented the research in modification of soy protein concentrate using glutaraldehyde and polyvinyl alcohol. The modified resin allow to process soy protein polymer without any plasticizer. The modified resin also showed increased tensile properties, improved thermal stability and reduced moisture resistance as compared to soy protein concentrate resin.

At present, the market has got an excess protein, especially protein hydrolysates from leather waste. Collagen belongs to the most important technical proteins, which enables more effective preparation of adhesive mixtures, Sedliačik [7, 8].

The aim of research was to develop a hardener for MEF adhesive mixtures. The mixtures could be used for woof gluing in bond quality grade 3, according to the standard EN 314-1, 2. The adhesive joint of grade 3 is applicable at outdoor conditions – at unlimited climatic influences. Non modified commercial MEF adhesives provide glued joint in grade 2. The MEF hardener was modified by biopolymers of animal origin. Various waste biopolymers (leather waste) could be secondary used.

2.2 EXPERIMENTAL PART

The experiments were carried out with the adhesive (KRONOCOL SM 10) and the particular hardener (hardener—product of Duslo Šala). Required hardener addition is 3%.

To prepare a modified hardener, biopolymers in the form of collagen substrates were used. Substrates were prepared by dechromation of chrome leather waste at two different temperatures and were specified as activator VIPOTAR I (prepared at 20°C) and activator VIPOTAR II (at 30°C). Substrates pH value was adapted to the value of 4.0. Solubility and hydrophobic improvement was assured by addition of lyotropic agent and hydrophobic agent (methylester of tannery fat MEKT). Commercial hardener was activated by addition of activators VIPOTAR I or VIPOTAR II in ratios 3.5%. Adhesive mixtures were tested in 3-layer beech plywood. Pressing temperature was 130°C, adhesive consumption 150 gm^{-2}.

Shear strength was measured and evaluated using a tensile testing machine LaborTech 4.050 with 5 kN head. Glued joint quality was tested according to the standard STN EN 314-1. Bond quality was expressed as grade 1, 2 or 3. Requirements for joint quality at plywood are determined by the standard STN EN 314-2.

2.3 DISCUSSION AND RESULTS

To test the effectiveness of activator VIPOTAR I, influence of various concentrations of the activator on shear strength of prepared plywood was tested. If activator was added in hardener (in adhesive mixture), shear strength of the joint was increased. The improvement was observed only under specific activator concentration. The optimal addition was determined as 3.5%. At higher ratio (5%), the shear strength was lower when compared to ratios 3.5% or 2.5%.

In Table 1, mean values of shear strength evaluated according the method for grade 3, together with individual measured minimal and maximal values, are shown. The mentioned tendency of shear strength is also evident in this detailed evaluation. Based on the experiments, further experiments were carried with addition of 3.5%.

TABLE 1　The shear strength of plywood specimens glued with the adhesive mixture with various amount of activator VIPOTAR I

Sample	Activator addition in hardener [%]	Required standard value of shear strength [MPa]	Average shear strength [MPa]	Minimal measured shear strength [MPa]	Maximal measured shear strength [MPa]
Reference	–	1.0	**1.1**	0.82	1.26
1	2.5	1.0	**1.6**	1.33	2.31
2	3.5	1.0	**1.9**	1.66	2.59
3	5.0	1.0	**1.3**	0.92	1.46

When preparing adhesive mixture for the experiments of water resistance, both activators VIPOTAR I and VIPOTAR II were used. Activators were added in the amount of 3.5%. Resulting shear strength values for plywood conditioned for grade 2 are listed in Table 2.

Table 2 The shear strength of preliminary conditioned plywood specimens (gluing in grade 2)

Sample /modifier/	Required standard value [MPa]	Average shear strength [MPa]	Standard deviation [MPa]	Coefficient of variation [%]	Minimal measured shear strength [MPa]	Maximal measured shear strength [MPa]	Number of samples
1- VIPO-TAR I	1.0	2.8	0.20	7.3	2.4	3.2	12
2 – VIPO-TAR I	1.0	2.5	0.23	9.2	2.2	2.9	12
3 –VIPOTAR II	1.0	2.5	0.28	11.1	2.0	3.0	12
4 –VIPOTAR II	1.0	2.4	0.26	10.9	2.1	3.0	12

All mean values of shear strength for grade 2 markedly exceeded required standard value of 1.0 [MPa], even all individual measured values were double than standard required value. The shear strength in comparison with the shear strength of the joint glued without the modifiers was significantly higher, more than doubled. Final shear strength values for plywood conditioned for grade 3 are listed in Table 3.

TABLE 3 The shear strength of preliminary conditioned plywood specimens (gluing in grade 3)

Sample /modifier/	Re-quired standard value [MPa]	Average shear strength [MPa]	Standard deviation [MPa]	Coefficient of varia-tion [%]	Minimal measured shear strength [MPa]	Maximal measured shear strength [MPa]	Number of samples [ks]
1 – VIPO-TAR I	1.0	2.4	0.29	11.7	1.7	2.9	15
2 – VIPO-TAR I	1.0	2.3	0.37	16.1	1.6	2.8	13
3 – VIPO-TAR II	1.0	2.3	0.22	9.7	1.8	2.8	15
4 – VIPO-TAR II	1.0	1.9	0.35	18.9	1.4	2.4	15

Similarly as for grade 2, all mean shear strength values of grade 3 exceeded required standard value of 1.0 MPa. Moreover, all individual measured values were the standard required value. Similarly as in the experiments for grade 2, the shear strength at grade 3 compared with the shear strength of the joint glued without the modifiers, was higher.

If we compare individual measured minimal and maximal values of shear strength for two various ways of conditioning of tested material, we can see that strength for grade 3 reached lower values when compared with grade 2. The same tendency was observed at mean values of shear strength. Such results can be expected, as preliminary conditioning for grading 3 is significantly more aggressive (longer total time of boiling interrupted with drying at higher temperature).

All tested adhesive mixtures and glued joints met the standard for grade 2 and grade 3, as well, and significantly exceeded the shear strength values of the reference sample. The findings confirmed the expected presumption, the strength and water resistance of adhesive bond is markedly influenced by the addition of a small amount of biopolymer (skin collagen). Commercial hardener was modified by activators VIPOTAR I and VIPOTAR II in the ratio 3.5%. If we consider the ratio of activators in all volume of adhesive mixture, the concentration of them is very low; nevertheless, their impact on the resulting strength and water resistance of adhesive joints is so marked.

2.4   CONCLUSION

The assumption that the addition of biopolymers in form of hydrolysates containing skin collagen can result in increased shear strength and increased water resistance, was confirmed. Collagen macromolecules dispersed in solution or in adhesive mixture have good adhesion to glued surface. In line with the results of other authors, we assume the right chemical reaction between the functional groups of protein and functional groups of the adhesive. From the results, it is visible that researched additives can become modifiers for adhesive mixtures based on MEF adhesives. The MEF adhesives used in praxis are graded as adhesives class 2. Glued joints graded as class 2 are suitable in the environments with higher moisture (for example sheltered exterior, outdoor conditions—short-time climatic influences, indoor conditions with higher moisture when compared with grade 1). Both of the tested collagen substrates significantly increased the shear strength of glued joint, and enabled to grade the bond as 3. Adhesive bonds graded as 3 are applicable at outdoor conditions—at unlimited climatic influences.

KEYWORDS

- **Biopolymer**
- **Gluing**
- **Hardener**
- **Melamine adhesive**
- **Plywood**
- **Water-resistance**

ACKNOWLEDGMENT

This publication was prepared as part of the project "Application of Knowledge-based Methods in Designing Manufacturing Systems and Materials" co-funded by the Ministry of Education, Science, Research, and Sport of the Slovak Republic within the granted stimuli for research and development from the State Budget of the Slovak Republic pursuant to Stimuli for Research and Development Act No. 185/2009 Coll. and the amendment of Income Tax Act No. 595/2003 Coll. in the wording of subsequent regulations in the wording of Act No. 40/2011 Coll.

REFERENCES

1. Cyr, P. L., Riedl, B., and Wang, X. M. Investigation of Urea-Melamine-Formaldehyde (UMF) resin penetration in Medium-Density Fiberboard (MDF) by High Resolution Confocal Laser ScanningMicroscopy. In: *Holz als Roh-und Werkstoff*, **66**, 129–134 (2007).
2. Tohmura, S., Inoue, A., and Sahari, S. H. Influence of the melamine content in melamine-urea-formaldehyde resins on formaldehyde emission and cured resin structure. *In J. Wood Sci.*, **47**, 451–457 (2001).
3. Dukarska, D. and Lecka J. Polyurethane foam scraps as MUPF and PF filler in the manufacture of exterior plywood. In: *Annals of Warsaw University of Life Sciences—SGGW, Forestry and Wood Technology*, Warszawa, **65**, 14–19 (2008).
4. Jozwiak, M. Possibility of gluing veneers with high moisture content with the use modified MUF adhesives resin. In: *Annals of Warsaw Agricultural University—SGGW. Forestry and Wood Technology*, Warszawa, **61** (2007).
5. Polus-ratajczak, I., Mazela, B., and Golinski, P. The chemical interaction of animal origin proteins with cellulose and lignin in wood preservation. In: *Annals of Warsaw Agricultural University—SGGW, Forestry and Wood Technology*, Warszawa, **53**, 296–299 (2003).
6. Chaba, S. and Netravali, N. A. Green composites. Part 2: Characterization of flax yarn and glutaraldehyde/poly (vinyl alcohol) modified soy protein concentrate composites. *In: Journal of materials science*, **40**, 6275–6282 (2005).
7. Sedliačik, J., and Sedliačiková, M. Innovation tendencies at application of adhesives in wood working industry. In: *Annals of Warsaw University of Life Sciences—SGGW. Forestry and Wood Technology*, Warszawa, **69**, 262–266 (2009) ISSN 1898-5912.
8. Sedliačik, J., Šmidriaková, M., and Jabloński, M. Obniženie energetycznych wymagań wytwarzania sklejek. *Przemysl drzewny* **4**, 24–26, (2008) ISSN 0373-9856.
9. STN EN 314-1: 2005. Preglejované dosky. Kvalita lepenia. Časť 1: Skúšobné metódy.
10. STN EN 314-2: 2005. Preglejované dosky. Kvalita lepenia. Časť 2: Požiadavky.

CHAPTER 3

POSSIBILITIES OF APPLICATION OF COLLAGEN COLOID FROM SECONDARY RAW MATERIALS AS MODIFIER OF POLYCONDENSATION ADHESIVES

JAN MATYASOVSKY, PETER JURKOVIC, PETER DUCHOVIC, PAVOL DUCHOVIC, JAN SEDLIACIK, and IGOR NOVÁK

CONTENTS

3.1   INTRODUCTION

This chapter presents the utilization of collagen jelly as one of several possibilities of leather waste reprocessing. In the frame of experimental research, soluble collagen was used as a modifier for polycondensation adhesives composition. Based on the results it can be said, that collagen has a significant influence on basic properties of urea-formaldehyde (UF) and phenol-formaldehyde (PF) adhesives and also on mechanical and physical properties of glued joints.

During the leather processing up to 25% mass of input raw material comes to chromium tanned waste. As presence of such large amount of waste presents economic and mainly ecologic problem for leather tanning industry, big effort is given to development of technologies for processing respectively disposal of chromium tanned waste in the world. Technologies based on different principles are result of this effort which enable to separate chromium from collagen. Application of these procedures in industry and also the scale of evaluation of chromium waste depend also on effective application of obtained products.

The importance of lowering of formaldehyde (fd) emission from hardened UF adhesives and lowering of the price of PF adhesives was solved in the project of the 5th Frame Program European Commission with the name: "Radical Environmentally Sustainable Tannery Operation by Resource Management (RESTORM)". The VIPO Partizanske—Slovak republic with UTB Zlin–Czech republic, CSIC Barcelona—Spain and UNC Northampton—Great Britain solved together described work packages.

The aim of the part of WP 6.4 was development of ecologic technology of dechromation of chromium shavings without oxidation Cr^{3+} to Cr^{6+}, with remained fibril structure with evaluation of the influence of pH, temperature and number of dechromation bathes on amount of removed chromium. Further, the influence of these parameters on looses of collagen by hydrolysis was followed and the degree of collagen destruction was evaluated by determination of isoelectric point and following of its increased solubility in the dependence on temperature of dechromation.

The aim of WP1 was to develop more valuable polycondensation adhesives with improved ecologic parameters by modification with natural non-toxic, biologically easy decomposable biopolymers. The influence of the amount of biopolymer in adhesive mixture was tested on lifetime, gel time, viscosity, lowering of formaldehyde emission, and mechanical and physical parameters of plywood.

3.2   EXPERIMENTAL PART

Used material:
1.   Chromium shavings,
2.   Chemicals and materials for modification of adhesive mixtures:
Resin from Chemko a.s., Strazske production:

Diakol M1—water solution of UF polycondensate determined for production of board materials, used under heat in combination with hardener.
- *Look:* milky liquid, dry content matter: 65% weight,
- *pH:* 7.4–8.5,

- *Gel time:* 60–80 sec,
- Content of free formaldehyde: maximum 0.35 mg fd/g.

Hardener R-60—water solution of ammonium nitrate, treated with formic acid to pH 4–5, concentration 57–60% weight,
- Technical flour,
- Collagen jelly,
- And beech veneer of thickness 1.8 mm.

3.3 DISCUSSION AND RESULTS

Following results were reached by experimental trials:
1. Development of environmentally friendly technology of chromium shavings dechromation.
2. Modification of commercially produced polycondensation UF adhesive with the application of collagen jelly prepared from chromium shavings.
- Technology of separation of soluble $Cr_2(SO_4)_3$ after alkali processing v $Ca(OH)_2$ was proposed for interruption of the bound Cr^{3+} — OOC – collagen – dechromation.
- In experimental work, there was followed the influence of acid concentration— pH of water solution, influence of temperature and number of dechromation bathes on the degree of dechromation and looses of collagen by increasing of its solubility.
- Collagen jelly was prepared for application into UF adhesives.

In experimental work, there was followed:

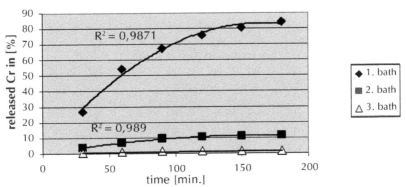

the first, second and third dechromation bath at the value of pH = 1.5 and temprature 20 °C

FIGURE 1 *(Continued)*

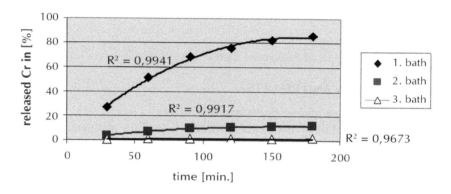

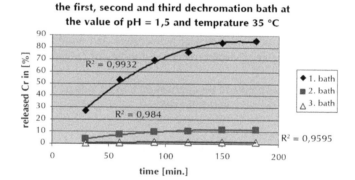

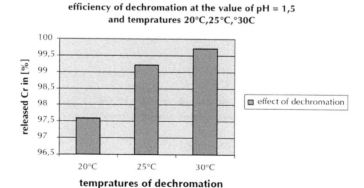

FIGURE 1 Experimentally determined dependencies of the released Cr in [%] in the first, second, and third dechromation bath at the value of pH 1.5 and temperatures of 20°C, 25°C, and 30°C.

FIGURE 2   Laboratory prepared dechromed shavings by proposed dechromation technology without oxidation of Cr^{3+} to Cr^{6+}.

3.3.1   THE INFLUENCE OF ADDITION OF COLLAGEN JELLY SAMPLES NO. 1, 2, AND 3 ON LIFETIME OF ADHESIVE MIXTURE

Obtained results of lifetime determination of UF adhesive mixtures with collagen jelly samples No. 1, 2, and 3 are presented in the Table 1. The lifetime of adhesive mixtures was followed 48 hr.

TABLE 1   Lifetime of UF adhesive mixture with collagen jelly samples No. 1, 2, and 3

No. of sample	lifetime of adhesive mixture [hr]
0	>48
1	>24< 48
2	>48
3	>48

From obtained results follow, that sample No. 1 has shortened lifetime >24 hr< 48 hr and the lifetime of samples No. 2 and 3 is comparative with the reference sample of UF adhesive >48 hr.

THE INFLUENCE OF ADDITION OF COLLAGEN JELLY SAMPLES NO. 1, 2, AND 3 ON GEL TIME OF ADHESIVE MIXTURE

Obtained results of gel time determination of UF adhesive mixtures with collagen jelly samples No. 1, 2, and 3 are presented in the Figure 3. The gel time was determined in the laboratory test-tube at the temperature of 100°C.

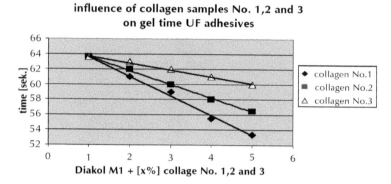

FIGURE 3   Experimentally determined dependency of the influence of collagen jelly samples No. 1, 2, and 3 on gel time of UF adhesive.

Determination of the gel time proved, that collagen jelly is suitable as a modifier of UF adhesive. Collagens No. 1, 2, and 3 lightly accelerate the condensation reaction. Collagen jelly No. 1 obtained from first dechromation bath—least hydrolysed, the most significantly accelerates polycondensation reaction in comparison with the reference sample. Hydrolysis of collagen is increased by the impact of temperature and vitriol acid is consumed by reaction in dechromation bath, what reduces the gel time.

3.3.2   THE INFLUENCE OF ADDITION OF COLLAGEN JELLY SAMPLES NO. 1, 2, 3 ON THE CONTENT OF FREE FORMALDEHYDE IN UF RESIN CONDENSED AT THE TEMPERATURE OF 120°C AND TIME 15 MIN

Obtained results of the influence of amount of collagen jelly samples No. 1, 2, and 3 on formaldehyde emission from UF adhesive mixtures are in the Figure 4.

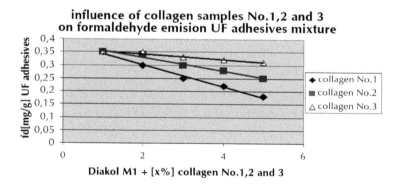

FIGURE 4   Experimentally determined dependency of the influence of collagen jelly samples No. 1, 2, and 3 on formaldehyde emission in UF adhesive mixture in [mg/g] condensed at the temperature of 120°C and time 15 min.

Determination of fd emission proved, that collagen jelly is suitable as modifier of UF adhesive for lowering of formaldehyde emission. Collagen jelly No. 1 obtained from first dechromation bath—least hydrolysed, the most significantly accelerates polycondensation reaction in comparison with the reference sample. Hydrolysis of collagen is increased by the impact of temperature, while comes to collagen hydrolysis and particular decay of amino-acids and also to lowering of amide nitrogen, what was confirmed by the decrease of isoelectric point. The decrease of reactive NH_2 groups in collagen No. 2 and 3 is expressed as lowered ability to bind of formaldehyde.

THE INFLUENCE OF ADDITION OF COLLAGEN JELLY SAMPLES NO. 1, 2, AND 3 ON THE VISCOSITY OF UF ADHESIVE MIXTURE

Obtained results of the influence of amount of collagen jelly samples No. 1, 2, and 3 on the change of viscosity UF adhesive mixture. Viscosity of adhesive mixtures was measured with Höppler viscometer in laboratory conditions at temperature of 20°C, results are in the Figure 5.

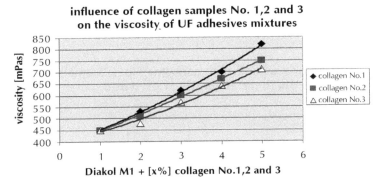

FIGURE 5   Experimentally determined dependency of the influence of collagen jelly samples No. 1, 2, and 3 on the viscosity of adhesive mixtures in [mPa.s].

Determination of the viscosity change of UF adhesive mixtures confirmed, that collagen jelly is suitable as modifier of UF adhesive for viscosity treatment. Collagen jelly can replace extenders as for example technical flour. Collagen jelly No. 1 obtained form first dechromation bath—least hydrolysed, most significantly treats viscosity in comparison with the reference sample. The decrease of mole weight of collagens No. 2 and 3 is expressed as lowered ability to treat the viscosity of UF adhesives.

3.3.3 STRENGTH PROPERTIES OF PLYWOOD

Strength properties of glued joints bonded with UF adhesive mixture with collagen jelly were tested on beech plywood. Plywood was prepared according to EN standardized procedure.

ADHESIVE MIXTURES

Diakol M1 + collagen jelly used for viscosity treatment, marked as 1, 2, and 3 added in amount of 20 weight parts on 100 weight parts of adhesive.

0—REFERENCE SAMPLE

Technical flour used for viscosity treatment in the rate of 20 weight parts on 100 weight parts of adhesive Diakol M1.

Results of shear strength of glued joints after dry climatisation and after soaking proved, that all samples fulfill required standard values.

- shear strength of glued joints partially decreased at collagen jelly samples No. 2 and 3 in comparison with the reference sample after soaking in water,
- shear strength of glued joints partially increased in comparison with the reference sample after dry climatisation at collagen jelly samples No. 1, 2, and 3. Results of measurements of influence of amount of added collagen jelly samples No. 1, 2, and 3 on shear strength of plywood bonded with UF adhesive mixtures are presented on the Figure 6.

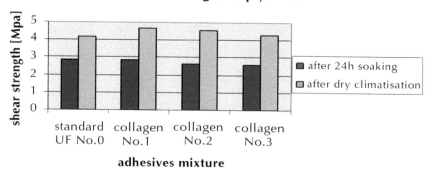

FIGURE 6   Results of tests of the influence of collagen jelly samples No. 1, 2, and 3 on shear strength of plywood after dry climatisation and after 24 hr soaking in water 20 ± 3°C in [MPa].

Testing of shear strength of UF adhesive mixtures proved that collagen jelly samples No. 1, 2, and 3 are suitable as modifiers – extenders for UF adhesives for viscosity treatment. There is possible to replace 100% of technical flour by collagen jelly.

3.4 CONCLUSION

Project RESTORM, parts WP 1 and WP 6 solved the evaluation of biopolymers from chromium shavings by their application into contemporary used polycondensation adhesives.

Specific conclusions following from this work:
- Obtained results bring new knowledge about possibilities of dechromation of chromium shavings without oxidation on toxic Cr^{6+} – perspective possibility to process leather tanning waste by proposed technology,
- Experimentally verified influence of collagen on properties of UF adhesive mixtures: lifetime, viscosity, gel time, emission of formaldehyde, and shear strength of plywood—perspective possibility to improve ecologic parameters of wood products.

KEYWORDS

- **Adhesive mixtures**
- **Collagen jelly**
- **Formaldehyde emission**
- **Gel time**
- **Hydrolysis of collagen**
- **Shear strength**

ACKNOWLEDGMENT

This chapter was processed in the frame of the APVV projects No. APVV-351-010 as the result of author's research at significant help of APVV agency, Slovakia.

REFERENCES

1. Blažej, A. et al. *Štruktúra a vlastnosti vláknitých bielkovín*, VEDA, Bratislava (1978).
2. Blažej, A. et al. *Technologie kůže a kožešin. Praha*, SNTL, ALFA, Bratislava, pp. 20–25, pp. 208–215 (1984).
3. Cabeza, F. L. Isolation of protein products from chrome leather waste. In: *Journal of the Society of Leather Technologists and Chemists*, **83**(1), 14–19.
4. Klásek, A. Způsoby odchromování usňových odpadů. In: *Kožedělné odpady a jejich ekonomické využití. Brno:* Zborník prednášok ČSVTS, pp. 25–33 (1983).
5. Kolomazník, K., Shánelová, K., and Dvořáčková, M. Modifikované aminoplasty proteínovými hydrolyzáty pro lepení dřeva. In: *Pokroky vo výrobe a použití lepidiel v drevopriemysle. Vinné*, TU Zvolen, p. 91 (1999) ISBN 80-228-0790-7.
6. Matyašovský, J. et al. Modification of polycondensation adhesives with animal proteins. Part II. In: Annals of Warsaw Agricultural University. *Forestry and Wood Technology*, **55**, 354–359 (2004) ISSN 028-5704.
7. Sedliačik, M. *Nové kompozície polykondenzačných lepidiel a ich aplikácie v drevárskom priemysle.* TU Zvolen, p. 202, (1992) ISBN 80-228-0207-7.

8.   Sedliačik, M. and Sedliačik, J. Technológia spracovania dreva II. *Lepidlá a pomocné látky*. TU Zvolen, p. 247, (1998) ISBN 80-228-0399-5.

9.   Sedliačik, J. Optimalizácia procesu lepenia hygienicky nezávadných preglejovaných materiálov. *Kandidátska práca*. TU Zvolen, p. 112 (2000).

10. Sedliačik, J. and Sedliačik, M. *Lepidlá a lepenie dreva*. TU Zvolen, p. 196, (2003) ISBN 228-1258-7.

CHAPTER 4

REUSE OF INDUSTRIAL WASTES AS CONSTRUCTION KEY MATERIAL

A. K. HAGHI and M. C. BIGNOZZI

CONTENTS

4.1 INTRODUCTION

Plastic waste management is one of the major environmental concerns in the world. Plastics can be separated into two types. The first type is thermoplastic which can be melted for recycling in the plastic industry, such as polyethyleneterephthalate (PET). The second type is thermosetting plastic. This plastic cannot be melted by heating (such as melamine). At present, these plastic wastes are disposed by either burning or burying. Therefore, both the ways contributing to the environmental problems. This chapter describes the use of thermosetting plastic waste as aggregate within lightweight concrete for building application and possibilities for reuse of thermosetting plastic waste in the concrete are described in detail.

Plastic materials production has reached global maximum capacities leveling at 260 Mt in 2007, where in 1990 the global production capacity was estimated at an 80 Mt. It is estimated that production of plastics worldwide is growing at a rate of about 5% per year. Its low density, strength, user-friendly designs, fabrication capabilities, long life, light weight, and low cost are the factors behind such phenomenal growth. Plastics also contribute to daily life functions in many aspects. With such large and varying applications, plastics contribute to an ever increasing because the molecular chains are bonded firmly with meshed crosslink [1-6].

The self weight of concrete elements is high and can represent a large proportion of the load on structure. Therefore, using lightweight concrete with a lower density can result in significant benefits such as superior load-bearing capacity of elements, smaller cross-sections and reduced foundation sizes. A lightweight structure is also desirable in earthquake prone areas. It is convenient to classify the various types of lightweight concrete by their method of production [3-6].

These are:

- By using porous lightweight aggregate of low apparent specific gravity, that is lower than 2.6, for example, pumice material. This type of concrete is known as lightweight aggregate concrete [7-9].
- By introducing large voids within the concrete or mortar mass, these voids should be clearly distinguished from the extremely fine voids produced by air entrainment such as aluminum powder. This type of concrete is variously known as aerated, cellular, and foamed or gas concrete [9-12].
- By omitting the fine aggregate from the mix so that a large number of interstitial voids is present, normal weight coarse aggregate is generally used. This concrete is known as no-fines concrete.

In this chapter, the technology of aerated concrete to produce lightweight concrete has been employed. Since the specific gravity of thermosetting plastic is about one-half of the typical fine and coarse aggregates, therefore possibilities for application of thermosetting plastic waste in the concrete are investigated.

4.2 EXPERIMENTAL

The materials used in present chapter are as follows:

Ordinary Portland Cement: Type I Portland cement conforming to ASTM C150-94.

Sand: Fine aggregate is taken from natural sand. Therefore, it was used after separating by sieve in accordance with the grading requirement for fine aggregate (ASTM C33-92). Table 1 presents the properties of the sand and its gradation is presented in Figure 1.

TABLE 1 Properties of sand and melamine aggregates

Properties	Sand	Melamine
density (g/cm^3)	2.60	1.574
bulk density (g/cm^3) (Melamine)	–	0.3–0.6
Water absorption (%)	1.64	7.2
Max size (mm)	4.75	1.77
Min size (mm)	–	0.45
Sieve 200 (%)	0.24	–
Flammability (Melamine)	non flammable	non flammable
Decomposition (Melamine)	–	at >280°C formation of NH$_3$

Thermosetting Plastic: Melamine is a widely used type of thermosetting plastic. Therefore, in the present work has been selected for application in the mixed design of composite. The mechanical and physical properties of melamine are shown in Table 1. The melamine waste was ground with a grinding machine. The ground melamine waste was separated under sieve analysis. Scanning electron micrographs (SEM) of melamine aggregates is shown in Figure 2. Those have irregular shape and rough surface texture. The grain size distributions were then plotted as shown in Figure 1. It was observed that the gradation curve of the combination of sand and plastic aggregates after sieve number 16 meets most of the requirements of ASTM C33-92.

The melamine aggregates were been saturated surface dry. Therefore, melamine aggregates immerse in water at approximately 21°C for 24 hr and removing surface moisture by warm air bopping.

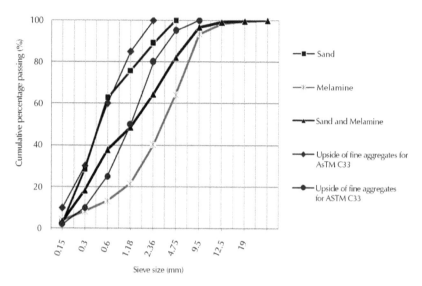

FIGURE 1 Gap-grading analysis of sand and melamine aggregates according to ASTM C33-92.

FIGURE 2 *(Continued)*

FIGURE 2 The SEM photographs of materials: (a) melamine aggregates and (b) silica fume.

Aluminum Powder: In the present chapter, aluminum powder was selected as an agent to produce hydrogen gas (air entrainment) in the cement. This type of lightweight concrete is then called aerated concrete. The following are possible chemical reactions of aluminum with water:

$$2Al + 6H_2O \qquad 2Al(OH)_3 + 3H_2 \qquad\qquad (1)$$
$$2Al + 4H_2O \qquad 2AlO(OH) + 3H_2 \qquad\qquad (2)$$
$$2Al + 3H_2O \qquad Al_2O_3 + 3H_2 \qquad\qquad (3)$$

The first reaction forms the aluminum hydroxide bayerite $(Al(OH)_3)$ and hydrogen, the second reaction forms the aluminum hydroxide boehmite $(AlO(OH))$ and hydrogen, and the third reaction forms aluminum oxide and hydrogen. All these reactions are thermodynamically favorable from room temperature past the melting point of aluminum (660°C). All are also highly exothermic. From room temperature to 280°C, $Al(OH)_3$ is the most stable product, while from 280–480°C, $AlO(OH)$ is most stable. Above 480°C, Al_2O_3 is the most stable product. The following reaction illustrates the combined effect of hydrolysis and hydration on tricalcium silicat.

$$3CaO.SiO_2 + water \qquad xCaO.ySiO_2(aq.) + Ca(OH)_2 \qquad (4)$$

In considering the hydration of Portland cement it is demonstrate that the more basic calcium silicates are hydrolysed to less basic silicates with the formation of calcium hydroxide or "slaked lime" as a by product. It is this lime which reacts with the aluminum powder to form hydrogen in the making of aerated concrete from Portland cement:

$$2Al + 3Ca(OH)_2 + 6H_2O \qquad 3CaO.Al_2O_3.6H_2O + 3H_2 \qquad (5)$$

Hydrogen gas creates many small air (hydrogen gas) bubbles in the cement paste. The density of concrete becomes lower than the normal weight concrete due to this air entrainment.

Silica Fume: In the present chapter, silica fume has been used. Its chemical compositions and physical properties are being given in Table 2 and Table 3, respectively. The SEM of silica fume is shown in Figure 2.

Superplasticizer: Premia 196 with a density of 1.055 ± 0.010 kg/m³ was used. It was based on modified polycarboxylate.

TABLE 2   Chemical composition of silica fume

Chemical composition	Silica fume
SiO_2 (%)	86–94
Al_2O_3 (%)	0.2–2
Fe_2O_3 (%)	0.2–2.5
C (%)	0.4–1.3
Na_2O (%)	0.2–1.5
K_2O (%)	0.5–3
MgO (%)	0.3–3.5
S (%)	0.1–0.3
CaO (%)	0.1–0.7
Mn (%)	0.1–0.2
SiC (%)	0.1–0.8

TABLE 3   Physical properties of silica fume

Items	Silica fume
Specific gravity (gr/cm³)	2.2–2.3
particle size (μm)	< 1
Specific surface area (m²/gr)	15–30
Melting point (°C)	1230
Structure	amorphous

4.2.1 MIX DESIGN

To determine the suitable composition of each material, the mixing proportions were tested in the laboratory, as shown in Table 4. In this chapter, the mix proportions were separated for five experimental sets. For each set, the cement and aluminum powder contents was specified as a constant proportion. The proportion of each of the remaining materials, that is sand, water, silica fume, aluminum powder, and melamine, was varied for each mix design.

4.2.2 EXPERIMENTAL TECHNIQUES

Mortar was mixed in a standard mixer and placed in the standard mold of $50 \times 50 \times 50$ mm according to ASTM C109-02. In the pouring process of mortar, an expansion of volume due to the aluminum powder reaction had to be considered. The expanded portion of mortar was removed until finishing. The fresh mortar was tested for slump according to ASTM C143-03. The specimens were cured by wet curing at normal room temperature. The hardened mortar was tested for dry density, compressive strength, water absorption, and voids for the curing age of 7 and 28 days. The test results for melamine, sand and water contents were reported for 7 days curing age for mix nos. 1–3, because these were very close to the results of 28 days. When silica fume was added in the latter mix nos. 4 and 5, the test results were presented for 28 days. This is because the presence of silica fume increases the duration for completion of the chemical reaction. The testing procedures of dry density, water absorption, and voids were performed according to ASTM C642-97 and compressive strength was performed according to ASTM C109-02.

TABLE 4   Mix proportions of melamine lightweight composites (by weight)

Mix no.	Cement	Aluminum powder	Sand	Water	Silica fume	Melamine	Super-plasti-cizer
1. Determination of melamine content (1st trial mix design)	1.0	0.004	1.0	0.35	-	1.0	-
						1.5	
						2.0	
						2.5	
						3.0	
2. Determination of sand content	1.0	0.004	1.0	0.35	-	1.0	-

TABLE 4 *Continued*

			1.2				
			1.4				
			1.6				
			1.8				
3. Determination of water content or water–cement ratio (w/c)	1.0	0.004	1.4	0.30	-	1.0	-
				0.35			
				0.40			
				0.45			
				0.50			
				0.55			
4. Determination of silica fume content	1.0	0.004	1.4	0.35	0.10	1.0	0.005
					0.15		0.007
					0.20		0.009
					0.25		0.012
					0.30		0.015
					0.35		0.020
5. Determination of melamine content (final mix design)	1.0	0.004	1.4	0.35	0.25	1.0	0.012
						1.2	0.011
						1.4	0.010
						1.6	0.009
						1.8	0.008
						2.0	0.007
						2.2	0.006

4.3 DISCUSSION AND RESULTS

4.3.1 MIX NUMBER 1 (DETERMINATION OF MELAMINE CONTENT FOR THE FIRST TRIAL MIX DESIGN)

Figure 3 present the variations in the compressive strength and dry density for 7 days age of mortars as a function of the value of melamine substitutes used. It can initially be seen, to increased melamine, the compressive strength and dry density of composites decreased. The reduction in the compressive strength is due to the addition of melamine aggregates or could be due to either a poor bond between the cement paste and the melamine aggregates or to the low strength that is characteristic of plastic aggregates.

TABLE 5 Specification of non-load-bearing lightweight concrete (ASTM C129)

Type	Compressive strength (MPa)		Density (kg/m^3)
	Average of three unit	Individual unit	
II	4.1	3.5	< 1680

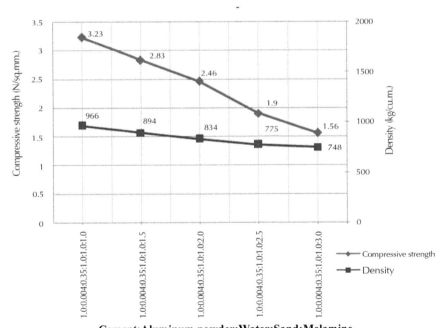

Cement:Aluminum powder:Water:Sand:Melamine

FIGURE 3 Compressive strength and density for varying melamine content (curing for 7 days).

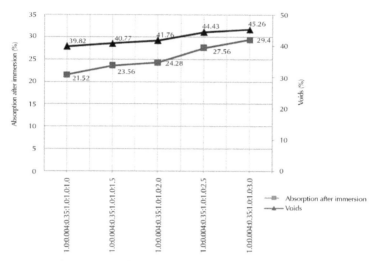

Cement:Aluminum powder:Water:Sand:Melamine

FIGURE 4   Absorption after immersion for varying melamine content (curing for 7 days).

The absorption is an indirect parameter to examine the inside porosity of mortar. The results showed that the absorption after immersion and voids of mortar increased as the melamine content increased (see Figure 4 and Figure 5). Therefore, to increased melamine plastic, the inside porosity of mortar increased. This might be other reason for the reduction in the compressive strength and density.

FIGURE 5   Optical photographs of samples containing varying melamine, right to left containing 1.0, 1.5, 2.0, 2.5, and 3.0 the weight percentage of melamine

4.3.2   MIX NUMBER 2 (DETERMINATION OF SAND CONTENT)

The results of compressive strength and dry density for 7 days age are shown in Figure 6. It can be seen that a reduction of sand leads to a reduction in the strength and dry density. The compressive strength and dry density for sand content equal to or greater than 1.4 exactly satisfy the standard value.

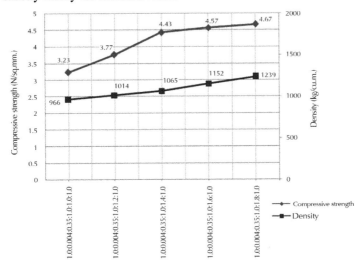

Cement:Aluminum powder:Water:Sand:Melamine

FIGURE 6   Compressive strength and density for the determination of the optimum sand content (curing for 7 days).

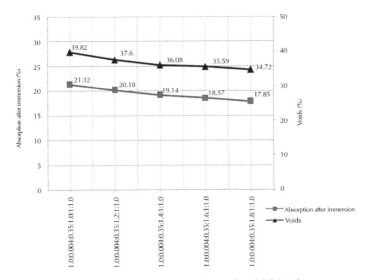

Cement:Aluminum powder:Water:Sand:Melamine

FIGURE 7   Absorption after immersion and voids for varying sand content (curing for 7 days).

Figure 7 present the variations in the absorption after immersion and voids as a function of the value of sand substitutes used. The results showed that the absorption after immersion and voids of mortar decreased as the sand content increased.

4.3.3 MIX NUMBER 3 (DETERMINATION OF WATER CONTENT)

The results of compressive strength and dry density for 7 days age are shown in Figure 8. The results showed that the compressive strength and dry density of mortar decreased as the water content increased.

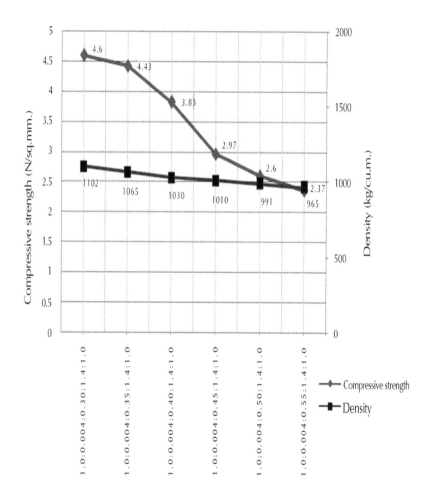

Cement:Aluminum powder:Water:Sand:Melamine

FIGURE 8 Compressive strength and density for the determination of the optimum water content (curing for 7 days).

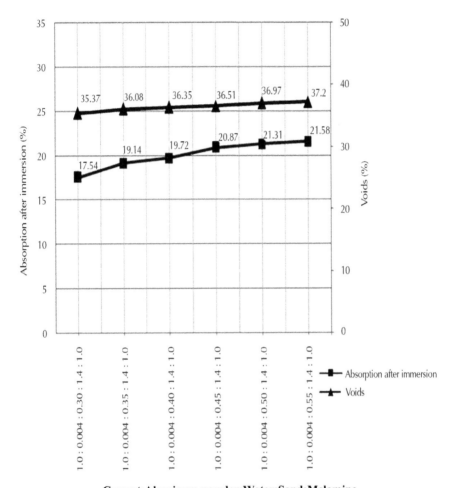

Cement:Aluminum powder:Water:Sand:Melamine

FIGURE 9 Absorption after immersion and voids for varying water content (curing for 7 days).

4.3.4 MIX NUMBER 4 (DETERMINATION OF SILICA FUME CONTENT)

Figure 10 present the variations in the compressive strength and dry density for 28 days age of mortars as a function of the value of silica fume substitutes used. It was found that the results of compressive strength for seven days age do not increase when compared with those without silica fume.

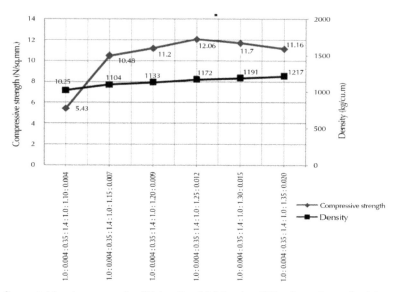

Cement:Aluminum powder:Water:Sand:Melamine:Silica fume:Superplasticizer

FIGURE 10 Compressive strength and density for the determination of the optimum silica fume content (curing for 28 days).

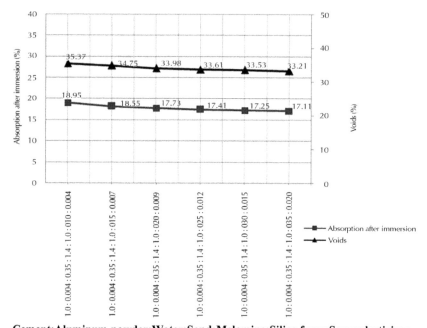

Cement:Aluminum powder:Water:Sand:Melamine:Silica fume:Superplasticizer

FIGURE 11 Absorption after immersion and voids for varying silica fume content (curing for 28 days).

4.3.5 MIX NUMBER 5 (DETERMINATION OF THE FINAL MELAMINE PLASTIC CONTENT)

Figure 12 present the variations in the compressive strength and dry density for 28 days age of mortars as a function of the value of melamine substitutes used. It was found that the presence of melamine caused a reduction in the dry density and compressive strength of concretes as discussed previously. Figure 13 show that the scanning electron microscopy analysis of composites reveals that cement paste-melamine aggregates adhesion is imperfect and weak. Therefore, the problem of bonding between plastic particles and cement paste is main reason to decrease of compressive strength. An optimum melamine content of 2.0 was selected. The results of compressive strength and dry density, which are 7.06 MPa and 887 kg/m³, are according to ASTM C129 Type II standard.

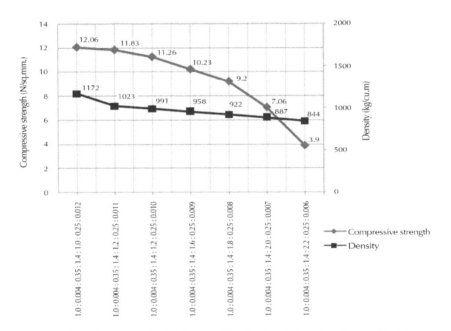

Cement:Aluminum powder:Water:Sand:Melamine:Silica fume:Superplasticizer

FIGURE 12 Compressive strength and density for the determination of the optimum melamine plastic content (curing for 28 days).

The results showed that the absorption after immersion and voids of mortar increased as the melamine content increased (see Figure 14). Also, the structure analysis of mortars by scanning electron microscopy has revealed a low level of compactness in mortars when the value of melamine plastic increased (see Figure 15). It was confirmed that to increased melamine plastic, the inside porosity of mortar increased.

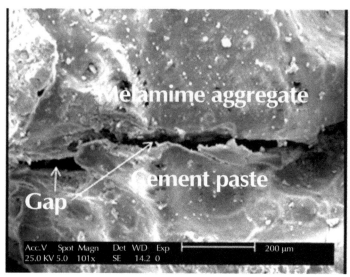

FIGURE 13   Microstructure of concrete containing 1.6 melamine by weight of cement, as obtained using SEM (enlargement: 101x).

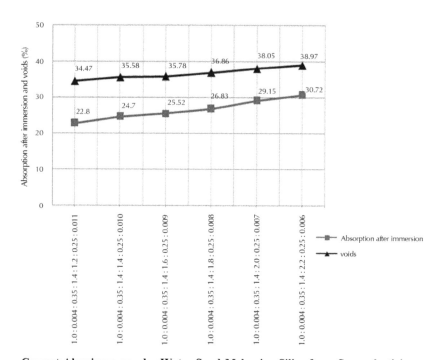

Cement:Aluminum powder:Water:Sand:Melamine:Silica fume:Superplasticizer

FIGURE 14   Absorption after immersion and voids for varying melamine plastic content (curing for 28 days).

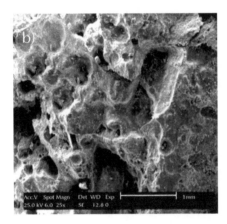

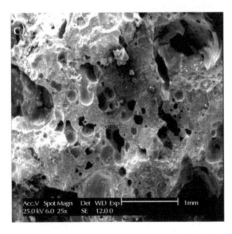

FIGURE 15 The SEM of various mortars containing melamine plastic aggregates. (a) 1.2 by weight of cement (enlargement: 25x), (b) 1.6 by weight of cement (enlargement: 25x), and (c) 2.0 by weight of cement (enlargement: 25x).

4.3.6 COMPARISON RESEARCH FINDINGS ON THE USE OF WASTE PLASTIC IN CONCRETE

The results of this chapter are in a perfectly agreement with the other research findings on the use of waste plastic in concrete (see Table 6).

4.4 CONCLUSION

Melamine plastic aggregates can be successfully and effectively utilized for non-load-bearing lightweight concrete according to ASTM C129 Type II standard. The following conclusions were drawn from the investigation:

1. With an increase of replacement ratio of materials of lightweight concrete by melamine plastic aggregates.
   - The compressive strength and densities of the lightweight concrete were reduced.
   - The absorption and volume of permeable voids of the lightweight concrete were increased.
2. The existed linear relationship between the compressive strength and density. Also, the existed inverse relationship between the compressive strength, density, and absorption, volume of permeable voids.
3. Density and compressive strength of mortar decreased as the water content increased.
4. The compressive strength increase as the addition of SF increases.
5. Utilization of other plastics in the mix proportion for non-load-bearing lightweight concrete according to ASTM C129 Type II standard and other standards are suggested for further studies.

A CASE STUDY

Industrial waste of different origins and nature can be used as unconventional constituents for the preparation of new blended cements. This case study collects different researches previously carried out with the aim to highlight the feasibility of this recycling route that can be considered highly rewarding for the cement industry. Chemical, physical and mechanical properties of the new blended cements are reviewed and compared with the requirements set by EN 197-1 for common cements.

The worldwide cement production in 2007 was 2.77 billion tons: Asia is the first producer (70%), followed by European Union countries (9.5%). Indeed, cement industry can be considered strategic: in fact, from one side, it produces an essential product in building and civil engineering for the construction of safe, reliable and long lasting buildings and infrastructures. On the other side it is very important from the economic point of view (for example Indian cement industry is playing a very import role in the economic development of the country). However, cement industry environment-wise is also responsible for a large use of not renewable raw materials (clays and calcium carbonate) and fossil fuels (for example clinker, the main cement constituent, is obtained at T = 1500°C) resulting in heavy emissions of carbon dioxide (CO_2) in atmosphere. In fact, in 2006, the European cement industry used an energy equivalent of about 26 Mt of coal for the production of 266 Mt of cement and it is estimated that 1 ton of CO_2 is emitted for each ton of cement produced. This induces cement industry to consider the possibility to introduce waste of different nature and origin in cement productive process. Two routes are currently taken into consideration: one involves the use of waste as alternative fuel, the other considers waste as a new cement constituent. However, both routes generate concerns. Using waste as an alternative fuel, it must be excluded the possibility that dangerous volatile compounds such as dioxins can be emitted into the environment during clinker production. Using waste as new cement constituent, it must be ascertained the absence of substances that can negatively interfere with cement reaction.

European standard EN 197-1, providing a complete classification of cements valid in Europe, confirms that a constituent of cement can also be a waste. In fact, except the ordinary Portland cement (OPC) constituted by 95% of clinker (CEM I), blended cements contain, besides clinker, other constituents coming from waste of different productive processes. Blast furnace slag, silica fume and fly ash respectively derive from iron ore processing for cast iron and steel production, from silicon and ferro-silicon alloys production and from coal combustion for electric energy production. Cement classification based on the different constituents is reported in Table 1. It has been proved that the addition of waste such as blast furnace slag, silica fume and fly-ash leads to several advantages for the relevant cements: (i) the lower content of clinker compare to OPC allows to classify these binders as low-heat cements; (ii) the development of cement mechanical strength, although it is slower at the early curing age, noteworthy increases with time, even after 28 days; (iii) durability behavior improves due to a less content of Portlandite, usually formed during OPC hydration reactions. Moreover, blended cements always need less clinker for their production thus involving a minor use of natural raw materials and fuels, less quarries exploitation and lower CO_2 emission. When the new cement constituents are waste, benefits as the safeguard of disposal sites and saving of natural raw materials must be added at the above quoted list.

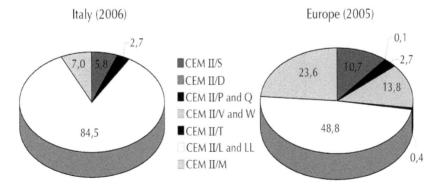

FIGURE 1   Italian and European market division (%) for the different types of CEM II.

TABLE 1 Cement classification according to EN 197-1 (wt%) (Minor additional constituents (0-5%) are not reported for brevity sake)

Main types	Notations		Main Constituents										
			Clinker	Blast Furnace slag	Silica fume	Pozzolana		Fly ash		Burnt shale	Limestone		
						nat	natural calcined	Sil.	Calc.				
			K	S	D	P	Q	V	W	T	L	LL	
CEM I	Portland Cement	CEM I	95–100	-	-	-	-	-	-	-	-	-	-
CEM II	Portland slag cement	CEM II/A-S	80–94	6–20	-	-	-	-	-	-	-	-	-
		CEM II/B-S	80–94	21–35	-	-	-	-	-	-	-	-	-
	Portland silica-fume cement	CEM II/A-D	90–94	-	6–10	-	-	-	-	-	-	-	-
		CEM II/A-P	80–94	-	-	6–20	-	-	-	-	-	-	-
		CEM II/B-P	65–79	-	-	21–35	-	-	-	-	-	-	-
	Portland Pozzolan cement	CEM II/A-Q	80–94	-	-	-	6–20	-	-	-	-	-	-
		CEM II/B-Q	65–79	-	-	-	21–35	-	-	-	-	-	-

TABLE 1 *(Continued)*

Type	Notation	Clinker	Slag (S)	Fly ash V	Fly ash W	Burnt shale (T)	Limestone L	Limestone LL	Composite (M)	Pozzolana
Portland fly-ash cement	CEM II/A-V	80–94	–	6–20	–	–	–	–	–	–
	CEM II/B-V	65–79	–	21–35	–	–	–	–	–	–
	CEM II/A-W	80–94	–	–	6–20	–	–	–	–	–
	CEM II/B-W	65–79	–	–	21–35	–	–	–	–	–
Portland burnt shale cement	CEM II/A-T	80–94	–	–	–	6–20	–	–	–	–
	CEM II/B-T	65–79	–	–	–	21–35	–	–	–	–
Portland limestone cement	CEM II/A-L	80–94	–	–	–	–	6–20	–	–	–
	CEM II/B-L	65–79	–	–	–	–	21–35	–	–	–
	CEM II/A-LL	80–94	–	–	–	–	–	6–20	–	–
	CEM II/B-LL	65–79	–	–	–	–	–	21–35	–	–
Portland composite cement	CEM II/A-M	80–94	–	–	–	–	–	–	6–20	–
	CEM II/B-M	65–79	–	–	–	–	–	–	21–35	–
CEM III Blast furnace cement	CEM III/A	35–64	36–65	–	–	–	–	–	–	–
	CEM III/B	20–34	66–80	–	–	–	–	–	–	–
	CEM III/C	5–19	81–95	–	–	–	–	–	–	–
CEM IV Pozzolan cement	CEM IV/A	65–89	–	–	–	–	–	–	–	11–35
	CEM IV/B	45–64	–	–	–	–	–	–	–	36–55
CEM V Composite cement	CEM V/A	40–64	18–30	–	–	–	–	–	–	18–30
	CEM V/B	20–38	31–50	–	–	–	–	–	–	31–50

Not all the cement types reported in Table 1 are usually manufactured. So far, as a consequence to produce less clinker, CEM II is the most produced cement with a percentage of 80.8 and 56.1 over the total production, in Italy and Europe respectively. Figure 1 shows how CEM II is fractioned on the market. Although Portland-limestone cement is the most popular, CEM II based on waste (blast furnace slag, silica fume, fly ash, with an amount ranging from 6 to 35%) is 48.2% in Europe, thus meaning that recycling is already successfully adopted in cement industry.

With the aim to pursue waste suitable to work as new cement constituents, several researches have been carried out specifically studying ground glass, matt waste, rice husk ash, municipal solid waste incinerator bottom ash, ceramic waste, ferroalloy industry waste, electric-arc furnace slag, and so on.

In this appendix industrial waste and relevant new blended cement constituted by 25 wt% waste and 75 wt% CEM I, separately investigated, are collectively reported and compared.

Their chemical, physical and mechanical characteristics are discussed in the framework of the requirements imposed by the European standard EN 197-1. The industrial waste considered is matt waste (MW), which derives from purification processes of cullet coming from separated glass waste collection, polishing (PR) and glazing (GR) residues, coming from porcelain stoneware polishing and glazing sludge, respectively, and ladle electric-arc furnace slag (LS), generated in the refining process of electric-arc furnace steels production. The four types of waste, coming from glass, ceramic and steel industry, are produced in large amount (6000 ÷ 25000 t/y), thus ensuring a noteworthy quantity of recyclable material for cement production.

THE CHEMICAL-PHYSICAL CHARACTERIZATION OF THE INVESTIGATED WASTE

A highly representative batch of each of the investigated waste was collected—MW was kindly supplied by Sasil, Gruppo Minerali (Novara, Italy), LS by Acciaieria di Rubiera (Casalgrande, Reggio Emilia, Italy), porcelain stoneware polishing and glazing sludge by S.A.T. S.p.A. Service for the Environment (Sassuolo, Modena, Italy) and by different glazing lines of a single Italian manufacturer, respectively. All the materials were submitted to some treatments before their use. Porcelain stoneware polishing and glazing sludge were dried (T = 105°C for 24–36 hr) to obtain solid residues (PR and GR) and ground in a laboratory ball mill to reach a grain size distribution close to that of commercial cement. The MW was also ground for the same reason. Ladle slag, although is produced as a very fine powder, was sieved to eliminate the fraction greater than 0.106 mm. Figure 2 collectively reports the grains size distributions of MW, LS, PR, GR, and that of a commercial OPC CEM I 52.5 R: the average size, ranging from 6 to 20 μm, increases in this order: LS >CEM I >GR >PR >MW.

A chemical and mineralogical characterization of the materials was then carried out to establish the main oxide constituents and mineralogical phases. The results are summarized in Table 2 and Table 3: SiO_2 is the main constituent of MW, GR, and PR but while MW has the typical chemical-mineralogical composition of soda-lime glass (about 10 and 13 wt% content of CaO and Na_2O, respectively, and amorphous silica phase), GR and PR are also rich in Al_2O_3 (15–20 wt%) and ZrO_2 (1–3 wt%), minerals such as zircon, quartz and albite deriving from glazing and porcelain stoneware body, respectively. The main constituents of LS are CaO (55 wt%), SiO_2 (24 wt%), and Al_2O_3 (13 wt%) with mineralogical crystalline phases as calcium silicates and aluminates.

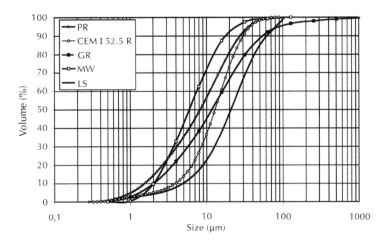

FIGURE 2   Grains size distribution of the investigated industrial waste.

TABLE 2   Chemical analysis (main oxide, wt%) of the investigated industrial waste (x-ray fluorescence spectrometer, XRF, PW 1414, Philips)

Oxide	MW (%)	LS (%)	GR (%)	PR (%)
SiO_2	71.04	23.85	52.36	62.19
Al_2O_3	2.02	13.69	19.37	15.75
TiO_2	-	0.20	0.45	0.34
Fe_2O_3	0.35	3.85	0.84	0.59
CaO	10.58	55.24	5.73	2.24
MgO	1.75	3.22	2.43	6.75
K_2O	0.75	0.22	1.32	1.46
Na_2O	13.52	0.31	3.90	3.71
ZrO_2	n.d.	<0.10	3.01	1.19
MnO	n.d.	0.32	<0.10	<0.10
ZnO	n.d.	<0.10	0.99	0.12
BaO	n.d.	<0.10	0.54	<0.10

TABLE 3   Main mineralogical phases of the investigated industrial waste (X-ray diffractometer with Ni-filtered Cu Kα (λ = 1.54 Å) radiation, XRD PW 1840, Philips)

	MW	LS	GR	PR
Crystalline phase	n.d.	olivine (γ–C_2S), ghelenite, mayenite, iron silicates, iron magnesium calcium silicate	quartz, zircon, albite	quartz, albite calcian
Amorphous phase	silica	n.d.	silica	silica

The EN 197-1 sets specific requirements concerning the amount of chloride and sulphate in cement, loss of ignition (LOI) must be also limited to a value≤ 5.0%. These limitations are necessary to ensure that deleterious reactions, such as steel bar corrosion and/or delayed ettringite formation, induced by Cl⁻ and SO_4^{-2} respectively, do not occur in cement base materials.

Table 4 reports the values determined according to EN 196-2 for the investigated materials and OPC CEM I 52.5R. MW, LS, and GR values agree with all the limits sets, whereas PR exhibits a higher Cl⁻ content and slightly overcomes the limit LOI value. Cl⁻ derives from the salts ($AlCl_3$, $FeCl_3$) used as flocculants in the separation process and from the magnesium chloride matrix of abrasive tools used for polishing. The LOI value is mainly related to the presence of calcite phase and SiC, coming from the abrasive tools. However, both the data can be acceptable as PR is going to be used mixed with CEM I 52.5 R (25 and 75% respectively), hence the overall Cl⁻% and LOI will be inside the limits required for cement.

Besides chemical requirements, physical properties such as setting time and soundness are very important to establish if a cement based binder is suitable to be adopted in the building industry. Setting time is a measurement of the time required by a cement paste to start hydration reactions thus leading to a loss of workability at the fresh state; soundness test allows to determine if volume expansion of cement paste occurs under accelerated curing conditions. Deleterious expansion phenomena are usually caused by the presence of free CaO and MgO in the binder. Table 5 reports the results obtained on the new blended cement based on the investigated waste, having the general composition—75 wt% CEM I 52.5 R + 25 wt% waste.

The results determined for 100 wt% CEM I 52.5 R are reported for comparison as well as the limits set by EN 197-1. Soundness results of all the new blended cement are comparable with that of CEM I and well below the limit; the same can be observed for the initial setting time, except for that of the binders containing MW and GR, which are slightly higher than 107 min. Such increase, although still acceptable for cement binder, has been ascribed by organic impurities ($\leq 1\%$) that leads to a slight slowdown of initial setting reactions.

The chemical and physical parameters obtained for MW, LS, GR, and PR allow their use as new constituent for cement production according to the restrictions reported by the European standard.

TABLE 4 Chemical analysis results of the investigated industrial waste and CEM I 52.5 R (average of 2 measurements). Limits set by EN 197-1 for cement are also reported

	MW	LS	GR	PR	CEM I 52.5 R	Limits set by EN 197-1 for cement
Chloride (wt %) EN 196-2	0.04	$\leq$ 0.01	≤ 0.01	0.24	0.04	≤ 0.1
Sulfate (SO_3 wt %) EN 196-2	$\leq$ 0.01	0.38	0.06	0.09	0.65	≤ 3.5
Loss of Ignition (wt %) EN 196-2	0.8	2.7	2.9	5.2	3.6	≤ 5.0

TABLE 5 Physical properties of the investigated binder based on industrial waste and CEM I 52.5 R (average of 2 measurements). Limits set by EN 197-1 for cement are also reported

Binder		Initial setting time (min)	Soundness (mm)
CEM I 52.5 R		107	0.2
75% CEM I 52.5 R + 25% MW		134	0.3
75% CEM I 52.5 R + 25% LS		106	0.4
75% CEM I 52.5 R + 25% GR		117	0.2
75% CEM I 52.5 R + 25% PR		105	0.2
Limits set by EN 197/1 for cement	32.5 R	$\geq$75	≤ 10
	42.5 R	$\geq$60	
	52.5 R	$\geq$45	

THE BEHAVIOR OF THE NEW BINDERS AT THE FRESH AND HARDENED STATE

Two parameters are extremely important for a binder involved in concrete production: workability and compressive mechanical strength determined at the fresh and hardened state respectively. For the binders under studying, workability was measured by mini slump test on paste sample prepared with a water/binder ratio equal to 0.5. The results, reported in Table 6, shows that workability increases in this order: CEM I >MW-binder >LS-binder >GR-binder >PR-binder. Workability is strictly related to waste average size, shape and its tendency to form particles agglomerates.

TABLE 6   Workability results at the fresh state (average of 2 measurements)

Binder	Water/binder ratio	Mini slump (mm)
CEM I 52.5 R	0.5	80
75% CEM I 52.5 R + 25% MW	0.5	70
75% CEM I 52.5 R + 25% LS	0.5	62
75% CEM I 52.5 R + 25% GR	0.5	55
75% CEM I 52.5 R + 25% PR	0.5	45

As far as concern mechanical requirements, EN 197-1 sets for each kind of cement reported in Table 1 a division based on cement mechanical strength—three classes (32.5, 42.5, and 52.5 N/mm² determined at 28 days of curing) are identified and each class can have a normal (N) or rapid/early (R) strength development (Table 7).

The compressive strength was determined on mortar samples prepared, by Hobart planetary mixer, according to the normalized mix-design (binder, sand and water in a weight ratio of 1:3:0.5) and procedure for cement mechanical strength determination (EN 196-1). Mortar samples were named M, followed by the acronym of the waste type used; mortar prepared with CEM I 52.5 R was identified as MREF. Figure 3 collectively reports the compressive strength at 2, 7, and 28 days and the following observation can be drawn—(i) all the new blended cements exhibit a compressive strength lower than that of CEM I 52.5 R at all curing times; (ii) according to the limits reported for cements, all the new blended binders overcome the threshold values at 2 and 28 days required for 42.5 R strength class; (iii) only the PR based binder reaches the limits required to be classified in the 52.5 R strength class.

It is usual that the addition of a constituent with a different chemical composition from clinker may decrease the development of mechanical strength in the first 28 days of curing, however pozzolan and blast furnace slag constituents usually lead to a continuative increase in the mechanical properties with curing, even after 28 days. A useful tool to understand if a new constituent is involved in mechanical strength development at long curing time is the determination of the activity index (AI). This index, reported in EN 450-1 for concrete fly ash, is the compressive strength ratio between samples containing 75% CEM I 52.5 R + 25% waste and reference mortar with 100% CEM I 52.5 R, cured at 28 and 90 days. Values of AI ≥75 and 85% at 28 and 90 days, respectively, mean that the new constituent is active in the strengthening development. The data, collected in Table 7, show that the activity index of the investigated waste increases with this order: PR >MW >GR >LS.

The LS and GR exhibit a lower/almost equal AI than the limit at 90 days, thus indicating that their action as cement constituent is limited to a filler effect. Activity index of MW and PR largely overcomes the limit, thus meaning that MW and PR actively participate in mechanical strength development.

TABLE 7 Cement classification based on the development of compressive strength

	Strength class	Compressive strength (MPa)		
		2 days	7 days	28 days
Limits set by EN 197-1	32.5 N		≥ 16.0	≥ 32.5
for cement	32.5 R	≥ 10.0	-	≤ 52.5
	42.5 N	≥ 10.0		≥ 42.5
	42.5 R	≥ 20.0	-	≤ 62.5
	52.5 N	≥ 20.0		≥ 52.5
	52.5 R	≥ 30.0	-	

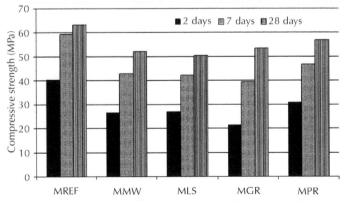

FIGURE 3 Compressive strength at different curing time of mortar samples prepared with the investigated binders.

Clearly, the different behavior of the industrial waste is strictly related to their nature: the amorphous phase of MW makes it similar to an artificial pozzolan constituent (fly ash, silica fume), whereas PR, containing both amorphous and crystalline phases, contributes to the formation of two different types of gels during hardening process, as previously ascertained. Both MW and PR are thus chemically involved in clinker hydration reactions, leading to matrices with very compact microstructures.

TABLE 8   Activity index for the investigated industrial waste

Binder	Activity index	
	28 days	90 days
75% CEM I 52.5 R + 25% MW	84	92
75% CEM I 52.5 R + 25% LS	80	67
75% CEM I 52.5 R + 25% GR	84	88
75% CEM I 52.5 R + 25% PR	90	101
Limits set by EN 450-1	≥75%	≥ 85%

4.5   SUMMARY

The investigated waste is chemically, physically and mechanically suitable as 25 wt% constituent for the production of new blended cements. PR blended cement can be classified as 52.5 R strength class, whereas MW, GR, and LS blended cement belong to 42.5 R strength class. The waste behavior is different:

- The GR and LS work more as inert addition (filler effect) than as active constituents of the relevant blended cements. The reason can be found in the crystalline phases characteristic of the two types of waste that are not able to take part/collaborate to the clinker hydration process.
- The PR and MW have an active role in the relevant blended cement—activity index of both waste at 90 days is close to 100, thus indicating that compressive strength of mortar samples increases with curing time, event after 28 days.

KEYWORDS

- **Aluminum powder**
- **Melamine plastic**
- **Ordinary portland cement**
- **Plastic waste management**
- **Thermosetting plastic**

REFERENCES

1. Albano, C., Camacho, N., Hernandez, M., Matheus, A., and Gutierrez. A. Influence of content and particle size of waste pet bottles on concrete behavior at different w/c ratios. *Waste Management*, **29**, 2707–2716 (2009).
2. Al-Salem, S. M., Lettieri, P., and Baeyens, J. The valorization of plastic solid waste (PSW) by primary to quaternary routes: From re-use to energy and chemicals. *Progress in Energy and Combustion Science*, **36**, 103–129 (2010).
3. Chan, Y. N. S. and Ji, X. Comparative study of the initial surface absorption and chloride diffusion of high performance zeolite, silica fume and PFA concretes. *Cement & Concrete Composites*, **21** 293–300 (1999).
4. Choi, Y. W., Moon, D. J., Kim, Y. J., and Lachemi, M. Characteristics of mortar and concrete containing fine aggregate manufactured from recycled waste polyethylene terephthalate bottles. *Construction and Building Materials*, **23**, 2829–2835 (2009).
5. Choi, Y. W., Moon, D. J., Chung, J. S., and Cho, S. K. Effects of waste PET bottles aggregate on the properties of concrete. *Cement and Concrete Research*, **35**, 776–781 (2005).
6. Ismail, Z. Z. and AL-Hashmi, E. A. Use of waste plastic in concrete mixture as aggregate replacement. *Waste Management*, **28**, 2041–2047 (2008).
7. Marzouk, O. Y., Dheilly, R. M., and Queneudec, M. Valorization of post-consumer waste plastic in cementitious concrete composites. *Waste Management*, **27**, 310–318 (2007).
8. Naik, T. R., Singh, S. S., Huber, C. O., and Brodersen, B. S. Use of postconsumer waste plastics in cement-based composites. *Cement and Concrete Research*, **26**(10), 1489–1492 (1996).
9. Panyakapo, P. and Panyakapo M.. Reuse of thermosetting plastic waste for lightweight concrete. *Waste Management*, **28**, 1581–1588 (2008).
10. Rao, G. A. Investigations on the performance of silica fume incorporated cement pastes and mortars. *Cement and Concrete Research*, **33** 1765–1770 (2003).
11. Siddique, R., Khatib, J., and Kaur, I. Use of recycled plastic in concrete: A review. *Waste Management*, **28**, 1835–1852 (2008).

CHAPTER 5

A NEW GENERATION OF COMPOSITE SOLID PROPELLANTS

MAYSAM MASOODI and MOHAMAD ALI DEHNAVI

CONTENTS

5.1   INTRODUCTION

The nitrate ester plasticized polyether (NEPE) is a new generation of composite solid propellants which its high energy content in comparing with conventional formulations, has attracted many researchers in the defense departments, all over the world. These composite propellant mixtures have been prepared by using ammonium perchlorate (AP), aluminum (Al) powder, and polyether polymer binder polyethylene glycol (PEG) by varying the percentage of plasticizer and addition tow particle size of energetic filer octagon or cyclotetramethylene-tetranitramine (HMX), and studied mechanical properties. The data on tow particle size of HMX indicate that particle size and ratio of the energetic filler affects the mechanical properties of the propellant. The modulus of elasticity is not dramatically as the particle size decreases and affects on tensile strength, modulus, and elongation drastically.

The NEPE propellants are high-energetic composite solid propellants that use a polyether, such as PEG or an ethylene oxide–tetrahydrofuran co-polyether, as a polymeric binder, and nitroglycerin (NG), as plasticizer. The large amounts of solid particles, such as Al powder, HMX and AP are contained in the propellants. This type of propellant integrates the advantages of double-base propellants and composite propellants, and adds excellent low temperature mechanical properties. Therefore, this type has been studied extensively and applied broadly in many countries since it was first developed in the USA in the 1970s [1, 2]. The general composition of the high energy propellant is about 22–27 wt% binder and about 73–78 wt% solids. The binder itself has several components—two polymers, a plasticizer, are a curative, and two stabilizers. The binder also contains a trace amount of catalyst. The solids component of the propellant is comprised of a metalized fuel, energetic filler, and an oxidizer [3, 4]. Table 1 illustrates four examples of preferred propellant compositions, by weight percent. Examples I and II exemplify the most preferred formulations. It is to be understood that the values specified in Table 1 are approximate, and some variations from those shown are still within the scope of this preparation.

5.2   EXPERIMENTAL

5.2.1   MATERIALS

The AP, procured from Ammonium Perchlorate Experimental Plant (APEP), was used in trimodal distribution having average particle size of 300 μm, 60 μm, and 6 μm, respectively. PEG having various molecular weight, was procured from trade. Al powder, having average particle size 15±3 μm, was procured from "The Metal Powder Company", HMX powder having tow particle size and used as such NG and Desmodur N-100 used as plasticizer and curative, respectively, were procured from trade. Other ingredients were also procured from trade and used as such.

The particle size of solid ingredients was determined by Malvern Particle Size Analyzer, model 2600C in non-aqueous medium. The viscosity build-up was determined by Brookfield Viscometer, model HBT dial type. The density of the cured propellant mixture was determined by Archimedes' principle using a density attachment with 0.1 mg precision balance of Mettler Toledo makes. The mechanical properties like tensile strength, % elongation and E-modulus of cured propellant mixtures were evaluated using dumb-bells on tensile testing machine, Instron 1185 conforming to ASTM D638 at a cross head speed of 50mm/min at ambient temperature. Solid strand burn rate (SSBR) was determined using acoustic emission technique in nitrogen atmosphere at 70 kg/cm^2 pressure. The calorimetric value (cal-val) of the mixture was evaluated by Parr Bomb Calorimeter using 1 g of sample.

5.2.2 PROCEDURE

PREPARATION OF NEPE PROPELLANT MIXTURE

All the experimental mixing of NEPE propellants were carried in a vertical planetary mixer. A general procedure for the preparation of composition is described in detail. PEG with other liquid ingredients (except curative) was charged into vertical planetary mixer. It was mixed well for half an hour followed by vacuum mixing for another half an hour to drive out entrapped air. After this, Al powder (15 ± 3 µm) was added in two steps. The AP (powder) was added in such a way that homogenous mixing could take place. The overall mixing temperature was maintained at $55°C \pm 1°C$. After addition of complete solid ingredients, the mixing of mixture was carried out under vacuum for half an hour. In the meantime, the temperature of the mixer was brought down to $37 \pm 1°C$. At this stage, Desmodur N-100 was added and it was further mixed for another 35 min. The mixture was cast into mould and cured at 50°C for 5 days and used for evaluation of its different properties.

TABLE 1 Formulation of NEPE propellant [5]

Components	Example I	Example II
Binder		
Polymers		
Polyethylene Glycol (PEG)	6.25	6.25
Cellulose Acetate Butyrate (CAB)	0.06	0.06
Plasticizer		
Nitroglycerin (NG)	19.02	19.02
Cross-linking Curative		
Desmodur N-100	0.88	0.88
Stabilizer		
2nitrodiphenyl amine (2NDPA)	0.19	0.19

TABLE 1   *(Continued)*

N-Methyl-p-nitroaniline (MNA)	0.60	0.60
Cure Catalyst		
Triphenyl Bismuth (TPB)	0.02	0.02
Solids		
Fuel		
Aluminum	18.00	18.00
Energetic Filler		
Cyclotetramethylene Tetranitramine (HMX)	47.00	46.00
Oxidizer		
Ammonium Perchlorate (AP)	8.00	9.00

5.3   DISCUSSION AND RESULTS

Examples I and II in the Table 1, which represent the most preferred embodiments of the present preparation, have the following mechanical properties:

TABLE 2   Mechanical properties of Examples I and II

Properties	Example I	Example II
Modulus(psi)	430	545
Tensile strength(psi)	92	99
Elongation(%)	270	273

These values are determined at a 2 in/min pull rate at 80° F. The above values were obtained from a 600 gal lon mixed batch [6]. The preferred polymers are PEG and cellulose acetate butyrate (CAB). The polymer PEG functions to give physical strength to the binder when cross-linked with itself and/or CAB. Poly(propylene glycol) (PPG) can be used for some PEG but the plasticizers are less soluble in it so that in most cases it is preferred to use PEG as the polyol polymer. PEG and CAB also serve as sources of fuel in the propellant CAB as a cross linker provides physical strength by improving tensile strength and the modulus of elasticity In other embodiments, such chemicals as cellulose acetate, cellulose butyrate, trimethylolpropane, or glycerin can be substituted in part or in whole for the CAB component of the inventive fuel. Preferably a mixture of cellulose acetate and cellulose butyrate are used in combination as substitutes for CAB rather than either alone [6, 7].

5.3.1 EFFECT OF PEG MOLECULAR WEIGHT IN MECHANICAL PROPERTIES

In the present formulation, the mechanical properties of the propellant can be modified by varying the molecular weight of the PEG employed in the propellant formulation. The molecular weight of the polyether and the modulus of elasticity vary inversely but the elongation varies directly. Thus, increasing the molecular weight of the polyether causes a steady drop in the modulus of elasticity and increases the elongation in the resultant fuel. The molecular weight of the PEG has a direct bearing on the cross-link density and gelling efficiency of the present propellant, particularly at the higher Pl:PO (that is plasticizer–polymer) levels due to the dilution of the polymer. The various effects of PEG molecular weight modulation can be seen in Table 3.

TABLE 3 Effects of PEG molecular weight

MW	s(psi)	e(%)	E(psi)
430	64	37	680
1020	70	56	728
1376	77	100	510
3240	65	292	296
4100	67	411	204

In alternate embodiments, other long chain polyols may be used in part or in whole in place of PEG. For instance, the mechanical properties can be modified by blending the PEG component of the fuel with various amounts of PPG to form block copolymers. This effect is demonstrated in Table 5. However, the use of PEG/PPG tends to yield a poor modulus. The inclusion of polypropylene for PEG tends to reduce the solubility of plasticizers and thus in most cases PEG alone is preferred and used [6, 8].

5.3.2 ENERGETIC PLASTICIZERS AND OTHER ENERGETIC MATERIAL THAT USE POLYMERIC PROPELLANTS

The preferred plasticizer is NG, a high energy compound. The mixing of CAB and NG results in a material that is formable and plastic. Its use in the present preparation results in improved mechanical properties and higher performance for the binder. In alternate embodiments, other nitrate esters generally may serve as suitable plasticizers. Butanethol trinitrate (BTTN), triethylene glycoldi nitrate (TEGDN), diethylene glycol dinitrate (DEGDN), and trimethylolethane trinitrate (TMETN) are examples of plasticizers that can be utilized within the purview of the present formulation. The cross-linking curative of the subject preparation is responsible for cross-linking the various components of the fuel. A polyfunctional isocyanate containing the ferred. It has an NCO functionality of at least 3. An especially suitable curative is desmodur N-100, commercially available from Mobay Chemical. Co. which is a complex mixture of biurets, uretediones, isocyanurates, and unreacted hexamethylene diisocyanate. The optimal cross-linking and mechanical properties are obtained when the stabilizer for NG and N-methyl-p-nitroaniline (MNA), acts to stabilize the complete propellant. In additional embodiments of the present preparation, other polyfunctional isocyanates may be successfully employed as cross-linking curatives. The pentaerythritol tetraisocyanate is a chemical which can be considered as an alternative curing agent. Difunctional isocyanates tend to increase the elasticity of the resulting fuel to a great degree, and so are generally less preferred [5, 7, 9].

The organo bismuth compounds are used as cure catalysts. Triphenyl bismuth (TPB) is the preferred cure catalyst in the present in preparation it functions to speed up the cross-linking process and helps to provide a very long pot life or working life (that is, until a viscosity of about 40 kilopoise is reached). Because TPB functions as a relatively slow reacting cure catalyst, it is particularly advantageous with large missiles where the set time for the fuel is longer. Competing and interfering 30 reactions are minimized in this system.

Other organo metal cure catalysts can be employed in place of TPB. Examples are trialkyl bismuths, such as triethyl bismuth. Other metals can be used but they can have adverse effects unrelated to cure. The choice depends on the final fuel qualities desired and accommodations to be made to the exigencies of the particular production process employed. The considerations involved include the size of the missile for which the fuel is being manufactured [6].

The metal used to form the metalized fuel used in the preferred embodiment of the present preparation is free metal Al. The Al reacts with the oxidizers primarily to provide heat as a product of the combustion process. This fuel is particularly useful in larger missiles and may be eliminated in smaller tactical missiles, or for minimum smoke missiles. In other embodiments, different metalized fuels can be employed with that of the preferred embodiment or as substitutes for it, depending on the nature of the fuel desired. Suitable other metals for use in metalized fuels are boron and beryllium (although the latter is quite toxic). The energetic filler employed in the preferred embodiment of the present preparation is HMX, which generates heat and gases. Also useful in this capacity is cyclotrimethylenetrinitramine (RDX) [10-13].

5.4 EFFECT OF PARTICLE SIZE AND RATIO OF ENERGETIC MATERIAL

As in Table 4, the particle size and ratio of the energetic filler can be varied to fit the needs of a particular fuel requirement. An increase in quantity enhances the energetic nature of the fuel. The HMX size affects the mechanical properties of the propellant. The modulus of elasticity is not dramatically as the particle size decreases. The particle size is varied to obtain a proper viscosity for processing, but the optimum mechanical properties are achieved with the finest possible HMX [6, 7].

TABLE 4   Effect of HMX size on mechanical properties

HMX Size(m)	4	2
Tensile strength(psi)	91	97
Elongation(%)	290	450
Modulus(psi)	430	370

The preferred oxidizer in the present preparation is AP, which functions to oxidize all of the hydrocarbons and the metal, Al, to generate heat and gases. AP is also used to control the burning rate of the propellant. The particle size and ratio of the oxidizer can be varied to fit the needs of a particular fuel requirement. The propellants provided in Examples I and II specified in Table I have the following typical properties [10, 11].

TABLE 5   Typical properties for two examples

	ExampleI	ExampleII
Density(lb/in^3)	0.0665	0.0665
Burning rate(1000psi,80° F., in/sec)	0.42	0.49
Specific impuls, (lbf.sec/lbm)	271.5	271.4

5.5   CONCLUSION

The NEPE propellants are high-energetic composite solid propellants that use energetic material such as NG, HMX, and AP in composition the propellants. This propellant with unique characteristics, high specific impulse high density, and volumetric specific impulse or high efficiency ratio has many significant advantages than solid propellant. The particle size and ratio of the energetic filler affects the mechanical properties of the propellant. The modulus of elasticity is not dramatically as the particle size decreases.

KEYWORDS

- **Butanethol trinitrate**
- **Cyclotetramethylene-tetranitramine**
- **Nitrate ester plasticized polyether**
- **Poly(propylene glycol)**
- **Triethylene glycol dinitrate**

ACKNOWLEDGMENT

The authors of this chapter appreciate the financial support and cooperation of Mosem research center of Engineering and Technology faculty of Imam Hosein comprehensive university.

REFERENCES

1. Yong, L., Luoxin, W., Xinlin, T., and Song Nnian, L. An SEM and EDS study of the microstructure of NEPE propellant, *J. Serb. Chem. Soc.*, **75**, 369–376 (2010).
2. Yong, L., Luoxin, W., Xinlin, T., SongNnian, L., and Weimin, Y. A study on the microstructure of a NEPE propellant dissolved in HCl and KOH solutions, *J. Serb. Chem. Soc.*, **75**, 987–996 (2010).
3. Li, S., Liu, Y., Tuo, X., and Wang, X. Mesoscale dynamic simulation on phase separation between plasticizer and binder in NEPE propellants, *Polymer*, **49**, 2775–2780 (2008).
4. Ping, H. Z., Ying Nie, H., Yuang, Z. Y., min, T. L., Li, Y. H.,and Gang, Ma. X. Migration kinetics and mechanisms of plasticizers, stabilizers at interfaces of NEPE propellant/ HTPB liner/EDPM insulation, *Journal of Hazardous Materials.*, 251–257 (2012).
5. Agrawal, J. P. *High energy materials: propellants, explosives, and pyrotechnics*, Weinheim: Wiley-VCH, (2010).
6. Hughes, Ch. W., Godsey, J. H., Keller, R. F.,and Cir. R. *High energy propellant formulation*, United States Statutory Invention Registration, H1341,2,(1994).
7. Davenas, A. *Solid Rocket Propulsion Technology*, Pergamum Press, Technology and Research Director, SNPE, France, (1990).
8. Sui, X., Wang, N., Wan, Q., and Bi Sh. Effects of Relaxed Modulus on the Structure Integrity of NEPE Propellant Grains during High Temperature Aging, *Propellants Explos. Pyrotech*, **35**, 535–539, (2010).
9. Taihua, Z., Yilong, B., Shiying, W., and Peide, L. Damage of a high-energy solid propellant and its effects on combustion, *Acta Mechanical Sinica (English Series)*, November, **17**, 529–533 (2001).
10. Li, Sh., Yajun, W., Shanshan, Z., and Shuqiong, Z. *Theory and practice of energetic materials*, International Autumn Seminar on Propellants, Explosives, and Pyrotechnics, (2009). Yan, W., Lei, G., Yanbin, L., and Zhixian W. Combustion Synthesis of La0.8Sr0.2MnO3 and Its Effect on HMX Thermal Decomposition, *Chinese Journal of Chemical Engineering*, **18**, 397–401 (2010).Sun, Y., Zhu, B.,and Li, Sh. Effect of Nitrate Ester on the Combustion Characteristics of PET/HMX-based Propellants, *Defense Science Journal*, **61**, 206–213 (2011).Nair, U. R., Asthana, S. N., Subhananda Rao, A., and Gandhe, B. R. Advances in High Energy Materials. *Defense Science Journal*, **60**, 137–151 (2010).

CHAPTER 6

KEY ELEMENTS ON SURFACE PROPERTIES OF POLYIMIDE COPOLYMERS

IGOR NOVÁK, PETER JURKOVIČ, JAN MATYAŠOVSKÝ, PETR SYSEL, MILENA ŠPIRKOVÁ, and LADISLAV ŠOLTES

CONTENTS

6.1   INTRODUCTION

Several sorts of block polyimide based copolymers, namely poly(imide-co-siloxane) (PIS) block copolymers containing siloxane blocks in their polymer backbone have been investigated. In comparison with pure polyimides the PIS block copolymers possess some improvements, for example, enhanced solubility, low moisture sorption, and their surface reaches the higher degree of hydrophobicity already at low content of polysiloxane (PSI) in PIS copolymer. This kind of the block copolymers are used as high-performance adhesives and coatings. The surface as well as adhesive properties of PIS block copolymers depends on the content and length of siloxane blocks. The surface properties of PIS block copolymers are strongly influenced by enrichment of the surface with siloxane-based segments. Micro phase separation of PIS block copolymers occurs due to the dissimilarity between the chemical structures of siloxane and imide blocks even at relatively low lengths of the blocks. The surface analysis of PIS block copolymers using various methods of investigation ,for example, contact angle measurements, scanning electron microscopy (SEM), transmission electron microscopy (TEM), atomic force microscopy (AFM), attenuated total reflectance-Fourier transform infrared spectroscopy (ATR-FTIR), and X-ray photoelectron spectroscopy (XPS), was performed, and the strength of the adhesive joint to more polar polymer was studied. The surface and adhesive properties are discussed in view of the varied composition of PIS block copolymers.

The polyimides present an important class of polymers, necessary in microelectronics, printed circuits construction, and aerospace investigation, mainly because their high thermal stability and good dielectric properties [1-4]. The PIS block copolymers containing siloxane blocks in their polymer backbone have been investigated [5, 6]. In comparison with pure polyimides the PIS block copolymers possess some improvements, for example, enhanced solubility, low moisture sorption, and their surface reaches the higher degree of hydrophobicity already at low content of PSI in PIS copolymer. This kind of the block copolymers are used as high-performance adhesives and coatings. The surface as well as adhesive properties of PIS block copolymers depends on the content and length of siloxane blocks. The surface properties of PIS block copolymers are strongly influenced by enrichment of the surface with siloxane segments [7]. The micro phase separation of PIS block copolymers occurs due to the dissimilarity between the chemical structures of both siloxane, and imide blocks.

6.2   EXPERIMENTAL

The 2-aminoterminated ODPA-BIS P polyimides with controlled molecular weight were synthesized by solution imidization (first step in NMP at room temperature for 24 hr, second step in NMP–BCB mixture at 180°C). The number-average molecular weights of products were in the range M_n = 2000–18,000 g/mol (by ^{1}H NMR spectroscopy). The surface morphology (height image) and local surface heterogeneities (phase image) were measured by AFM. All the measurements were performed under ambient conditions using a commercial atomic force microscope (NanoScope™ Dimension IIIa, MultiMode Digital Instruments, USA) equipped with the PPP-NCLR tapping-mode probe (Nanosensors™ Switzerland; spring constant 39 N/m, resonant frequency 160 kHz). The surface energy of PIS block copolymer was determined via measurements of contact angles of a set of testing liquids (that is re-distilled water, ethylene glycol, formamide, methylene iodide, and 1-bromo naphthalene) using surface energy evaluation (SEE) system completed with a web camera (Masaryk University, Czech Republic) and necessary PC software. The drop of the testing liquid (V = 3 µl) was placed with a micropipette (0–5 µl, Biohit, Finland) on the polymer surface, and a contact angle of the testing liquid was measured. The peel strength of adhesive joint (P_{peel}) to polyacrylate was measured by 90° peeling of adhesive joint using universal testing machine Instron 4301 (Instron, England) with 100 N measuring cell. The adhesive joints for peel tests were fixed in aluminum peeling circle.

6.3 DISCUSSION AND RESULTS

The AFM measurements of the PIS copolymers are shown in Figure 1. Its measurements of the surface topography (height image) and tip-sample interaction (phase image) of the samples containing 0–33 wt% of siloxane monomer revealed differences in both characteristics. Only characteristic samples, that is 0, 10, 20, and 33 wt% of siloxane are shown in the Figure 1, sample containing 30 wt% of siloxane is very similar in height and phase images to the sample with 33 wt% siloxane.

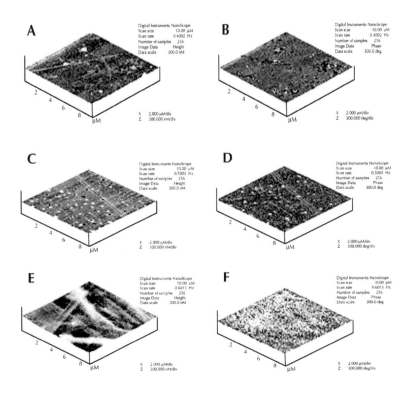

FIGURE 1 *(Continued)*

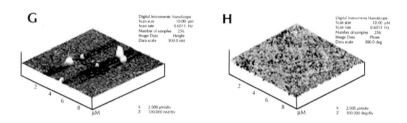

FIGURE 1 AFM images of PIS block copolymers films—pure polyimide (A, B), 10 wt% of siloxane (C, D), 20 wt% of siloxane (E, F), and 33 wt% of siloxane (G, H) Height images (A, C, E, G), and phase (B, D, F, H) images, respectively.

The comparison of height images—samples containing 20% (Figure 1 (E)) and 30% (not shown in figure) have rugged and funicular surface relief. On the other hand, surfaces of pure polyimide (Figure 1 (A)), 10% copolymer (Figure 1 (C)) and 33% copolymer (Figure 1 (D)) contain individual formations on the surfaces –"hills" of different size and height (tens–hundreds nm) and furthermore holes (tens of nm size) on 10% sample. Moreover, funicular formations are shadowed also in the Figure 1 (A) and Figure 1 (C). The comparison of phase images—Figure 1 (B) versus Figure 1 (D), and Figure 1 (F) versus Figure 1 (H) exhibit mutually similar relief. If compared the phase images with the relevant topography images, that is Figure 1 (A) versus Figure 1 (C) and Figure 1 (E) versus Figure 1 (G), it is evident while height images are similar for first couple as well, significant differences for second couple exist.

Figure 2 shows the contact angles of re-distilled water deposited on PIS block copolymer surface versus content of siloxane in copolymer. The contact angles of water by Figure 2 increased by growth of siloxane content and/or Si/N ratio in copolymer. The contact angles of PIS block copolymer increase from 76° for pure polyimide, to 95° for 10% of siloxane in copolymer up to 102° for 30% of siloxane in copolymer. Micro phase separation in PIS block copolymer occurs even at relatively low block lengths due to dissimilarity between the chemical structures of the siloxane, and imide blocks.

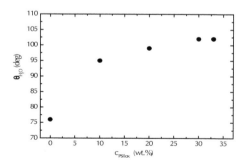

FIGURE 2 Contact angles of water versus siloxane content in PIS block copolymer.

The dependencies of the surface energy and its polar component of PIS block copolymer determined by Owens-Wendt-Rabel-Kaelble (OWRK) method [7] versus content of siloxane in copolymer are shown in Figure 3. The surface energy of PIS block copolymer decreases significantly with the concentration of siloxane from 46.0 mJ.m^{-2} (pure polyimide) to 34.2 mJ.m^{-2} (10% of siloxane), and to 30.2 mJ.m^{-2} (30% of siloxane). The polar component of the surface energy reached the value 22.4 mJ.m^{-2} (pure polyimide), which decreases with content of siloxane in PIS copolymer to 4.6 mJ.m^{-2} (10% of siloxane) and 0.8 mJ.m^{-2} (30% of siloxane) The surface energy of pure polyimide is 46 mJ.m^{-2}, while the value of the surface energy of poly (dimethyl siloxane) is only 20.9 mJ.m^{-2}. At room temperature the siloxane molecules are above their glass temperature, their segments are capable to migrate to the polymeric surface, so making it more hydrophobic. The surface of the PSI copolymer films should be covered with PSI segments having their thickness in molecular order.

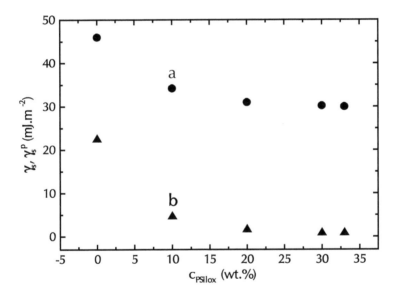

FIGURE 3 Surface energy and its polar component of PSI block copolymer versus siloxane content.

Figure 4 shows the dependence of the peel strength of adhesive joint PSI block copolymer to epoxy versus content of siloxane. It is seen that the peel strength of adhesive joint PIS copolymer-epoxy decreases with growth in siloxane content in the whole concentration range. The fact that the strength of the adhesive joints decreases with increase in siloxane content reflects the increases hydrophobicity of the polymeric surface. The peel strength of adhesive joint to epoxy adhesive diminished from 1.2 MPa (pure polyimide), to 1.05 MPa (10% of siloxane), and to 0.65 MPa (30% of siloxane). This decrease of peel strength of adhesive joint is relatively steady for all investigated content of siloxane in block copolymer. Comparing polyimide with PSI block copolymer containing 30% of siloxane shows that the peel strength of adhesive joint to epoxy decreased more than two times. The presence of siloxane in PSI block copolymer caused the more hydrophobic surface of copolymer (surface energy of co-polymer containing 10% of siloxane was 34.2 mJ.m^{-2}).

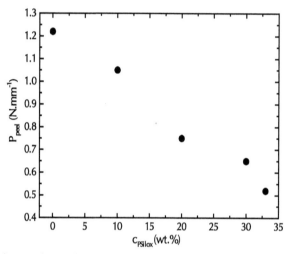

FIGURE 4   Peel strength of adhesive joint PSI block copolymer-epoxy versus concentration of siloxane.

6.4   CONCLUSION

The morphology of PIS block copolymer has been changed due segregation of silox-ane segments, constitution of polyimide continuous phase in copolymer was affirmed. A significant increase of roughness of PSI copolymer surface, if the content of siloxane is growing, was observed. The values of contact angles of water extremely increased by rising of siloxane content in PSI block copolymer and at higher composition were leveled off. The content of siloxane in copolymer increased, the surface energy, and its polar component of PSI copolymer diminished, the dispersive component of the surface energy on opposite increased, and if the content of siloxane in PIS copolymer rises up, strength of adhesive joint to epoxy decreased almost linearly.

KEYWORDS

- **Atomic force microscopy**
- **Hydrophobicity**
- **Poly(imide-co-siloxane)**
- **Siloxane**
- **Surface energy evaluation**

ACKNOWLEDGMENT

This publication was prepared as part of the project "Application of Knowledge-based Methods in Designing Manufacturing Systems and Materials" co-funded by the Ministry of Education, Science, Research and Sport of the Slovak Republic within the granted stimuli for research and development from the State Budget of the Slovak Republic pursuant to Stimuli for Research and Development Act No. 185/2009 Coll. and the amendment of Income Tax Act No. 595/2003 Coll. in the wording of subsequent regulations in the wording of Act. No. 40/2011 Coll.

REFERENCES

1. Niklaus, F., Enoksson, P., Kälvesten, E., and Stemme, G. J. Micromech. Microeng., 11, 100 (2001).
2. Chan-Park, M. B. and Tan, S. S. Int. J. Adhes. Adhesives, 22, 471 (2002).
3. Engel, Chen, J., and Liu, C. J. Micromech. Microengn, 13, 359 (2003).
4. Sysel, P., Hobzová, R., Šindelář, V., and Brus, J. Polymer, 42, 10079 (2001).
5. Sysel, P. and Oupicky, D. Polym. Intern., 40, 275 (1996).
6. McGrath, J. E., Dunson, D. L., Mecham, S. J., and Hedrick, J. L. Adv. Polym. Sci., 140, 61 (1999).

CHAPTER 7

A CFD MODEL FOR POLYMER FUEL CELL

MIRKAZEM YEKANI, MEYSAM MASOODI, NIMA AHMADI,
MOHAMAD SADEGHI AZAD, and KHODADAD VAHEDI

CONTENTS

7.1   INTRODUCTION

The fuel cell is a device that converts chemical energy directly into the electrical energy. Its main components are cell electrolyte, anode electrode, and cathode electrode. The electrochemical reactions performing around the electrodes and generating the electric potential lead to electric current. The oxygen reduction reaction is three time slower than hydrogen oxidation reaction. In fuel cell system with theoretical viewpoint, electrical energy can be generated regularly while the oxidant and fuel are injecting into the electrodes. Depreciation, corrosion and unsuited operation occur due to loss of the life time of the fuel cell. In this system, fuel, this is mostly hydrogen, flows in anode side and oxidant gases such as air or pure oxygen pass in cathode side.

The most attractive choice among variety of fuel cell is Polymer electrolyte membrane fuel cell (PEMFC). In PEMFC, there is a solid polymer electrolyte conducting protons between two platinum impregnated porous electrodes, the present electrodes are cast as thin films and bonded to the membrane. William T. Grubbs [1] originally performed the use of organic cation exchange membrane polymers in fuel cells in 1959. One of the best electrolytes used in PEMFC is Nafion. Water is produces as liquid instead of steam in a polymer electrolyte fuel cell (PEFC). The first three-dimensional model was presented by Dutta et al. in 2000 [4] Berning, Djilali and Li et al. have also developed steady-state, three-dimensional, no isothermal models for PEMFCs recently [5] proton exchange membrane fuel cell (PEMFC) have been extensively suggested for future power supply in the vehicles, by developing the power generation systems and electronically applications.

Although research and investigations in fuel cell systems have been developed a lot, but these systems and applications are still very expensive and complicated so they are not suitable for commercial uses. The complete fuel cell systems have been demonstrated for a number of transportation applications including public transit buses and passenger automobiles. One of the most important goals of recent development has been cost reduction and high volume manufacture for the catalyst, membranes, and bipolar plates. This issue will come off by ongoing research to increase power density, improve water management, operate at ambient conditions, tolerate reformed fuel, and extend stack life. In recent years various research and experiments on PEMFC, by various geometries have been developed. One of the main requirements of these cells is maintaining a high water content in the electrolyte to ensure high ionic conductivity.

During the reactive mode of operation water content in the cell will be determined by the balance of water or its transport. The contributing factors to water transport are the water drag through the cell, back diffusion from the cathode, and the diffusion of any water in the fuel stream through the anode.

Water transport is function of cell current and the characteristics of the membrane and the electrodes. Water drag refers to the amount of water that pulled by osmotic action along with the proton. Water management has a noticeable impact on cell performance, because at high current densities mass transport issues associated with water formation and distribution limit cell output. Without sufficient water management, an imbalance will happen between water production and evaporation within the cell. Against effects, include dilution of reactant gases by water vapor, flooding of the electrodes, and dehydration of the solid polymer membrane. If dehydration occurs the adherence of the membrane to the electrode also will be adversely affected.

As there is no free liquid electrolyte to form a conducting bridge so intrinsic contact between the electrodes and the electrolyte membrane is important. If water was exhausted more than produced, thus it is important to humidify the incoming anode gas. If there is too much humidification, however, the electrode floods, which causes problems with diffusing the gas to the electrode. A smaller current, larger reactant flow, lower humidity, higher temperature, or lower pressure will result in a water deficit. A higher current, smaller reactant flow, higher humidity, lower temperature, or higher pressure will lead to a water surplus.

There have been attempts to use external wicking connected to the membrane to either drain or supply water by capillary action in order to control the water in the cell. The ionic conductivity of the electrolyte is higher when the membrane is fully saturated, and this offers a low resistance to current flow and increases efficiency. Operating temperature has a significant influence on PEFC performance. If the temperature increases the internal resistance of the cell will decrease, mainly by downfall of the ohmic resistance of the electrolyte. In addition, mass transport limitations get reduced at higher temperatures. The overall result is an improvement in cell performance. The experimental data [1, 2] suggest a voltage gain in the range of 1.1 mV to 2.5 mV for each degree (°C) of temperature increase. The other advantage of Operating at higher temperatures is reducing the chemisorptions of CO as this reaction is exothermic. Improving the cell performance through an increase in temperature, however, limited by the high vapor pressure of water in the ion exchange membrane. This is due to the membrane's susceptibility to dehydration and the subsequent loss of ionic conductivity. Operating pressure also influences cell performance.

An increase in the oxygen pressure from 3 to 10.2 atmospheres produces an increase of 42 mV in the cell voltage at 215 mA/cm². According to the Nernst equation, the increase in the reversible cathode potential that expected for this increase in oxygen pressure is about 12 mV, which is considerably less than the measured value. These results demonstrate that an increase in the pressure of oxygen results in a significant reduction in polarization at the cathode [1, 2] Performance improvements due to increased pressure must be balanced against the energy required to pressurize the reactant gases. The overall system must be optimized according to output, efficiency, cost, and size. Operating at pressure above ambient conditions would most likely be reserved for stationary power applications.

In this work, a three-dimensional, steady state, mathematical model was developed. Because of no material found by zero permeability, some researches and investigations have been done by mathematical and numerical modeling of fuel cell, which respect bipolar plates as another parts of PEMFC as gas diffusion layers (GDLs). The GDLs that currently made from carbon fiber or carbon cloth and functions to wick away liquid water, transport reactants of H_2 and O_2 and conduct electrons. Its thickness is normally between 200 and 300 μm. [7, 8] Design of a flow channel is very complicated and difficult to optimize geometry, shape and size of flow fields like that the gas flow channels and bipolar plates since there are many parameters which affect the fuel cell operation such as different materials use in cathode and anode bipolar plates [8] channel/rib ratio [9], and channel path length [10] Catalyst layer which is the region where the membrane and the electrodes overlap and the H_2 oxidation or O_2 reduction reaction occur. It allows electron and ion conduction at the same time. There are other parameters, which affect cell performance such as operating temperature, pressure and humidification of the gas in the cell. It is necessary to understand these parameter and their effects on cell performance. For this reasons we understand it is a point to set up these design factors at optimum values in order to increase the PEMFC operation performance. In this model, major transport phenomena in PEMFC and the effects of prominent GDLs on cell performance and output cell voltage were studied. The modeling data for base case validate with experimental data.

Using a three-dimensional computation model, numerical simulations are applied to compare the performance characteristics of PEMFCs with conventional straight gas flow channel and PEMFCs with prominent GDLs. The simulations focus especially on the effect of prominent GDLs on cell performance, the electrochemical reaction efficiency and the electrical performance of the PEMFCs. The numerical simulations reveal that prominent GDLs improves the transport of the reactant gases through the porous layers; it is due to increase the efficiency of mentioned fuel cell, and prominent GDLs yields appreciably higher voltage and power density. Finally the numerical results of proposed CFD model (base case) are compared with the available experimental data that represent good agreement.

7.2   EQUATIONS

7.2.1   MATHEMATICAL MODEL

Figure 1 shows a schematic of a single cell of a PEMFC. It is made of two porous electrodes, a polymer electrolyte membrane, two-catalyst layer and two-gas distributor plates. The membrane is sandwiched between the gas channels.

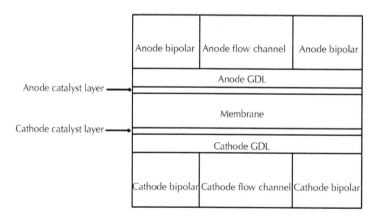

FIGURE 1   Schematic of a single straight-channel of PEMFC.

7.2.2 GOVERNING EQUATIONS

In this numerical simulation, a single domain model formation was used for the governing equations. These governing equations consist of mass conservation, momentum and species equations, which can be written as:

$$(\nabla.\rho u) = 0 \tag{1}$$

$$\frac{1}{\left(\varepsilon^{eff}\right)^2} \nabla.(\rho uu) = -\nabla P + \nabla.(\mu \nabla u) + S_u \tag{2}$$

$$\nabla.(uC_K) = \nabla.\left(D_K^{eff} \nabla C_K\right) + S_K \tag{3}$$

$$\nabla.\left(\kappa_e^{eff} \nabla \Phi_e\right) + S_\Phi = 0 \tag{4}$$

In Equation 1, ρ is the density of gas mixture. According to model assumption, mass source and sink term neglected. ε is the effective porosity inside porous mediums, and μ is the viscosity of the gas mixture in the momentum equation is shown as Equation 2. The momentum source term, Su, is used to describe Darcy's drag for flow through porous GDLs and catalyst layers [11] as:

$$S_u = -\frac{\mu}{K}u \tag{5}$$

K is the gas permeability inside porous mediums. D_K^{eff} in the species equation shown as Equation 3, is the effective diffusion coefficient of species k (for example hydrogen, oxygen, nitrogen, and water vapor) and is defined to describe the effects of porosity in the porous GDLs and catalyst layers by the Bruggeman correlation [12] as:

$$D_K^{eff} = \left(\varepsilon^{eff}\right)^{1.5} D_K \tag{6}$$

Transport properties for species are given in Table 1:

TABLE 1　Transport properties [13]

Property	Value
H_2 Diffusivity in the gas channel, D^0_{H2}	1.10×10^{-04} m²/s
O_2 Diffusivity in the gas channel, D^0_{o2}	3.20×10^{-05} m²/s
H_2O Diffusivity in the gas channel, D^0_{H2o}	7.35×10^{-05} m²/s
H_2 Diffusivity in the membrane, D^{mem}_{H2}	2.59×10^{-10} m²/s
O_2 Diffusivity in the membrane, D^{mem}_{o2}	1.22×10^{-10} m²/s

Additionally, diffusion coefficient is function of temperature and pressure [13] by next equation:

$$D_K = D^\circ{}_K \left(\frac{T}{T_\circ}\right)^{\frac{3}{2}} \left(\frac{P_\circ}{P}\right) \tag{7}$$

The charge conservation equation is shown as Equation 4 and κ_e is the ionic conductivity in the ion metric phase and has been incorporated by Springer et al. [14] as:

$$\kappa_e = \exp\left[1268\left(\frac{1}{303} - \frac{1}{T}\right)\right] \times (0.005139\lambda - 0.00326) \tag{8}$$

Moreover, in recent equation, λ is defined as the number of water molecules per sulfonate group inside the membrane. The water content can be assumed function of water activity, a, is defined according to experimental data [15]:

$$\lambda = 0.3 + 6a\left[1 - \tanh(a - 0.5)\right] + 3.9\sqrt{a}\left[1 + \tanh\left(\frac{a - 0.89}{0.23}\right)\right] \tag{9}$$

Water activity, a, is defined by:

$$a = \frac{C_w RT}{P_w^{sat}} \tag{10}$$

The proton conductivity in the catalyst layers by introducing the Bruggeman correlation [16] is defined as:

$$\kappa_e^{eff} = \varepsilon_m^{1.5} \kappa_e \tag{11}$$

In recent equation ε_m is the volume fraction of the membrane-phase in the catalyst layer. The source and sink term in Equation 3 and Equation 4 are given in Table 2.

Local current density in the membrane can be calculated by:

$$I = -\kappa_e \nabla \Phi_e \tag{12}$$

Then the average current density is calculated as follow:

$$I_{ave} = \frac{1}{A} \int_{A_{mem}} IdA \tag{13}$$

Where A is the active area over the MEA.

7.2.3 WATER TRANSPORT

Water molecules in PEM fuel cell are transported via electro-osmotic drag due to the properties of polymer electrolyte membrane in addition to the molecular diffusion. H^+ protons transport water molecules through the polymer electrolyte membrane and this transport phenomenon is called electro-osmotic drag. In addition to the molecular diffusion and electro-osmotic drag, water vapor is also produced in the catalyst layers due to the oxygen reduction reaction.

Water transport through the polymer electrolyte membrane is defined by:

$$\nabla \cdot \left(D_{H_2O}^{mem} \nabla C_{H_2O}^{mem} \right) - \nabla \cdot \left(\frac{n_d}{F} i \right) = 0 \tag{14}$$

TABLE 2 Source/sink term for momentum, species, and charge conservation equations for individual regions

	Momentum	Species	Charge
Flow channels	$S_u = 0$	$S_K = 0$	$S_\Phi = 0$
Bipolar plates	$S_u = -\dfrac{\mu}{K} u$	$S_K = 0$	$S_\Phi = 0$
GDLs	$S_u = -\dfrac{\mu}{K} u$	$S_K = 0$	$S_\Phi = 0$

TABLE 2   *(Continued)*

Catalyst layers	$S_u = 0$	$S_K = -\nabla \cdot \left(\dfrac{n_d}{F} I \right) - \dfrac{S_K j}{nF}$	$S_\Phi = j$
Membrane	$S_u = 0$	$S_K = -\nabla \cdot \left(\dfrac{n_d}{F} I \right)$	$S_\Phi = 0$

where nd and $D_{H_2O}^{mem}$ are defined as the water drag coefficient from anode to cathode and the diffusion coefficient of water in the membrane phase, respectively.

The number of water molecules transported by each hydrogen proton H$^+$ is called the water drag coefficient. It can be determined from the following equation [15]:

$$n_d = \begin{cases} 1 & \lambda < 9 \\ 0.117\lambda - 0.0544 & \lambda \geq 9 \end{cases} \tag{15}$$

The diffusion coefficient of water in the polymer membrane is dependent on the water content of the membrane and is obtained by the following fits of the experimental expression [17]:

$$D_w^{mem} = \begin{cases} 3.1 \times 10^{-7} \lambda \left(e^{0.28\lambda} - 1 \right) e^{\left(\frac{-2346}{T} \right)} & 0 < \lambda \leq 3 \\ 4.17 \times 10^{-8} \lambda \left(1 + 161 e^{-\lambda} \right) e^{\left(\frac{-2346}{T} \right)} & Otherwise \end{cases} \tag{16}$$

The terms are therefore related to the transfer current through the solid conductive materials and the membrane. The transfer currents or source terms are non-zero only inside the catalyst layers. The transfer current at anode and cathode can be described by Tafel equations as follows:

$$R_{an} = j_{an}^{ref} \left(\frac{[H_2]}{[H_2]_{ref}} \right)^{\gamma_{an}} \left(e^{\alpha_{am} F \eta_{an} / RT} - e^{-\alpha_{cat} F \eta^{an} / RT} \right) \tag{17}$$

$$R_{cat} = j_{an}^{ref} \left(\frac{[O_2]}{[O_2]_{ref}} \right)^{\gamma_{cat}} \left(-e^{\alpha_{am} F \eta_{cat} / RT} + e^{-\alpha_{cat} F \eta^{cat} / RT} \right) \tag{18}$$

According to the Tafel equation, the current densities in the anode and cathode catalysts can be expressed by the exchange current density, reactant concentration, temperature, and over-potentials according to the Tafel equations. Where the surface over potential is defined as:

The difference between proton potential and electron potential.

$$\eta_{an} = \varphi_{sol} - \varphi_{mem} \tag{19}$$

$$\eta_{cat} = \phi_{sol} - \phi_{mem} - V_{oc} \tag{20}$$

The open circuit potential at the anode is assumed to be zero, while the open circuit potential at the cathode becomes a function of a temperature as:

$$V_{oc} = 0.0025T + 0.2329 \tag{21}$$

The protonic conductivity of membrane is dependent on water content, where σ_m is the ionic conductivity in the ionomeric phase and has been correlated by Springer et al. [23]:

$$\sigma = (0.005139\lambda - 0.00326)\exp\left[1268\left(\frac{1}{303} - \frac{1}{T}\right)\right] \tag{22}$$

Energy equation given by Equation 23:

$$\nabla.(\rho u T) = \nabla.(\lambda_{eff}\nabla T) + S_T \tag{23}$$

where, λ_{eff} is the effective thermal conductivity, and the source term of the energy equation, S_T, is defined with the following equation:

$$S_T = I^2 R_{ohm} + h_{reaction} + \eta_a i_a + \eta_c i_c \tag{24}$$

In this equation, R_{ohm}, is the ohmic resistance of the membrane, $h_{reaction}$, is the heat generated thorough the chemical reactions, η_a and η_c, are the anode and cathode over potentials, which are calculated as:

$$R_{ohm} = \frac{t_m}{\sigma_e} \tag{25}$$

Here, t_m is the membrane thickness.

$$\eta_a = \frac{RT}{\alpha_a F} \ln \left[\frac{IP}{j_{0_a} P_{O_{H_2}}} \right] \tag{26}$$

$$\eta_C = \frac{RT}{\alpha_c F} \ln \left[\frac{IP}{j_{0_c} P_{0_{O_2}}} \right] \tag{27}$$

where, α_a and α_c are the anode and cathode transfer coefficients $P_0 0_2$, is the partial pressure of hydrogen and oxygen, and j_0 is the reference exchange current density.

The fuel and oxidant fuel rate, u, is given by following equations:

$$u_{in,a} = \frac{\xi_a I_{ref} A_{mem}}{2 C_{H_2,in} F A_{ch}} \tag{28}$$

$$u_{in,c} = \frac{\xi_c I_{ref} A_{mem}}{4 C_{O_2,in} F A_{ch}}$$

In present equation, I_{ref} is the reference current density and ξ is a stoichiometric ratio, which is defined as the ratio between the amount supplied and the amount required of the fuel based on the reference current density. The species concentrations of flow inlets are assigned by the humidification conditions of both the anode and cathode inlets.

7.2.4 BOUNDARY CONDITION

Equation 1–Equation 4 form the complete set of governing equations for the traditional mathematical model. Boundary conditions are dispensed at the external boundaries. The constant mass flow rate at the channel inlet and constant pressure condition at the channel outlet, the no-flux conditions are executed for mass, momentum, species, and potential conservation equations at all boundaries except for inlets and outlets of the anode and cathode flow channels.

7.3 DISCUSSION AND RESULTS

7.3.1 MODEL VALIDATION

To validate the numerical simulation model used in this study, the performance curves of voltage and current density for base case compared with the experimental data of Wang et al [18].

The fuel cell operating condition and geometric parameters are shown in Table 3. It is used fully humidified inlet condition for anode and cathode. The transfer current at anode and cathode can be described by Tafel equations.

The polarization and power density curve of present model is shown in Figure 2. This curve signifies the good agreement between present model (for base case) with experimental data.

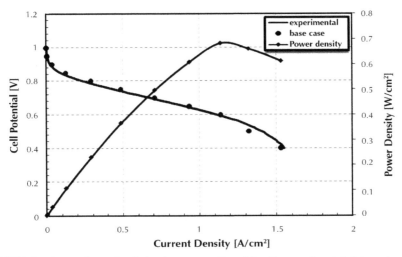

FIGUIRE 2 Comparison of polarization curve of model with experimental data and power density curve at 1.5 (A/m²).

TABLE 3 Geometric Parameters and operating condition of PEMFC [18]

Parameter	Value
Gas channel length	7.0×10^{-2} m
Gas channel width and depth	1.0×10^{-3} m
Bipolar plate width	5.0×10^{-4} m
Gas diffusion layer thickness	3.0×10^{-4} m
Catalyst layer thickness	1.29×10^{-5} m
Membrane thickness	1.08×10^{-4} m
Cell temperature	70°C
Anode pressure	3 atm
Cathode pressure	3 atm

According to Figure 3 and Figure 4, oxygen mass fraction is high at the entrance and then decrease along the fuel cell length due to consumption of oxygen by water formation. On the other hands, water increase along the cell length, this increase of water mass fraction was related with the fact that the water was formed by electrochemical reaction along the channel and water was transported from anode side by electro-osmotic drag coincidently.

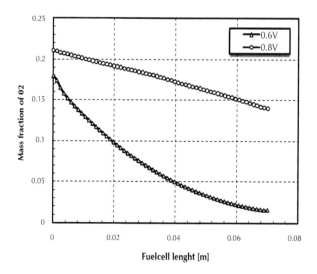

FIGURE 3   Comparison mass fraction of oxygen at the interface of cathode GDL and cathode catalyst layer for two different cell voltages, along the cell.

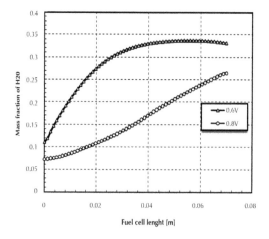

FIGURE 4   Comparison mass fraction of water at the interface of cathode GDL and cathode catalyst layer for two different cell voltages, along the cell.

In addition, it can be anticipated that the mass fraction of oxygen and water will be decreased and increased respectively, by increasing of cell voltage due to increasing the electrochemical reactions. The current density on the catalyst layer is shown in Figure 5. The current density at the inlet was the highest and decreased along the cell. The highest value of current flux density at inlet is probably because of the high concentration of hydrogen and oxygen and high electro-osmotic mass flux at the inlet region.

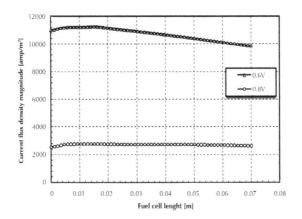

FIGURE 5   Comparison current flux density at the Interface of cathode GDL and cathode catalyst layer for two different cell voltages, along the cell.

In this work, also the effect of prominent GDLs on cell performance and efficiency was studied and compared with base case. Geometrical specification of case with prominent GDLs given in Table 4. Figure 6(a,b) and Figure 7 compare polarization curve and power density curve of two numerical cases respectively.

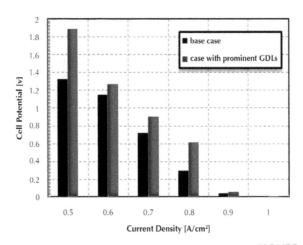

FIGURE 6   *(Continued)*

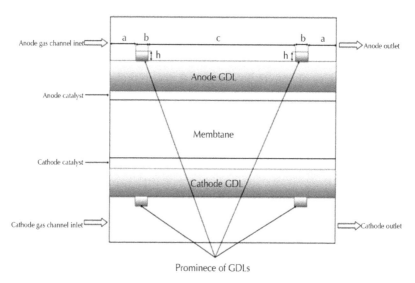

FIGURE 6   (a) Comparison polarization curve of two numerical models (b) Cross sectional schematic of case with prominent GDLs.

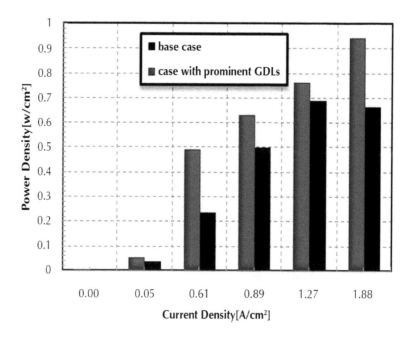

FIGURE 7   Comparison power density curve of two numerical models.

Figure 8 shows the cross sectional schematic of base case which validated with experimental data.

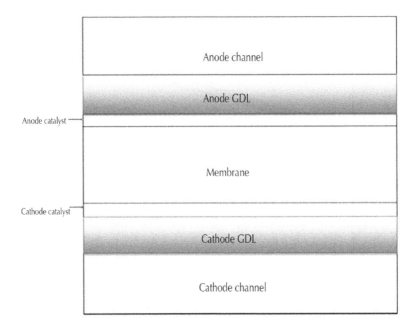

FIGURE 8   Cross sectional schematic of base case.

TABLE 4   Geometrical specification of case with prominent GDLs

Symbol	Value
a	15 mm
·b	5 mm
c	30 mm
h	0.25 mm

It is notice that case with prominent GDLs produces more current density than base case at same cell voltage. Figure 9 illustrates the comparison of distribution of velocity at cathode gas channel. It is clear that the case with prominent GDLs yields a notable increase in velocity. Prominence of GDLs increases the velocity by decreasing the cross sectional area of gas flow at gas channels. This increase of velocity provides the reactant gases to the catalyst layers thus the efficiency of catalytic reaction increases and therefore the performance of PEMFC improves. Also prominences of GDLs improve the flow of reaction hence reduces the membrane drawing effect. Thus the performance of fuel cell especially at the higher current densities improves. Figure 10 and Figure 11 show the contours of velocity distribution for two cases and confirm the mentioned reasons. The comparison of protonic conductivity at membrane and cathode catalyst layer interface for entry region is shown in Figure 12.

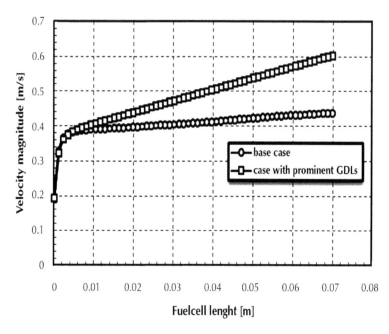

FIGURE 9 Comparison velocity magnitude of two numerical cases at cathode gas flow channel along the fuel cell.

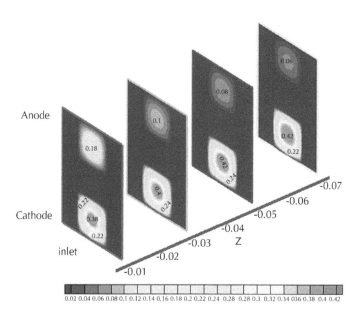

FIGURE 10 Velocity magnitude for base case.

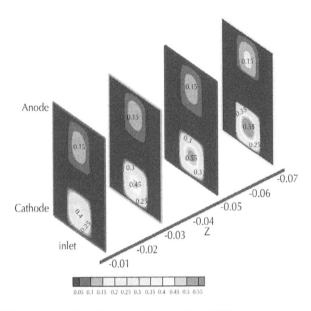

FIGURE 11   Velocity magnitude for case with prominent GDLs.

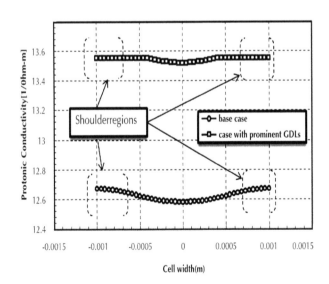

FIGURE 12   Comparison protonic conductivity at the Interface of membrane and cathode catalyst layer for two different numerical model at entry region ($L/L_0 = 0.1428$).

Figure 13 illustrates the distribution of oxygen mass fraction at membrane-cathode catalyst interface, for two different cases at the entry region of fuel cell. In case with prominent GDLs since the current density generative the cell higher than base case therefore high current density due to high accelerate in the electrochemical reaction rate and oxygen consumption. Lack of oxygen at the shoulder region of cell causes to higher mass fraction losses.

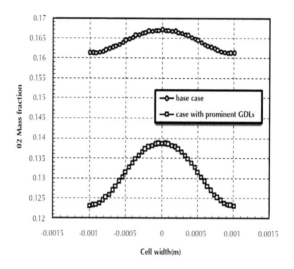

FIGURE 13   Comparison mass fraction of oxygen at the Interface of membrane and cathode catalyst layer for two different numerical model at entry region ($L/L_0 = 0.1428$).

For downstream region of channel high mass fraction losses becomes worse due to diminution of the reactant with moving downstream which shown in Figure 14 (a,b).

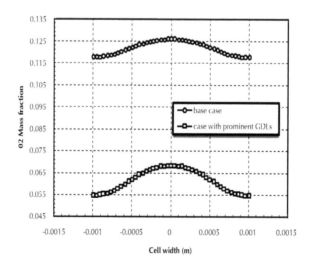

FIGURE 14   *(Continued)*

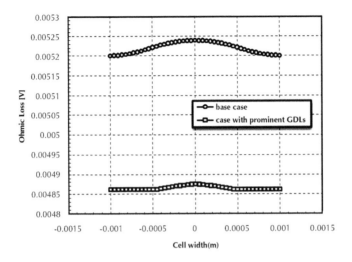

FIGURE 14 Comparison mass fraction of oxygen at the Interface of membrane and cathode catalyst layer for two different numerical model at exit region ($L/L_0 = 0.8571$).

The contour of oxygen mass fraction at interface of membrane and cathode catalyst layer is shown in Figure 15. This shows oxygen mass fraction decreases gradually along the flow channel due to the consumption of oxygen at the catalyst layer.

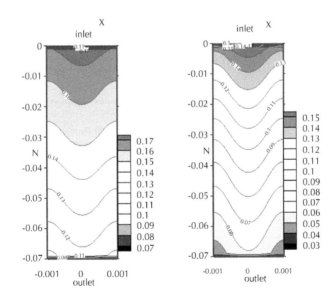

FIGURE 15 Mass fraction of oxygen at the Interface of membrane and cathode catalyst layer for base case (left) and case with prominent GDLs (right).

At the catalyst layer, the concentration of oxygen is balanced by consuming the oxygen and the amount of oxygen that diffuses towards the catalyst layer, driven by the concentration gradient. The lower diffusivity of the oxygen along the flow with the low concentration of oxygen in ambient air results in noticeable oxygen diminution along the fuel cell.

Hydrogen at the anode provides a proton, releasing an electron in the process that must pass through an external circuit to reach the cathode. The proton, which remains solvated with a certain number of water molecules, diffuses through the membrane to the cathode to react with Oxygen and the returning electron. Water successively produced at the cathode.

The comparison of mass fraction of water at the cathode side is shown in Figure 16 and Figure 17. Respectively for entry and exit region of cell for two cases. Water concentration at the cathode membrane and catalyst layer interface increases along the flow channel. This increase of water concentration associates with the phenomenon that the Water composes by electrochemical reaction along the channel and transports from anode side by electro-osmotic drag coincidently. In case with prominent GDLs higher current density due to higher reaction rate. Then the mass fraction of water for base case lower than case with prominent GDLs. Figure 18 shows the distribution of water along the cell at membrane and cathode GDL interface.

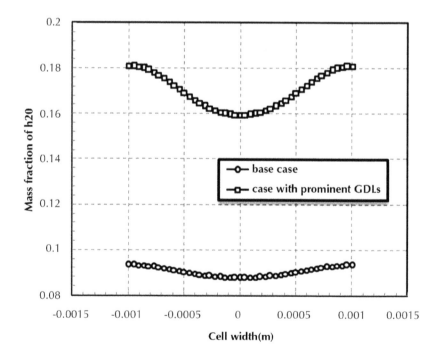

FIGURE 16   Comparison mass fraction of water at the Interface of membrane and cathode catalyst layer for two different numerical model at entry region (L/L$_0$ = 0.1428).

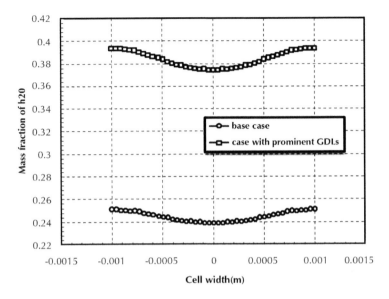

FIGURE 17 Comparison mass fraction of water at the interface of membrane and cathode catalyst layer for two different numerical model at exit region (L/L$_0$ = 0.8571).

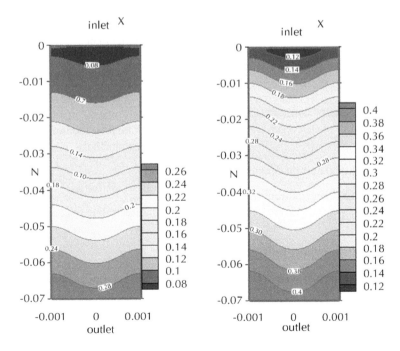

FIGURE 18 Mass fraction of water at the Interface of membrane and cathode catalyst layer for base case (left) and case with prominent GDLs (right).

7.4 CONCLSIONS

In this chapter a full three-dimensional computational fluid dynamics model based on finite volume method of a proton exchange membrane fuel cell (PEMFC) has been developed to study the effect of prominent GDLs on cell performance such as cell current density, power density and efficiency of fuel cell. In this work we repose the prominence of GDLs at entry and exit region of gas flow channel both anode and cathode side. By comparison the result of this case such as polarization curve, power density curve, velocity distribution and species mass fraction by result of base case it can be seen that the prominences of GDLs increase the flow velocity which it is due to improvement of the gas flow along the channel which provides the reactant gases to the catalyst layers. Also gas flow improvement is due to reduce the membrane draw-ing effect thus the efficiency of catalytic reaction increases and more current density achieves with comparison by the base case at same cell potential. Therefore the per-formance of PEMFC improves. Thus it is clear that using the GDLs with prominence in PEMFC can be useful in increasing the efficiency of the cell.

NOMENCLATURE

a	Water activity	λ	Water content in the membrane
C m³)	Molar concentration (mol/	ζ	Stoichiometric ratio
D	Mass diffusion coefficient (m²/s)	η	Over potential (v)
F	Faraday constant (C/mol)		
		λ_{eff}	Effective thermal conductivity
I	Local current density (A/m²)	(w/m-k)	
J	Exchange current density (A/m²)	ϕ_e	Electrolyte phase potential (v)
K	Permeability (m²)		
M	Molecular weight (kg/mol)	Subscripts and superscripts	
		a	Anode
n_d	Electro-osmotic drag coefficient		
P	Pressure (Pa)	c	Cathode

R	Universal gas constant (J/mol-K)	ch	Channel
T	Temperature (K)	k	Chemical species
t	Thickness	m	Membrane
$\vec{u}$	Velocity vector	MEA	Membrane electrolyte assembly
V_{cell}	Cell voltage	ref	Reference value
V_{oc}	Open-circuit voltage	sat	saturated
W	Width	w	Water
X	Mole fraction		

Greek letters

α	Water transfer coefficient
ε^{eff}	Effective porosity
ρ	Density (kg/m³)
μ	Viscosity (kg/m-s)
σ_e	Membrane conductivity (1/ohm-m)

KEYWORDS

- **Catalyst layer**
- **CFD**
- **Geometrical configuration**
- **Modeling and simulation**
- **PEM fuel cell**

REFERENCES

1. Fuel Cell Handbook EG&G Services, Parsons, Inc.Science Applications International Corporation.
2. Amphlett J.C., et al., *The Operation of a Solid Polymer Fuel Cell: A Parametric Model*, Royal Military College of Canada.
3. Ledjeff K., et al. *Low Cost Membrane Fuel Cell for Low Power Applications*, Fraunhofer-Institute for Solar Energy Systems, Program and Abstracts, Fuel Cell Seminar (1992).
4. Dutta S., Shimpalee S., and Van Zee J. W. Three-dimensional numerical simulation of straight channel PEM fuel cells, *Journal of Applied Electrochemistry*, **30**, 135–146 (2000).
5. Gurau V., Liu H., and Kakac S. *Two-dimensional model for proton exchange membrane fuel cells*, AICHE Journal, **44**(11), 2410–2422 (1998).
6. Seung Chi Lee and Chul S. Y. *Korean J. Chem. Eng.*,**21** 1153–1160 (2004).
7. Chi Seung Lee, Chang Hyun Yun, and Byung Moo Kim,Se Chan Jang and Sung Chul Yi. *J. Ceramic Processing Research*, **6**(2), 188–195 (2005).
8. Kumar A. and Reddy R. G., *Power J. Sources*, **114**, 54–62 (2003).
9. Shimpalee S. and Van Zee J. W. *Int. Hydrog. J. Energy*, **32**(7), 842–856 (2007).
10. Shimpalee S, Greenway S., and Van Zee J.W. *J. Power Sources*, **160**, 398–406 (2006).
11. Garau V., Liu H., and Kakac S. *AIChE J.*, **44**(11), 2410–2422 (1998).
13. Meredith R. E. and Tobias C. W. *in Advances in Electrochemistry and Electrochemical Engineering 2*, Tobias, C. W. (Ed.), Interscience Publishers, New York, (1960).
14. Byron Bird R., Warren E. Stewart, and Edwin N.L, *in Transport Phenomena* John Wiley & Sons, Inc, (1960).
15. Springer T. E., Zawodzinski T. A., and Gottesfeld S. *Electrochem J. Soc.*, **138** 2334–2342 (1991).
16. Kuklikovsky A. A. *Electrochem J. Soc.*, **150**(11) A1432–A1439 (2003).
17. Meredith R. E. and Tobias C. W. *in Advances in Electrochemistry and Electrochemical Engineering 2*, Tobias, C. W. (Ed.), Interscience Publishers, New York, (1960).
18. Yeo S. W. and Eisenberg A. *Appl J. Polym. Sci.*, **21** 875 (1997).
19. Wang L., Husar A., Zhou T., and Liu H. *Int. J. Hydrog. Energy*, **28**(11) 1263–1272 (2003).
20. Jung Ch. Y., Kim J. J., Lim S. Y., and Yi S. Ch., Journal of Ceramic Processing Research. Vol. 8, No. 5, pp. 369–375 (2007)
21. Patankar S. V. *Numerical Heat Transfer and Fluid Flow*, Hemisphere Publishing Corporation (1980).
22. Ahmed D. H and Sung H. J, *Journal of Power Sources*, **162**, 327–339 (2006).
23. Maher A. R. and Albaghdadi S. *Renewable Energy*, **33**, 1334–1345 (2008).
24. Springer T. E., Zawodinski T. A., and Gottesfeld S. *J. Electrochem. Soc.*, **136**, 2334, (1991).

CHAPTER 8

POLYHYDROXYBUTYRATE–CHITOSAN MIXED COMPOSITIONS UNDER EXTERNAL INFLUENCES CHANGES IN THE STRUCTURAL PARAMETERS AND MOLECULAR DYNAMICS

S. G. KARPOVAA, A. L. IORDANSKIIB, A. A. POPOVA,
S. M. LOMAKINA, and N. G. SHILKINA

CONTENTS

8.1   INTRODUCTION

The differential scanning calorimetry (DSC), electron paramagnetic resonance (EPR) probe analysis, large angle X-ray diffraction (XRD), and ultraviolet (UV) spectroscopy are used to study the molecular dynamics and structure of hydroxybutyrate (PHB) copolymer, chitosan, and mixed compositions thereof upon thermal treatment in an aquatic medium.

It is shown that, in mixed compositions, starting from 30% polyhydroxybutyrate (PHB), the correlation time increases by an order of magnitude, indicative of a sharp slowdown of the molecular mobility of the probe, and concurrently, the degree of crystallinity decreases abruptly, as evidenced by DSC and XRD analyses. The diffusion coefficient of rifampicin in mixed compositions also decreases with increasing PHB content. A short term (1 hr) thermal treatment (at 70°C) in water results in an increase in the molecular mobility of the probe in the system. The crystallinity changes in complex ways. Biodegradable compositions based on natural polymers are widely used in practice. Because of mixing, new physicochemical properties of compositions can arise, which are not inherent to the original components. Innovative technologies use biodegradable system for producing clean construction materials and packagings [1-5]. A typical representative of polyhydroxyalkanoates [6] is PHB, along with its useful properties has some undesirable ccharacteristics—high cost and fragility. To overcome these limitations its copolymers with poly-(3 hydroxyvalerate) (PHBV), as well as compositions with other biomedical polymers, in particular chitosan. The chitosan is a natural nontoxic biodegradable polymer, produced by deacetylation of chitin, is the second most common biopolymer in nature after cellulose. It is used in film coatings and membranes to extend the shelf life of foodstuffs. However, a high sensitivity chitosan to moisture limits its applications. This disadvantage can be overcome by mixing chitosan with moisture proof polymers. In this case, however, the material remains biodegradable.

Varying the composition of the PHB/chitosan mixture and thus affecting its morphology and crystallinity makes it possible to prepare composite materials with various physicochemical characteristics, such as permeability, water solubility, the rate, and mechanism of degradation, and so on. The DSC and NMR spectroscopy measurements showed that the crystallization of PHB in blends with chitosan is increasingly suppressed with increasing concentration of polysaccharides [8]. The authors supposed that can form hydrogen bonds between carbonyl groups of PHB and amide groups of chitosan. However, there is evidence [9] on the changes of PHB in PHB/chitosan blends that contradicts the results [8], which is likely associated with mixture preparation conditions.

An effective way to assess the state of the amorphous and crystalline phases of original polymers and mixtures thereof is to use a combination of dynamic and structural methods. In this study we used EPR spectroscopy (probe analysis) DSC, and large angle XRD, and UV spectroscopy. This combination of structural and dynamic methods enable to obtain a more complete assessment of the structural evolution of the PHB/chitosan mixture in an aqueous medium over short time intervals (a few hours), which precedes the hydrolytic decomposition of the polymer system.

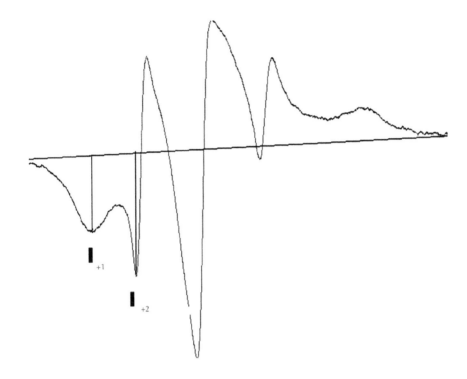

FIGURE 1   EPR spectrum of a mixed composition containing 80% chitosan and 20% PHB.

8.2   OBJECTS AND METHODS

We studied mixtures of biodegradable polymers, PHB, and chitosan. The films were prepared from natural biodegradable polymer PHB of series 16F, produced by microbiological synthesis (BIOMER®, Germany). The original polymer was a fine white powder. The PHB has a molar mass of $Mw = 2.06 \times 10^5$ g/mol, density of $d = 1.248$ g/cm^3, $Tm = 177°C$, and degree of crystallinity of 65%. Chitosan, of domestic production (Bioprogress, Shchelkovo), an infusible polysaccharide, was in the form of a fine powder. The molar weight of this polymer is $Mw = 4.4 \times 10^5$ g/mol and the degree of deacetylation, 82.3%. In casting the films, we used the following solvents—chloroform and dioxane (reagent grade) (ZAO Ekos, Russia) for PHB and acetic acid (reagent grade) for chitosan. The films were prepared by mixing a solution of chitosan in an aqueous acidic medium and a PHB solution in dioxane.

The chitosan solution was prepared by dissolving its powder in acetic acid. The molecular mobility was studied by the paramagnetic probe method, by measuring the correlation time τ that characterizes the rotational mobility of the probe as described in [10]. The probe was the 2,2,6,6-tetramethylpipe-ridine-1-oxyl (TEMPO) radical, introduced into the film in the form of its vapors.

The calorimetric studies of the samples were performed on a DSK 204 F1 instrument (Netzsch, Germany) in an inert atmosphere of argon at a heating rate of 10 K/min. The average measurement error was 2%. The X-ray diffraction analysis of the samples was carried out on RU 200 Rotaflex instrument (Rigaku, Japan) with a 12 kW generator with a rotating copper anode in the transmission mode (40 kV, 140 mA, Cu($K\alpha$)radiation ($\lambda = 0.1542$ nm, Ni filter).

The kinetics of the release of rifampicin was studied using a DU-65 UV spectrophotometer (Backman USA) with output to a chart recorder. For this purpose, a rifampicin film was immersed in a container with a phosphate buffer solution (pH 6.86), and the kinetics of rifampicin release was monitored by periodic sampling. The measurement error was 5%.

8.3   DISCUSSION AND RESULTS

8.3.1   STUDYING THE INTERACTION OF THE AMORPHOUS PHASE OF PHB AND ITS COMPOSITIONS WITH CHITOSAN WITH WATER

The EPR spectra of chitosan–PHB mixed compositions and of the individual component are superposition of two spectra corresponding to radicals with correlation times τ_1 and τ_2: τ_2, which characterize molecular mobility in loose amorphous regions and in a more dense amorphous phase, respectively (Figure 1). All calculations of EPR spectra were performed using the stochastic Liouville equation (random trajectory method) [11]. Let and be the intensities of the first peaks of the triplets corresponding hindered and faster motion, respectively.

We plotted the dependence of the ratio of the intensities of the first lines of the spectra, on the composition of the mixture (Figure 2). This parameter characterizes the ratio of the fraction of amorphous phase with slow molecular mobility to that of amorphous phase with fast mobility. It is seen that, in pure PHB, the fraction of slow component is small, increasing by almost an order of magnitude in mixed compositions. It should be noted that, in pure chitosan and mixed compositions with 10% and 20% PHB, this ratio is even higher. These data are indicative of a heterogeneous structure, which is especially pronounced in mixed compositions and chitosan.

An experiment was performed to determine the number of radicals in identical samples of PHB, chitosan, and mixtures thereof. Figure 2 shows that the number of radicals in pure PHB is almost two orders of magnitude larger than that in chitosan and compositions. These results indicate a low permeability of interphase regions for the radical. Note also that, starting from 30% PHB, the number of radicals in samples (Figure 2b), is almost 10 times higher than in mixtures with 10% and 20% PHB, which means that chitosan and mixtures containing up to 20% PHB (inclusive) have the densest structure of amorphous regions.

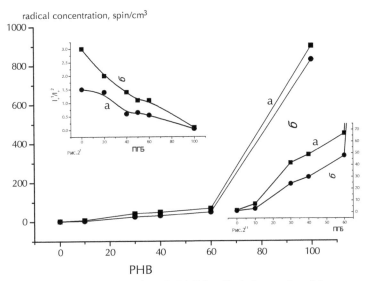

FIGURE 2   Dependence of the concentration of TEMPO radicals in samples of the same mass on the composition of the mixture, (a) Ratio of the first peaks of the spectra with correlation times τ_1 and τ_2 (1) before and (2) annealing. (b) Concentration of the radical in mixed compositions of the same mass with PHB content of up to 60%.

The shape of the spectra of mixed compositions and chitosan is very different from the spectrum of PHB. While the first spectrum of PHB, representing hindered motion, is almost superimposed on the spectrum corresponding to fast motion, the respective spectra of chitosan and mixed compositions are clearly separated (Figure 1). These data suggest that the continuous phase in mixed compositions is chitosan and that the interphase layers have a relatively high density. Figure 3 (curve 1) shows how the correlation time τ_2 depends on the composition of the mixture. As can be seen, in individual chitosan and in compositions with 10% and 20% PHB, the correlation time is relatively short $(9.5 \div 19) \times 10^{-10}$s, while compositions with 30% PHB are characterized by a correlation time of 280×10^{-10}s (for PHB, $\tau_2 = 65 \times 10^{-10}$s). The other compositions also have long correlation times. According to DSC data, the glass transition temperature of chitosan is ≈40°C, whereas a mixture with 30% PHB features no glass transition peak in the entire temperature range covered (from –20°C to 200°C). The results suggest that, when mixed with PHB, chitosan experiences a transition from the glassy to a highly elastic state, so that the radical, penetrating into dense regions of the phase, provides high values of the correlation time τ.

The samples were exposed to water at 70°C for 1 hr. Curve 2 in Figure 3 shows how the correlation time depends on the composition of the mixture. It can be seen that the behavior of the dependence is the same—only τ_2 increases, indicating a decline in molecular mobility in the loose amorphous phase. The ratio of the intensities of the first lines in the spectrum (/) with correlation times τ_1 and τ_2 after annealing in water (Figure 2a, curve 2) indicates an increase in the fraction of amorphous regions with low mobility. Only PHB showed no difference. Note also that the number of radicals sorbed in compositions after exposure to water at 70°C for 1 hr was smaller than in the initial polymers, which also indicates an increase in chain rigidity upon thermal treatment in water (Figure 2).

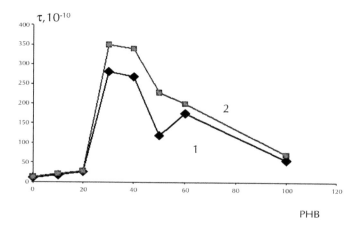

FIGURE 3 Dependence of the correlation time on the composition of the PHB–chitosan mixture (1) before and (2) after annealing at 70°C for 1 hr.

The values of the activation energy for all the studied compositions are given in the Table 1. Figure 4 shows the temperature dependence of the correlation time for three compositions. As can be seen, for PHB, the dependence is linear, whereas those for the mixed compositions exhibit a kink at ≈60°C. This means, that, at this temperature, molecular mobility in dense regions of chitosan is "unfrozen", since the activation energy E is almost two times higher than that for PHB.

TABLE 1 Activation energies for mixed compositions of chitosan and PHB

Content of chitosan,%	E_1 kJ/mol	E_2 kJ/mol
100%	7,5	
90%	8,8	42
80%	10	38
70%	33	50
60%	31	45
50%	9	50
40%	25	60
0%	27	27

Note: The experimental error is 5%.

Note that, in the initial segment, the value of E for chitosan and mixed compositions with 10% and 20% of PHB is small, increasing almost 2.5 fold for compositions with a high content of PHB (except for a composition with 50% PHB). This also shows that, starting from 30% PHB, the structure of the amorphous phase of mixed compositions becomes more rigid due to the transformation of the crystal structure into an amorphous, and the radical, accumulating in these dense regions, indicates a slowdown of molecular motion.

We performed experiments aimed at studying the kinetics of the release of rifampicin from films. These studies are of great interest for solving problem concerning the diffusion of drugs. The rate of these processes largely depends on the structure of the amorphous phase of the polymer. The diffusion coefficients of rifampicin in mixtures was calculated by the equation $D = \pi l^2 (tg\alpha)^2 / 16$, where D is the diffusion coefficient, α is the angle of slope of the kinetic curves for the release of rifampicin from the film, and l is the film thickness. Figure 5 shows the dependence of D on the content of PHB in the composition. It is seen that, with increasing content of PHB in the composition of the system, the diffusion mobility declines, which may be due to a slowdown of the molecular mobility, as shown in Figure 2. Note that, just after the PHB content in the mixture becomes larger than 20%, the diffusion coefficient decreases sharply, whereas for compositions with 30% and 40% PHB, these values are close to each other (Figure 5).

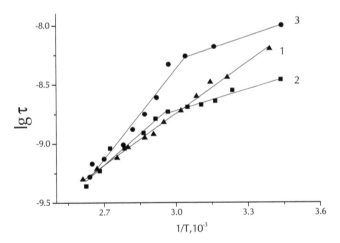

FIGURE 4 Temperature dependence of the correlation time for (*1*) PHB and mixed compositions with (*2*) 20% and (*3*) 50% PHB.

Thus, studying the amorphous phase in PHB–chitosan compositions showed that a mixture containing more than 20% of PHB is characterized by the absence of the glassy component and has a relatively high density. The diffusion coefficient decreases with increasing PHB content in the composition, especially sharply in passing from 20% to 30% PHB. Exposure to water at 70°C leads to an increase in molecular mobility due to compaction of the amorphous phase of the composition.

8.3.2 A STUDY OF THE CRYSTALLINE PHASE OF PHB, CHITOSAN, AND MIXTURES THEREOF

The crystalline phase of mixed compositions was studied using DSC (Figure 6). Characteristically, the bimodal shape of the peak of PHB melting transforms into a single peak for mixed compositions, which indirectly indicates the merger of the two forms of the crystalline state of PHB. The high-temperature peak at 175°C manifests the melting of the well organized crystal structure, whereas the low temperature peak at 161°C corresponds to the melting of the less ordered crystalline structure of PHB. The DSC data showed that, with increasing PHB content in the mixture, its crystallinity and the melting temperature of the crystallites decrease. Thus, in compositions with 10% PHB, the degree of crystallinity is as high as 82% (Figure 6), but then, with increasing the PHB content, it decreased to 60% (for PHB, $\chi = 61\%$). Note that the degree of crystallinity also decreases abruptly. While the crystallinity of a mixture with 20% PHB is 79%, in a 30% PHB mixture, drops to $\chi = 60\%$, decreasing further $\chi = 56.7$ at 60% PHB. According to DSC data, the degree of crystallinity of chitosan is low, only 6.5%, with no crystalline phase detected in mixtures. The melting point of PHB in a mixed composition with 10% PHB is 172.7°C, lowing to 171°C at 30% and 60% PHB.

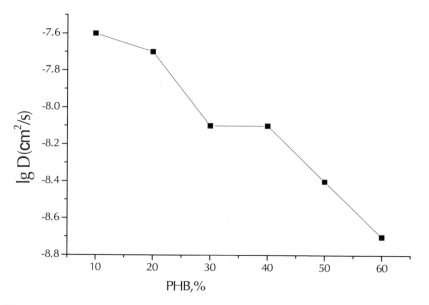

FIGURE 5   Dependence of the diffusion coefficient D of rifampicin on the composition of the PHB–chitosan mixture.

These data indicate that, when mixed with chitosan, PHB acquires a less ordered crystalline structure. This can be attributed to the interaction between chitosan and PHB via hydrogen bonds between ester groups of PHB and amino groups of the polysaccharide. As a result, the structure of chitosan becomes looser, and the crystalline phase is nonexistent in mixed compositions.

It is important to note that, in mixed compositions with 10% and 20% PHB, the crystallinity of the latter is extremely high, probably because of a low degree of entanglement of chains, a factor that facilitates crystallization, while the fraction hydrogen bonds remains very small. Similar results were obtained by using XRD. This method showed that the degree of crystallinity in chitosan is 38% and decreases with increasing PHB content. For example, in a mixed composition with 20% PHB, it is 30%, decreasing to 15% at 30% PHB and to zero at 40% PHB (Figure 7).

The crystallinity of PHB also decreases as its content in the mixed composition increase. The crystallinity of pure PHB is 62%, whereas in mixtures chitosan, its crystallinity decreases to 54%, 44%, and 40% at 20%, 30%, and 40% PHB, respectively. These data also show that the crystallinity of PHB in mixtures decreases sharply staring from its content of 30%.

Thus, with increasing PHB fraction (up to 60%) in mixtures with chitosan, the degree of crystallinity of both chitosan and PHB decline, indicating an increase in the fraction of dense regions in the amorphous phase. Note that, while the crystallinity of chitosan decreased by 5% relative to pure chitosan upon mixing with 20% PHB, in a mixture with 30% PHB, the decrease reaches 23%, with these changes being reflected in the correlation time.

Note that, starting namely from 30% PHB in the composition, the correlation time begins to increase sharply. This suggests that a fraction of the crystallites transforms into an amorphous state, and probe radicals in these mixtures, accumulating in these dense formations, provide a high correlation time.

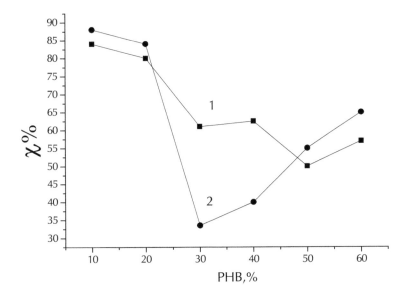

FIGURE 6 Dependence of the degree of crystallinity on the composition of the PHB–chitosan mixture (1) before and(2) after annealing at 70°C for 1 hr.

According to DSC, exposure to water at 70°C for 1 hr results in an increase in the degree of crystallinity of mixed compositions with 10%, 20%, 50%, and 60% PHB. In samples with 30% and 40% PHB and in chitosan, the crystallinity decreased by almost half (Figure 6). According to XRD analysis, after exposure of mixed compositions with 10%, 20%, 30%, and 40% PHB to water at 70°C for 1 hr, no crystalline phase in chitosan was detected. After such treatment, the melting point of PHB increased by 6–7°C for all the studied polymers. That the crystallinity of polymers increases after annealing is well known [12, 13], especially in view of the plasticizing effect of water, which accelerates crystallization. At the same time, the decrease of crystallinity with increasing PHB in the mixture is indicative of the interaction of chitosan with PHB via the formation of hydrogen bonds between ester group of PHB and amino groups of chitosan. Note that, after exposure to water at 70°C, chitosan in mixed compositions shows no crystallinity, as evidenced by XRD data.

8.4   CONCLUSION

Thus, it was shown that the structure of mixed compositions changes drastically stating from a 30% PHB content (70% chitosan), which results in a significant increase in the correlation time. This is indicative of an increase in the fraction of structures with tightly packed chains, as confirmed by data on the diffusion of drugs. The degree of crystallinity also reduces sharply with increasing content of PHB in the composition, as demonstrated by DSC and XRD data. Thermal treatment at 70°C for 1 hr causes an increase in the correlation time, indicative of a decrease in molecular mobility. At the same time, the degree of crystallinity changes in a complex manner.

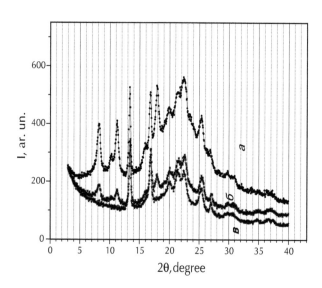

FIGURE 7   X-Ray diffractograms for PHB–chitosan mixtures with (1) 20, (2) 30, and (3) 40% PHB.

KEYWORDS

- **Chitosan**
- **Differential scanning calorimetry**
- **Electron paramagnetic resonance**
- **Polyhydroxybutyrate**
- **Rifampicin**

REFERENCES

1. Yu, L., Dean, K., and Li, L. *Prog. Polym. Sci.*,**31**, 576 (2006).
2. *Polymer Blends, Handbook Vol. 1*, L. A. Utracki (Ed.). Kluwer Academic, Dodrecht, (2002).
3. Freier, T., Sternberg, S., Behrend, D., and Schmitz, K. P. *In Health Issues of Biopolymers. Polyhydroxybutyrate Series of Biopolymers*, A. V. Steinbuchel (Ed.). Wiley, Weinheim, **10**, p. 247 (2003).
4. Iordanskii, A. L., Rogovina, S. Z., Kosenko, R. Yu., Ivantsova, E. L., and Prut, E. V. *Dokl. Phys. Chem.*, **431**, 60 (2010).
5. Ivantsova, E. L., Iordanskii, A. L., Kosenko, R. Yu., et al. *Khim.-Farm. Zh.*, **45**, 39 (2011).
6. Ivantsova, E. L., Kosenko, R. Yu., Iordanskii, A. L., et al. *Polymer Sci., Ser. A*, **54**, 87 (2012).
7. Nikolenko, Yu. M., Kuryavyi, V. G., Sheveleva, I. V., Zemskova, L. A., and Sergienko, V. I. *Inorg. Mater.*, **46**, 221 (2010).
8. Ikejima, T. and Inoue, Y. *Carbohydr. Polym.*, **41**, 351 (2000).
9. Ikejima, T., Yagi, K., and Inoue, Y. *Macromol. Chem. Phys.*, **200**, 413 (1999).
10. Buchachenko, A. L. and Vasserman, A. M. *Stable Radicals* ,Khimiya, Moscow, [in Russian] (1973).
11. Antsiferova, L. I., Vasserman, A. M., Ivanova, A. I., Livshits, V. A., and Nazemets, N. S. *Spectra Atlas*, Nauka, Moscow, [in Russian] (1977).
12. Artsis, M. I., Bonartsev, A. P., Iordanskii, A. L., Bonartseva, G. A., and Zaikov G. E. *Mol. Cryst. Liq. Cryst.*, **523**, 21 (2010).
13. Shibryaeva, L. S., Shatalova, O. V., Krivandin, A. V., et al. *Polymer Sci., Ser. A*, **45**, 244 (2003).

CHAPTER 9

KEY CONCEPTS ON TRANSFORMING MAGNETIC PHOTOCATALYST TO MAGNETIC DYE-ADSORBENT CATALYST

SATYAJIT SHUKLA

CONTENTS

9.1 INTRODUCTION

The magnetic dye adsorbent catalyst has been synthesized *via* hydrothermal process-ing of the magnetic photocatalyst followed by typical washing and thermal treatments. The magnetic dye adsorbent catalyst consists of a core-shell nanocomposite with the core of a magnetic ceramic particle (such as mixed cobalt ferrite ($CoFe_2O_4$), hematite (Fe_2O_3), and pure cobalt ferrite ($CoFe_2O_4$)) and the shell of nanotubes of dye adsorbing material (such as hydrogen titanate). The samples have been characterized for deter-mining the phase structure, morphology, size, and magnetic properties using the X-ray and selected area electron diffraction (SAED), transmission electron microscope (TEM), and vibrating sample magnetometer (VSM). The photocatalytic activity under the ultraviolet (UV) radiation exposure and the dye adsorption under the dark condi-tion have been measured using the methylene blue (MB) as a model catalytic dye-agent. It has been demonstrated that, the transformation of the magnetic photocatalyst to the magnetic dye adsorbent catalyst is accompanied by a change in the mechanism of dye-removal from an aqueous solution from the photocatalytic degradation to the surface adsorption under the dark condition. It has been also shown that, due to its magnetic nature, the magnetic dye adsorbent catalyst can be separated from the treated solution using an external magnetic field and the previously adsorbed dye can be re-moved from the surface of nanotubes *via* typical surface cleaning treatment, which make the recycling of the magnetic dye-adsorbent catalyst possible.

The organic synthetic dyes find applications in various fields including textile, leather tanning, paper production, food technology, agricultural research, light har-vesting arrays, photo electrochemical cells, and hair coloring. Due to the large scale production, extensive use, and subsequent discharge of colored waste waters contain-ing the toxic and non-biodegradable pollutants such as organic synthetic dyes, the latter are considered to be environmentally unfriendly and health hazardous. More-over, they affect the sunlight penetration and the oxygen solubility in the water bodies, which in turn affect the underwater photosynthetic activity and life sustainability. In addition to this, due to their strong color even at lower concentrations, the organic synthetic dyes generate serious aesthetic issues in the waste water disposal [1-5].

As a consequence, powerful oxidation/reduction methods are required to ensure the complete decolorization and degradation of the organic synthetic dyes and their metabolites present in the waste water effluents. Over the last two decades, photoca-talysis has been the area of rapidly growing interest for the removal organic synthetic dyes from the industrial effluents, which involves the use of semiconductor particles as photocatalyst for the initiation of the redox chemical reactions on their surfaces [6-9]. When the semiconductor oxide particle is illuminated with the radiation having energy comparable to its band-gap energy, it generates highly active oxidizing/reducing sites, which can potentially oxidize/reduce large number of organic-wastes. Metal oxide and metal sulfide semiconductors, such as titania (TiO_2) [6-9], zinc oxide (ZnO) [10], tin oxide (SnO_2) [11], zinc sulfide (ZnS) [12], and cadmium sulfide (CdS) [13] have been successfully applied as photocatalyst for the removal of highly toxic and non-biodegradable pollutants commonly present in air and waste water. Among them, TiO_2 is believed to be the most promising one since it is cheaper, environmentally friendly, non-toxic, highly photocatalytically active, and stable to chemical and photo-corrosion. However, its effective application as a photocatalyst is hindered due to some of its major limitations. First, TiO_2 nanocrystallites trend to aggregate (or agglomerate) into large-sized nanoparticles, which affects its performance as a photocatalyst due to the decreased specific surface area. Secondly, it has lower absorption in the visible region, which makes it less effective in using the readily available solar energy. Third, the separation of photocatalyst from the treated effluent, *via* traditional sedimentation and coagulation approaches, has been difficult and time consuming.

The technologies such as adsorption on inorganic or organic matrices and microbiological or enzymatic decomposition have also been developed for the removal of organic synthetic dyes from the waste water to decrease their impact on the environment [5, 14-16]. However, the treatment of waste water containing organic synthetic dyes using these techniques is very costly and has lower efficiency in the color removal and mineralization. The adsorption method results in the generation of large amount of sludge, which causes further difficulties in the recovery of the photocatalyst for recycling the product as a catalyst. Therefore, further development of these techniques for the effective waste water treatment is essential.

In the literature, to overcome the major drawbacks associated with the photocatalytic degradation and adsorption mechanisms, the magnetic photocatalyst has been developed [17-26], which consists of a "core-shell" composite particle with a ceramic magnetic particle as a core and TiO_2 based photocatalyst particles as shell. Such magnetic nanocomposite, which possesses both the photocatalytic and magnetic properties, can be effectively separated from the treated solution using an external magnetic field. However, even the magnetic photocatalyst has been associated with several drawbacks.

First, they show limited photocatalytic activity due to the presence of a core magnetic ceramic particle, which reduces the volume fraction of the photocatalyst available for the photo degradation. Second, the total time of dye decomposition using the magnetic photocatalyst is substantially higher (few hours). Third, the dye removal using the magnetic photocatalyst is based predominantly on the photocatalytic degradation mechanism. Forth, being an energy dependent mechanism (that is, requiring an exposure to the UV, visible, or solar radiation), the photocatalytic degradation is relatively an expensive process. Fifth, the dye removal *via* other mechanism(s) such as the surface adsorption, which is not an energy dependent process (that is, it can be carried out in the dark) has never been utilized for the magnetic photocatalyst. This has been mainly due to the non-suitability of the magnetic photocatalyst for the surface adsorption mechanism as a result of its lower specific surface area. Sixth, the techniques to enhance the specific surface area of the magnetic photocatalyst are not yet reported.

From these points of view, we demonstrate here the conversion of a magnetic photocatalyst having lower specific surface area to a "magnetic dye adsorbent catalyst", having higher specific surface area, consisting a composite structure with the core of a magnetic ceramic particle and the shell of nanotubes of a dye adsorbent material [27-29]. It is demonstrated here that, such conversion is accompanied by a concurrent change in the organic dye removal mechanism from the photocatalytic degradation under the radiation exposure to the surface adsorption under the dark condition, which offers several advantages over the former.

9.2 EXPERIMENTAL

9.2.1 PROCESSING OF MAGNETIC CERAMIC PARTICLES

A mixed $CoFe_2O_4$ and Fe_2O_3 (CFH), and pure $CoFe_2O_4$ magnetic ceramic powders are first processed *via* polymerized complex technique [28, 30]. 36.94 g of citric acid was dissolved in 40 ml of ethylene glycol as complexing agents and stirred to get a clear solution. 17 g of cobalt(II) nitrate $(Co(NO_3)_2\, 6H_2O, 98 + \%)$ and 47.35 g of iron(III) nitrate $(Fe(NO_3)_3\, 9H_2O, 99.99 + \%)$ (Sigma-Aldrich, India) were added and the solution was stirred for 1 hr followed by heating at 80°C for 4 hr. The yellowish gel obtained was charred at 300°C for 1 hr in a vacuum furnace. A black colored solid precursor was obtained which, after grinding, was heated at 600°C for 6 hr to obtain a mixed $CoFe_2O_4$-Fe_2O_3 magnetic powder. Further calcination at 900°C for 4 hr resulted in the formation of pure-$CoFe_2O_4$ magnetic powder. The selection of mixed $CoFe_2O_4$-Fe_2O_3 or pure-$CoFe_2O_4$ powder as a core magnetic material for the photocatalytic and dye adsorption measurements was as per convenience.

9.2.2 PROCESSING OF MAGNETIC PHOTOCATALYST

An insulating layer of silica (SiO_2) was deposited on the surface of magnetic ceramic particles using the Stober process [28, 31]. To 2 g suspension of magnetic powder dispersed in 250 ml of 2-propanol (S.D. Fine-Chem Ltd., India), 1 ml of ammonium hydroxide $(NH_4OH, 25\ wt\%,$ Qualigens Fine Chemicals, India) solution was slowly added. This was followed by the drop wise addition of 7.3 ml of tetraethylorthosilicate (TEOS),(98%, Sigma-Aldrich, India) and the resulting suspension was allowed to settle after stirring for 3 hr. The clear top solution was decanted and the powder was washed with 100 ml of 2-propanol and distilled water followed by drying in an oven at 60°C overnight.

   In order to deposit nanocrystalline TiO_2, 2 g of SiO_2 coated $CoFe_2O_4$-Fe_2O_3 magnetic powder was suspended in a clear solution of prehydrolized titanium(IV) iso-propoxide $(Ti(OC_3H_7)_4)$, (98 %, Sigma-Aldrich, India) precursor (4.73 g) dissolved in 125 ml of 2-propanol (The prehydrolized precursor was obtained due to the reaction of pure $Ti(OC_3H_7)_4$ with the atmospheric moisture over a prolonged period of time). To this suspension, a clear solution consisting 1.5 ml of distilled water ($R = 5$, defined as the ratio of molar concentration of water to that of the precursor), dissolved in 125 ml of 2-propanol, was added drop wise. The suspension was stirred for 10 hr, and after settling, the top solution was decanted. The powder was washed with 100 ml of 2-propanol and then dried in an oven at 80°C overnight followed by calcination at 600°C for 2 hr. The above procedure was utilized for R = 10 and 20 using the pure alkoxide precursor, with a reduced concentration (0.5 g), and pure $CoFe_2O_4$ as a core magnetic ceramic particle. In this case, the coating process was repeated twice to obtain relatively thicker TiO_2-coating.

9.2.3   HYDROTHERMAL TREATMENT OF MAGNETIC PHOTOCATALYST

The magnetic photocatalyst, as processed in this chapter, was subjected to the hydrothermal treatment under highly alkaline condition followed by typical washing and thermal treatments [27, 28]. 0.5 g of conventional magnetic photocatalyst was suspended in a highly alkaline aqueous solution containing 10 M sodium hydroxide (NaOH, 97%, S.D. Fine-Chem Ltd., India) filled up to 84 vol.% of Teflon-beaker placed in a 200 ml stainless steel (SS 316) vessel. The process was carried out in an autoclave (Amar Equipment Pvt. Ltd., Mumbai, India) at 120°C for 30 hr under an autogenous pressure. The hydrothermal product was washed once using 100 ml of 1 M hydrochloric acid (HCl, Qualigens Fine Chemicals, India) solution for 1 hr followed by washing multiple times with distilled water till the final pH of the filtrate was in between 5–7. The washed powder was dried in an oven at 110°C overnight (dried-sample) and then calcined at 400°C for 1 hr (calcined-sample).

9.2.4   CHARACTERIZATION

The morphology of different samples at the nanoscale was examined using the TEM (Tecnai G2, FEI, The Netherlands) operated at 300 kV. The SAED patterns were obtained to confirm the crystallinity and the structure of different samples. The crystalline phases present were determined using the X-ray diffraction (XRD) (XRD, PW1710, Phillips, The Netherlands). The broad scan analysis was typically conducted within the $2-\theta$ range of 10-80° using the Cu $K\alpha$ (λ = 1.542 Å) X-radiation. The magnetic properties of different samples were measured using a VSM attached to a physical property measurement system (PPMS). The pristine samples were subjected to different magnetic field strengths (H) and the induced magnetization (M) was measured at 270 K. The external magnetic field was reversed on saturation and the hysteresis loop was traced [28].

9.2.5   DYE ADSORPTION AND PHOTOCATALYTIC ACTIVITY MEASUREMENTS

The dye adsorption measurements in the dark were conducted using the MB (>96%, S. D. Fine-Chem Ltd., India) as a model catalytic dye agent. A 75 ml of aqueous suspension was prepared by dissolving 7.5 µmol L^{-1} of MB dye and then dispersing 1.0 g L^{-1} of the catalyst powder in pure distilled water. The suspension was stirred in the dark and 3 ml sample suspension was separated after each 30 min time interval for total 180 min. The catalyst powder was separated using a centrifuge (R23, Remi Instruments India Ltd.) and the solution was used to obtain the absorption spectra using the UV visible absorption spectrophotometer (UV-2401 PC, Shimadzu, Japan). The normalized concentration of surface adsorbed MB dye was calculated using the equation of the form:

$$\%MB_{adsorbed} = \left(\frac{C_0 - C_t}{C_0}\right)_{MB} \times 100 \qquad (1)$$

which is equivalent of the form:

$$\%MB_{adsorbed} = \left(\frac{A_0 - A_t}{A_0}\right)_{MB} \times 100 \qquad (2)$$

where, C_0 and C_t correspond to the MB dye concentration at the start and after string time 't', under the dark condition, with the corresponding absorbance of A_0 and A_t. In few experiments, the powder was used for the successive cycles of dye adsorption, under the dark condition, to demonstrate its reusability as a catalyst. The dye adsorption experiments were typically conducted under two different solution-pH (6.4 (neutral) and 10).

During the measurement of photocatalytic activity, the aqueous suspension of MB dye and the catalyst powder was stirred in the dark for 1 hr to stabilize the surface adsorption of the former on the surface of latter. The suspension is then exposed to the UV radiation, having the wavelength within the range of 200–400 nm peaking at 360 nm, in a Rayonet photoreactor (The Netherlands) and 3 ml sample suspension was separated after each 10 min time interval for total 60 min. The powder was separated using a centrifuge and the solution was used to obtain the absorption spectra. The normalized residual MB dye concentration was calculated using the equation of the form:

$$\%MB_{adsorbed} = \left(\frac{C_t}{C_0}\right)_{MB} \times 100 \qquad (3)$$

which is equivalent of the form:

$$\%MB_{adsorbed} = \left(\frac{A_t}{A_0}\right)_{MB} \times 100 \qquad (4)$$

where, C_0 and C_t correspond to the MB dye concentration just at the beginning of the UV radiation exposure (that is, after stirring the suspension under the dark condition for 1 hr) and after the UV radiation exposure time of 't' with the corresponding absorbance of A_0 and A_t.

9.3   DISCUSSION AND RESULTS

The XRD broad scan spectra, as obtained using the mixed $CoFe_2O_4$-Fe_2O_3 and pure-$CoFe_2O_4$ magnetic ceramic powders, are presented in Figure 1(a) and Figure 1(b). The major peaks corresponding to $CoFe_2O_4$ and Fe_2O_3 are identified after comparison with the Joint Committee on Powder Diffraction Standards (JCPDS) card numbers 22-1086 and 33-663. The formation of magnetic ceramic powders consisting mixed $CoFe_2O_4$-Fe_2O_3 and pure-$CoFe_2O_4$ is, thus, confirmed via broad-scan XRD analyses.

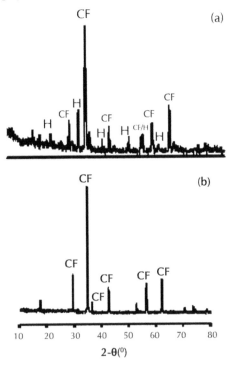

FIGURE 1 XRD patterns as obtained for the mixed $CoFe_2O_4$-Fe_2O_3 (CFH) (a) and pure-$CoFe_2O_4$ (b) magnetic ceramic particles.

The TEM image of a magnetic photocatalyst particle, exhibiting a "core-shell" structure, with a core of magnetic ceramic particle and the shell of a sol-gel derived nanocrystalline coating of anatase-TiO_2 particles, is shown in Figure 2(a). The SAED pattern as obtained from the core is shown as an inset in Figure 2(a), which confirms the crystalline nature of the mixed $CoFe_2O_4$-Fe_2O_3 magnetic ceramic particle. The presence of anatase-TiO_2 in the shell has also been confirmed via XRD analysis as reported elsewhere [27, 28]. As observed in Figure 2(b), the hydrothermal treatment results in the morphological transformation within the shell involving the conversion of nanocrystalline anatase-TiO_2 particles into the nanotubes of hydrogen titanate ($H_2Ti_3O_7$) as identified via high magnification images of the shell presented in Figure 2(c) and Figure 2(d), and the corresponding SAED pattern shown as an inset in Figure 2(c). The "core-shell" magnetic nanocomposite, Figure 2(b), with the core of a magnetic ceramic particle and the shell of nanotubes of $H_2Ti_3O_7$ is termed as a "magnetic dye adsorbent catalyst", which possesses the magnetic, dye adsorption (in the dark), and catalytic properties.

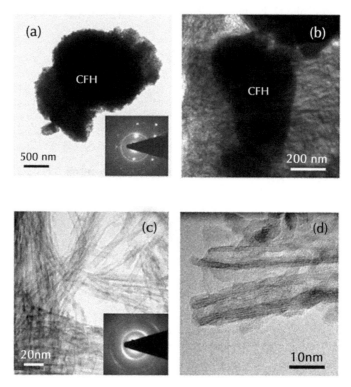

FIGURE 2 TEM images of the magnetic photocatalyst (a) and the magnetic dye-adsorbent catalyst (b). The high magnification images of the shell of magnetic dye adsorbent catalyst are presented in (c) and (d). The SAED pattern corresponding to the core of magnetic photocatalyst is shown as an inset in (a); while, that corresponding to the shell of magnetic dye-adsorbent catalyst is shown as an inset in (c). The samples are processed with R = 5.

The comparison of the morphologies of magnetic photocatalyst and the magnetic dye adsorbent catalyst is schematically shown in Figure 3. The formation mechanism of magnetic dye adsorbent catalyst *via* hydrothermal treatment of the magnetic photo-catalyst, under highly alkaline condition, can be explained using the model originally proposed for the free standing powder [32-34], which is applied here for the similar conversion in the form of coating. When the anatase-TiO_2 particles, present in the coating form on the surface of magnetic ceramic particle, Figure 3(a), are subjected to the hydrothermal treatment under highly alkaline condition, an exfoliation of single layer nanosheets of sodium titanates ($Na_2Ti_3O_7$) results from the bulk anatase-TiO_2 structure, which continuously undergo the dissolution and crystallization processes [34].

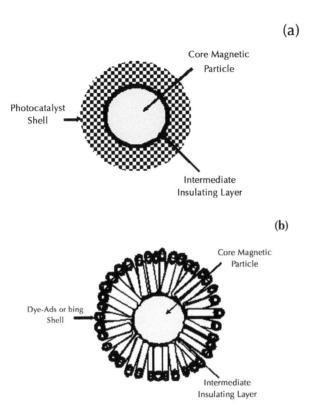

FIGURE 3   Schematic representation of the morphologies of magnetic photocatalyst (a) and magnetic dye-adsorbent catalyst (b).

$$3TiO_2 + 2NaOH \rightarrow Na_2Ti_3O_7 + H_2O \qquad (5)$$

$$Na_2Ti_3O_7 \longleftrightarrow 2Na^+ + Ti_3O_7^{2-} \qquad (6)$$

Due to their higher surface area to volume ratio and the presence of dangling bonds along the two long-edges, the nanosheets of $Na_2Ti_3O_7$ have a strong drive to rollup. However, this rollup tendency is opposed by the repulsive force produced by the charge on the nanosheets created by the presence of Na^+ ions. These Na^+ ions are easily replaced *via* ion-exchange mechanism with H^+ ions in the subsequent washing steps in an acidic aqueous solution and pure water, which reduces the repulsive force to rollup.

$$Na_2Ti_3O_7 + 2HCl \rightarrow H_2Ti_3O_7 + 2NaCl \qquad (7)$$

$$Na_2Ti_3O_7 + 2H_2O \rightarrow H_2Ti_3O_7 + 2NaOH \qquad (8)$$

This results in the formation of nanotubes of $H_2Ti_3O_7$, which are normally transformed to those of anatase-TiO_2 following the calcination treatment at higher temperature.

$$H_2Ti_3O_7 \xrightarrow{\Delta} 3TiO_2 + H_2O \qquad (9)$$

The XRD pattern of the magnetic dye adsorbent catalyst (calcined sample), however, did not reveal the presence of anatase-TiO_2 on the surface [27, 28]. Since the $H_2Ti_3O_7$ to anatase TiO_2 transformation has been observed earlier in the powder form [35], it appears that, such transformation is possibly retarded in the coating form due to the substrate effect.

The *M-H* graphs, as obtained for the magnetic photocatalyst and the magnetic dye adsorbent catalyst (calcined sample), are presented in Figure 4. The presence of a hysteresis loop is noted, which suggests the ferromagnetic nature of these composite particles. For the magnetic photocatalyst, the obtained values of saturation magnetization, remanence magnetization, and coercivity are 59 emu g^{-1}, 24 emu g^{-1}, 1410 Oe; while, those for the magnetic dye adsorbent catalyst (calcined sample) are 45 emu g^{-1}, 15 emu g^{-1}, 578 Oe. It is noted that, the magnetic dye adsorbent catalyst (calcined sample) shows reduced saturation magnetization, remanence magnetization, and coercivity relative to those observed for the magnetic photocatalyst. This has been attributed to the combined effect of decrease in the volume fraction and increase in the average particle size and crystallinity of the core magnetic ceramic particle following the hydrothermal, drying, and calcination treatments [28]. Nevertheless, the hydrothermal product (calcined-sample) also possesses the magnetic property and is suitable for its separation, after the photocatalytic and dye adsorption experiments, using an external magnetic field.

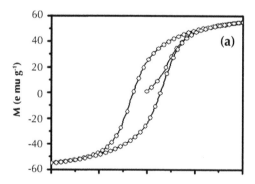

FIGURE 4 *(Continued)*

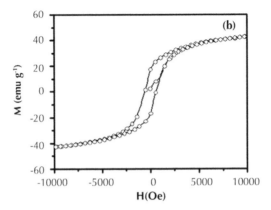

FIGURE 4   Variation in the M as a function of external magnetic field strength (H) as obtained for the magnetic photocatalyst (a) and magnetic dye adsorbent catalyst (calcined-sample) (b). The samples are processed with R = 5.

The variation in the normalized residual MB dye concentration as a function of stirring time, under the UV radiation exposure, as obtained for the pure and Pd-deposited [36] magnetic photocatalyst is shown in Figure 5(a). It has been demonstrated earlier that, the sol-gel derived pure nanocrystalline anatase-TiO_2 particles completely remove the MB dye *via* photocatalytic degradation mechanism under similar test conditions within the UV radiation exposure time of 1 hr [37]. It hence, appears that the magnetic photocatalyst possesses very slow MB dye degradation kinetics. The comparison of the kinetics of MB dye removal *via* surface adsorption mechanism, under the dark condition, using the magnetic photocatalyst and the magnetic dye adsorbent catalyst is presented in Figure 5(b). Due to its higher specific surface area, the magnetic dye adsorbent catalyst (dried-sample) exhibits >99% of dye-adsorption, which is larger than that (~90%) of the calcined-sample. This is attributed here to some loss in the specific surface-area of the magnetic dye-adsorbent catalyst as a result of the calcination treatment. Nevertheless, the magnetic dye adsorbent catalysts (dried and calcined-samples) show significantly higher amount of dye adsorption on the surface, under the dark condition, relative to that (~50–55%) shown by the magnetic photocatalyst. It is, thus, successfully demonstrated that under the dark condition, the magnetic dye adsorbent catalyst removes an organic dye from an aqueous solution predominantly *via* surface adsorption mechanism. On the other hand, the magnetic photocatalyst cannot completely remove the MB dye neither *via* photocatalytic degradation mechanism (under the UV radiation exposure) nor *via* surface adsorption mechanism (under the dark condition). This is further supported by the comparison of the qualitative variation in the color of the MB dye solution as a function of stirring time, as observed for the magnetic photocatalyst and the magnetic dye adsorbent catalyst (dried-sample), Figure 6(a). In Figure 6(b), the qualitative variation in the color of the $H_2Ti_3O_7$ nanotubes (without the core magnetic ceramic particle) is presented under different conditions. It is noted that, the initial white color of the nanotubes changes to blue after the surface adsorption of the MB dye under the dark condition. After a typical surface cleaning treatment [27], the surface adsorbed MB dye can be decomposed, which is suggested by the disappearance of the blue color. This produces surface cleaned $H_2Ti_3O_7$ nanotubes which can be recycled for the next cycle of dye adsorption under the dark condition with the dye adsorption capacity comparable with the original powder. (Since the surface cleaning treatment is effective with pure-$H_2Ti_3O_7$ nanotubes, it would also be effective in the presence of the core magnetic ceramic particle). As a result, it seems that, the magnetic dye adsorbent can be recycled and used as a catalyst for the dye removal application.

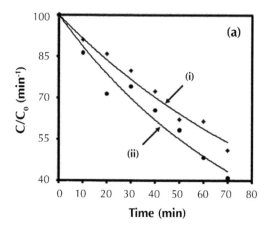

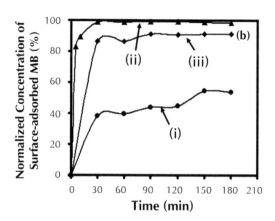

FIGURE 5 (a) Variation in the normalized residual MB dye concentration as a function of stirring time under the UV radiation exposure as obtained for the pure (i) and palladium (Pd)-deposited (ii) magnetic photocatalyst. (b) Variation in the normalized concentration of surface adsorbed MB as a function of stirring time under the dark condition as obtained for the magnetic photocatalyst (i) and magnetic dye adsorbent catalyst, dried (ii) and calcined (iii) samples. The samples are processed with R = 5.

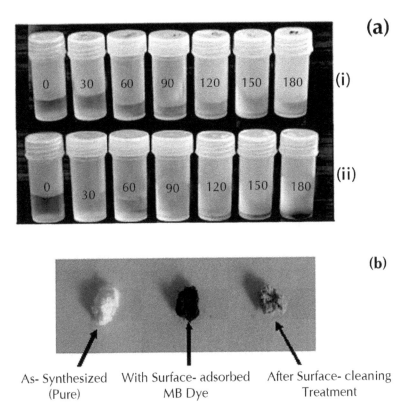

FIGURE 6 (a) Qualitative variation in the color of the MB dye solution as a function of stirring time under the dark-condition as obtained for the magnetic photocatalyst (i) and magnetic dye adsorbent catalyst (dried sample) (ii). The samples are processed with R = 5. (b) Qualitative variation in the color of the $H_2Ti_3O_7$ nanotubes powder (without the core magnetic ceramic particle) under different conditions.

The increased surface adsorption of the MB dye, under the dark condition, as observed earlier in Figure 5(b) for $R = 5$ following the hydrothermal treatment of the magnetic photocatalyst, is also shown by those processed with larger R-values (10 and 20), Figure 7(a) and Figure 7(b). Interestingly, comparison of Figure 8 and Figure 7(b) shows that, at higher solution pH in the basic region (pH = 10), both the magnetic photocatalyst as well as the magnetic dye adsorbent catalyst exhibit high and comparable surface adsorption of the MB dye under the dark condition. However, when the same catalyst is used for the repetitive dye adsorption cycles, at higher solution-pH in the basic region (pH = 10), the magnetic photocatalyst rapidly loses its maximum dye adsorption capacity, Figure 9(a), during each successive cycles. On the other hand, under similar test conditions, the magnetic dye adsorbent catalyst (calcined-sample) retains its maximum dye adsorption capacity, Figure 9(b).

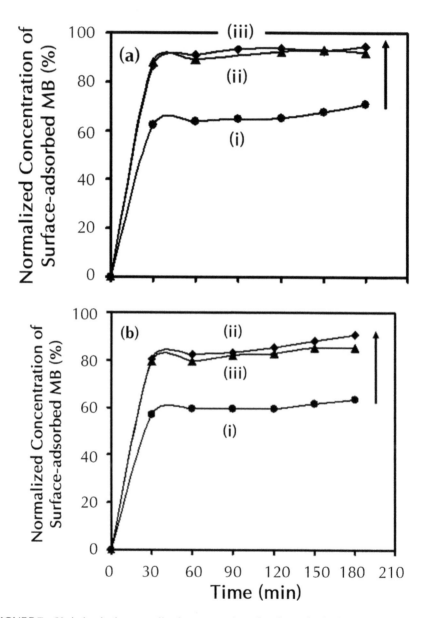

FIGURE 7 Variation in the normalized concentration of surface adsorbed MB as a function of stirring time under the dark condition as obtained for different samples—magnetic photocatalyst (i) and magnetic dye adsorbent catalyst, dried (ii) and calcined (iii) samples. The samples are processed with R = 10 (a) and R = 20 (b), and the dye adsorption measurements are conducted at the neutral solution-pH (6.4).

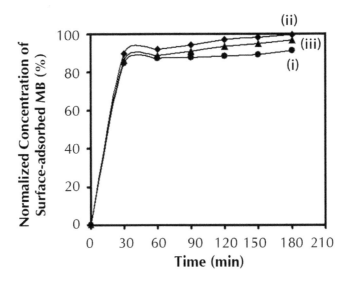

FIGURE 8 Variation in the normalized concentration of surface-adsorbed MB as a function of stirring time under the dark condition as obtained for different samples—magnetic photocatalyst (i) and magnetic dye adsorbent catalyst, dried (ii) and calcined (iii) samples. The samples are processed with R = 20 and the dye adsorption measurements are conducted at the basic solution-pH (10).

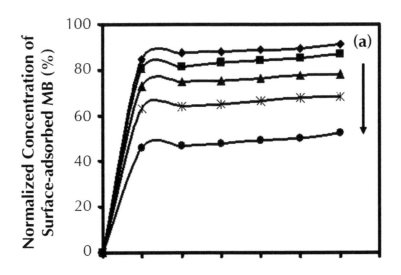

FIGURE 9 *(Continued)*

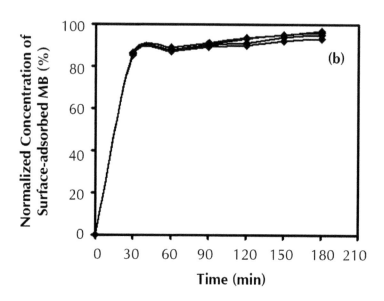

FIGURE 9   Variation in the normalized concentration of surface adsorbed MB as a function of stirring time under the dark condition, for the successive dye adsorption cycles, as obtained for different samples—magnetic photocatalyst (a) and magnetic dye adsorbent catalyst (calcined-sample) (b). The samples are processed with R = 20 and the dye adsorption measurements are conducted at the basic solution-pH (10).

As a result, overall it appears that, the transformation of magnetic photocatalyst to the magnetic dye adsorbent catalyst is accompanied by a concurrent change in the dye removal mechanism from the photocatalytic degradation under the UV radiation exposure to the surface adsorption under the dark condition.

9.4   CONCLUSION

The magnetic photocatalyst has been successfully converted to the magnetic dye adsorbent catalyst *via* hydrothermal followed by typical washing and thermal treatments. The latter consists of a core-shell nanocomposite with the core of a magnetic ceramic particle, such as mixed $CoFe_2O_4$-Fe_2O_3 and pure-$CoFe_2O_4$, and the shell of nanotubes of dye adsorbing material such as $H_2Ti_3O_7$. The magnetic dye adsorbent catalyst has been successfully utilized for the removal of organic dye from an aqueous solution, under the dark condition, *via* surface adsorption mechanism. Due to their magnetic nature, the magnetic dye adsorbent catalyst can be separated from the treated solution using an external magnetic field. The successful removal of previously adsorbed dye from the surface of nanotubes, *via* typical surface treatment, suggests that the magnetic dye adsorbent catalyst can be recycled.

KEYWORDS

- Cobalt ferrite
- Hematite
- Tetraethylorthosilicate
- Vibrating sample magnetometer
- X-ray and selected area electron diffraction

ACKNOWLEDGMENT

The author thanks CSIR, India for funding the photocatalysis and nanotechnology research at NIIST-CSIR, India through the Projects # NWP0010 and P81113.

REFERENCES

1. Forgacsa, E., Cserhati, T., and Oros, G. *Environ. International*, **30**, 953–971 (2004).
2. Gupta, G. S., Shukla, S., Prasad, G., and Singh, V. N. *Environ. Technol.*, **13**, 925–936 (1992).
3. Shukla, S. P. and Gupta, G. S. *Ecotoxicol. Environ. Saf.*, **24**,155–163 (1992).
4. Sokolowska-Gajda, J., Freeman, H. S., and Reife, A. *Dyes Pigments*, **30**, 1–20 (1996).
5. Robinson, T., McMullan, G., Marchant, R., and Nigam, P. *Bioresource Technol.*, **77**, 247–255 (2001).
6. Carp, O., Huisman, C. L., and Reller, A. *Prog. Solid State Ch.*, **32**, 33–177 (2004).
7. Tachikawa, T., Fujitsuka, M., and Majima, T. *J. Phys. Chem. C*, **111**, 5259–5275 (2007).
8. Chen, X. and Mao, S. S. *Chem. Rev.*, **107**, 2891–2959 (2007).
9. Fujishima, A., Zhang, X., and Tryk, D. A. *Surf. Sci. Rep.*, **63**, 515–582 (2008).
10. Marto, J., Marcos, P. S., Trindade, T., and Labrincha, J. A. *J. Haz. Mat.*, **163**, 36–42 (2009).
11. Pan, S. S., Shen, Y. D., Teng, X. M., Zhang, Y. X., Li, L., Chu, Z. Q., Zhang, J. P., Li, G. H., and Hub, X. *Mater. Res. Bull.*, **44**, 2092–2098 (2009).
12. Feng, S., Zhao, J., and Zhu, Z. *Mater. Sci. Eng. B*, **150**, 116–120 (2008).
13. Datta, A., Priyama, A., Bhattacharyya, S. N., Mukherjee, K. K., Saha, A. *J. Colloid Interf Sci.*, **322**, 128–135 (2008).
14. Shaul, G. M., Holdsworth, T. J., Dempsey, C. R.,and Dostal, K. A. *Chemosphere*, **22**, 107–119 (1991).
15. Gupta, V. K. and Suhas. *J. Environ. Manage.*, **90**, 2313–2342 (2009).
16. Crini, G. *Bioresource Technol.*, **97**, 1061–1085 (2006).
17. Beydoun, D., Amal, R., Scott, J., Low, G., and Evoy, S. M. *Chem. Eng. Technol.*, **24**, 745–748 (2001).
18. Song, X., Gao, L. *J. Am. Ceram. Soc.*, **90**, 4015–4019 (2007).
19. Gao, Y., Chen, B., Li, H., and Ma, Y. *Mater. Chem. Phys.*, **80**, 348–355 (2003).
20. Xiao, H. M., Liu, X. M., and Fu, S. Y. *Compos. Sci. Technol.*, **66**, 2003–2008 (2006).
21. Rana, S., Rawat, J., Sorensson, M. M., and Misra, R. D. K. *Acta Biomater.*, **2**, 421–432 (2006).
22. Lee, S. W., Drwiega, J., Mazyck, D., Wu, C. Y., and Sigmund, W. M. *Mater. Chem. Phys.*, **96**, 483–488 (2006).
23. Fu, W., Yang, H., Li, M., Li, M., Yang, N., and Zou, G. *Mater. Lett.*, **59**, 3530–3534 (2005).

24. Jiang, J., Gao, Q., Chen, Z., Hu, J., and Wu, C. *Mater. Lett.*, **60**, 3803–3808 (2006).
25. Beydoun, D., Amal, R., Low, G., and McEvoy, S. *J. Mol. Catal. A*, **180**, 193–200 (2002).
26. Siddiquey, I. A., Furusawa, T., Sato, M., and Suzuki, N. *Mater. Res. Bull.*, **43**, 3416–24 (2008).
27. Shukla, S., Warrier, K. G. K., Vaarma, M. R., Lajina, M. T., Harsha, N., and Reshmi, C. P. PCT Application No. PCT/IN2010/000198.
28. Lajina, T., Shereef, A., Shukla, S., Pattelath, R., Varma, M. R., Suresh, K. G., Patil, K., and Warrier, K. G. K. *J. Am. Ceram. Soc.* (No. DOI: 10.1111/j.1551-2916.2010.03949.x).
29. Shukla, S., Varma, M. R., Suresh, K. G., and Warrier, K. G. K. *Magnetic Dye-Adsorbent Catalyst: A "Core-Shell" Nanocomposite*, In: Proceedings of NanoTech Conference and Expo, TechConnect World Summit Conferences and Expo 2010, Anaheim, California, U.S.A., **1**, 830–833, June 21–24, (2010).
30. Varma, P. C. R., Manna, R. S., Banerjee, D., Varma, M. R., Suresh, K. G., and Nigam, A. K. *J. Alloy Compd.*, **453**, 298–303 (2008).
31. Lee, S. W., Drwiega, J., Mazyck, D., Wu, C. Y., and Sigmund, W. M. *Mater. Chem. Phys.*, **96**, 483–488 (2006).
32. Kasuga, T., Hiramatsu, M., Hoson, A., Sekino, T., and Niihara, K. *Langmuir*, **14**, 3160–3163 (1998).
33. Kasuga, T., Hiramatsu, M., Hoson, A., Sekino, T., and Niihara, K. *Adv. Mater.*, **11**, 1307–1311 (1999).
34. Bavykin, D. V., Parmon, V. N., Lapkina, A. A., and Walsh, F. C. *J. Mater. Chem.*, **14**, 3370–3377 (2004).
35. Sun, X. and Li, Y. *Chem. Eur. J.*, **9**, 2229–2238 (2003).
36. Harsha, N., Ranya, K. R., Shukla, S., Biju, S., Reddy, M. L. P., and Warrier, K. G. K. *J. Nanosci. Nanotech*, (In Press).
37. Baiju, K. V., Shukla, S., Sandhya, K. S., James, J., and Warrier, K. G. K. *J. Phys. Chem. C*, **111**, 7612–7622 (2007).

CHAPTER 10

NEW TYPES OF ETHYLENE COPOLYMERS ON THE BASE NANOCOMPOSITE

IGOR NOVÁK, PETER JURKOVIČ, JÁN MATYAŠOVSKÝ, and LADISLAV ŠOLTÉS

CONTENTS

10.1 INTRODUCTION

This chapter deals with adhesive and mechanical properties study of nanocomposites based on ethylene-acrylic acid copolymer during aluminum bonding. The main objective was to describe the changes of copolymer properties during increasing of the nanofiller's concentration. Based on executed experiments it was found out, that the properties of tested nanocomposite system were mostly improved depending on the contents of the nanofiller in the system. The optimum concentration of nanofiller Aerosil 130 SLP in the composite was 2.5 wt% for cohesive mechanical properties of the system and 3.5 wt% for adhesive ones. Thermal properties of the composite system showed their maximum within concentration of 4.5 wt% of nanofiller.

When compared with other types of composites, thermoplastics have some advantages.
- They are solvent-free and non-toxic (in most cases).
- They are characterized by short time of creation of adhesive bond respectively foil.
- They are applicable at low temperatures.
- They ensures high adhesion to different material and high impact strength of the joint.
- They ensure suitable initial strength of adhesive joints.
- They have good storage stability.
- They are proper for gluing automation and increasing labor productivity
- No undesirable moisture is brought into the materials—it means that there is not necessary the long-term storage of products in conditioned environment.

Nowadays, adhesives based on EAA (ethylene–acrylic acid) copolymers, EVA (ethylene–vinyl acetate) copolymers, thermoplastic polymers, polyamide, polyesters, polyethylene, and cellulose [1-6] belong to most often used composites. By addition of a proper type of filler, mentioned properties can be even improved. The aim of this contribution is to evaluate the influence of nanofiller on the properties of EAA copolymer.

10.2 EXPERIMENTAL PART

As a polymer, EAA copolymer MICHEM Adhesive 20 EAA, with the ratio of 20% wt. of acrylic acid and the ratio of 80% wt. of ethylene, was used. Characteristic properties of the product are:
- Appearance: slight turbidity, almost transparent polymer,
- Density: 1.3 g/cm^3,
- Melt flow: 1.8 g/10min,
- Content of volatiles: less than 0.1 wt.%.

Aerosil 130 SLP (Degussa comp.) was used as filler into nanocomposite system. Aerosil is a flame-patterned silica oxide with an average particle size from 40 to 50 nm. Figure 1 shows the microscopic image of used filler. As we can see, the structure of the filler is spherical with a minimal difference in particles size and non-porous/solid surface.

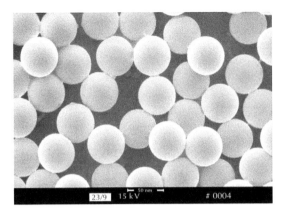

FIGURE 1   The detail of Aerosil 130 SLP particles.

For preparation of nanocomposite system, EAA copolymer was used as the base for copolymer matrix which was blended with the filler in concentrations 0, 0.5, 1, 1.5, 2.5, 3.5, and 4.5% wt. To mix the mixture, we used Plastograf Brabender PLE 331 heated by silicone oil in fully filled tank W-50-hr (volume 50 cm³). The temperature at mixing of nanocomposite was adjusted to 180°C by a thermostat containing tempered silicone medium. Mixing was at 35 rpm⁻¹ for 10 min at predetermined temperature. Considering the properties of individual components, it was preferable to use a triangular blade.

At measurement of adhesive characteristics, the aluminum sheet with thickness of 2 mm and chemical composition listed in Table 1 was used.

TABLE 1   Chemical composition of adherends

Elements	Al	Cu	Fe	Mg	Mn	Ni	Si	Zn
Content (wt. %)	99.5	0.0025	0.32	0.002	0.0035	0.013	0.12	0.007

To measure the peeling strength of adhesive joint, the aluminum foil AlMgSi 0.5 with thickness of 0.1 mm were used.

Before gluing, the surface of adherents was grinded with 120 grit sandpaper and then scratches were aligned with 1000 grit sandpaper. Afterwards, the surface was cleaned of grease and other dirtiness with a mixture of benzene and toluene (volume ratio 1:1). To ensure a constant spacing between bonded adherents and an equal thickness of adhesive, two distant wires with diameter φ 0.15 mm were placed parallel on the bottom board.

The surface of aluminum foil used in the peeling test was only ungreased with a mixture of benzene and toluene. To measure cohesive characteristics, it was necessary to make test blades according to Figure2.

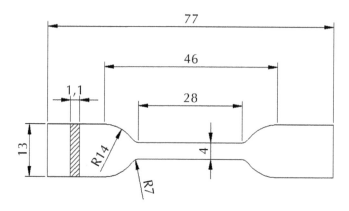

FIGURE 2 Specimen for testing of tensile strength.

To make them, first boards from filled and unfilled systems (dimensions of 74 × 100 × 1.1) were prepared in a shape in hydraulic press at 180°C, pressure 250 kPa, for 5 min. After cooling of them in a mechanical press, test blades were scissored.

For preparing the samples for testing of adhesive properties (Figure 3), thin layer of hot-melt adhesive was inserted between two cleaned and ungreased aluminum boards with distant wires ϕ 0.15 mm. Lap joint was foil-wrapped into teflon foil and the whole sample was fixed with aluminum foil and put between press plates tempered at 180°C. At pressure of 100 kPa during 5 min, lap joint was formed. The specimens for peeling test were made similarly.

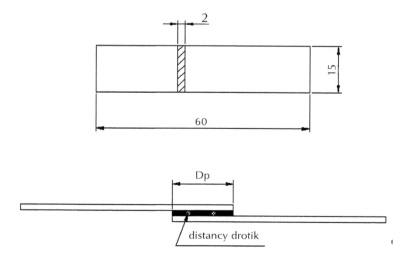

FIGURE 3 Lap adhesive joint.

Methods of testing included mechanical tests (measurement of cohesive proper-ties and hardness), adhesive tests (measurement of shear strength of adhesive joint at loading by tension [2], measurement of strength of adhesive joint at peeling [3], measurement of surface properties, thermo gravimetric analysis, and measurement of thermal properties.

Measurement of cohesive characteristics included the loading the test blade by ten-sile (Figure 1) at rate of separation of the jaws 50 mm/min with machine Instron 4301 (Instron, England), when following characteristics were evaluated—maximal tensile strength (MPa), maximal elongation (%), elongation at rupture (%), tensile strength at rupture (MPa), Young module of elasticity (MPa), yield strength (MPa), and elonga-tion at yield.

Measurement of hardness in ShD was done according to ASTM D 2122-2. Equip-ment D Scale Durometer PTC 307—L designed for plastics and react-plastics was used.

To measure adhesive characteristics, the test machine Instron 4301 was used (rate of separation of the jaws 50 mm/min). Following characteristics were evaluated:
- Shear strength (MPa)
- Relative elongation (%)
- Young module of elasticity (MPa)
- Energy of destruction of adhesive joint (J)

At peel test, the tested specimen was fixed in testing machine Instron 4301. Board A1 was fixed in the low jaw and aluminum foil was fixed in the upper movable jaw. Rate of separation was slower, only 10 mm/min. The values evaluated were:
- Strength of the joint at maximal loading (MPa)
- Average peel power (N)
- Average tear tension (N/mm)

Besides, also thermographic analysis was done with a thermogravimeter TG-1 (Perkin Elmer, USA).

10.3 DISCUSSION AND RESULTS

The Figure 4 presents the dependence of maximal tensile strength (R_{max}) and tensile strength at breaking (R_r) on the content of filler in composite adhesive. From measured results follow, that with increasing content of nanoparticles of filler in EAA, the maxi-mal tensile strength of composite is non-linearly increased. It can be assumed, that further filling will increase the value of maximal tensile strength but only for certain concentration. At this concentration, EAA composite will be saturated with Aerosil 130 SLP, what causes insufficient wetting of surface of filler particles and following lowering of max. tensile strength.

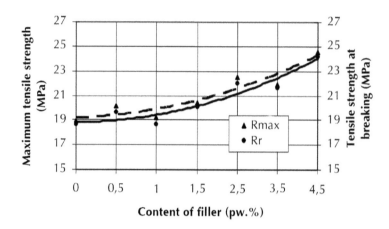

FIGURE 4   Dependence of max. tensile strength (R_{max}) and tensile strength at breaking (R_r) on the content of filler.

The dependence of adhesive shear strength of joint on the content of filler is on the Figure 5. Considering the high specific surface of nanoparticle filler (130 m²/g), intense change of investigated parameter occurs already at low concentrations of filler. Increased dispersion of measured values can be justified by the possible presence of nonhomogeneous in the composite system, as well as the deteriorative wetting of the aluminum substrate in the growth of filler content. Substantially is worsened the spreading of copolymer melt adhesive on the glued surface due to an increase melt viscosity of hot melt glue, which deteriorates the surface wetting.

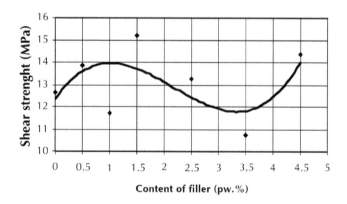

FIGURE 5   Dependence of adhesion joint shear strength on the filler content.

Character of dependence of average peeling stress is parabolic with the maximum at the content of filler 3.5 wt% (Figure 6). Also in this case, measured values show higher variance, similarly as at measurement of adhesive shear strength of joint.

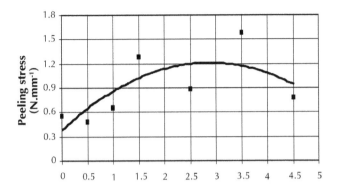

FIGURE 6 Dependence of peeling stress on adhesive concentration.

Thermo gravimetric analysis confirmed, that temperature of 10% weight loss and temperature of sudden weight loss (Figure 7) had after initial decrease increasing tendency. With the increase of filler particles, the temperature of loss 10% weight is increasing from 360 to 385°C, which represents a rise up to 8%. The reason is higher absorption of heat with Aerosil 130 SLP. Temperatures of sudden loss reach lower values (342 up to 374°C) in comparison with the temperature of loss 10% weight.

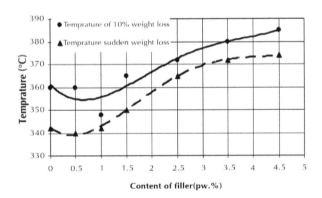

FIGURE 7 The dependence of temperature of 10% weight loss and temperature of sudden weight loss on the content of Aerosil 130 SLP.

10.4 CONCLUSION

On the base of realized experiments it can be concluded, that nanoparticle filler Aerosil 130 SLP influences individual properties of filled EAA system differently. The filler has positive impact to improve the cohesion and adhesion strength, heat resistance, peeling tension, and surface properties of the system. On the other hand, reduces the relative extension, factors of heat and thermal conductivity, and specific volume heat capacity. The cohesive mechanical parameters of the system can be stated as an optimal concentration of nanofiller Aerosil 130 SLP 2.5 wt%, the adhesion properties of 3.5 wt%. Nanoparticles composite systems showed the highest heat resistance in filler concentration from 3.5 to 4.5 wt%. For practical application of filled EAA nanocomposite systems is therefore necessary to know how to use, environment, application temperature and method of stress, and accordingly select the optimal concentration nanofiller.

KEYWORDS

- **Composite**
- **EAA copolymers,**
- **Hot-melt adhesives**
- **Nanofillers**

ACKNOWLEDGMENT

This contribution is the result of the project implementation Research of the Application Potential of Renewable and Recycled Materials and Information Technologies in the Rubber Industry" (project code ITMS: 26220220173) supported by the Research and Development Operational Program funded by the ERDF.

REFERENCES

1. Novák, I. and Pollák, V. Adhezíva využívané na lepenie v automobilovom priemysle. In: Chemagazín, No.4, ročník XII, pp. 4–5 (2002).
2. STN EN 1465. Stanovenie pevnosti v šmyku preplátovaného lepeného spoja pri namáhaní vťahu (2000).
3. STN EN 28510-2. Skúška odlupovania lepeného spoja skúšobného telesa z ohybného a tuhého adherendu. Časť 2: Odlupovanie pod uhlom 180 (2000).
4. Novák, I., Florián, Š., Pollák, V., and Žigo, O. Tlakovo-citlivé elektricky vodivé adhezíva. In: Chemagazín, No. 4, ročník XX, pp. 22–23 (2010).
5. Kinloch, A. J. Adhesion and Adhesives. Chapman and Hall, UK (1994).
6. Lu, D. and Wong, C. P. Intern. J. Adhesion and Adhesives, 20, p.189 (2000).

CHAPTER 11

HYPERBRANCHED POLYMERS: A PROMISING NEW CLASS OF KEY ENGINEERING MATERIALS

RAMIN MAHMOODI, TAHEREH DODEL, TAHEREH MOIENI, and MAHDI HASANZADEH

CONTENTS

11.1   INTRODUCTION

In the chapter, a hyperbranched, functional, water-soluble, and amine terminated polymer is synthesized by melt polycondensation reaction of methyl acrylate and diethylenetriamine. The polymer is then characterized by Fourier transform infrared spectroscopy (FTIR). The color characteristics of the dyed samples are evaluated using CIELAB method. The result showed that the color strength of hyperbranched polymers (HBPs) treated PA6 fabrics is more than pure PA6 fabric due to the presence of terminal primary amino groups in the molecular structure of the HBPs. Moreover, the washing fastness of the dyed treated PA6 fabrics were also good compared with that obtained by conventional dyeing.

Dendritic polymers are highly branched polymers with tree like branching having an overall spherical or ellipsoidal shape. These macromolecules consist of three subsets namely dendrimers, dendrigraft polymers, and HBPs [1]. The HBPs are highly branched, polydisperse and three-dimensional macromolecules synthesized from a multifunctional monomer to produce a molecule with dendritic structure [2]. These polymers have remarkable properties, such as low melt and solution viscosity, low chain entanglement, and high solubility, as a result of the large amount of functional end groups and globular structure, so they are excellent candidates for use in random applications, particularly for modifying fibers [3, 4].

Recently, the application of HBPs in textile industry has been developed especially for improvement of dye uptake of several fibers. For instance, Burkinshaw et al. [5] investigated the dyeability of polypropylene (PP) fibers modified by HBP. The results showed that the incorporation of HBP prior to fiber spinning considerably improved the color strength of PP fiber with C.I. Disperse Blue 56 and has no significant effect on physical properties of the PP fibers. In the study on applying HBP to cotton fabric by Zhang et al. [6-9], it was demonstrated that HBP treatment on cotton fabrics has no undesirable effect on mechanical properties of fabrics. Dyeing of treated cotton fabrics with direct and reactive dyes in the absence of electrolyte showed that the color strength of treated samples was better than untreated cotton fabrics. Furthermore, it has shown that the application of HBP to cotton fabrics can reduce UV transmission and impart antibacterial properties. In another study, the dyeability of modified polyethylene terephthalate (PET) fabrics by amine-terminated HBP was investigated by Hasanzadeh et al. [10]. They reported that the dye uptake of HBP-treated PET fabrics is significantly greater than that of untreated PET ones due to the presence of terminal primary amino groups in the molecular structure of the HBP, that will protonate in the liquid phase and give rise to positive charge at lower pH values.

Literature review showed that there has not been a previous report regarding the application of HBP on PA6 fabric and dyeing with acid dyes. In this work, new synthesized HBP with an amine terminal group was used to improve the dyeability of PA6 fabrics with acid dye. For this purpose, amine-terminated AB_2-type HBP was synthesized from methyl acrylate and diethylenetriamine by melt polycondensation. The obtained HBP was characterized using FTIR. Then the HBP was applied to PA6 fabric and the dyeability of treated samples and untreated one with C.I. Acid Red 131 was investigated.

11.2   EXPERIMENTAL

11.2.1   MATERIALS

Methyl acrylate (molecular weight = 86.09 g/mol) and diethylenetriamine (molecular weight = 103.17 g/mol) were purchased from Merck Co. and used as received. The PA6 knitted fabrics was used throughout this work and before using, the samples were scoured by a 2 g/L of anionic detergent at 70°C for 30 min and the procedure was followed by washing in cold distilled water and ambient drying. The selected acid dye for this work was C.I. Acid Red 131 provided by the Ciba Ltd. (Iran).

11.2.2   MEASUREMENTS

The FTIR spectrum of AB_2-type monomer and corresponding HBP were recorded by Nicolet 670 FTIR spectrophotometer in the wave number range of 500–4000 cm^{-1} which nominal resolution for all spectra was 4 cm^{-1}. The spectral reflectance of the dyed samples was determined using a spectrophotometer. As shown by Equation (1), the Kubelka–Munk single-constant theory was employed to calculate the color strength as reflectance function (K/S value) [11].

$$(K/S)_\lambda = \frac{\left(1-R_\lambda\right)^2}{2R_\lambda} \tag{1}$$

where R_λ is the reflectance value of dyed sample at the wavelength of maximum absorbance (λ_{max}), K is the absorption coefficient, and S is the scattering coefficient. Washing fastness test was performed according to ISO/R 105/IV, Part 8.

11.2.3   SYNTHESIS OF AMINE-TERMINATED HYPERBRANCHED POLYMER

For synthesis of amine-terminated hyperbranched polymer two steps were employed. The first step was the synthesis of AB_2-type monomers and the second one was the preparation of HBP using a melt polycondensation reaction. Diethylenetriamine (0.5 mol and 52 ml) was added in a three-necked flask equipped with a constant voltage dropping funnel, condenser, and a nitrogen inlet. The flask was placed in an ice bath and solution of methyl acrylate (0.5 mol and 43 ml) in methanol (100 ml) was added dropwise into the flask. The reaction mixture was stirred with a magnetic stirrer. Then the mixture was removed from the ice bath and left to react with a flow of nitrogen at room temperature (Scheme 1). After stirring for 14 hr, the nitrogen flow was removed and AB_2 type monomer was obtained.

The obtained light yellow and viscous mixture was transferred to an eggplant shaped flask for an automatic rotary vacuum evaporator to remove the methanol under low pressure. Then the temperature was raised to 150°C using an oil bath and condensation reaction was carried out for 6 hr. A pale yellow viscous amine-terminated hyperbranched polymer was obtained.

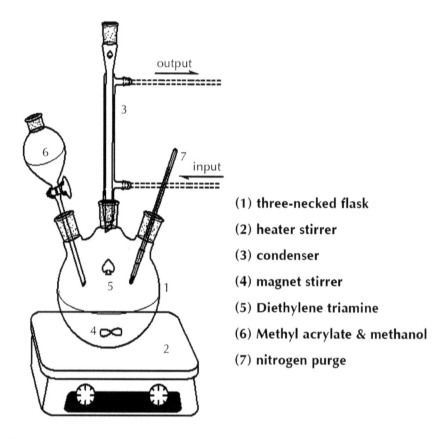

output

input

(1) **three-necked flask**

(2) **heater stirrer**

(3) **condenser**

(4) **magnet stirrer**

(5) **Diethylene triamine**

(6) **Methyl acrylate & methanol**

(7) **nitrogen purge**

SCHEME 1   Schematic of reactor for monomer synthesis.

11.2.4 MODIFICATION OF PA FABRICS USING HBP

The PA6 fabrics were immersed into baths containing 10% HBP and 0.5% nan-ionic surfactant (Irgasol NA) with the liquor to goods ratio of 60:1 at the temperature 40°C for 30 min. The pH was adjusted to 5.5 with acetic acid prior to adding the fabric. After exhaustion, the samples (PA6-HBP) were allowed to dyeing process.

11.2.5 DYEING PROCEDURE

The dyeing process of PA6-HBP samples was carried out in the same bath. Acid dye (0.5, 0.75, 1% owf) was added to the bath at the temperature 40°C. Then the temperature was raised up to boiling point at a rate of 2°C/min. Dyeing was then continued for 60 min with occasional stirring. Dyeing process was carried out using a liquor ratio of 60:1, and the procedure was followed by washing in warm water at about 50°C and then with cold water. The same procedure was also carried out for pure PA6 fabrics. Figure 1 show the dyeing profile.

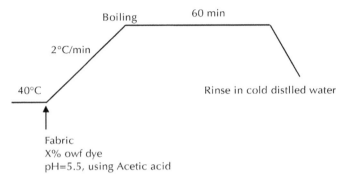

FIGURE 1 Method and recipe for dyeing of samples with Acid Red 131.

11.3 DISCUSSION AND RESULTS

11.3.1 *SYNTHESIS AND CHARACTERIZATION OF HBP*

Amine-terminated hyperbranched polymer were synthesized in two step reaction comprise preparation of AB_2 type monomers (1 and 2) by Michael addition reaction of methyl acrylate and diethylenetriamine and synthesis of HBP by polycondensation reaction respectively (Scheme 2).

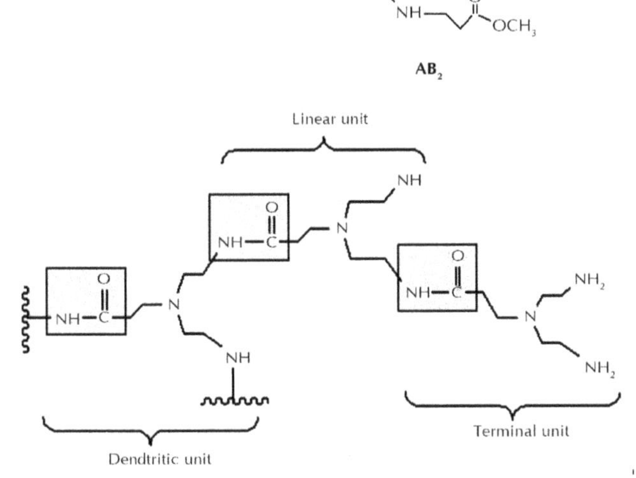

SCHEME 2 Chemical structure of amine terminated hyperbranched polymer.

In order to confirm the polycondensation reaction progress, FTIR of AB$_2$-type monomer and HBP were studied (Figure 2). Spectral differences between AB$_2$ type monomer and HBP are observed in the fingerprint region between 1800 and 900 cm^{-1}. The absorption bands in FTIR spectra are assigned according to literature [12].

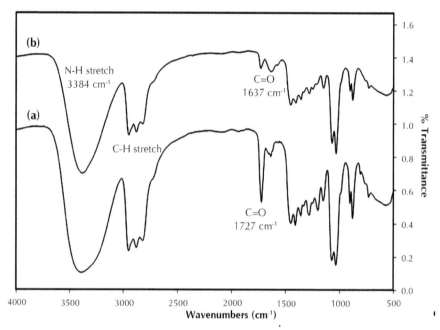

FIGURE 2   FTIR spectra of (a) AB$_2$ type monomer and (b) hyperbranched polymer.

The FTIR spectra of both monomer and polymer show that the peak positions are at 3384, 2950, 2880, and 2819 cm^{-1}. The band at above 3300 cm^{-1} is due to N–H stretching vibration, while the band at 1463 cm^{-1} reflects the CH$_2$ group bending vibration. In Figure 2 (a), the absorption at 1727 cm^{-1} is attributed to C=O stretching vibration of esters. This peak is generally weak in FTIR spectrum of HBP (Figure 2 (b)). Moreover, the band at 1637 cm^{-1} corresponds to C=O stretching vibration of amides. Melt-polycondensation reaction and synthesis can be responsible for this change in the band intensity [6].

11.3.2   DYEING PROPERTIES OF PA6-HBP FABRICS

The effect of HBP treatment on dye absorbance behavior (dyeability) of PA6 fabrics was evaluated by measuring its optical properties. Figure 3 shows the reflectance spectra of dyed PA6 and PA6-HBP fabrics. It can be seen that the reflectance spectrum of PA6-HBP samples is generally lower than PA6 samples. Therefore, the treatment of PA6 fabrics by HBP significantly reduced the reflectance.

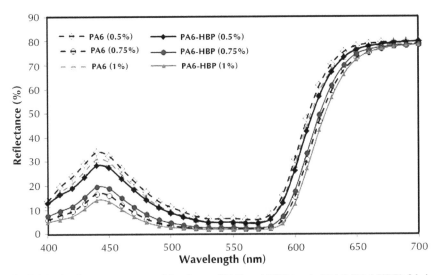

FIGURE 3   Reflectance spectra of dyed pure (PA6) and HBP treated PA6 (PA6-HBP) fabrics.

The color strength (K/S value) results of the PA6 and PA6-HBP fabrics dyed with Acid Red 131 using a competitive dyeing method are shown in Figure 4. A higher K/S value of PA6-HBP samples indicated greater dye uptake than that of PA6 sample. The increase in the color strength is due to the positively charged amino groups on the PA6-HBP fabrics. Moreover, the CIELAB color coordinates (L*, a*, b*, C*, and h°) of samples are shown in Table 1. It can be concluded that the HBP treatment has significant effects on the color coordinates of PA6 fabrics.

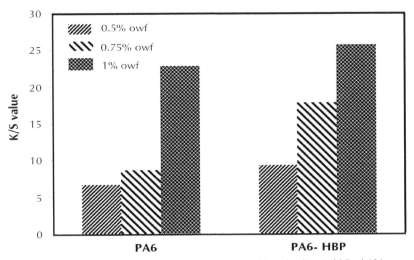

FIGURE 4   Effect of HBP treatment on color strength achieved using Acid Red 131.

TABLE 1 Color coordinates of dyed samples

Fabrics	Dye (% owf)	L*	a*	b*	C*	h°
PA6	0.5	48.140	55.798	−17.867	58.58879	−0.30989
	0.75	36.914	60.330	−7.471	60.79083	−0.12321
	1	45.958	57.430	−17.393	60.00601	−0.29407
PA6-HBP	0.5	44.995	55.988	−15.714	58.15141	−0.27363
	0.75	38.949	59.685	−10.187	60.54811	−0.16905
	1	33.099	55.476	−4.346	55.64597	−0.07818

Table 2 shows the washing fastness of PA6 and PA6-HBP fabric dyed with acid dye. Compared with the pure PA6 fabrics, the treated ones have good washing fastness. This result showed that, the washing fastness of acid dye was largely unaffected by HBP treatment. Furthermore, the color strength of samples before and after washing was shown in Figure 5. It is clear that the extent of color strength change was relatively small for treated and pure PA6 fabrics.

TABLE 2 Color fastness of pure (PA6) and HBP treated PA6 (PA6-HBP) fabrics

Fabrics	Dye (% owf)	K/S	Washing fastness	
			Fading	Staining
PA6	0.5	6.78	4–5	4–5
	0.75	8.75	4–5	4–5
	1	22.93	4	4
PA6-HBP	0.5	9.31	5	5
	0.75	17.81	5	5
	1	25.75	4–5	4–5

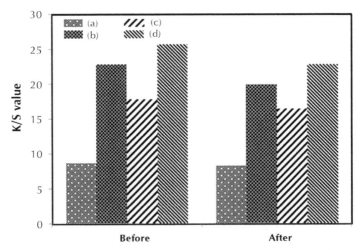

FIGURE 5 Effect of washing fastness test on color strength of (a), (b) PA6 fabric dyed with 0.75% and 1% owf, respectively, (c) and (d) PA6-HBP fabric dyed with 0.75% and 1% owf, respectively.

11.4 CONCLUSION

In this chapter, a hyperbranched, functional ,and water-soluble polymer was success-fully synthesized by melt polycondensation reaction of methyl acrylate and diethyl-enetriamine. The FTIR spectroscopy measurement of HBP indicated that this polymer comprised terminal amine group. The HBP was applied to PA6 fabric using the ex-haustion process. The study of dyeability of treated samples with C.I. Acid Red 131 indicated that the color strength of HBP treated PA6 fabrics is more than pure PA6 fabric due to the presence of terminal primary amino groups in the molecular structure of the HBP, that will protonate in the liquid phase and give rise to positive charge at lower pH values. Moreover, the washing fastness of acid dye was largely unaffected by HBP treatment.

KEYWORDS

- **Dendritic polymers**
- **FTIR spectrum**
- **HBP treatment**
- **Hyperbranched polymers**
- **PA6 fabrics**

ACKNOWLEDGMENT

The author would like to express their grateful thanks to Dr. M. Ghanbar Afjeh from Amirkabir University of Technology for the color measurements.

REFERENCES

1. Jikei, M. and Kakimoto, M. Hyperbranched polymers: a promising new class of materials. *Progress in Polymer Science*, **26**, 1233–1285 (2001) doi: 10.1016/S0079-6700(01)00018-1.
2. Gao, C. and Yan, D. Hyperbranched polymers: from synthesis to applications. *Progress in Polymer Science*, **29**, 183–275 (2004) doi: 10.1016/j.progpolymsci.2003.12.002.
3. Schmaljohann, D., Pötschke, P., Hässler, R., Voit, B. I., Froehling, P. E., Mostert, B., and Loontjens, J. A. Blends of amphiphilic, hyperbranched polyesters and different polyolefins. *Macromolecules*, **32**, 6333–6339 (1999) doi: 10.1021/ma9902504.
4. Seiler, M. Hyperbranched polymers: Phase behavior and new applications in the field of chemical engineering. *Fluid Phase Equilibria*, **241**, 155–174 (2006) doi: 10.1016/j.fluid.2005.12.042.
5. Burkinshaw, S. M., Froehling, P. E., and Mignanelli, M. The effect of hyperbranched polymers on the dyeing of polypropylene fibres. *Dyes and Pigments*, **53**, 229–235 (2002) doi: 10.1016/S0143-7208(02)00006-2.
6. Zhang, F., Chen, Y., Lin, H., and Lu, Y. Synthesis of an amino-terminated hyperbranched polymer and its application in reactive dyeing on cotton as a salt-free dyeing auxiliary. *Coloration Technology*, **123**, 351–357 (2007) doi: 10.1111/j.1478-4408.2007.00108.x.
7. Zhang, F., Chen, Y., Lin, H., Wang, H., and Zhao, B. HBP-NH$_2$ grafted cotton fiber: Preparation and salt-free dyeing properties. *Carbohydrate Polymer*, **74**, 250–256 (2008) doi: 10.1016/j.carbpol.2008.02.006.
8. Zhang, F., Chen, Y. Y., Lin, H., and Zhang, D. S. Performance of cotton fabric treated with an amino-terminated hyperbranched polymer. *Fibers and Polymers*, **9**, 515–520 (2008).
9. Zhang, F., Chen, Y., Ling, H., and Zhang, D. Synthesis of HBP-HTC and its application to cotton fabric as an antimicrobial auxiliary. *Fibers and Polymers*, **10**, 141–147 (2009) doi: 10.1007/s12221-009-0141-6.
10. Hasanzadeh, M., Moieni, T., and Hadavi Moghadam, B. Synthesis and characterization of an amine terminated AB$_2$-type hyperbranched polymer and its application in dyeing of poly(ethylene terephthalate) fabric with acid dye, *Advances in Polymer Technology*, **32**, 20 (2013) doi: 10.1002/adv.21345.
11. McDonald, R. *Color physics for industry*. England: Society of dyers and colorists (1997).
12. Pavia, D. L., Lampman, G. M., Kriz, G. S., and Vyvyan, J. R. *Introduction to spectroscopy*. US: Brooks/Cole (2001).

CHAPTER 12

CURRENT STATUS AND BIOMEDICAL APPLICATION SPECTRUM OF POLY (3-HYDROXYBUTYRATE) AS A BACTERIAL BIODEGRADABLE POLYMER

A. L. IORDANSKII, G. A. BONARTSEVA, YU. N. PANKOVA,
S. Z. ROGOVINA, K. Z. GUMARGALIEVA, G. E. ZAIKOV,
and A. A. BERLIN

CONTENTS

12.1 INTRODUCTION

The biodegradable poly(3-hydroxybutyrate) (PHB) as member of natural polymer family has been widely used and advanced in the innovative biomedical areas owing to relevant combination of biocompatibility, transport characteristics, and mechanical behavior. This review presents an extensive description to hydrolysis and enzymatic degradation during long-time period preferably under laboratory condition (*in vitro*) and as medical implant (*in vivo*). Besides, the focus of review is devoted to the literature comparison of different works in PHB degradation and its biomedical application including work of authors.

Currently, the intensive development of biodegradable and biocompatible materials for medical implication provokes comprehensive interdisciplinary studies on biopolymer structures and functions. The well-known and applicable biodegradable polymers are polylactic acid (PLA) and polyglycolides (PGA), and their copolymers, poly-ε-caprolactone, poly(orthoesters), poly-β-malic acid, poly(propylene fumarate), polyalkylcyanoacrylates, polyorthoanhydrides, polyphosphazenes, some natural polysaccharides (starch, chitosan, alginates, agarose, dextrane, chondroitin sulfate, and hyaluronic acid), and proteins (collagen, silk fibroin, spidroin, fibrin, gelatin, and albumin). Since some of these polymers should be synthesized through chemical stages (for example, *via* lactic and glycolic acids) it is not quite correct to define them as the biopolymers.

Besides biomedicine applications, the biodegradable biopolymers attract much attention as perspective materials in wide areas of industry, nanotechnologies, farming, and packaging owing to the relevant combination of biomedical, transport, and physical-chemical properties.

It is worth to emphasis that only medical area of these biopolymers includes implants and prosthesis, tissue engineering scaffolds, novel drug dosage forms in pharmaceutics, novel materials for dentistry and others.

Each potentially applicable biopolymer arranges a wide multidisciplinary network, which usually includes tasks of searching for efficacy ways of biosynthesis reactions; economical problems associated with large-scale production; academic studies of mechanical, physicochemical, and biochemical properties of the polymer and material of interest, technology of preparation and using this biopolymer, preclinical and clinical trials of these materials and products, a market analysis and perspectives of application of the developed products and many other problems.

The poly((R)-3-hydroxybutyrate) (P3HB) is an illustrative example for the one of centers for formation of the mentioned scientific-technological network and a basis for the development of various biopolymer systems [1–2]. In recent decades an intense development of biomedical application of bacterial PHB in producing of biodegradable polymer implants and controlled drug release systems [3-6] needs for comprehensive understanding of the PHB biodegradation process. The examination of PHB degradation process is also necessary for development of novel friendly environment polymer packaging [7-9]. It is generally accepted that biodegradation of PHB both in living systems and in environment occurs *via* enzymatic and non-enzymatic processes that take place simultaneously under natural conditions. It is, therefore, important to understand both processes [6, 10]. Opposite to other biodegradable polymers (for example, PGA and PLGA), PHB is considered to be moderately resistant to degradation *in vitro* as well as to biodegradation in biological media. The rates of degradation are influenced by the characteristics of the polymer, such as chemical composition, crystallinity, morphology, and molecular weight [11, 12]. In spite of that PHB application *in vitro* and *in vivo* has been intensively investigated, the most of the available data are often incomplete and sometimes even contradictory. The presence of conflicting data can be partially explained by the fact that biotechnologically produced PHB with standardized properties is relatively rare and is not readily available due to a wide variety of its biosynthesis sources and different manufacturing processes.

The inconsistencies can be explained also by excess applied trend in PHB degradation research. At most of the papers observed in this review, PHB degradation process has been investigated in the narrow framework of development of specific medical devices. Depending on applied biomedical purposes biodegradation of PHB was investigated under different geometry: films and plates with various thickness [13-16], cylinders [17-19], monofilament threads [20-22] and micro- and nanospheres [23, 24]. At these experiments PHB was used from various sources, with different molecular weight and crystallinity.

Besides, different technologies of PHB devices manufacture affect such important characteristics as polymer porosity and surface structure [14, 15]. Reports regarding the complex theoretical research of mechanisms of hydrolysis, enzymatic degradation and biodegradation *in vivo* of PHB processes are relatively rare [13-15, 16, 25-27] that attaches great value and importance to these investigations. Nevertheless, the effect of thickness, size, and geometry of PHB device, molecular weight and crystallinity of PHB on the mechanism of PHB hydrolysis and biodegradation was not yet well clarified.

12.2 HYDROLYSIS AND ENZYMATIC DEGRADATION OF PHB

12.2.1 NONENZYMATIC HYDROLYSIS OF PHB IN VITRO

The examination of hydrolytic degradation of natural P3HB *in vitro* is a very important step for understanding of PHB biodegradation. There are several very profound and careful examinations of PHB hydrolysis that was carried out for 10–15 years [25-28]. The hydrolytic degradation of PHB was usually examined under standard experimental conditions simulating internal body fluid, in buffered solutions with pH = 7.4 at 37°C but at seldom cases the higher temperature (55°C, 70°C, and more) and other values of pH (from 2 to 11) were selected.

The classical experiment for examination of PHB hydrolysis in comparison with hydrolysis of other widespread biopolymer, PLA, was carried out by Koyama N. and Doi Y. [25]. They compared films (10×10 mm size, 50 μm thickness, and 5 mg initial mass) from PHB ($M_n = 300000$ and $M_w = 650000$) with PLA ($M_n = 9000$ and $M_w = 21000$) prepared by solvent casting and aged for 3 weeks to reach equilibrium crystallinity. They have shown that hydrolytic degradation of natural PHB is the slow process. The mass of PHB film remained unchanged at 37°C in 10 mM phosphate buffer (pH = 7.4) over a period of 150 days, while the mass of the PLA film rapidly decreased with time and reached 17% of the initial mass after 140 days. The rate of decrease in the M_n of the PHB was also much slower than the rate of decrease in the M_n of PLA. The M_n of the PHB decreased approximately to 65% of the initial value after 150 days, while the M_n of the PLA decreased to 20% (M_n = 2000) at the same time.

As PLA used at this research was with low molecular weight it is worth to compare these data with the data of hydrolysis investigation with the same initial M_n as for PHB. In other work the mass loss of two polymer films (PLA and PHB) with the same thickness (40 μm) and molecular weight (M_w = 450000) was studied *in vitro*. It was shown that the mass of PLA film decreased to 87%, whereas the mass of PHB film remained unchanged at 37°C in 25 mM phosphate buffer (pH = 7.4) over a period of 84 days, but after 360 days the mass of PHB film was 64.9% of initial one [29-31].

The cleavage of polyester chains is known to be catalyzed by the carboxyl end groups, and its rate is proportional to the concentrations of water and ester bonds that on the initial stage of hydrolysis are constant owing to the presence of a large excess of water molecules and ester bonds of polymer chains. Thus, the kinetics of nonenzymatic hydrolysis can be expressed by a simplified equation [32-33]:

$$\ln M_n = \ln M_n^0 - k_h t \qquad (1)$$

where M_n and M_n^0 are the number-average molecular weights of a polymer component at time t and zero, respectively and k_h is effective hydrolysis constant.

The average number of bond cleavage per original polymer molecule, N, is given by Equation 2:

$$N = (M_n^0/M_n) - 1 = k_d M_m P_n^0 t \qquad (2)$$

where k_d is the effective rate constant of hydrolytic depolymerization, and P_n^0 is the initial number-average degree of polymerization at time zero, M_m is constant molecular mass of monomer. Thus, if the chain scission is completely random, the value of N is linearly dependent on time.

The molecular weight decrease with time is the distinguishing feature of mechanism under nonenzymatic hydrolysis condition in contrast to enzymatic hydrolysis condition of PHB when M_n values remained almost unchanged. It was supposed also that water soluble oligomers of PHB with molecular mass about 3 kDa may accelerate the hydrolysis rate of PHB homopolymer [25]. In contrast, Freier T. et al. [14] showed that PHB hydrolysis was not accelerated by the addition of pre-degraded PHB, the rate of mass and M_w loss of blends (70/30) from high-molecular PHB (M_w = 641000) and low-molecular PHB (M_w = 3000) was the same with degradation rate of pure high-molecular PHB.

Meanwhile, the addition of amorphous atactic PHB (at-PHB) (M_w = 10000) to blend with high-molecular PHB caused significant acceleration of PHB hydrolysis, the relative mass loss of PHB/at-PHB blends was 7% in comparison with 0% mass loss of pure PHB, the decrease of M_w was 88% in comparison with 48% M_w decrease of pure PHB [14, 34]. We have showed that the rate of hydrolysis of PHB films depends on M_w of PHB. The PHB films of high molecular weight (M_w = 450000 and 1000000) degraded slowly as it was described whereas, films from PHB of low molecular weight (M_w = 150000 and 300000 kDa) lost weight relatively gradually and more rapidly [29-31].

To enhance the hydrolysis of PHB a higher temperature was selected for degradation experiments, 55°C, 70°C and more [25]. It was showed by the same research team that the weights of films (12 mm diameter, 65 μm thick) from PHB (M_n = 768 and 22 kDa, M_w = 1460 and 75 kDa) were unchanged at 55°C in 10 mM phosphate buffer (pH = 7.4) over a period of 58 days. The M_n value decreased from 768kDa to 245 kDa for 48 days.

The film thickness increased from 65 to 75 μm for 48 days, suggesting that water permeated the polymer matrix during the hydrolytic degradation. The examination of the surface and cross-section of PHB films before and after hydrolysis showed that surface after 48 days of hydrolysis was apparently unchanged, while the cross-section of the film exhibited a more porous structure (pore size< 0.5 μm). It was shown also that the rate of hydrolytic degradation is not dependent upon the crystallinity of PHB film.

The observed data indicates that the nonenzymatic hydrolysis of PHB in the aqueous media proceeds *via* a random bulk hydrolysis of ester bonds in the polymer chain films and occurs throughout the whole film, since water permeates the polymer matrix [25-26]. Moreover, as over the whole degradation time the first-order kinetics was observed and the molecular weight distribution was unimodal, a random chain scission mechanism is very probable both on the crystalline surfaces and in the amorphous regions of the biopolymer [14, 35]. For synthetic amorphous at-PHB it was shown that its hydrolysis is the two-step process. First, the random chain scission proceeds that accompanying with a molecular weight decrease. Then, at a molecular weight of about 10000, mass loss begins [28].

The analysis of literature data shows a great spread in values of rate of PHB hydrolytic degradation *in vitro*. It can be explained by different thickness of PHB films or geometry of PHB devices used for experiment as well as by different sources, purity degree, and molecular weight of PHB (Table 1). At 37°C and pH = 7.4 the weight loss of PHB (unknown M_w) films (500 μm thick) was 3% after 40 days incubation [38], 0% after 52 weeks (364 days) and after 2 years (730 days) incubation (640 kDa PHB, 100 μm films) [14-15], 0% after 150 days incubation (650 kDa PHB, 50 μm film) [25], 7.5% after 50 days incubation (279 kDa PHB, unknown thickness of films) [37], 0% after 3 months (84 days) incubation (450 kDa PHB, 40 μm films), 12% after 3 months (84 days) incubation (150 kDa PHB, 40 μm films) [29-30], 0% after 180 days incubation of monofilament threads (30 μm in diameter) from PHB (470 kDa) [22-23]. The molecular weight of PHB dropped to 36% of the initial values after 2 years (730 days) of storage in buffer solution [15], to 87% of the initial values after 98 days [38], to 58% of the initial values after 84 days [29-30] (Table 1).

TABLE 1   Nonenzymatic hydrolysis of PHB *in vitro* (data for comparison)

Type of device	Initial M_w of PHB, kDa	Size/Thickness, μm	Conditions	Relative mass loss of PHB, %	Relative loss of PHB M_w, %	Time and Days	Links
Film	650	50	37°C, pH = 7.4	0	35	150	25
Film	640	100	37°C, pH = 7.4	0	64	730	15
Film	640	100	37°C, pH = 7.4	0	45	364	14
Film	450	40	37°C, pH = 7.4	0	42	84	29-30
Film	150	40	37°C, pH = 7.4	12	63	84	29-30
Film	450	40	37°C, pH = 7.4	35,1	-	360	31
Film	279	-	37°C, pH = 7.4	7.5	-	50	36
Plate	-	500	37°C, pH = 7.4	3	-	40	30
Plate	380	1000	37°C, pH = 7.4	0	-	28	43
Plate	380	2000	37°C, pH = 7.4	0	8	98	43
Thread	470	30	37°C, pH = 7.0	0	-	180	23
Thread	-	-	37°C, pH = 7.2	0	-	182	22
Microspheres	50	250-850	37°C, pH = 7.4	0	0	150	42
Thread	470	30	37°C, pH = 5.2	0	-	180	22
Film	279	-	37°C, pH = 10	100	-	28	36
Film	279	-	37°C, pH = 13	100	-	19	36
Film	650	50	55°C, pH = 7.4	0	68	150	25
Plate	380	2000	55°C, pH = 7.4	0	61	98	43
Film	640	100	70°C, pH = 7.4	-	55	28	14
Film	150	40	70°C, pH = 7.4	39	96	84	29-30
Film	450	40	70°C, pH = 7.4	12	92	84	29-30
Microspheres	50	250-850	85°C, pH = 7.4	50	68	150	42
Microspheres	600	250-850	85°C, pH = 7.4	25	-	150	42

In acidic or alkaline aqueous media PHB degrades more rapidly, 0% degradation after 140 days incubation in 0.01M NaOH (pH = 11) (200 kDa, 100 µm film thickness) with visible surface changing [39], 0% degradation after 180 days incubation of PHB threads in phosphate buffer (pH = 5.2 and 5.9) [23], complete PHB films biodegradation after 19 days (pH = 13) and 28 days (pH = 10) [37]. It was demonstrated that after 20 weeks of exposure to NaOH solution, the surfaces of PHB samples became rougher, along with an increased density in their surface layers. From these results, one may surmise that the non-enzymatic degradation of PHAs progresses on their surfaces before noticeable weight loss occurs as illustrated in Figure 1 [39].

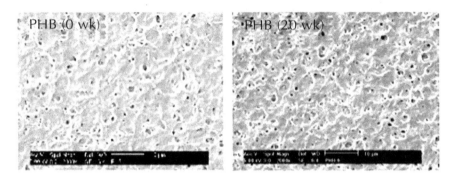

FIGURE 1   Scanning electron microscopy photographs of PHB films both before (0 wk, panels on the left) and after 20 weeks (panels on the right) of non-enzymatic hydrolysis in 0.01 N NaOH solution (scale bars, 10 µm).

It was shown also that treatment of PHB film with 1M NaOH caused a reduction in pore size on film surface from 1–5 µm to around 1 µm that indicates a partially surface degradation of PHB in alkaline media [40-41]. At higher temperature no weight loss of PHB films and threads was observed after 98 and 182 days incubation in phosphate buffer (pH = 7.2) at 55°C and 70°C, respectively [22], 12% and 39% of PHB (450 and 150 kDa, respectively) films after 84 days incubation at 70°C [35, 40], 50% and 25% after 150 days incubation of microspheres (250–850 µm diameter) from PHB (50 kDa and 600 kDa, respectively) [42].

During degradation of PHB monofilament threads, films and plates the change of mechanical properties was observed under different conditions *in vitro* [22, 43]. It was shown that a number of mechanical indices of threads became worse, load at break lost 36%, strain at break lost 33%, Young's modulus did not change, tensile strength lost 42% after 182 days incubation in phosphate buffer (pH = 7.2) at 70°C. But at 37°C the changes were more complicated, at first load at break increased from 440 g to 510 g (16%) at 90th day and then decreased to the initial value on 182th day, strain at break increased rapidly from 60% to 70% (in 17%) at 20th day and then gradually increased to 75% (in 25%) at 182th day, Young's modulus did not change [22]. For PHB films it was demonstrated a gradual 32% decrease in Young's modulus and 77% fall in tensile strength during 120 days incubation in phosphate buffer (pH = 7.4) at

37°C [43]. For PHB plates more complicated changes were observed, at first tensile strength dropped in 13% for 1st day and then increased to the initial value at 28th day, Young's modulus dropped in 32% for 1st day and then remain unchanged up to 28th day, stiffness decreased sharply also in 40% for 1st day and then remain unchanged up to 28th day [44].

12.2.2 ENZYMATIC DEGRADATION OF PHB IN VITRO

The examination of enzymatic degradation of PHB *in vitro* is the following important step for understanding of PHB performance in animal tissues and in environment. The most papers observed degradation of PHB by depolymerases of its own bacterial producers. The degradation of PHB *in vitro* by depolymerase was thoroughly examined and mechanism of enzymatic PHB degradation was perfectly clarified by Doi Y. [25-26]. At these early works it was shown that 68–85% and 58% mass loss of PHB (Mw = 650–768 and 22 kDa, respectively) films (50–65 µm thick) occurred for 20 hr under incubation at 37°C in phosphate solution (pH = 7.4) with depolymerase (1.5–3 µg/ml) isolated from *A. faecalis*. The rate (k_e) of enzymatic degradation of films from PHB (M_n = 768 kDa and 22 kDa) was 0.17 mg/hr and 0.15 mg/hr, respectively. The thickness of polymer films dropped from 65 to 22 µm (32% of initial thickness) during incubation. The scanning electron microscopy examination showed that the surface of the PHB film after enzymatic degradation was apparently blemished by the action of PHB depolymerase , while no change was observed inside the film. Moreover, the molecular weight of PHB remained almost unchanged after enzymatic hydrolysis, the M_n of PHB decreased from 768 to 669 kDa or unchanged (22 kDa) [25-26].

The extensive literature data on enzymatic degradation of PHB by specific PHB depolymerases was collected in detail in review of Sudesh K., Abe H., and Doi Y. [45]. We would like to summarize some the most important data. But at first it is necessary to note that PHB depolymerase is very specific enzyme and the hydrolysis of polymer by depolymerase is the unique process. But in animal tissues and even in environment the enzymatic degradation of PHB is occurred mainly by nonspecific esterases [24, 46]. Thus, in the frameworks of this review, it is necessary to observe the fundamental mechanisms of PHB enzymatic degradation.

The rate of enzymatic erosion of PHB by depolymerase is strongly dependent on the concentration of the enzyme. The enzymatic degradation of solid PHB polymer is heterogeneous reaction involving two steps, namely, adsorption and hydrolysis. The first step is adsorption of the enzyme onto the surface of the PHB material by the binding domain of the PHB depolymerase, and the second step is hydrolysis of polyester chains by the active site of the enzyme. The rate of enzymatic erosion for chemosynthetic PHB samples containing both monomeric units of (R)- and (S)-3-hydrohybutyrate is strongly dependent on both the stereo composition and on the tacticity of the sample as well as on substrate specificity of PHB depolymerase. The water-soluble products of random hydrolysis of PHB by enzyme showed a mixture of monomers and oligomers of (R)-3-hydrohybutirate. The rate of enzymatic hydrolysis for melt-crystallized PHB films by PHB depolymerase decreased with an increase in the crystallinity of the PHB film, while the rate of enzymatic degradation for PHB chains in an amorphous state was approximately 20 times higher than the rate for PHB chains in a crystalline state. It was suggested that the PHB depolymerase predominantly hydrolyzes polymer chains in the amorphous phase and then, subsequently, erodes the crystalline phase. The surface of the PHB film after enzymatic degradation was apparently blemished by the action of PHB depolymerase, while no change was observed inside the film. Thus, depolymerase hydrolyses of the polyester chains in the surface layer of the film and polymer erosion proceeds in surface layers, while dissolution, the enzymatic degradation of PHB are affected by many factors as monomer composition, molecular weight, and degree of crystallinity [45]. The PHB polymer matrix ultrastructure [47] plays also very important role in enzymatic polymer degradation [48].

At the next step it is necessary to observe enzymatic degradation of PHB under the conditions that modeled the animal tissues and body fluids containing nonspecific esterases. *In vitro* degradation of PHB films in the presence of various lipases as nonspecific esterases was carried out in buffered solutions containing lipases [18, 49-50], in digestive juices (for example, pancreatin) [14], simulated body fluid [51] biological media (serum, blood, and so on) [23] and crude tissue extracts containing a mixture of enzymes [24] to examine the mechanism of nonspecific enzymatic degradation process. It was noted that a Ser..His..Asp triad constitutes the active center of the catalytic domain of both PHB depolymerase [52] and lipases [53]. The serine is part of the pentapeptide Gly X1-Ser-X2-Gly, which has been located in all known PHB depolymerases as well as in lipases, esterases and serine proteases [52].

On the one hand, it was shown that PHB was not degraded for 100 days with a quantity of lipases isolated from different bacteria and fungi [49-50]. On the other hand, the progressive PHB degradation by lipases was shown [18, 40-41]. The PHB enzymatic biodegradation was studied also in biological media, it was shown that with pancreatin addition no additional mass loss of PHB was observed in comparison with simple hydrolysis [14], the PHB degradation process in serum and blood was demonstrated to be similar to hydrolysis process in buffered solution [31], whereas, progressive mass loss of PHB sutures was observed in serum and blood, 16% and 25%, respectively, after 180 days incubation [23], crude extracts from liver, muscle, kidney, heart, and brain showed the activity to degrade the PHB, from 2 to 18% mass loss of PHB microspheres after 17 hr incubation at pH 7.5 and 9.5 [24]. The weight loss of PHB (M_w = 285000) films after 45 days incubation simulated body fluid was about 5% [51]. The degradation rate in solution with pancreatin addition, obtained from the decrease in M_w of pure PHB, was accelerated about threefold, 34% decrease in M_w after incubation for 84 days in pancreatin (10 mg/ml in Sorensen buffer) versus 11% decrease in M_w after incubation in phosphate buffer [14]. The same data was obtained for PHB biodegradation in buffered solutions with porcine lipase addition, 72% decrease in M_w of PHB (M_w = 450000) after incubation for 84 days with lipase (20 U/mg, 10 mg/ml in tris-buffer) versus 39% decrease in M_w after incubation in phosphate buffer [18]. This observation is in contrast to enzymatic degradation by PHB depolymerases which was reported to proceed on the surface of the polymer film with an almost unchanged molecular weight [24, 25]. It has been proposed that for depolymerases the relative size of the enzyme compared with the void space in solvent cast films is the limiting factor for diffusion into the polymer matrix [54] whereas, lipases can penetrate into the polymer matrix through pores in PHB film [40, 41]. It was shown that lipase (0.1 g/l in buffer) treatment for 24 hr caused significant morphological change in PHB film surface, transferring from native PHB film with many pores ranging from 1 to 5 μm in size into a pore free surface without producing a quantity of hydroxyl groups on the film surface. It was supposed that the pores had a fairly large surface exposed to lipase, thus it was degraded more easily (Figure 2) [40, 41]. It indicates also that lipase can partially penetrate into pores of PHB film but the enzymatic degradation proceeds mainly on the surface of the coarse polymer film achievable for lipase. Two additional effects reported for depolymerases could be of importance. It was concluded that segmental mobility in amorphous phase and polymer hydrophobicity play an important role in enzymatic PHB degradation by nonspecific esterases [14]. The significant impairment of the tensile strength and other mechanical properties were observed during enzymatic biodegradation of PHB threads in serum and blood. It was shown that load at break lost 29%, Young's modulus lost 20%, and tensile strength did not change after 180 days of threads incubation, the mechanical properties changed gradually [23].

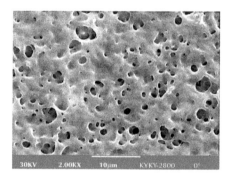 

FIGURE 2   Scanning electron microscopy photographs documented the surface structure of PHB polymer films: (a) PHB film; (b) PHB film treated with lipase (0.1 g/l at 30°C and pH = 7.0 for 24 hr).

12.2.3   BIODEGRADATION OF PHB BY SOIL MICROORGANISMS

The polymers exposed to the environment are degraded through their hydrolysis, mechanical, thermal, oxidative, and photochemical destruction, and biodegradation [7, 38, 55, 56]. One of the valuable properties of PHB is its biodegradability, which can be evaluated using various field and laboratory tests. The requirements for the biodegradability of PHB may vary in accordance with its applications. The most attractive property of PHB with respect to ecology is that it can be completely degraded by microorganisms finally to CO_2 and H_2O. This property of PHB allows manufacturing biodegradable polymer objects for various applications (Figure 3) [4].

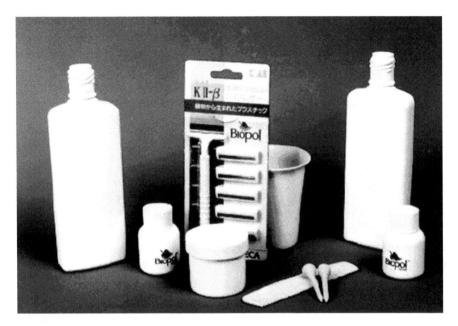

FIGURE 3 Molded PHB objects for various applications. In soil burial or composting experiments, such objects biodegrade in about three months.

The degradation of PHB and its composites in natural ecosystems, such as soil, compost, and bodies of water, was described in a number of publications [4, 38, 55, 56]. Maergaert et al. isolated from soil more than 300 microbial strains capable of degrading PHB *in vitro* [55]. The bacteria detected on the degraded PHB films were dominated by the genera *Pseudomonas, Bacillus, Azospirillum, Mycobacterium,* and *Streptomyces* and so on. The samples of PHB have been tested for fungicides and resistance to fungi by estimating the growth rate of test fungi from the genera *Aspergillus, Aureobasidium, Chaetomium, Paecilomyces, Penicillum,* and *Trichoderma* under optimal growth conditions. The PHB film did not exhibit neither fungicide properties, nor the resistance to fungal damage, and served as a good substrate for fungal growth [57].

It was studied biodegradability of PHB films under aerobic, microaerobic, and anaerobic condition in the presence and absence of nitrate by microbial populations of soil, sludge from anaerobic and nitrifying/denitrifying reactors, and sediment of a sludge deposit site, as well as to obtain active denitrifying enrichment culture degrading PHB (Figure 4) [58]. Changes in molecular mass, crystallinity, and mechanical properties of PHB have been studied. A correlation between the PHB degradation degree and the molecular weight of degraded PHB was demonstrated. The most degraded PHB exhibited the highest values of the crystallinity index. As it has been shown by Spyros et al., PHAs contain amorphous and crystalline regions, of which the former are much more susceptible to microbial attack [59]. If so, the microbial degradation of PHB must be associated with a decrease in its molecular weight and an increase in its crystallinity, which was really observed in the experiments. Moreover, microbial degradation of the amorphous regions of PHB films made them more rigid. However, further degradation of the amorphous regions made the structure of the polymer much looser [58].

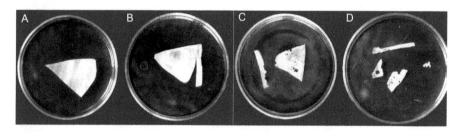

FIGURE 4   Undegraded PHB film (A) and PHB films with different degrees of degradation after 2 months incubation in soil suspension: anaerobic conditions without nitrate (B), microaerobic conditions without nitrate (C), and microaerobic conditions with nitrate (D).

The PHB biodegradation in the enriched culture obtained from soil on the medium used to cultivate denitrifying bacteria (Gil'tai medium) has been also studied. The dominant bacterial species, *Pseudomonas fluorescens* and *Pseudomonas stutzeri*, have been identified in this enrichment culture. Under denitrifying conditions, PHB films were completely degraded for seven days. Both the film weight and M_w of PHB decreased with time. In contrast to the data of Doi et al. [26] who found that M_w of PHB remained unchanged upon enzymatic biodegradation in an aquatic solution of PHB- depolymerase from *Alcaligenes faecalis,* in our experiments, the average viscosity molecular weight of the higher and lower molecular polymers decreased gradually from 1540 to 580 kDa and from 890 to 612 kDa, respectively. As it was shown at single PHB crystals [47] the "exo" type cleavage of the polymer chain, that is a successive removal of the terminal groups, is known to occur at a higher rate than the "endo" type cleavage, that is, a random breakage of the polymer chain at the enzyme-binding sites. Thus, the former type of polymer degradation is primarily responsible for changes in its average molecular weight. However the "endo" type attack plays the important role at the initiation of biodegradation, because at the beginning, a few polymer chains are oriented so that their ends are accessible to the effect of the enzyme [60]. The biodegradation of the lower-molecular polymer, which contains a higher number of terminal groups, is more active, probably, because the "exo" type degradation is more active in lower than in higher molecular polymer [58, 61].

12.2.4   BIODEGRADATION OF PHB IN VIVO IN ANIMAL TISSUES

The first scientific works on biodegradation of PHB *in vivo* in animal tissues were carried out 15–20 years ago by Miller N.D. et al. and Saito T. et al. [22, 24]. They are high-qualitative researches that disclosed many important characteristics of this process. As it was noted the both enzymatic and non-enzymatic processes of biodegradation of PHB *in vivo* can occur simultaneously under normal conditions. But it does not mean that polymer biodegradation *in vivo* is a simple combination of non-enzymatic hydrolysis and enzymatic degradation. Moreover, *in vivo* the biodegradation (decrease of molecular weight and mass loss) of PHB) is a controversial subject in the literature. As it was noted above for *in vitro* PHB hydrolysis, the main reason for the controversy is the use of samples made by various processing technologies and the incomparability of different implantation and animal models.

The most of researches on PHB biodegradation was carried out with use of prototypes of various medical devices on the base of PHB, solid films and plates [13, 16, 18, 31, 62], porous patches [14, 15], porous scaffolds [63], electrospun microfiber mats [64], nonwoven patches consisted of fibers [65-69], screws [31], cylinders as nerve guidance channels and conduits [16, 20-21], monofilament sutures [22-23], cardiovascular stents [70], and microspheres [24, 71]. *In vivo* biodegradation was studied on various laboratory animals, rats [14, 18, 20-24, 64], mice [16, 75], rabbits [13, 62, 70, 72], minipigs [15], cats [20], calves [65], sheep [66-68], and even at clinical trials on patients [69]. It is obviously that these animals differ in level of metabolism very much, for example, only weight of these animals differs from 10–20 g (mice) to 50 kg (calves). The implantation of devices from PHB was carried out through different ways, subcutaneously [13, 16, 18, 22, 23, 72], intraperitoneally on a bowel [14], subperiostally on the osseus skull [15-62], nerve wrap-around [19-21], intramuscularly [71, 72], into the pericardium [66-69], into the atrium [65], and intravenously [24]. The terms of implantation were also different: 2.5 hr, 24 hr, 13 days, 2 months [24]; 7, 14, 30 days [21], 2, 7, 14, 21, 28, 55, 90, 182 days [22]; 1, 3, 6 months [13, 16, 19]; 3, 6, 12 months [65]; 6, 12 months [66]; 6, 24 months [69]; 3, 6, 9, 12, 18, 24 months [68].

The most entire study of PHB *in vivo* biodegradation was fulfilled by Gogolewski S. et al. and Qu X.-H. et al. [13, 16]. It was shown that PHB lost about 1.6% (injection-molded film, 1.2 mm thick, M_w of PHB = 130 kDa) [16] and 6% (solvent-casting film, 40 μm thick, M_w = 534 kDa) [13] of initial weight after 6 months of implantation. But the observed small weight loss was partially due to the leaching out of low molecular weight fractions and impurities present initially in the implants. The M_w of PHB decreased from 130000 to 74000 (57% of initial M_w) [16] and from 534000 to 216000 (40% of initial Mw) [13] after 6 months of implantation. The polydispersity of PHB polymers narrowed during implantation. The PHB showed a constant increase in crystallinity (from 60.7 to 64.8%) up to 6 months [16] or an increase (from 65.0 to 67.9%) after 1 month and a fall again (to 64.5%) after 6 month of implantation [13] which suggests the degradation process had not affected the crystalline regions. This data is in accordance with data of PHB hydrolysis [25] and enzymatic PHB degradation by lipases *in vitro* [14] where, M_w decrease was observed. The initial biodegradation of amorphous regions of PHB *in vivo* is similar to PHB degradation by depolymerase [45].

Thus, the observed biodegradation of PHB showed coexistence of two different degradation mechanisms in hydrolysis in the polymer: enzymatically or non-enzymatically catalyzed degradation. Although non-enzymatical catalysis occurred randomly in homopolymer, indicated by M_w loss rate in PHB, at some point in a time, a critical molecular weight is reached whereupon enzyme-catalyzed hydrolysis accelerated degradation at the surface because easier enzyme/polymer interaction becomes possible. However considering the low weight loss of PHB, the critical molecular weight appropriate for enzymes predominantly does not reach, yet resulting low molecular weight and crystallinity in PHB could provide some sites for the hydrolysis of enzymes to accelerate the degradation of PHB [13, 16]. Additional data revealing the mechanism of PHB biodegradation in animal tissues was obtained by Kramp B. et al. in long-term implantation experiments. A very slow, clinically not recordable degradation of films and plates was observed during 20 month (much more than in experiments mentioned above).

A drop in the PHB weight loss evidently took place between the 20th and 25th month. Only initial signs of degradation were to be found on the surface of the implant until 20 months after implantation but no more test body could be detected after 25 months [62]. The complete biodegradation *in vivo* in the wide range from 3 to 30 months of PHB was shown by other researches [65, 67-69, 73], whereas, almost no weight loss and surface changes of PHB during 6 months of biodegradation *in vivo* was shown [16, 22].

The residual fragments of PHB implants were found after 30 months of the patches implantation [66, 68]. A reduction of PHB patch size in 27% was shown in patients after 24 months after surgical procedure on pericardial closure with the patch [69]. Significantly more rapid biodegradation *in vivo* was shown by other researches [13, 20, 23, 43, 65]. It was shown that 30% mass loss of PHB sutures occurred gradually during 180 days of *in vivo* biodegradation with minor changes in the microstructure on the surface and in volume of sutures [23]. It was shown that PHB nonwoven patches (made to close atrial septal defect in calves) was slowly degraded by polynucleated macrophages, and 12 months postoperatively no PHB device was identifiable but only small particles of polymer were still seen.

The absorption time of PHB patches was long enough to permit regeneration of a normal tissue [65]. The PHB sheets progressive biodegradation was demonstrated qualitatively at 2, 6 and 12 months after implantation as weakening of the implant surface, tearing/cracking of the implant, fragmentation, and a decrease in the volume of polymer material [21, 43, 72]. The complete biodegradation of PHB (M_w = 150–1000 kDa) thin films (10–50 μm) for 3–6 months was shown and degradation process was described. The process of PHB biodegradation consists of several phases. At initial phase PHB films was covered by fibrous capsule.

At second phase capsulated PHB films very slowly lost weight with simultaneous increase of crystallinity and decrease of M_w and mechanical properties of PHB. At third phase PHB films were rapidly disintegrated and then completely degraded. At 4th phase empty fibrous capsule resolved (Figure 5) [18, 31]. Interesting data were obtained for biodegradation *in vivo* of PHB microspheres (0.5–0.8 μm in diameter). It was demonstrated indirectly that PHB loss about 8% of weight of microspheres accumulated in liver after 2 month of intravenous injection. It was demonstrated also a presence of several types PHB degrading enzymes in the animal tissues extracts [24].

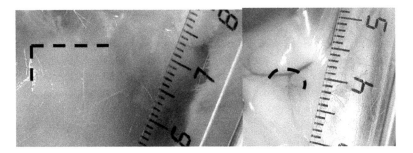

FIGURE 5 Biodegradation of PHB films *in vivo*. Connective-tissue capsule with PHB thin films (outlined with broken line) 2 weeks (98% residual weight of the film) (left photograph) and 3 months (0% residual weight of the film) (left photograph) after subcutaneous implantation.

Some researchers studied a biodegradation of PHB threads with a tendency of analysis of its mechanical properties *in vivo* [22, 23]. It was shown that at first load at break index decreased rapidly from 440 to 390 g (12%) at 15th day and then gradually increased to the initial value at 90th and remain almost unchanged up to 182nd day [22] or gradually decreased in 27% during 180 days [23], strain at break decreased rapidly from 60 to 50% (in 17% of initial value) at 10th day and then gradually increased to 70% (in 17% of initial value) at 182nd day [22] or did not change significantly during 180 days [23].

It was demonstrated that the primary reason of PHB biodegradation *in vivo* was a lysosomal and phagocytic activity of polynucleated macrophages and giant cells of foreign body reaction. The activity of tissue macrophages and nonspecific enzymes of body liquids made a main contribution to significantly more rapid rate of PHB biodegradation *in vivo* in comparison with rate of PHB hydrolysis *in vitro*. The PHB material was encapsulated by degrading macrophages. The presence of PHB stimulated uniform macrophage infiltration, which is important for not only the degradation process but also the restoration of functional tissue. The long absorption time produced a foreign-body reaction, which was restricted to macrophages forming a peripolymer layer [23, 65, 68, 72].

Very important data that clarifies the tissue response that contributes to biodegradation of PHB was obtained by Lobler M. It was demonstrated a significant increase of expression of two specific lipases after 7 and 14 days of PHB contact with animal tissues. Moreover, liver specific genes were induced with similar results. It is striking that pancreatic enzymes are induced in the gastric wall after contact with biomaterials [46]. Saito T. et al. suggested the presence of at least two types of degradative enzymes in rat tissues: liver serine esterases with the maximum of activity in alkaline media (pH = 9.5) and kidney esterases with the maximum of activity in neutral media [24]. The mechanism of PHB biodegradation by macrophages was demonstrated at cultured macrophages incubated with particles of low-molecular weight PHB [74]. It was shown that macrophages and, to a lesser level, fibroblasts have the ability to take up (phagocytize) PHB particles (1–10 μm). At high concentrations of PHB particles (>10 μg/mL) the phagocytosis is accompanied by toxic effects and alteration of the functional status of the macrophages but not the fibro-blasts. This process is accompanied by cell damage and cell death. The elevated production of nitric oxide (NO) and tumor necrosis factor alfa (TNF-α) by activated macrophages was observed. It was suggested that the cell damage and cell death may be due to phagocytosis of large amounts of PHB particles; after phagocytosis, polymer particles may fill up the cells, and cause cell damage and cell death. It was demonstrated also that phagocytized PHB particles disappeared in time due to an active PHB biodegradation process (Figure 6) [74].

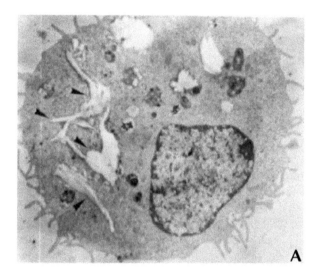

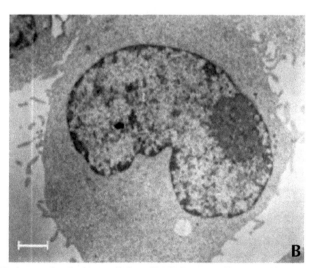

FIGURE 6   Phagocytosis of microparticles of PHB in macrophages. TEM analysis of cultured macrophages in the presence (A) or absence (B) of 2 µg PHB microparticles/mL for 24 hr. Bar in B represents 1 µm, for A and B.

12.3   APPLICATION OF PHB

12.3.1   MEDICAL DEVICES ON THE BASE OF PHB AND PHB IN VIVO BIOCOMPATIBILITY

The perspective area of PHB application is development of implanted medical devices for dental, cranio-maxillofacial, orthopaedic, cardiovascular, hernioplastic, and skin surgery [3, 6]. A number of potential medical devices on the base of PHB: bioresorbable surgical sutures [6,22-23, 31, 75, 76], biodegradable screws and plates for cartilage and bone fixation [6, 24, 31, 62], biodegradable membranes for periodontal treatment [6, 31, 77, 78], surgical meshes with PHB coating for hernioplastic surgery [6, 31], wound coverings [79], patches for repair of a bowel, pericardial and osseous defects [14, 15, 65-69], nerve guidance channels and conduits [20, 21], cardiovascular stents [80], and so on was developed (Figure 7).

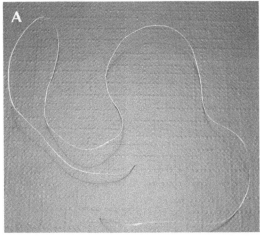

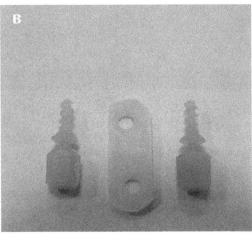

FIGURE 7   *(Continued)*

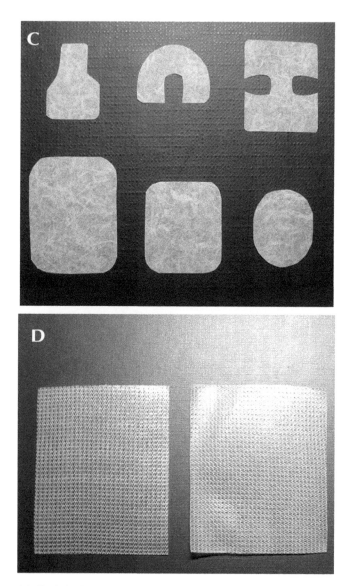

FIGURE 7 Medical devices on the base of PHB. (A) bioresorbable surgical suture, (B) biodegradable screws and plate for cartilage and bone fixation, (C) biodegradable membranes for periodontal treatment, and (D) surgical meshes with PHB coating for hernioplastic surgery, pure (left) and loaded with antiplatelet drug, dipyridamole (right).

The tissue reaction *in vivo* to implanted PHB films and medical devices was studied. In most cases a good biocompatibility of PHB was demonstrated. In general, no acute inflammation, abscess formation, or tissue necrosis was observed in tissue surrounding of the implanted PHB materials. In addition, no tissue reactivity or cellular mobilization occurred in areas remote from the implantation site [13, 16, 31, 71]. On the one hand, it was shown that PHB elicited similar mild tissue response as PLA did [16], but on the other hand the use of implants consisting of PLA, polyglicolic acid, and their copolymers is not without a number of sequelae related with the chronic inflammatory reactions in tissue [81-85].

The subcutaneous implantation of PHB films for 1 month has shown that the samples were surrounded by a well-developed, homogeneous fibrous capsule of 80–100 μm in thickness. The vascularized capsule consists primarily of connective tissue cells (mainly, round, immature fibroblasts) aligned parallel to the implant surface. A mild inflammatory reaction was manifested by the presence of mononuclear macrophages, foreign body cells, and lymphocytes. Three months after implantation, the fibrous capsule has thickened to 180–200 μm due to the increase in the amount of connective tissue cells and a few collagen fiber deposits. A substantial decrease in inflammatory cells was observed after 3 months, tissues at the interface of the polymer were densely organized to form bundles. After 6 months of implantation, the number of inflammatory cells had decreased and the fibrous capsule, now thinned to about 80–100 μm, consisted mainly of collagen fibers, and a significantly reduced amount of connective tissue cells. A little inflammatory cells effusion was observed in the tissue adherent to the implants after 3 and 6 months of implantation [13, 16]. The biocompatibility of PHB has been demonstrated *in vivo* under subcutaneous implantation of PHB films. Tissue reaction to films from PHB of different molecular weight (300kDa, 450kDa, and 1000 kDa) implanted subcutaneously was relatively mild and did not change from tissue reaction to control glass plate [18, 31].

At implantation of PHB with contact to bone the overall tissue response was favorable with a high rate of early healing and new bone formation with some indication of an osteogenic characteristic for PHB compared with other thermoplastics, such as polyethylene. Initially there was a mixture of soft tissue, containing active fibroblasts, and rather loosely woven osteonal bone seen within 100 μm of the interface. There was no evidence of a giant cell response within the soft tissue in the early stages of implantation. With time this tissue became more orientated in the direction parallel to the implant interface. The dependence of the bone growth on the polymer interface is demonstrated by the new bone growing away from the interface rather than towards it after implantation of 3 months. By 6 months post-implantation the implant is closely encased in new bone of normal appearance with no interposed fibrous tissue. Thus, PHB-based materials produce superior bone healing [43].

The regeneration of a neointima and a neomedia, comparable to native arterial tissue, was observed at 3–24 months after implantation of PHB nonwoven patches as transannular patches into the right ventricular outflow tract and pulmonary artery. In the control group, a neointimal layer was present but no neomedia comparable to native arterial tissue. Three layers were identified in the regenerated tissue: neointima with an endothelium-like lining, neomedia with smooth muscle cells, collagenous and elastic tissue, and a layer with polynucleated macrophages surrounding istets of PHB, capillaries and collagen tissue. The lymphocytes were rare. It was concluded that PHB nonwoven patches can be used as a scaffold for tissue regeneration in low-pressure systems. The regenerated vessel had structural and biochemical qualities in common with the native pulmonary artery [68]. The biodegradable PHB patches implanted in atrial septal defects promoted formation of regenerated tissue that macroscopically and microscopically resembled native atrial septal wall. The regenerated tissue was found to be composed of three layers: monolayer with endothelium-like cells, a layer with fibroblasts and some smooth-muscle cells, collagenous tissue and capillaries, and a third layer with phagocytizing cells isolating and degrading PHB. The neointima contained a complete endothelium-like layer resembling the native endothelial cells. The patch material was encapsulated by degrading macrophages. There was a strict border between the collagenous and the phagocytizing layer. The presence of PHB seems to stimulate uniform macrophage infiltration, which was found to be important for the degradation process and the restoration of functional tissue. Lymphocytic in-filtration as foreign-body reaction, which is common after replacement of vessel wall with commercial woven Dacron patch, was wholly absent when PHB. It was sug-gested that the absorption time of PHB patches was long enough to permit regenera-tion of a tissue with sufficient strength to prevent development of shunts in the atrial septal position [65]. The prevention of postoperative pericardial adhesions by closure of the pericardium with absorbable PHB patch was demonstrated. The regeneration of mesothelial layer after implantation of PHB pericardial patch was observed. The complete regeneration of mesothelium, with morphology and biochemical activity similar to findings in native mesothelium, may explain the reduction of postoperative pericardial adhesions after operations with insertion of absorbable PHB patches [67]. The regeneration of normal filament structure of restored tissues was observed by im-munohistochemical methods after PHB devices implantation [66]. The immunohisto-chemical demonstration of cytokeratine, an intermediate filament, which is constituent of epithelial, and mesodermal cells, agreed with observations on intact mesothelium. Heparin sulfate proteoglycan, a marker of basement membrane, was also identified [66]. However, in spite of good tissue reaction to implantation of cardiovascular PHB patches, PHB endovascular stents in the rabbit iliac arteria caused intensive inflamma-tory vascular reactions [80].

The PHB patches for the gastrointestinal tract were tested using animal model. The patches made from PHB sutured and PHB membranes were implanted to close experimental defects of stomach and bowel wall. The complete regeneration of tissues of stomach and bowel wall was observed at 6 months after patch implantation without strong inflammatory response and fibrosis [14, 86].

Recently an application of biodegradable nerve guidance channels (conduits) for nerve repair procedures and nerve regeneration after spinal cord injury was demon-strated. Polymer tubular structures from PHB can be modulated for this purpose. Suc-cessful nerve regeneration through a guidance channel was observed as early as after 1 month. Virtually all implanted conduits contained regenerated tissue cables centrally located within the channel lumen and composed of numerous myelinated axons and Schwann cells. The inflammatory reaction had not interfered with the nerve regenera-tion process. Progressive angiogenesis was present at the nerve ends and through the walls of the conduit. The results demonstrate good guality nerve regeneration in PHB guidance channels [21, 87].

The biocompatibility of PHB was evaluated by implanting microspheres from PHB (M_w = 450 kDa) into the femoral muscle of rats. The spheres were surrounded by one or two layers of spindle cells, and infiltration of inflammatory cells and mononuclear cells into these layers was recognized at 1 week after implantation. After 4 weeks, the number of inflammatory cells had decreased and the layers of spindle cells had thickened. No inflammatory cells were seen at 8 weeks, and the spheres were encapsulated by spindle cells. The toxicity of PHB microspheres was evaluated by weight change and survival times in L1210 tumor-bearing mice. No differences were observed in the weight change or survival time compared with those of control. These results suggest that inflammation accompanying microsphere implantation is temporary as well as toxicity to normal tissues is minimal [71].

The levels of tissue factors, inflammatory cytokines, and metabolites of arachidonic acid were evaluated. Growth factors derived from endothelium and from macrophages were found. These factors most probably stimulate both growth and regeneration occurring when different biodegradable materials were used as grafts [46, 65, 67, 86]. The positive reaction for thrombomodulin, a multifunctional protein with anticoagulant properties, was found in both mesothelial and endothelial cells after pericardial PHB patch implantation. The prostacycline production level, which was found to have cytoprotective effect on the pericardium and prevent adhesion formation, in the regenerated tissue was similar to that in native pericardium [65, 67]. The PHB patch seems to be highly biocompatible, since no signs of inflammation were observed macroscopically and also the level of inflammation associated cytokine messanger ribonucleic acid (mRNA) did not change dramatically, although a transient increase of interleukin-1β and interleukin-6 mRNA through days 1–7 after PHB patch implantation was detected. In contrast, tumor necrosis factor-α mRNA was hardly detectable throughout the implantation period, which agrees well with an observed moderate fibrotic response [46, 86].

12.3.2 *PHB AS TISSUE ENGINEERING MATERIAL AND PHB IN VITRO BIOCOMPATIBILITY*

The biopolymer PHB is promising material in tissue engineering due to high biocompatibility *in vitro*. Cell cultures of various origins including murine and human fibroblasts [15, 40, 78, 88-92], human mesenchymal stem cells [93], rabbit bone marrow cells (osteoblasts) [36, 89, 94], human osteogenic sarcoma cells [95], human epithelial cells [90, 95], human endothelial cells [96, 97], rabbit articular cartilage chondrocytes [98-99] and rabbit smooth muscle cells [100], human neurons (schwann cells) [101] in direct contact with PHB when cultured on polymer films and scaffolds exhibited satisfactory levels of cell adhesion, viability, and proliferation. Moreover, it was shown that fibroblasts, endothelium cells, and isolated hepatocytes cultured on PHB films exhibited high levels of cell adhesion and growth (Figure 8) [102]. A series of 2D and 3D PHB scaffolds was developed by various methods: polymer surface modification [97], blending [36, 78, 88, 90, 93, 99, 103], electrospinning [104-106], salt leaching [36, 94, 107, 108], microspheres fusion [109], forming on porous mold [110], laser cutting [111].

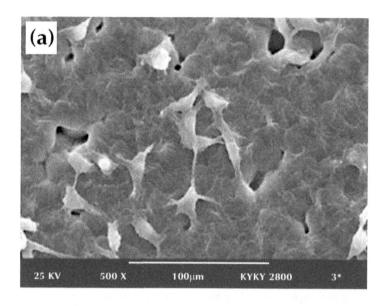

FIGURE 8 Scanning electron microscopy image of 2 days growth of fibroblast cells on films made of (a) PHB; (e) PLA; (500x). Cell density of fibroblasts grown on PHB film is significantly higher versus cell density of fibroblasts grown on PLA film.

It was shown also that cultured cells produced collagen II and glycosaminogly-can, the specific structural biopolymers formed the extracellular matrix [95, 98, 99]. A good viability and proliferation level of macrophages and fibroblasts cell lines was obtained under culturing in presence of particles from short-chain low-molecular PHB [74]. However it was shown that cell growth on the PHB films was relatively poor: the viable cell number ranged from 1×10^3 to 2×10^5 [40, 89, 99]. An impaired interaction between PHB matrix and cytoskeleton of cultured cells was also demonstrated [95]. It was reported that a number of polymer properties including chemical composition, surface morphology, surface chemistry, surface energy, and hydrophobicity play important roles in regulating cell viability and growth [112]. The investigation showed that this biomaterial can be used to make scaffolds for *in vitro* proliferating cells [40, 89, 98].

The most widespread methods to manufacture the PHB scaffolds for tissue engineering by means of improvement of cell adhesion and growth on polymer surface are change of PHB surface properties and microstructure by salt-leaching methods and enzymatic/chemical/physical treatment of polymer surface [40, 89, 98, 113]. Adhesion to polymer substrates is one of the key issues in tissue engineering, because adhesive interactions control cell physiology. One of the most effective techniques to improve adhesion and growth of cells on PHB films is treatment of polymer surface with enzymes, alkali or low pressure plasma [40, 113]. Lipase treatment increases the viable cell number on the PHB film from 100 to 200 times compare to the untreated PHB film. The NaOH treatment on PHB film also indicated an increase of 25 times on the viable cell number compared with the untreated PHB film [40]. It was shown that treatment of PHB film surface with low pressure ammonia plasma improved growth of human fibroblasts and epithelial cells of respiratory mucosa due to increased hydrophylicity (but with no change of microstructure) of polymer surface [90]. It was suggested that the improved hydrophilicity of the films after PHB treatment with lipases, alkali and plasma allowed cells in its suspension to easily attach on the polymer films compared to that on the untreated ones. The influence of hydrophilicity of biomaterial surface on cell adhesion was demonstrated earlier [114].

But a microstructure of PHB film surface can be also responsible for cell adhesion and cell growth [115-117]. Therefore, noticed above modification of polymer film surface after enzymatic and chemical treatment (in particular, reduced pore size and a surface smoothing) is expected to play an important role for enhanced cell growth on the polymer films [40]. Different cells prefer different surface. For example, osteoblasts preferred rougher surfaces with appropriate size of pores [115, 116] while fibroblast prefer smoother surface, yet epithelial cells only attached to the smoothest surface [117]. This appropriate roughness affects cell attachment as it provides the right space for osteoblast growth, or supplies solid anchors for filapodia. A scaffold with appropriate size of pores provided better surface properties for anchoring type II collagen filaments and for their penetration into internal layers of the scaffolds implanted with chondrocytes. This could be illuminated by the interaction of extracellular matrix proteins with the material surface. Moreover, the semicrystalline surface PHB ultrastructure can be connected with protein adsorption and cell adhesion [91, 92, 118]. The appropriate surface properties may also promote cell attachment and proliferation by providing more spaces for better gas/nutrients exchange or more serum protein adsorption [36, 94, 98]. Additionally Sevastianov et al. found that PHB films in contact with blood did not activate the hemostasis system at the level of cell response, but they did activate the coagulation system and the complement reaction [119].

The high biocompatibility of PHB may be due to several reasons. First of all, PHB is a natural biopolymer involved in important physiological functions both prokaryotes and eukaryotes. The PHB from bacterial origin has property of stereospecificity that is inherent to biomolecules of all living things and consists only from residues of D(-)-3-hydrohybutyric acid [120-122]. Low molecular weight PHB (up to 150 resides of 3-hydrohybutyric acid), complexed to other macromolecules (cPHB), was found to be a ubiquitous constituent of both prokaryotic and eukaryotic organisms of nearly all phyla [123-127]. The complexed cPHB was found in a wide variety of tissues and organs of mammals (including human): blood, kidney, vessels, nerves, vessels, eye, brain, as well as in organelles, membrane proteins, lipoproteins, and plaques. The cPHB concentration ranged from 3–4 µg/g wet tissue weight in nerves and brain to 12 µg/g in blood plasma [128, 129]. In humans, total plasma cPHB ranged from 0.60 to 18.2 mg/l, with a mean of 3.5 mg/l. [129]. It was shown that cPHB is a functional part of ion channels of erythrocyte plasma membrane and hepatocyte mitochondria membrane [130, 131]. The singular ability of cPHB to dissolve salts and facilitate their transfer across hydrophobic barriers defines a potential physiological niche for cPHB in cell metabolism [125]. However, a mechanism of PHB synthesis in eukaryotic organisms is not well clarified that requires additional studies. Nevertheless, it could be suggested that cPHB is one of products of symbiotic interaction between animals and gut microorganisms. It was shown, for example, that E.coli is able to synthesize low molecular weight PHB and cPHB plays various physiological roles in bacteria cell [127, 132].

The intermediate product of PHB biodegradation, D(-)-3-hydroxybutyric acid is also a normal constituent of blood at concentrations between 0.3 and 1.3 mM and contains in all animal tissues [133, 134]. As it was noted that PHB has a rather low degradation rate in the body in comparison to, for example,, poly(lactic-co-glycolic) acids (PLGA), that prevent from increase of 3-hydroxybutyric acid concentration in surrounding tissues [13, 16], whereas, PLA release, following local pH decrease in implantation area and acidic chronic irritation of surrounding tissues is a serious problem in application of medical devices on the base of PLGA [135, 136]. Moreover, chronic inflammatory response to polylactic and polyglycolic acids that was observed in a number of cases may be induced by immune response to water-soluble oligomers that released during degradation of synthetic polymers [136-138].

12.3.3 NOVEL DRUG DOSAGE FORMS ON THE BASE OF PHB

An improvement of medical materials on the base of biopolymers by encapsulating different drugs opens up the wide prospects in applications of new devices with pharmacological activity. The design of injection systems for sustained drug delivery in the forms of microspheres or microcapsules prepared on the base of biodegradable polymers is extremely challenging in the modern pharmacology. The fixation of pharmacologically active component with the biopolymer and following slow drug release from the microparticles provides an optimal level of drug concentration in local target organ during long-term period (up to several months). At curative dose the prolonged delivery of drugs from the systems into organism permits to eliminate the shortcomings in peroral, injectable, aerosol, and the other traditional methods of drug administration. Among those shortcomings hypertoxicity, instability, pulsative character of rate delivery, and ineffective expenditure of drugs should be pointed out. Alternatively, applications of therapeutical polymer systems provide orderly and purposefully the deliverance for an optimal dose of agent that is very important at therapy of acute or chronic diseases [1, 3, 6, 139]. An ideal biodegradable microsphere formulation would consist of a free-flowing powder of uniform-sized microspheres less than 125 μm in diameter and with a high drug loading. In addition, the drug must be released in its active form with an optimized profile. The manufacturing method should produce the microspheres which are reproducible, scalable, and benign to some often delicate drugs, with high encapsulation efficiency [140, 141].

The PHB as a biodegradable and biocompatible is a promising material for producing of polymer systems for controlled drug release. A number of drugs with various pharmacological activities were used for development of polymer controlled release systems on the base of PHB and its copolymers: model drugs (2,7-dichlorofluorescein [142], Dextran-FITC [143], methyl red (MR) [122, 144, 145], and 7-hydroxethylthe-ophylline [146, 147]), antibiotics and antibacterial drugs (rifampicin [148,149], tetracycline [150], cefoperazone and gentamicin [151], sulperazone and duocid [152-155], sulbactam and cefoperazone [156], fusidic acid [157], and nitrofural [158]), anticancer drugs (5-fluorouracil [159], 2',3'-diacyl-5-fluoro-2'-deoxyuridine [71], paclitaxel [160, 161], rubomycin [162], chlorambucil, and etoposide [163]), anti-inflammatory drug (indomethacin [164], flurbiprofen [165], and ibuprofen [166]), analgesics (tramadol [167], vasodilator and antithrombotic drugs (dipyridamole [168, 169], NO donor [170, 171], nimodipine [174], felodipine [175]), and proteins (hepatocyte growth factor [176], mycobacterial protein for vaccine development [177], bone morphogenetic protein 7 [178]). The various methods for manufacture of drug-loaded PHB matrices and microspheres were used: films solvent casting [144-147], emulsification and solvent evaporation [148-164], spray drying [172], layer-by-layer self-assembly [165], and supercritical antisolvent precipitation [173]. The biocompatibility and pharmacological activity of some of these systems was studied [71, 148, 154-156, 160, 161, 167, 171]. But only a few drugs were used for production of drug controlled release systems on the base of PHB homopolymer: 7-hydroxethyltheophylline, MR, 2',3'-diacyl-5-fluoro-2'-deoxyuridine, rifampicin, tramadol, indomethacin, dipyridamole, and paclitaxel [71, 154-156, 160-171]. The latest trend in PHB drug delivery systems study is PHB nanoparticles development loaded with different drugs [179-180].

The first drug sustained delivery system on the base of PHB was developed by Korsatko W. et al., who observed a rapid release of encapsulated drug, 7-hydroxethyltheophylline, from tablets of PHB (M_w = 2000 kDa), as well as weight losses of PHB tablets containing the drug after subcutaneous implantation. It was suggested that PHB with molecular weight greater than 100 kDa was undesirable for long-term medication dosage [146].

Pouton C.W. and Akhtar S. describing the release of low molecular drugs from PHB matrices reported that the latter have the tendencies to enhance water penetration and pore forming [181]. The entrapment and release of the model drug, MR, from melt-crystallized PHB was found to be a function of polymer crystallization kinetics and morphology whereas, overall degree of crystallinity was shown to cause no effect on drug release kinetics. The MR released from PHB films for 7 days with initial phase of rapid release ("burst effect") and second phase with relatively uniform release. The release profiles of PHB films crystallized at 110°C exhibited a greater burst effect when compared to those crystallized at 60°C. This was explained by better trapping of drug within polymeric spherulites with the more rapid rates of PHB crystallization at 110°C [144, 145].

Kawaguchi T. et al. showed that chemical properties of drug and polymer molecular weight had a great impact on drug delivery kinetics from PHB matrix. Microspheres (100–300 μm in diameter) from PHB of different molecular weight (65000, 135000, and 450000) were loaded with prodrugs of 5-fluoro-2'deoxyuridine (FdUR) synthesized by esterification with aliphatic acids (propionate, butyrate, and pentanoate). The prodrugs have different physicochemical properties, in particular, solubility in water (from 70 mg/ml for FdUR to 0.1 mg/ml for butyryl- FdUR). The release rates from the spheres depended on both the lipophilicity of the prodrug and the molecular weight of the polymer. Regardless of the polymer, the relative release rates were propionyle- FdUR >butyryl- FdUR >pentanoyl- FdUR. The release of butyryl- FdUR and pentanoyl- FdUR from the spheres consisting of low-molecular-weight polymer (M_w = 65000) was faster than that from the spheres of higher molecular weight (M_w = 135000 or 450000). The effect of drug content on the release rate was also studied. The higher the drug content in the PHB microspheres, the faster was the drug release. The release of FdUR continued for more than 5 days [71].

Kassab A.C. developed a well-managed technique for manufacture of PHB microspheres loaded with drugs. Microspheres were obtained within a size of 5–100 μm using a solvent evaporation method by changing the initial polymer/solvent ratio, emulsifier concentration, stirring rate, and initial drug concentration. The drug overloading of up to 0,41 g rifampicin/g PHB were achieved. Drug release was rapid; the maximal duration of rifampicin delivery was 5 days. Both the size and drug content of PHB microspheres were found to be effective in controlling the drug release from polymer microspheres [149].

The sustained release of analgesic drug, tramadol, from PHB microspheres was demonstrated by Salman M.A. et al. It was shown that 58% of the tramadol (the initial drug content in PHB matrix = 18%) was released from the microspheres (7.5 μm in diameter) in the first 24 hr. Drug release decreased with time. From 2 to 7 days the drug release was with zero-order rate. The entire amount of tramadol was released after 7 days [167].

The kinetics of different drug release from PHB micro- and nanoparticles loaded with dipyridamole, indomethacin and paclitaxel was studied [160, 161, 164, 168, 169]. It was found that the release occurs *via* two mechanisms, diffusion and degradation, and operating simultaneously. The vasodilator and antithrombotic drug (dipyridamole) and anti-inflammatory drug (indomethacin) diffusion processes determine the rate of the release at the early stages of the contact of the system with the environment (the first 6-8 days). The coefficient of the release diffusion of a drug depends on its nature, the thickness of the PHB films containing the drug, the weight ratio of dipyridamole and indomethacin in polymer, and the molecular weight of PHB. Thus, it is possible to regulate the rate of drug release by changing of molecular weight of PHB, for example [164]. The biodegradable microspheres on the base of PHB designed for controlled release of dipyridamole and paclitaxel were kinetically studied. The profiles of release from the microspheres with different diameters present the progression of nonlinear and linear stages. The diffusion kinetic equation describing both linear (PHB hydrolysis) and nonlinear (diffusion) stages of the dipyridamole and paclitaxel release profiles from the spherical subjects has been written down as the sum of two terms: desorption from the homogeneous sphere in accordance with diffusion mechanism and the zero-order release. In contrast to the diffusivity dependence on microsphere size, the constant characteristics of linearity are scarcely affected by the diameter of PHB microparticles. The view of the kinetic profiles as well as the low rate of dipyridamole and paclitaxel release are in satisfactory agreement with kinetics of weight loss measured *in vitro* for the PHB films and observed qualitatively for PHB microspheres. Taking into account kinetic results, it was supposed that the degradation of PHB microspheres is responsible for the linear stage of dipyridamole and paclitaxel release profiles (Figure 9) [24, 160, 161, 168, 169].

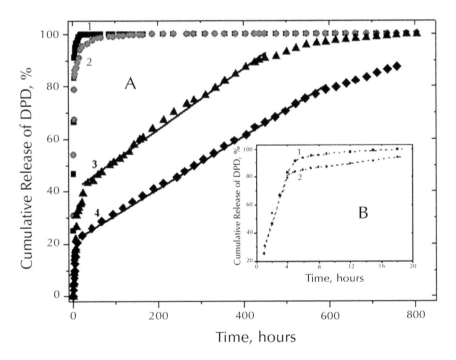

FIGURE 9 Kinetics profiles of DPD release from PHB microspheres *in vitro* (phosphate buffer, 37°C). A: General view of kinetic curves for the microspheres with different diameters: 4(1), 19(2), 63(3), and 92(4) μm. The lines show the second stage of release following the zero-order equation. B: Details of the curves for the microspheres with the smaller diameters: 4(1), 19(2).

The biocompatibility and pharmacological activity of advanced drug delivery systems on the base of PHB was studied [71, 148, 167-169]. It was shown that implanted PHB microspheres loaded with paclitaxel caused the mild tissue reaction. The inflammation accompanying implantation of PHB matrices is temporary and additionally toxicity relative to normal tissues is minimal [169]. No signs of toxicity were observed after administration of PHB microspheres loaded with analgesic and tramadol [167]. A single intraperitoneal injection of PHB microspheres containing anticancer prodrugs, butyryl- FdUR, and pentanoyl- FdUR, resulted in high antitumor effects against P388 leukemia in mice over a period of five days [71]. Embolization with PHB microspheres *in vivo* at dogs as test animals has been studied by Kasab et al. Renal angiograms obtained before and after embolization and also the histopathological observations showed the feasibility of using these microspheres as an alternative chemoembolization agent [148]. Epidural analgesic effects of tramadol released from PHB microspheres were observed for 21 hr, whereas, an equal dose of free tramadol was effective for less than 5 hr. It was suggested that controlled release of tramadol from PHB microspheres *in vivo* is possible, and pain relief during epidural analgesia is prolonged by this drug formulation compared with free tramadol [167].

The observed data indicate the wide prospects in applications of drug-loaded medical devices and microspheres on the base of PHB as implantable and injectable therapeutic systems in medicine for treatment of various diseases such as cancer, cardiovascular diseases, tuberculosis, osteomyelitis, arthritis, and so on[6].

12.4 CONCLUSION

The natural PHB is unique biodegradable thermoplastics of considerable commercial importance. With this review, we have attempted to systematically evaluate the impact of physicochemical factors on the hydrolysis and the biodegradation of natural PHB both *in vitro* and *in vivo*. Clearly, the degradation behavior observed is very dependent upon both physicochemical conditions. The geometry and structural and microbial properties. If these conditions of (bio)degradation are known, the systems on the base PHB can be designed in such biomedicine areas as medical devices, tissue scaffolds in bioengineering and development of novel biodegradable therapeutic systems for drug delivery.

KEYWORDS

- **Arachidonic acid**
- **Depolymerization**
- **Poly(3-hydroxybutyrate)**
- **Polylactic acid**
- **Polylactides**

ACKNOWLEDGMENT

The work was supported by the Russian Foundation for Basic Research (grant no. 13-03-00405-a) and the Russian Academy of Sciences under the program "Construction of New Generation Macromolecular Structures" (03/OC-13).

REFERENCES

1. Rice, J. J., Martino, M. M., De Laporte, L., Tortelli, F., Briquez, P. S., and Hubbell, J. A. Engineering the Regenerative Microenvironment with Biomaterials. *Adv. Healthcare Mater.*, **2**, 57–71 (2013), DOI: 10.1002/adhm.201200197.
2. Khademhosseini, A. and Peppas, N. A. Micro- and Nanoengineering of Biomaterials for Healthcare Applications. *Adv. Healthcare Mater.*, **2**, 10–12 (2013), DOI: 10.1002/adhm.201200444.
3. Chen, G. Q. and Wu, Q. The application of polyhydroxyalkanoates as tissue engineering materials. *Biomaterials*, **26**(33), 6565–6578 (2005).
4. Lenz, R. W. and Marchessault, R. H. Bacterial Polyesters: Biosynthesis, Biodegradable Plastics, and Biotechnology. *Biomacromolecules*, **6**(1), 1–8 (2005).
5. Anderson, A. J. and Dawes, E. A. Occurrence, metabolism, metabolic role, and industrial uses of bacterial polyhydroxyalkanoates. *Microbiological Reviews*, **54**(4), 450–472 (1990).
6. Bonartsev, A. P., Bonartseva, G. A., Shaitan, K. V., and Kirpichnikov, M. P. Poly(3-hydroxybutyrate) and poly(3-hydroxybutyrate)-based biopolymer systems. *Biochemistry (Moscow) Supplement Series B: Biomedical Chemistry*, **5**(1), 10–21 (2011).
7. Jendrossek, D. and Handrick, R. Microbial degradation of polyhydroxyalkanoates. *Annu Rev Microbiol.*, **56**, 403–432 (2002).
8. Fabra, M. J., Lopez-Rubio, A., and Lagaron, J. M. High barrier polyhydroxyalcanoate food packaging film by means of nanostructured electrospun interlayers of zein. *Food Hydrocolloids*, **32**, 106–114 (2013). DOI: http://dx.doi.org/10.1016/j.foodhyd.2012.12.007.
9. Kim, D. Y. and Rhee, Y. H. Biodegradation of microbial and synthetic polyesters by fungi. *Appl. Microbiol. Biotechnol.*, **61**, 300–308 (2003).
10. Marois, Y., Zhang, Z., Vert, M., Deng, X., Lenz, R., and Guidoin, R. Hydrolytic and enzymatic incubation of polyhydroxyoctanoate (PHO): a short-term in vitro study of a degradable bacterial polyester. *J. Biomater. Sci. Polym. Ed.*, **10**, 483–499 (1999).
12. Abe, H. and Doi, Y. Side-chain effect of second monomer units on crystalline morphology, thermal properties, and enzymatic degradability for random copolyesters of (R)-3-hydroxybutyric acid with (R)-3-hydroxyalkanoic acids. *Biomacromolecules*, **3**(1), 133–138 (2002).
13. Renstad, R., Karlsson, S., and Albertsson, A. C. The influence of processing induced differences in molecular structure on the biological and non-biological degradation of poly(3-hydroxybutyrate-co-3-hydroxyvalerate), P(3-HB-co-3-HV). *Polym. Degrad. Stab.*, **63**, 201–211 (1999).
14. Qu, X. H., Wu, Q., Zhang, K. Y., and Chen, G. Q. In vivo studies of poly(3-hydroxybutyrate-co-3-hydroxyhexanoate) based polymers: biodegradation and tissue reactions. *Biomaterials*, **27**(19), 3540–3548 (2006).
15. Freier, T., Kunze, C., Nischan, C., Kramer, S., Sternberg, K., Sass, M., Hopt, U. T., and Schmitz, K. P. In vitro and in vivo degradation studies for development of a biodegradable patch based on poly(3-hydroxybutyrate). *Biomaterials*, **23**(13), 2649–2657 (2002).

16. Kunze, C., Edgar Bernd, H., Androsch, R., Nischan, C., Freier T., Kramer, S., Kramp, B., and Schmitz, K. P. In vitro and in vivo studies on blends of isotactic and atactic poly (3-hydroxybutyrate) for development of a dura substitute material. *Biomaterials*, **27**(2), 192–201 (January, 2006).

17. Gogolewski, S., Jovanovic, M., Perren, S. M., Dillon, J. G., and Hughes, M. K. Tissue response and in vivo degradation of selected polyhydroxyacids: polylactides (PLA), poly(3-hydroxybutyrate) (PHB), and poly(3-hydroxybutyrate-co-3-hydroxyvalerate) (PHB/VA). *J. Biomed. Mater. Res.*, **27**(9), 1135–1148 (1993).

18. Boskhomdzhiev, A. P., Bonartsev, A. P., Ivanov, E. A., Makhina, T. K., Myshkina, V. L., Bagrov, D. V., Filatova, E. V., Bonartseva, G. A., and Iordanskii, A. L. Hydrolytic degradation of biopolymer systems based on poly(3-hydroxybutyrate. Kinetic and structural aspects. *International Polymer Science and Technology*, **37**(11), 25–30 (2010).

19. Boskhomdzhiev, A. P., Bonartsev, A. P., Makhina, T. K., Myshkina, V. L., Ivanov, E. A., Bagrov, D. V., Filatova, E. V., Iordanskiĭ, A. L., and Bonartseva, G. A. Biodegradation kinetics of poly(3-hydroxybutyrate)-based biopolymer systems. *Biochemistry (Moscow) Supplement Series B: Biomedical Chemistry*, **4**(2), 177–183 (2010).

20. Borkenhagen, M., Stoll, R. C., Neuenschwander, P., Suter, U. W., and Aebischer, P. In vivo performance of a new biodegradable polyester urethane system used as a nerve guidance channel. *Biomaterials*, **19**(23), 2155–2165 (1998).

21. Hazari, A., Johansson-Ruden, G., Junemo-Bostrom, K., Ljungberg, C., Terenghi, G., Green, C., and Wiberg, M. A new resorbable wrap-around implant as an alternative nerve repair technique. Journal of Hand Surgery, British and European Volume, **24B**(3), 291–295 (1999).

22. Hazari, A., Wiberg, M., Johansson-Rudén, G., Green, C., and Terenghi, G. A. Resorbable nerve conduit as an alternative to nerve autograft. *British Journal of Plastic Surgery*, **52**, 653–657 (1999).

23. Miller, N. D and Williams, D. F. On the biodegradation of poly-beta-hydroxybutyrate (PHB) homopolymer and poly-beta-hydroxybutyrate-hydroxyvalerate copolymers. *Biomaterials*, **8**(2), 129–137 (March, 1987).

24. Shishatskaya, E. I., Volova, T. G., Gordeev, S. A., and Puzyr, A. P. Degradation of P(3HB) and P(3HB-co-3HV) in biological media. *J Biomater Sci Polym Ed.*, **16**(5), 643–657 (2005).

25. Bonartsev, A. P., Livshits, V. A., Makhina, T. A., Myshkina, V. L., Bonartseva, G. A., and Iordanskii1, A. L. Controlled release profiles of dipyridamole from biodegradable microspheres on the base of poly(3-hydroxybutyrate) (PHB). *eXPRESS Polymer Letters.*, **1**(12), 797–803 (2007). DOI: 10.3144/expresspolymlett.2007.110.

26. Koyama, N. and Doi, Y. Morphology and biodegradability of a binary blend of poly((R)-3-hydroxybutyric acid) and poly((R,S)-lactic acid). *Can. J. Microbiol.*, **41**(1), 316–322 (1995).

27. Doi, Y., Kanesawa, Y., Kunioka, M., and Saito, T. Biodegradation of microbial copolyesters: poly(3-hydroxybutyrate-co-3-hydroxyvalerate) and poly(3-hydroxybutyrate-co-4-hydroxybutyrate). *Macromolecules*, **23**, 26–31 (1990a).

28. Holland, S. J., Jolly, A. M., Yasin, M., and Tighe, B. *J. Polymers for biodegradable medical devices.* II. Hydroxybutyrate-hydroxyvalerate copolymers: hydrolytic degradation studies. *Biomaterials*, **8**(4), 289–295 (1987).

29. Kurcok, P., Kowalczuk, M., Adamus, G., Jedlinrski, Z., and Lenz, R. W. Degradability of poly (b-hydroxybutyrate)s. Correlation with chemical microstucture. *JMS-Pure Appl. Chem.*, **A32**, 875–880 (1995).

30. Bonartsev, A. P., Boskhomodgiev, A. P., Iordanskii, A. L., Bonartseva, G. A., Rebrov, A. V., Makhina, T. K., Myshkina, V. L., Yakovlev, S. A., Filatova, E. A., Ivanov, E. A., Bagrov, D. V., and Zaikov, G. E. Hydrolytic Degradation of Poly(3-hydroxybutyrate), Polylactide and their Derivatives: Kinetics, Crystallinity, and Surface Morphology. *Molecular Crystals and Liquid Crystals*, **556**(1), 288–300 (2012).

31. Bonartsev, A. P., Boskhomodgiev, A. P., Voinova, V. V., Makhina, T. K., Myshkina, V. L., Yakovlev, S. A., Zharkova, I. I., Zernov, A. L., Filatova, E. A., Bagrov, D. V., Rebrov, A. V., Bonartseva, G. A., and Iordanskii, A. L. Hydrolytic degradation of poly(3-hydroxybutyrate) and its derivates: characterization and kinetic behavior. *Chemistry and Chemical Technology*, **6**(4), 385–392 (2012).

32. Bonartsev, A. P., Myshkina, V. L., Nikolaeva, D. A., Furina, E. K., Makhina, T. A., Livshits, V. A., Boskhomdzhiev, A. P., Ivanov, E. A., Iordanskii, A. L., and Bonartseva, G. A. Biosynthesis, biodegradation, and application of poly(3-hydroxybutyrate) and its copolymers - natural polyesters produced by diazotrophic bacteria. Communicating Current Research and Educational Topics and Trends in Applied Microbiology, A. Méndez-Vilas (Ed.), Formatex, Spain, V.1, p. 295–307 (2007).

33. Cha, Y. and Pitt, C. G. The biodegradability of polyester blends. *Biomaterials*, **11**(2), 108–112 (1990).

34. Schliecker, G., Schmidt, C., Fuchs, S., Wombacher, R., and Kissel, T. Hydrolytic degradation of poly(lactide-co-glycolide) films: effect of oligomers on degradation rate and crystallinity. *Int. J. Pharm.*, **266**(1–2), 39–49 (2003).

35. Scandola, M., Focarete, M. L., Adamus, G., Sikorska, W., Baranowska, I., Swierczek, S., Gnatowski, M., Kowalczuk, M., and Jedlinr ski, Z. Polymer blends of natural poly(3-hydroxybutyrate-co-hydroxyvalerate) and a synthetic atactic poly(3-hydroxybutyrate). Characterization and biodegradation studies. *Macromolecules*, **30**, 2568–2574 (1997).

36. Doi, Y., Kanesawa, Y., Kawaguchi, Y., and Kunioka, M. Hydrolytic degradation of microbial poly(hydroxyalkanoates). *Makrom. Chem. Rapid. Commun.*, **10**, 227–230 (1989).

37. Wang, Y. W., Yang, F., Wu, Q., Cheng, Y. C., Yu, P. H., Chen, J., and Chen, G. Q. Evaluation of three-dimensional scaffolds made of blends of hydroxyapatite and poly(3-hydroxybutyrate-co-3-hydroxyhexanoate) for bone reconstruction. *Biomaterials*, **26**(8), 899–904 (a) (2005).

38. Muhamad, I. I., Joon, L. K., and Noor, M. A. M. Comparing the degradation of poly-β-(hydroxybutyrate), poly-β-(hydroxybutyrate-co-valerate)(PHBV) and PHBV/Cellulose triacetate blend. *Malaysian Polymer Journal*, **1**, 39–46 (2006).

39. Mergaert, J., Webb, A., Anderson, C., Wouters, A., and Swings, J. *Microbial degradation of poly(3-hydroxybutyrate) and poly(3-hydroxybutyrate-co-3-hydroxyvalerate) in soils Applied and environmental microbiology*, **59**(10), 3233–3238 (1993).

40. Choi, G. G., Kim, H. W., and Rhee, Y. H. Enzymatic and non-enzymatic degradation of poly(3-hydroxybutyrate-co-3-hydroxyvalerate) copolyesters produced by Alcaligenes sp. MT-16. *The Journal of Microbiology*, **42**(4), 346–352 (2004).

41. Yang, X., Zhao, K., and Chen, G. Q. Effect of surface treatment on the biocompatibility of microbial polyhydroxyalkanoates. *Biomaterials*, **23**(5), 1391–1397 (2002).

42. Zhao, K., Yang, X., Chen, G. Q., and Chen, J. C. Effect of lipase treatment on the biocompatibility of microbial polyhydroxyalkanoates. *J. material science: materials in medicine.*, **13**, 849–854 (2002).

43. Wang, H. T., Palmer, H., Linhardt, R. J., Flanagan, D. R., and Schmitt, E. Degradation of poly(ester) microspheres. *Biomaterials*, **11**(9), 679–685 (1990).

44. Doyle, C., Tanner, E. T., and Bonfield, W. In vitro and in vivo evaluation of polyhydroxy-butyrate and of polyhydroxybutyrate reinforced with hydroxyapatite. *Biomaterials*, **12**, 841–847 (1991).

45. Coskun, S., Korkusuz, F., and Hasirci, V. Hydroxyapatite reinforced poly(3-hydroxybutyr-ate) and poly(3-hydroxybutyrate-co-3-hydroxyvalerate) based degradable composite bone plate. *J. Biomater. Sci. Polymer Edn.*, **16**, No. 12, pp. 1485–1502 (2005).

46. Sudesh, K., Abe, H., and Doi, Y. Synthesis, structure and properties of polyhydroxyal-kanoates: biological polyesters. *Prog. Polym. Sci.*, **25**, 1503–1555(2000).

47. Lobler, M., Sass, M., Kunze, C., Schmitz, K. P., and Hopt, U. T. Biomaterial patches su-tured onto the rat stomach induce a set of genes encoding pancreatic enzymes. *Biomateri-als*, **23**, 577–583 (2002).

48. Bagrov, D. V., Bonartsev, A. P., Zhuikov, V. A., Myshkina, V. L., Makhina, T. K., Zhar-kova, I. I., Yakovlev, S. G., Voinova, V. V., Boskhomdzhiev, A. P., Bonartseva, G. A., and Shaitan, K. V. Amorphous and semicrystalline phases in ultrathin films of poly(3-hydroxybutirate). TechConnect World NTSI-Nanotech 2012 Proceedings, ISBN 978-1-4665-6274-5, **1**, 602–605 (2012).

49. Kikkawa, Y., Suzuki, T., Kanesato, M., Doi, Y., and Abe, H. Effect of phase structure on enzymatic degradation in poly(L-lactide)/atactic poly(3-hydroxybutyrate) blends with dif-ferent miscibility. *Biomacromolecules*, **10**(4), 1013–1018 (2009).

50. Tokiwa, Y., Suzuki, T., and Takeda, K. Hydrolysis of polyesters by Rhizopus arrhizus lipase. *Agric. Biol. Chem.*, **50**, 1323–1325 (1986).

51. Hoshino, A. and Isono, Y. Degradation of aliphatic polyester films by commercially avail-able lipases with special reference to rapid and complete degradation of poly(L-lactide) film by lipase PL derived from Alcaligenes sp. *Biodegradation*, **13**, 141–147, (2002).

52. Misra, S. K., Ansari, T., Mohn, D., Valappil, S. P., Brunner, T. J., Stark, W. J., Roy, I., Knowles, J. C., Sibbons, P. D., Jones, E. V., Boccaccini, A. R., and Salih, V. Effect of nanoparticulate bioactive glass particles on bioactivity and cytocompatibility of poly(3-hydroxybutyrate) composites. *J. R. Soc. Interface.*, **7**(44), 453–465 (2010).

53. Jendrossek, D., Schirmer, A., and Schlegel, H. G. Biodegradation of polyhydroxyalkanoic acids. *Appl. Microbiol. Biotechnol.*, **46**, 451–463 (1996).

54. Winkler, F. K., D'Arcy, A., and Hunziker, W. *Structure of human pancreatic lipase, Na-ture*, **343**, 771–774 (1990).

55. Jesudason, J. J., Marchessault, R. H., and Saito, T. Enzymatic degradation of poly([R,S] β-hydroxybutyrate). *Journal of environmental polymer degradation.*, **1**(2), 89–98 (1993).

56. Mergaert, J., Anderson, C., Wouters, A., Swings, J., and Kersters, K. Biodegradation of polyhydroxyalkanoates. *FEMS Microbiol. Rev.*, **9**(2–4), 317–321 (1992).

57. Tokiwa, Y. and Calabia, B. P. Degradation of microbial polyesters. *Biotechnol. Lett.*, **26**(15), 1181–1189 (2004).

58. Mokeeva, V., Chekunova, L., Myshkina, V., Nikolaeva, D., Gerasin, V., and Bonartseva, G. Biodestruction of poly(3-hydroxybutyrate) by microscopic fungi: tests of polymer on resistance to fungi and fungicidal properties. *Mikologia and Fitopatologia*, **36**(5), 59–63 (2002).

59. Bonartseva, G. A., Myshkina, V. L., Nikolaeva, D. A., Kevbrina, M. V., Kallistova, A. Y., Gerasin, V. A., Iordanskii, A. L., and Nozhevnikova, A. N. Aerobic and anaerobic mi-crobial degradation of poly-beta-hydroxybutyrate produced by Azotobacter chroococcum. *Appl Biochem Biotechnol.*, **109**(1–3), 285–301 (2003).

60. Spyros, A., Kimmich, R., Briese, B. H., and Jendrossek, D. 1H NMR Imaging Study of Enzymatic Degradation in Poly(3-hydroxybutyrate) and Poly(3-hydroxybutyrate-co-3-hy-droxyvalerate). Evidence for Preferential Degradation of the Amorphous Phase by PHB

Depolymerase B from Pseudomonas lemoignei. *Macromolecules*, **30**(26), 8218–8225 (1997).

61. Hocking, P. J., Marchessault, R. H., Timmins, M. R., Lenz, R. W., and Fuller, R. C. Enzymatic Degradation of Single Crystals of Bacterial and Synthetic Poly(-hydroxybutyrate) *Macromolecules*, **29**(7), 2472–2478 (1996).

62. Bonartseva, G. A., Myshkina, V. L., Nikolaeva, D. A., Rebrov, A. V., Gerasin, V. A., and Makhina, T. K. The biodegradation of poly-beta-hydroxybutyrate (PHB) by a model soil community: the effect of cultivation conditions on the degradation rate and the physicochemical characteristics of PHB. *Mikrobiologiia*, **71**(2), 258–263 (2002). Russian

63. Kramp, B., Bernd, H. E., Schumacher, W. A., Blynow, M., Schmidt, W., Kunze, C., Behrend, D., and Schmitz, K. P. Poly-beta-hydroxybutyric acid (PHB) films and plates in defect covering of the osseus skull in a rabbit model. *Laryngorhinootologie*, **81**(5), 351–356 (2002). [Article in German].

64. Misra, S. K., Ansari, T. I., Valappil, S. P., Mohn, D., Philip, S. E., Stark, W. J., Roy, I., Knowles, J. C., Salih, V., and Boccaccini, A. R. Poly(3-hydroxybutyrate) multifunctional composite scaffolds for tissue engineering applications. *Biomaterials*, **31**(10), 2806–2815 (2010).

65. Kuppan, P., Vasanthan, K. S., Sundaramurthi, D., Krishnan, U. M., and Sethuraman, S. Development of poly(3-hydroxybutyrate-co-3-hydroxyvalerate) fibers for skin tissue engineering: effects of topography, mechanical, and chemical stimuli. *Biomacromolecules*, **12**(9), 3156–3165 (2011).

66. Malm, T., Bowald, S., Karacagil, S., Bylock, A., and Busch, C. A new biodegradable patch for closure of atrial septal defect. An experimental study. *Scand J Thorac Cardiovasc Surg.*, **26**(1), 9–14 (a) (1992).

67. Malm, T., Bowald, S., Bylock, A., Saldeen, T., and Busch, C. Regeneration of pericardial tissue on absorbable polymer patches implanted into the pericardial sac. An immunohistochemical, ultrastructural and biochemical study in the sheep. *Scandinavian Journal of Thoracic and Cardiovascular Surgery*, **26**(1), 15–21 (b) (1992).

68. Malm, T., Bowald, S., Bylock, A., and Busch, C. Prevention of postoperative pericardial adhesions by closure of the pericardium with absorbable polymer patches. An experimental study. *The Journal of Thoracic and Cardiovascular Surgery*, **104**, 600–607 (c) 1992.

69. Malm, T., Bowald, S., Bylock, A., Busch, C., and Saldeen, T. Enlargement of the right ventricular outflow tract and the pulmonary artery with a new biodegradable patch in transannular position. *European Surgical Research*, **26**, 298–308 (1994).

70. Duvernoy, O., Malm, T., Ramström, J., and Bowald, S. A biodegradable patch used as a pericardial substitute after cardiac surgery: 6- and 24-month evaluation with CT. *Thorac Cardiovasc Surg.*, **43**(5), 271–274 (Oct, 1995).

71. Unverdorben, M., Spielberger, A., Schywalsky, M., Labahn, D., Hartwig, S., Schneider, M., Lootz, D., Behrend, D., Schmitz, K., Degenhardt, R., Schaldach, M., and Vallbracht, C. A polyhydroxybutyrate biodegradable stent: preliminary experience in the rabbit. *Cardiovasc Intervent Radiol.*, **25**(2), 127–132 (2002).

72. Kawaguchi, T., Tsugane, A., Higashide, K., Endoh, H., Hasegawa, T., Kanno, H., Seki, T., Juni, K., Fukushima, S., and Nakano, M. Control of drug release with a combination of prodrug and polymer matrix: antitumor activity and release profiles of 2',3'-Diacyl-5-fluoro-2'-deoxyuridine from poly(3-hydroxybutyrate) microspheres. *Journal of Pharmaceutical Sciences*, **87**(6), 508–512 (1992).

73. Baptist, J. N. (Assignor to W.R. Grace Et Co., New York), US Patent No. 3 225 766, (1965).

74. Holmes, P. Biologically produced (R)-3-hydroxy-alkanoate polymers and copolymers. In: D. C. Bassett (Ed.) Developments in crystalline polymers. London, Elsevier, Vol. 2, 1–65 (1988).

75. Saad, B., Ciardelli, G., Matter, S., Welti, M., Uhlschmid, G. K., Neuenschwander, P., and Suterl, U. W. Characterization of the cell response of cultured macrophages and fibroblasts td particles of short-chain poly[(R)-3-hydroxybutyric acid]. *Journal of Biomedical Materials Research*, **30**, 429–439 (1996).

76. Fedorov, M., Vikhoreva, G., Kildeeva, N., Maslikova, A., Bonartseva, G., and Galbraikh, L. Modeling of surface modification process of surgical suture. Chimicheskie volokna, (6), 22–28 (2005). [Article in Russian]

77. Rebrov, A. V., Dubinskii, V. A., Nekrasov, Y. P., Bonartseva, G. A., Shtamm, M., Antipov, E. M. Structure phenomena at elastic deformation of highly oriented polyhydroxybutyrate. Vysokomol. Soedin. (Russian), **44**, 347–351 (2002) [Article in Russian].

78. Kostopoulos, L. and Karring, T. Augmentation of the rat mandible using guided tissue regeneration. *Clin Oral Implants Res.*, **5**(2), 75–82 (1994).

79. Zharkova, I. I., Bonartsev, A. P., Boskhomdzhiev, A. P., Efremov Iu.M., Bagrov, D. V., Makhina, T. K., Myshkina, V. L., Voinova, V. V., Iakovlev, S. G., Filatova, E. V., Zernov, A. L., Andreeva, N. V., Ivanov, E. A., Bonartseva, G. A., and Shaĭtan, K. V. The effect of poly(3-hydroxybutyrate) modification by poly(ethylene glycol) on the viability of cells grown on the polymer films. *Biomed. Khim.*, **58**(5), 579–591 (2012).

80. Kil'deeva, N. R., Vikhoreva, G. A., Gal'braikh, L. S., Mironov, A. V., Bonartseva, G. A., Perminov, P. A., and Romashova, A. N. Preparation of biodegradable porous films for use as wound coverings. *Prikl. Biokhim. Mikrobiol.*, **42**(6), 716–720 (2006). [Article in Russian].

81. Unverdorben, M., Spielberger, A., Schywalsky, M., Labahn, D., Hartwig, S., Schneider, M., Lootz, D., Behrend, D., Schmitz, K., Degenhardt, R., Schaldach, M., and Vallbracht, C. A polyhydroxybutyrate biodegradable stent: preliminary experience in the rabbit. *Cardiovasc. Intervent. Radiol.*, **25**, 127–132 (2002).

82. Solheim, E., Sudmann, B., Bang, G., and Sudmann, E. Biocompatibility and effect on osteogenesis of poly(ortho ester) compared to poly(DL-lactic acid). *J. Biomed. Mater. Res.*, **49**(2), 257–263 (2000).

83. Bostman, O. and Pihlajamaki, H. Clinical biocompatibility of biodegradable orthopaedic implants for internal fixation: a review. *Biomaterials*, **21**(24), 2615–2621 (2000).

84. Lickorish, D., Chan, J., Song, J., Davies, J. E. An in-vivo model to interrogate the transition from acute to chronic inflammation. *Eur. Cell. Mater.*, **8**, 12–19 (2004).

85. Khouw, I. M., van Wachem P. B., de Leij L.F., van Luyn M.J. Inhibition of the tissue reaction to a biodegradable biomaterial by monoclonal antibodies to IFN-gamma. *J. Biomed. Mater. Res.*, **41**, 202–210 (1998).

86. Su, S. H., Nguyen, K. T., Satasiya, P., Greilich, P. E., Tang, L., and Eberhart, R. C. Curcumin impregnation improves the mechanical properties and reduces the inflammatory response associated with poly(L-lactic acid) fiber. *J. Biomater. Sci. Polym. Ed.*, **16**(3), 353–370 (2005).

87. Lobler, M., Sass, M., Schmitz, K. P., and Hopt, U. T. Biomaterial implants induce the inflammation marker CRP at the site of implantation. *J. Biomed. Mater. Res.*, **61**, 165–167 (2003).

88. Novikov, L. N., Novikova, L. N., Mosahebi, A., Wiberg, M., Terenghi, G., and Kellerth, J. O. A novel biodegradable implant for neuronal rescue and regeneration after spinal cord injury. *Biomaterials*, **23**, 3369–3376 (2002).

89. Cao, W., Wang, A., Jing, D., Gong, Y., Zhao, N., and Zhang, X. Novel biodegradable films and scaffolds of chitosan blended with poly(3-hydroxybutyrate). *J. Biomater. Sci. Polymer Ed.*, **16**(11), 1379–1394 (2005).

90. Wang, Y. W., Yang, F., Wu, Q., Cheng, Y. C., Yu, P. H., Chen, J., Chen, G. Q. Effect of composition of poly(3-hydroxybutyrate-co-3-hydroxyhexanoate) on growth of fibroblast and osteoblast. *Biomaterials.*, **26**(7), 755–761 (b) (2005).

91. Ostwald, J., Dommerich, S., Nischan, C., and Kramp, B. In vitro culture of cells from respiratory mucosa on foils of collagen, poly-L-lactide (PLLA) and poly-3-hydroxy-butyrate (PHB). *Laryngorhinootologie*, **82**(10), 693–699 (2003) [Article in Germany].

92. Bonartsev, A. P., Yakovlev, S. G., Boskhomdzhiev, A. P., Zharkova, I. I., Bagrov, D. V., Myshkina, V. L., Mahina, T. K., Charitonova, E. P., Samsonova, O. V., Zernov, A. L., Zhuikov, V. A., Efremov, Yu. M., Voinova, V. V., Bonartseva, G. A., and Shaitan, K. V. The terpolymer produced by Azotobacter chroococcum 7B: effect of surface properties on cell attachment. PLoS ONE 8(2): e57200.

93. Bonartsev, A. P., Yakovlev, S. G., Zharkova, I. I., Boskhomdzhiev, A. P., Bagrov, D. V., Myshkina, V. L., Makhina, T. K., Kharitonova, E. P., Samsonova, O. V., Voinova, V. V., Zernov, A. L., Efremov, Yu. M., Bonartseva, G. A., and Shaitan, K. V. Cell attachment on poly(3-hydroxybutyrate)-poly(ethylene glycol) copolymer produced by Azotobacter chroococcum 7B. *BMC Biochemistry*, (2013). (in press)

94. Wollenweber, M., Domaschke, H., Hanke, T., Boxberger, S., Schmack, G., Gliesche, K., Scharnweber, D., and Worch, H. Mimicked bioartificial matrix containing chondroitin sulphate on a textile scaffold of poly(3-hydroxybutyrate) alters the differentiation of adult human mesenchymal stem cells. *Tissue Eng.*, **12**(2), 345–359 (February, 2006).

95. Wang, Y. W, Wu, Q., and Chen, G. Q. Attachment, proliferation and differentiation of osteoblasts on random biopolyester poly(3-hydroxybutyrate-co-3-hydroxyhexanoate) scaffolds. *Biomaterials*, **25**(4), 669-675 (2004).

96. Nebe, B., Forster, C., Pommerenke, H., Fulda, G., Behrend, D., Bernewski, U., Schmitz, K. P., and Rychly, J. Structural alterations of adhesion mediating components in cells cultured on poly-beta-hydroxy butyric acid. *Biomaterials*, **22**(17), 2425–2434 (2001).

97. Qu, X. H., Wu, Q., and Chen, G. Q. In vitro study on hemocompatibility and cytocompatibility of poly(3-hydroxybutyrate-co-3-hydroxyhexanoate) *J. Biomater. Sci. Polymer Ed.*, **17**(10), 1107–1121 (a) (2006).

98. Pompe, T., Keller, K., Mothes, G., Nitschke, M., Teese, M., Zimmermann, R., and Werner C. Surface modification of poly(hydroxybutyrate) films to control cell-matrix adhesion. *Biomaterials*, **28**(1), 28–37 (2007).

99. Deng, Y., Lin, X. S., Zheng, Z., Deng, J. G., Chen, J. C., Ma, H., and Chen, G. Q. Poly(hydroxybutyrate-co-hydroxyhexanoate) promoted production of extracellular matrix of articular cartilage chondrocytes in vitro. *Biomaterials*, **24**(23), 4273–4281 (2003).

100. Zheng, Z., Bei, F. F., Tian, H. L., and Chen, G. Q. Effects of crystallization of polyhydroxyalkanoate blend on surface physicochemical properties and interactions with rabbit articular cartilage chondrocytes, *Biomaterials*, **26**, 3537–3548 (2005).

101. Qu, X. H., Wu, Q., Liang, J., Zou, B., and Chen, G. Q. Effect of 3-hydroxyhexanoate content in poly(3-hydroxybutyrate-co-3-hydroxyhexanoate) on in vitro growth and differentiation of smooth muscle cells. *Biomaterials*, **27**(15), 2944–2950 (May, 2006).

102. Sangsanoh, P., Waleetorncheepsawat, S., Suwantong, O., Wutticharoenmongkol, P., Weeranantanapan, O., Chuenjitbuntaworn, B., Cheepsunthorn, P., Pavasant, P., and Supaphol, P. In vitro biocompatibility of schwann cells on surfaces of biocompatible polymeric electrospun fibrous and solution-cast film scaffolds. *Biomacromolecules*, **8**(5), 1587–1594 (2007).

103. Shishatskaya, E. I. and Volova, T. G. A comparative investigation of biodegradable poly-hydroxyalkanoate films as matrices for in vitro cell cultures. *J. Mater. Sci-Mater. M.*, **15**, 915–923 (2004).
104. Iordanskii, A. L., Ol'khov, A. A., Pankova, Yu. N., Bonartsev, A. P., Bonartseva, G. A., and Popov, V. O. Hydrophilicity impact upon physical properties of the environmentally friendly poly(3-hydroxybutyrate) blends: modification *via* blending. Fillers, Filled Polymers and Polymer Blends, Willey-VCH, 2006 r., 233, p. 108–116.
105. Suwantong, O., Waleetorncheepsawat, S., Sanchavanakit, N., Pavasant, P., Cheepsunthorn, P., Bunaprasert, T., and Supaphol, P. In vitro biocompatibility of electro-spun poly(3-hydroxybutyrate) and poly(3-hydroxybutyrate-co-3-hydroxyvalerate) fiber mats. *Int J Biol Macromol.*, **40**(3), 217–23 (2007).
106. Heidarkhan, T. A., Zadhoush, A., Karbasi, S., and Sadeghi, A. H. Scaffold percolative efficiency: in vitro evaluation of the structural criterion for electrospun mats. *J. Mater. Sci. Mater. Med.*, **21**(11), 2989–2998 (2010).
107. Masaeli, E., Morshed, M., Nasr-Esfahani, M. H., Sadri, S., Hilderink, J., van Apeldoorn A., van Blitterswijk C. A., and Moroni, L. Fabrication, characterization and cellular compatibility of poly(hydroxy alkanoate) composite nanofibrous scaffolds for nerve tissue engineering. *PLoS One.*, **8**(2), e57157 (2013).
108. Zhao, K., Deng, Y., Chun Chen J., and Chen, G. Q. Polyhydroxyalkanoate (PHA) scaffolds with good mechanical properties and biocompatibility. *Biomaterials*, **24**(6), 1041–1045 (2003).
109. Cheng, S. T., Chen, Z. F., and Chen, G. Q. The expression of cross-linked elastin by rabbit blood vessel smooth muscle cells cultured in polyhydroxyalkanoate scaffolds. *Biomaterials*, **29**(31), 4187–4194 (2008).
110. Francis, L., Meng, D., Knowles, J. C., Roy, I., and Boccaccini, A. R. Multi-functional P(3HB) microsphere/45S5 Bioglass-based composite scaffolds for bone tissue engineering. *Acta Biomater.*, **6**(7), 2773–2786 (2010).
111. Misra, S. K., Ansari, T. I., Valappil, S. P., Mohn, D., Philip, S. E., Stark, W. J., Roy, I., Knowles, J. C., Salih, V., and Boccaccini, A. R. Poly(3-hydroxybutyrate) multifunctional composite scaffolds for tissue engineering applications. *Biomaterials*, **31**(10), 2806–2815 (2010).
112. Lootz, D., Behrend, D., Kramer, S., Freier, T., Haubold, A., Benkiesser, G., Schmitz, K. P., and Becher, B. Laser cutting: influence on morphological and physicochemical properties of polyhydroxybutyrate. *Biomaterials*, **22**(18), 2447–2452 (2001).
113. Fischer, D., Li, Y., Ahlemeyer, B., Kriglstein, J., and Kissel, T. In vitro cytotoxicity testing of polycations: influence of polymer structure on cell viability and hemolysis. *Biomaterials*, **24**(7), 1121–1131 (2003).
114. Nitschke, M., Schmack, G., Janke, A., Simon, F., Pleul, D., and Werner, C. Low pressure plasma treatment of poly(3-hydroxybutyrate): toward tailored polymer surfaces for tissue engineering scaffolds. *J. Biomed. Mater. Res.*, **59**(4), 632–638 (2002).
115. Chanvel-Lesrat, D. J., Pellen-Mussi, P., Auroy, P., and Bonnaure-Mallet, M. Evaluation of the in vitro biocompatibility of various elastomers. *Biomaterials*, **20**, 291–299 (1999).
116. Boyan, B. D., Hummert, T. W., Dean, D. D., and Schwartz, Z. Role of material surfaces in regulating bone and cartilage cell response. *Biomaterials*, **17**, 137–146 (1996).
117. Bowers, K. T., Keller, J. C., Randolph, B. A., Wick, D. G., and Michaels, C. M. Optimization of surface micromorphology for enhanced osteoblasts responses in vitro. *Int. J. Oral. Max. Impl.*, 7, 302–310 (1992).
118. Cochran, D., Simpson, J., Weber, H., and Buser, D. Attachment and growth of periodontal cells on smooth and rough titanium. *Int. J. Oral. Max. Impl.*, **9**, 289–297 (1994).

119. Bagrov, D. V., Bonartsev, A. P., Zhuikov, V. A , Myshkina, V. L., Makhina, T. K., Zharkova, I. I., Yakovlev, S. G., Voinova, V. V., Boskhomdzhiev, A. P., Bonartseva, G. A., and Shaitan, K. V. Amorphous and semicrystalline phases in ultrathin films of poly(3-hydroxybutirate) Technical Proceedings of the 2012 NSTI Nanotechnology Conference and Expo, NSTI-Nanotech, 602–605 (2012).

120. Sevastianov, V. I., Perova, N. V., Shishatskaya, E. I., Kalacheva, G. S., and Volova, T. G. Production of purified polyhydroxyalkanoates (PHAs) for applications in contact with blood. *J. Biomater. Sci. Polym. Ed.*, **14**, 1029–1042 (2003).

121. Seebach, D., Brunner, A., Burger, H. M., Schneider, J., and Reusch, R. N. Isolation and 1H-NMR spectroscopic identification of poly(3-hydroxybutanoate) from prokaryotic and eukaryotic organisms. Determination of the absolute configuration (R) of the monomeric unit 3-hydroxybutanoic acid from Escherichia coli and spinach. *Eur. J. Biochem.*, **224**(2), 317–328 (1994).

122. Myshkina, V. L., Nikolaeva, D. A., Makhina, T. K., Bonartsev, A. P., and Bonartseva, G. A. Effect of growth conditions on the molecular weight of poly-3-hydroxybutyrate produced by Azotobacter chroococcum 7B. *Applied biochemistry and microbiology*, **44**(5), 482–486 (2008).

123. Myshkina, V. L., Ivanov, E. A., Nikolaeva, D. A., Makhina, T. K., Bonartsev, A. P., Filatova, E. V., Ruzhitskiĭ, A. O., and Bonartseva, G. A. Biosynthesis of poly-3-hydroxy-butyrate-3-hydroxyvalerate copolymer by Azotobacter chroococcum strain 7B. *Applied biochemistry and microbiology*, **46**(3), 289–296 (2010).

124. Reusch, R. N. Poly-β-hydroxybutryate/calcium polyphosphate complexes in eukaryotic membranes. *Proc. Soc. Exp. Biol. Med.*, **191**, 377–381 (1989).

125. Reusch, R. N. Biological complexes of poly-β-hydroxybutyrate. *FEMS Microbiol. Rev.*, **103**, 119–130 (1992).

126. Reusch, R. N. Low molecular weight complexed poly(3-hydroxybutyrate): a dynamic and versatile molecule in vivo. *Can. J. Microbiol.*, **41**(Suppl. 1), 50–54 (1995).

127. Müller, H. M. and Seebach, D. Polyhydroxyalkanoates: a fifth class of physiologically important organic biopolymers? *Angew Chemie*, **32**, 477–502 (1994).

128. Huang, R. and Reusch, R. N. Poly(3-hydroxybutyrate) is associated with specific proteins in the cytoplasm and membranes of Escherichia coli. *J. Biol. Chem.*, **271**, 22196–22201 (1996).

129. Reusch, R. N., Bryant, E. M., and Henry, D. N. Increased poly-(R)-3-hydroxybutyrate concentrations in streptozotocin (STZ) diabetic rats. *Acta Diabetol.*, **40**(2), 91–94 (2003).

130. Reusch, R. N., Sparrow, A. W., and Gardiner, J. Transport of poly-β-hydroxybutyrate in human plasma. *Biochim. Biophys. Acta*, **1123**, 33–40 (1992).

131. Reusch, R. N., Huang, R., and Kosk-Kosicka, D. Novel components and enzymatic activities of the human erythrocyte plasma membrane calcium pump. *FEBS Lett.*, **412**(3), 592–596 (1997).

132. Pavlov, E., Zakharian, E., Bladen, C., Diao, C. T. M., Grimbly, C., Reusch, R. N., and French R. J. A large, voltage-dependent channel, isolated from mitochondria by water-free chloroform extraction. *Biophysical Journal*, **88**, 2614–2625 (2005).

133. Theodorou, M. C., Panagiotidis, C. A., Panagiotidis, C. H., Pantazaki, A. A., and Kyriakidis, D. A. Involvement of the AtoS-AtoC signal transduction system in poly-(R)-3-hydroxybutyrate biosynthesis in Escherichia coli. *Biochim. Biophys. Acta.*, **1760**(6), 896–906 (2006).

134. Wiggam, M. I., O'Kane, M. J., Harper, R., Atkinson, A. B., Hadden, D. R., Trimble, E. R., and Bell, P. M. Treatment of diabetic ketoacidosis using normalization of blood 3-hy-

droxy-butyrate concentration as the endpoint of emergency management, *Diabetes Care*, **20**, 1347–1352 (1997).

135. Larsen, T. and Nielsen, N. I. Fluorometric determination of beta-hydroxybutyrate in milk and blood plasma. *J. Dairy Sci.*, **88**(6), 2004–2009 (2005).

136. Agrawal, C. M. and Athanasiou, K. A. Technique to control pH in vicinity of biodegrading PLA-PGA implants. *J. Biomed. Mater. Res.*, **38**(2), 105–114 (1997).

137. Ignatius, A. A. and Claes, L. E. *In vitro biocompatibility of bioresorbable polymers: poly(l, dl-lactide) and poly(l-lactide-co-glycolide)*, **17**(8), 831–839 (1996).

138. Rihova, B. Biocompatibility of biomaterials: Hemocompatibility, immunocompatibility and biocompatibility of solid polymeric materials and soluble targetable polymeric carriers. *Adv Drug. Delivery Rev.*, **21**, 157–176 (1996).

139. Ceonzo, K., Gaynor, A., Shaffer, L., Kojima, K., Vacanti, C. A., and Stahl, G. L. Polyglycolic acid-induced inflammation: role of hydrolysis and resulting complement activation. *Tissue Eng.*, **12**(2), 301–308 (2006).

140. Chasin, M. and Langer, R. (eds), Biodegradable Polymers as Drug Delivery Systems, New York, Marcel Dekker, 1990.

141. Johnson, O. L. and Tracy, M. A. Peptide and protein drug delivery. In: E. Mathiowitz, (Ed.) Encyclopedia of Controlled Drug Delivery. Hoboken, NJ: John Wiley and Sons, **2**, 816–832 (1999).

142. Jain, R. A. The manufacturing techniques of various drug loaded biodegradable poly(lactide-co-glycolide) (PLGA) devices. *Biomaterials*, **21**, 2475–2490 (2000).

143. Gursel, I. and Hasirci, V. Properties and drug release behaviour of poly(3-hydroxybutyric acid) and various poly(3-hydroxybutyrate-hydroxyvalerate) copolymer microcapsules. *J. Microencapsul.*, **12**(2), 185–193 (1995).

144. Li, J., Li, X., Ni, X., Wang, X., Li, H., and Leong, K. W. Self-assembled supramolecular hydrogels formed by biodegradable PEO–PHB–PEO triblock copolymers and a-cyclodextrin for controlled drug delivery. *Biomaterials*, **27**( ), 4132–4140 (2006).

145. Akhtar, S., Pouton, C. W., and Notarianni, L. J. Crystallization behaviour and drug release from bacterial polyhydroxyalkanoates. *Polymer*, **33**(1), 117–126 (1992).

146. Akhtar, S., Pouton, C. W., and Notarianni, L. J. The influence of crystalline morphology and copolymer composition on drug release from solution cast and melting processed P(HB-HV) copolymer matrices. *J. Controlled Release*, **17**, 225–234 (1991).

147. Korsatko, W., Wabnegg, B., Tillian, H. M., Braunegg, G., and Lafferty, R. M. Poly-D-hydroxybutyric acid-a biologically degradable vehicle to regard release of a drug. *Pharm. Ind.*, **45**, 1004–1007 (1983).

148. Korsatko, W., Wabnegg, B., Tillian, H. M., Egger, G., Pfragner, R., and Walser, V. Poly D(-)-3-hydroxybutyric acid (poly-HBA)-a biodegradable former for long-term medication dosage. 3. Studies on compatibility of poly-HBA implantation tablets in tissue culture and animals. *Pharm. Ind.*, **46**, 952–954 (1984).

149. Kassab, A. C., Piskin, E., Bilgic, S., Denkbas, E. B., and Xu, K. Embolization with polyhydroxybutyrate (PHB) micromerspheres: in vivo studies, *J. Bioact. Compat. Polym.*, **14**, 291–303 (1999).

150. Kassab, A. C., Xu, K., Denkbas, E. B., Dou, Y., Zhao, S., and Piskin, E. Rifampicin carrying polyhydroxybutyrate microspheres as a potential chemoembolization agent, *J. Biomater. Sci. Polym. Ed.*, **8**, 947–961 (1997).

151. Sendil, D., Gursel, I., Wise, D. L., and Hasirci, V. Antibiotic release from biodegradable PHBV microparticles. *J. Control. Release*, **59**, 207–17 (1999).

152. Gursel, I., Yagmurlu, F., Korkusuz, F., and Hasirci, V. In vitro antibiotic release from poly(3-hydroxybutyrate-co-3-hydroxyvalerate) rods. *J. Microencapsul.*, **19**, 153–164 (2002).

153. Turesin, F., Gursel, I., and Hasirci, V. Biodegradable polyhydroxyalkanoate implants for osteomyelitis therapy: in vitro antibiotic release. *J. Biomater. Sci. Polym. Ed.*, **12**, 195–207 (2001).

154. Turesin, F., Gumusyazici, Z., Kok, F. M., Gursel, I., Alaeddinoglu, N. G., and Hasirci, V. Biosynthesis of polyhydroxybutyrate and its copolymers and their use in controlled drug release. *Turk. J. Med. Sci.*, **30**, 535–541 (2000).

155. Gursel, I., Korkusuz, F., Turesin, F., Alaeddinoglu, N. G., and Hasirci, V. In vivo application of biodegradable controlled antibiotic release systems for the treatment of implant-related osteomyelitis. *Biomaterials*, **22**(1), 73–80 (2001).

156. Korkusuz, F., Korkusuz, P., Eksioglu, F., Gursel, I., and Hasirci, V. In vivo response to biodegradable controlled antibiotic release systems. *J. Biomed. Mater. Res.*, **55**(2), 217–228 (2001).

157. Yagmurlu, M. F., Korkusuz, F., Gursel, I., Korkusuz, P., Ors, U., and Hasirci, V. Sulbactam-cefoperazone polyhydroxybutyrate-co-hydroxyvalerate (PHBV) local antibiotic delivery system: In vivo effectiveness and biocompatibility in the treatment of implantrelated experimental osteomyelitis. *J. Biomed. Mater. Res.*, **46**, 494–503 (1999).

158. Yang, C., Plackett, D., Needham, D., and Burt, H. M. PLGA and PHBV microsphere formulations and solid-state characterization: possible implications for local delivery of fusidic acid for the treatment and prevention of orthopaedic infections. *Pharm. Res.*, **26**(7), 1644–1656 (2009).

159. Kosenko, R. Yu., Iordanskii, A. L., Markin, V. S., Arthanarivaran, G., Bonartsev, A. P., and Bonartseva, G. A. Controlled release of antiseptic drug from poly(3-hydroxybutyrate)-based membranes. combination of diffusion and kinetic mechanisms. *Pharmaceutical Chemistry Journal*, **41**(12), 652–655 (2007).

160. Khang, G., Kim, S. W., Cho, J. C., Rhee, J. M., Yoon, S. C., and Lee, H. B. Preparation and characterization of poly(3-hydroxybutyrate-co-3-hydroxyvalerate) microspheres for the sustained release of 5-fluorouracil. *Biomed. Mater. Eng.*, **11**, 89–103 (2001).

161. Bonartsev, A. P., Yakovlev, S. G., Filatova, E. V., Soboleva, G. M., Makhina, T. K., Bonartseva, G. A., Shaïtan, K. V., Popov, V. O., and Kirpichnikov, M. P. Sustained release of the antitumor drug paclitaxel from poly(3-hydroxybutyrate)-based microspheres. Biochemistry (Moscow) Supplement Series B: Biomedical Chemistry, **6**(1), 42–47 (2012).

162. Yakovlev, S. G., Bonartsev, A. P., Boskhomdzhiev, A. P., Bagrov, D. V., Efremov, Yu. M., Filatova, E. V., Ivanov, P. V., Mahina, T. K., and Bonartseva, G. A. In vitro cytotoxic activity of poly(3-hydroxybutyrate) nanoparticles loaded with antitumor drug paclitaxel. Technical Proceedings of the 2012 NSTI Nanotechnology Conference and Expo, NSTI-Nanotech, pp. 190–193 (2012).

163. Shishatskaya, E. I., Goreva, A. V., Voinova, O. N., Inzhevatkin, E. V., Khlebopros, R. G., and Volova, T. G. Evaluation of antitumor activity of rubomycin deposited in absorbable polymeric microparticles. *Bull. Exp. Biol. Med.*, **145**(3), 358–361 (2008).

164. Filatova, E. V., Yakovlev, S. G., Bonartsev, A. P., Mahina, T. K., Myshkina, V. L., and Bonartseva, G. A. Prolonged release of chlorambucil and etoposide from poly-3-oxybutyrate-based microspheres. *Applied Biochemistry and Microbiology*, **48**(6), pp 598–602 (2012).

165. Bonartsev, A. P., Bonartseva, G. A., Makhina, T. K., Mashkina, V. L., Luchinina, E. S., Livshits, V. A., Boskhomdzhiev, A. P., Markin, V. S., and Iordanskiĭ, A. L. New poly-(3-

hydroxybutyrate)-based systems for controlled release of dipyridamole and indomethacin. *Applied biochemistry and microbiology*, **42**(6), 625–630 (2006).

166. Coimbra, P. A., De Sousa, H. C., Gil, M. H. Preparation and characterization of flurbiprofen-loaded poly(3-hydroxybutyrate-co-3-hydroxyvalerate) microspheres. *J. Microencapsul.*, **25**(3), 170–178 (2008).

167. Wang, C., Ye, W., Zheng, Y., Liu, X., and Tong, Z. Fabrication of drug-loaded biodegradable microcapsules for controlled release by combination of solvent evaporation and layer-by-layer self-assembly. *Int. J. Pharm.*, **338**(1–2), 165–173 (2007).

168. Salman, M. A., Sahin, A., Onur, M. A., Oge, K., Kassab, A., and Aypar, U. Tramadol encapsulated into polyhydroxybutyrate microspheres: in vitro release and epidural analgesic effect in rats. *Acta Anaesthesiol. Scand.*, **47**, 1006–1012 (2003).

169. Bonartsev, A. P., Livshits, V. A., Makhina, T. A., Myshkina, V. L., Bonartseva, G. A., and Iordanskii, A. L. Controlled release profiles of dipyridamole from biodegradable microspheres on the base of poly(3-hydroxybutyrate). *Express Polymer Letters*, **1**(12), 797–803 (2007).

170. Livshits, V. A., Bonartsev, A. P., Iordanskii, A. L., Ivanov, E. A., Makhina, T. A., Myshkina, V. L., and Bonartseva, G. A. Microspheres based on poly(3-hydroxy)butyrate for prolonged drug release. *Polymer Science Series B*, **51**(7–8), 256–263 (2009).

171. Bonartsev, A. P., Postnikov, A. B., Myshkina, V. L., Artemieva, M. M., and Medvedeva, N. A. A new system of NO donor prolonged delivery on basis of controlled-release polymer, polyhydroxybutyrate. *American Journal of Hypertension*, **18**(5A), p.A (2005).

172. Bonartsev, A. P., Postnikov, A. B., Mahina, T. K., Myshkina, V. L., Voinova, V. V., Boskhomdzhiev, A. P., Livshits, V. A., Bonartseva, G. A., and Iorganskii, A. L. A new in vivo model of prolonged local NO action on arteries on basis of biocompatible polymer. *The Journal of Clinical Hypertension*, Suppl. A., **9**(5), A152 (c) (2007).

173. Stefanescu, E. A., Stefanescu, C., and Negulescu, I. I. Biodegradable polymeric capsules obtained via room temperature spray drying: preparation and characterization. *J. Biomater. Appl.*, **25**(8), 825–849 (2011).

174. Costa, M. S., Duarte, A. R., Cardoso, M. M., and Duarte, C. M. Supercritical antisolvent precipitation of PHBV microparticles. *Int J Pharm.*, **328**(1), 72–77 (2007).

175. Riekes, M. K., Junior, L. R, Pereira, R. N., Borba, P. A., Fernandes, D., and Stulzer, H. K. Development and evaluation of poly(3-hydroxybutyrate-co-3-hydroxyvalerate) and poly-caprolactone microparticles of nimodipine. *Curr Pharm Des.*, (March 12, 2013).

176. Bazzo, G. C., Caetano, D. B., Boch, M. L., Mosca, M., Branco, L. C., Zétola, M., Pereira, E. M., and Pezzini, B. R. Enhancement of felodipine dissolution rate through its incorporation into Eudragit® E-PHB polymeric microparticles: in vitro characterization and investigation of absorption in rats. *J. Pharm. Sci.*, **101**(4), 1518–1523 (2012).

177. Zhu, X. H., Wang, C. H., and Tong, Y. W. In vitro characterization of hepatocyte growth factor release from PHBV/PLGA microsphere scaffold. *J. Biomed. Mater. Res A.*, **89**(2), 411–423 (2009).

178. Parlane, N. A., Grage, K., Mifune, J., Basaraba, R. J., Wedlock, D. N., Rehm, B. H., and Buddle, B. M. Vaccines displaying mycobacterial proteins on biopolyester beads stimulate cellular immunity and induce protection against tuberculosis. *Clin. Vaccine Immunol.*, **19**(1), 37–44 (2012).

179. Yilgor, P., Tuzlakoglu, K., Reis, R. L., Hasirci, N., and Hasirci, V. Incorporation of a sequential BMP-2/BMP-7 delivery system into chitosan-based scaffolds for bone tissue engineering. *Biomaterials*, **30**(21), 3551–3559 (2009).

180. Errico, C., Bartoli, C., Chiellini, F., and Chiellini, E. Poly(hydroxyalkanoates)-based polymeric nanoparticles for drug delivery. *J. Biomed. Biotechnol.*, 2009, 2009, 571702.

181. Althuri, A., Mathew, J., Sindhu, R., Banerjee, R., Pandey, A., and Binod, P. Microbial synthesis of poly-3-hydroxybutyrate and its application as targeted drug delivery vehicle. *Bioresour Technol.*, pii: S0960-8524(13)00129-6 (2013).
182. Pouton, C. W. and Akhtar, S. Biosynthetic polyhydroxyalkanoates and their potential in drug delivery. *Adv. Drug Deliver. Rev.*, **18**:133–162 (1996).

CHAPTER 13

KEY CONCEPTS ON GROWTH AND CHARACTERIZATION OF METAL NANO-SIZED BRANCHED STRUCTURES

GUOQIANG XIE, MINGHUI SONG, KAZUO FURUYA, and AKIHISA INOUE

CONTENTS

13.1   INTRODUCTION

Metal nanostructures are of great importance in nanotechnology due to their potential applications as building blocks in optoelectronic devices, catalysis, chemical sensors, and other areas [1-3]. The formation of nanostructures with controlled size and morphology has been the focus of intensive research in recent years. Such nanostructures are important in the development of nanoscale devices and in the exploitation of the properties of nanomaterials [4-7]. Many fabrication methods including chemical vapor decomposition [8], the arc-discharge method [9], evaporation [10], hydrothermal reactions [11], and so on, have been developed for the production of nanomaterials with controlled sizes and shapes. However, up to date, fabrication of position controllable nanostructures at selected positions on a substrate is still a challenge.

Among the methods, electron-beam-induced deposition (EBID) process is one of the most promising techniques to fabricate small-sized structures on substrates. An advantage of this process is that the deposited position can be controlled. In this approach, an electron beam in a high vacuum chamber is focused on a substrate surface on which precursor molecules, containing the element to be deposited (for example organometallic compound or hydrocarbon), are adsorbed. As a result of complex beam-induced surface reactions, the precursor molecules absorbed in or near to the irradiated area, are dissociated into non-volatile and volatile parts by energetic electrons. The non-volatile materials are deposited on surface of the substrate, while the volatile components are pumped away. Due to the controllability of electron beam, fabrication of position controllable nanostructures can be realized. Using the technique, a variety of nanometer-sized structures, such as nanodots, nanowires, nanotubes, nanopatterns, two or three dimensional nano-objects, and so on, has been fabricated [12-22]. Due to easy to receive a stable fabrication condition, conductive substrates are generally used in these fabrications [16]. As a result, compact structures are usually fabricated with this process.

In the case using nonconductive substrates, the nanofabrication with the EBID technique is also very important for the technique to be applied in technology. The accumulation of charges readily occurs on an insulator substrate during the EBID process. The deposition of novel structures may occur under this condition. The growth of carbon fractal-like structures has already been observed on insulator substrates under electron beam irradiation due to the existence of residual pump oil remaining in the atmosphere of the vacuum chamber [23-25], but the growth of metallic nanostructures on insulator substrates using EBID process was first reported recently by Song et al. [26]. By using insulator substrates such as SiO_2 [27-30] and Al_2O_3 [26, 31-35], characteristic morphologies of nanostructures, such as arrays of nanowhiskers (or nanowires), arrays of nanodendrites, and fractal treelike structures, have been fabricated in transmission electron microscopes (TEMs) by the EBID process [26-35]. The typical size of the diameter of a nanowhisker, the tip of a nanodendrite, and the tip of a nanotree is about 3 nm, and is almost completely independent of the size of the electron beam. In the present review we reported the fabrication and characterization of the metal nano-sized branched structures and the composite nanostructures grown on insulator substrates by the EBID process.

Using an EBID process in a TEM, we fabricated self-standing metal nano-sized branched structures including nanowire arrays, nanodendrites, and nanofractal-like trees, as well as their composite nanostructures with controlled size and position on insulator (SiO_2, Al_2O_3) substrates. The fabricated nanostructures were characterized with high resolution transmission electron microscopy and X-ray energy dispersive spectroscopy. The growth mechanism was discussed. Effect of the electron beam accelerating voltage on crystallization of the nanostructures was investigated. The nanostructures of the different morphologies were obtained by controlling the intensity of the electron beam during the EBID process. A mechanism for the growth and morphology of the nanostructures was proposed involving charge up produced on the surface of the substrate, and the movement of the charges to and charges accumulation at the convex surface of the substrate and the tips of the deposits. High-energy electron irradiation enhanced diffusion of the metallic atoms in the nanostructures and hence promoted crystallization. More crystallized metal nano-branched structures were achieved by the EBID process using high energy electron beams.

13.2 FABRICATION AND CHARACTERIZATION OF NANO-SIZED BRANCHED STRUCTURES

Thin films of SiO_2 and Al_2O_3 were irradiated by an electron beam in TEM with organometallic precursor gases at room temperature. The EBID process was carried out using JEM-2010 or JEM-ARM1000 TEMs made by JEOL Co., Ltd. The energy of the electron beam used was from 200 to 1000 keV. The pressure in the specimen chambers of the TEMs was on the order of 10^{-5}–10^{-6} Pa. Insulator SiO_2 and Al_2O_3 substrates suitable for TEM observation were used. An organometallic precursor gas was introduced near the substrate using a special designed system comprising a nozzle with a diameter smaller than 0.1 mm and a reservoir containing the precursor [26]. Tungsten hexacarbonyl ($W(CO)_6$) or (Methylcyclopentadienyl)trimethylplatinum ($Me_3MeCpPt$) powders were used as the precursors for fabricating tungsten (W)-containing, or platinum (Pt)-containing nanostructures, respectively. The vapor pressures of these precursors are several Pa at room temperature. These precursors have been previously typically used to fabricate small objects on conductive substrates by EBID process [16]. The intensity of the electron beam for the EBID process was from 3.2 to 111.7×10^{18} e cm^{-2} sec^{-1} (current density: 0.51–17.9 A cm^{-2}), which was estimated by measuring the total intensity and the size of the electron beam under operating conditions. The fabricated structure was characterized *in situ* or after fabrication in a TEM. The experiments were performed at room temperature.

Figure 1 shows a series of micrographs of growth process of the W-nanodendrite structures using the TEM at an accelerating voltage of 1000 kV during the electron beam irradiation with an electron beam current density of 1.6 A cm^{-2} [29]. Figure 1(a) presents a micrograph before the electron beam irradiation. No deposits are observed at the SiO_2 substrate. Figure 1(b) shows a micrograph taken 3 sec after the beginning of the irradiation at fluence of 3.0×10^{19} e cm^{-2}. The electron beam irradiated the specimen from a direction perpendicular to the plane of the micrograph. Whisker-like deposits begin to nucleate and grow on the surface of the substrate. The deposits are about 3 nm in diameter and about 5 nm in length. The structures grow self-standing at positions separated from each other at a distance of several nanometers, and in parallel and nearly perpendicular to the surface of the substrate within the irradiated area. The whisker-like deposits grow longer and also denser under further electron beam irradiation, and new nucleation and growth deposits are not observed. Figure 1(c) shows a micrograph taken 3 min after the beginning of irradiation at fluence of 1.8×10^{21} e cm^{-2}. Similarity in length and an almost even thickness of all deposits are observed. Branching is observed at the tips of the deposits. The fabricated deposits have a nanodendrite structure. With further electron beam irradiation, new branches grew from the tips of the grown branches. This process continued as long as the irradiation continued. Figure 1(d) shows a micrograph taken 10 min after the beginning of irradiation at fluence of 6.0×10^{21} e cm^{-2}. The nanodendrite structures grow at both the tip and trunk. The diameter of the nanodendrite structures increased with the decrease in distance from the substrate but is about 3 nm at the tips. The average length of the nanodendrite structures increased to about 51.4 nm.

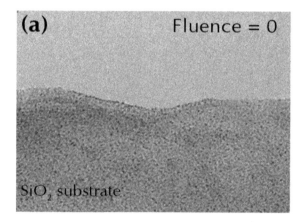

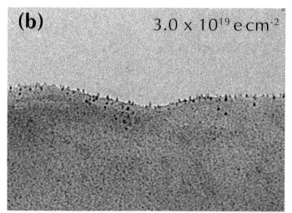

FIGURE 1   *(Continued)*

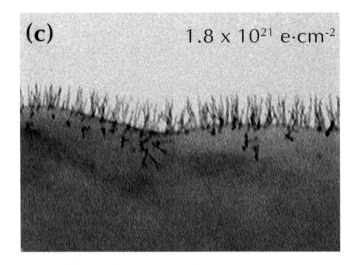

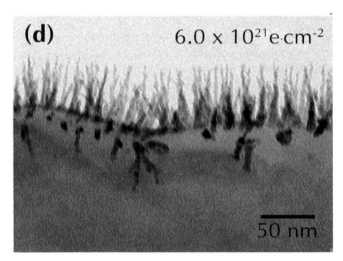

FIGURE 1 Bright-field TEM images of W-nanodendrite structures grown on surface of SiO_2 substrate by the EBID process at an electron beam accelerating voltage of 1000 kV with a current density of 1.6 A cm^{-2} after beginning of electron beam irradiation. (a) 0 sec, (b) 3 sec, (c) 3 min, and (d) 10 min [29].

Figure 2 shows the length of nanodendrite structures grown at various electron beam irradiation fluences (namely, electron beam irradiation time). The length of nanodendrite structures increases approximated linearly with electron beam irradiation fluence [29].

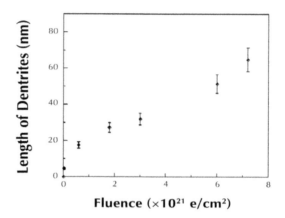

FIGURE 2 Relationship between length of the W-nanodendrite structures grown on surface of SiO$_2$ substrate by the EBID process at an electron beam accelerating voltage of 1000 kV with a current density of 1.6 A cm^{-2} and electron beam irradiating fluence [29].

Furthermore, the morphology of the nanostructures grown can also be controlled by the intensity of the electron beam [26, 36]. Figure 3(a) shows a micrograph of an array of W-nanowhiskers (or nanowires) fabricated on an Al$_2$O$_3$ substrate irradiated with a 200 keV electron beam at an intensity of 4.7×10^{18} e cm^{-2} sec^{-1} (0.75 A cm^{-2}) to about 120 sec [26]. The growth speed was very low. By increasing the current density to a higher value such as 1.6 A cm^{-2}, the growth speed was increased. The nanowhiskers were longer and also denser than those grown at the lower current density with an almost even thickness. Branching is observed to take place at the tips of the nanowhiskers and the morphology becomes dendritic structures [27, 29]. A further increase in the current density to 17.9 A cm^{-2} resulted in extensive branching and the formation of complicated nanotree structures (Figure 3(c)) [26].

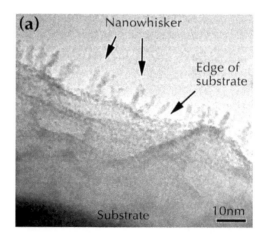

FIGURE 3   *(Continued)*

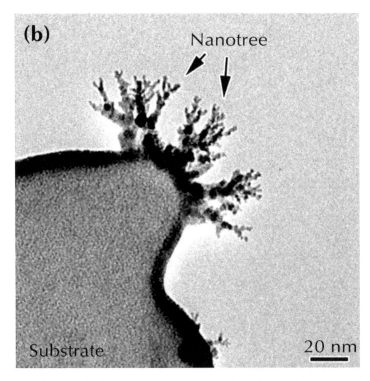

FIGURE 3 (a) W nanowhisker arrays grown on the surface of an Al_2O_3 substrate by the EBID process at an accelerating voltage of 200 kV. The current density of the electron beam was 0.75 A cm^{-2} and the irradiation time was 120 sec. (b) Fractal-like W nanotrees grown on the surface of an Al_2O_3 substrate by the EBID process. The current density of the electron beam was 17.9 A cm^{-2} and the irradiation time was 10 sec [26].

The composition of the fabricated nanodendrites is important, since they are re-lated to the physical and chemical properties of deposits. The chemical composition of the nanodendrites was examined using an X-ray energy dispersive spectroscopy (EDS). Figure 4 shows the spectra taken using EDS from an as-deposited nanodendrite fabricated at an electron beam irradiating fluence of 6.0×10^{21} e cm^{-2} after detaching the precursor source [29].

Figure 4(a) presents a spectrum taken from the tips of a nanodendrite structure, and Figure 4(b) shows that taken from its trunk. The size of the electron beam for the analysis using EDS is about 10 nm. The peaks of tungsten dominate these spectra. No obvious differences in the composition between the tip and the trunk of the nanoden-drite structure were observed. On the basis of the analyses using EDS, a relative con-tent of 89.4% W compared with 10.6% C for the spectrum from the tips is obtained. The relative content of W is higher than the reported values for W-deposits fabricated using the same precursor but at a lower voltage of the electron beam (25 kV) by EBID process, which was 75% [17].

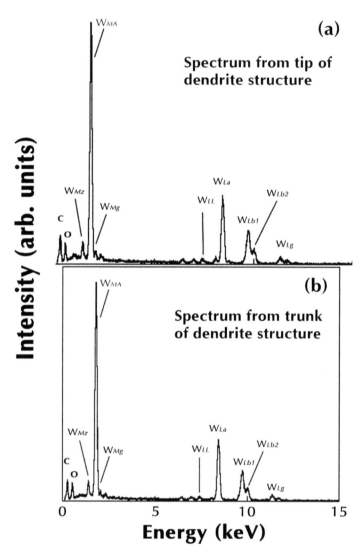

FIGURE 4   The EDS spectra taken from the nanodendrite grown on surface of SiO$_2$ substrate by the EBID process at an accelerating voltage of 1000 kV with a current density of 1.6 A cm^{-2} to an electron beam irradiating fluence of 6.0 × 10^{21} e cm^{-2}. (a) at tip; and (b) at trunk [29].

Using (Methylcyclopentadienyl)trimethylplatinum (Me$_3$MeCpPt) powder as a precursor, one can obtain Pt nanodendritic structures by the EBID process. Figure 5 shows TEM micrographs of Pt nanodendrite-like structures grown on an Al$_2$O$_3$ substrate in a 200 kV TEM [34]. The electron beam is defocused to a size of about 600 nm, corresponding to a current density of 0.52 A cm^{-2}. The irradiated time is 1 min. The irradiating fluence of the electron beam is 2.0 × 10^{20} e cm^{-2}. The edge of the electron beam is indicated by arrows in Figure 5(a). The contrast inside the irradiated area is obviously dark. The size of dendrite structures is much smaller than the diameter of the electron beam. The dendrites show a tendency to grow at the edge of the substrate. The dendrites have branches at the tip, as observed in Figure 5(b). The diameter of the nanodendrites become thicker near the substrate, which implies that the deposition takes place at both tip and trunk part. The typical thickness of the tips is less than 3 nm, as observed in Figure 5(b). The TEM observation, diffraction pattern analyses, and EDS analyses indicated that a nanodendrite structure with a high Pt content was formed.

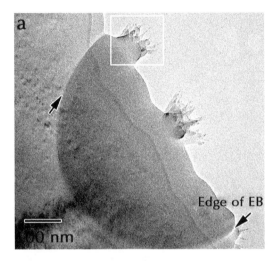

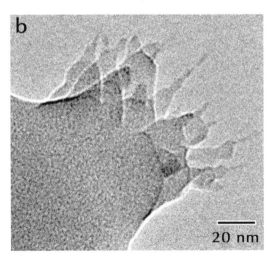

FIGURE 5 (a) Bright-field TEM micrographs of Pt-nanodendrite structures grown on the surface of an Al_2O_3 substrate in a 200 kV TEM with an EBID process 1 min after the start of the electron beam irradiation to a fluence of 2.0×10^{20} e cm^{-2}. (b) Enlargement of the square area in (a), showing the nanodendrites in more detail [34].

Figure 6 shows a high resolution TEM (HRTEM) micrograph of the fabricated Pt nanodendrite [34]. It is confirmed that fcc Pt nanoparticles and an amorphous part are contained in the structure. The morphology of the Pt nanodendrite is considerably different from that of the W-nanodendrite fabricated on an Al_2O_3 substrate, suggesting that the growth of a nanodendrite depends largely on the properties of the precursor.

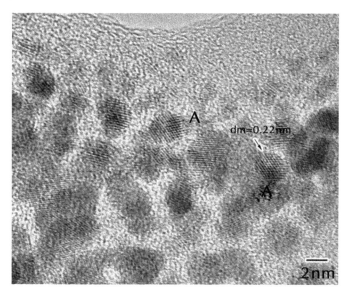

FIGURE 6   An HRTEM micrograph of a nanodendrite structure fabricated on an electron beam irradiating fluence of 2.6×10^{21} e cm^{-2} [34].

13.3   GROWTH MECHANISM OF NANO-SIZED BRANCHED STRUCTURES

It is well known that the EBID process is caused by the dissociation of molecules adsorbed to a surface by energetic electron beam. In this approach, a molecule of a precursor is first adsorbed on surface of a substrate and then decomposed into volatile and non-volatile parts by further irradiation of the energetic beam. The non-volatile fraction accumulates to form a deposit, while the volatile component is pumped away by the vacuum system. The dissociation mechanism is complex and not fully understood until the present time because of the huge number of excitation channels available even for small molecules. The details involving the decomposition have been argued to relate to secondary electrons on surface of a specimen produced by incident electron beam and backscattered electrons in an EBID process [17, 19].

If an EBID process is carried out on the surface of an electric grounded conductive substrate, the molecules absorbed may move or not move, but distribute randomly and be decomposed on the surface. Therefore a compact deposit is usually formed. On the other hand, in a case that the deposition is conducted on an insulator substrate, charge up may take place on the surface due to emission of secondary electrons under energetic beam irradiation. When specimens are exposed to electron bombardments, the molecules absorbed to a surface of a substrate or near the surface in the irradiated area may be polarized by irradiation of an incident electron beam. The irradiated area on the insulator is easily charged, forms local electric potential. It is reasonable to consider that the distribution of charges due to charge up on the surface may be not even in a nanometer scale.

This unevenness may be resulted from the nanoscaled unflatness or atomic steps on surface. Charges may accumulate at some places to some extent. Figure 7(a) shows the growth mechanism of a nanowhisker array [26]. The intensity of the electron beam is weak, but the accumulation of charges occurs. The distribution of the charges on the surface is assumed to be uneven at a nanometer scale. Charges may accumulate at some places (charge centers), where an electric field is generated. The dark dots represent the non-volatile fraction of the precursor molecules of which the deposits are composed, while the brighter dots represent the volatile fraction. Because the intensity of the electron beam is weak, the adsorbed molecules can move around considerably before being decomposed.

The precursor molecules adsorbed on the surface may be attracted to the charge centers since the molecules are easily polarized due to the weak bonding between the atoms of the molecules. The precursor molecules then decompose and form a deposit. After a deposit is formed, the charges on the surface of the substrate or the deposit tend to move and accumulate at the tip of the deposit, since W deposits have been reported to be conductive [17, 19]. Thus, molecules are attracted to the tip, and the deposit increases in length upon further electron beam irradiation. The formation of W nanodot deposits on the surface of a SiO_2 substrate and the growth of W nanowhiskers from these nanodots [30] are consistent with the above mechanism.

Figures 7(b) and (c) schematically show the growth conditions of nanodendrites and fractal-like nanotrees, respectively [26]. When the current density of the electron beam is increased to a definite extent, an adsorbed molecule on the surface may be decomposed in a very short period before it can move. Hence, a compact layer is deposited on the surface. This results in the entire irradiated area becoming conductive. After this stage, the charges on the surface can move a long distance to the substrate edge, particularly when the surface is convex. A strong electric field is thus generated near the convex surface. Once a deposit grows away from the surface at a point, charges then accumulate there.

This deposit thus grows preferentially because the charges attract precursor molecules. When the current density is moderate, the electric field generated near a tip may not strongly affect the trajectory of a molecule. Therefore, the deposit grows at both its tip and its main body because the molecules arrive at the tip and the main body at the same time. Therefore, the main body of the deposit increases in thickness with time and develops a dendritic morphology. On the other hand, if the current density of the electron beam is very strong, the generated electric field will become sufficiently strong to affect the trajectory of a molecule near the tip of a deposit. The molecules are thus likely to be attracted to the tip and are not easily adsorbed on the main body of the deposit. Therefore, a fractal-like tree morphology is formed.

a

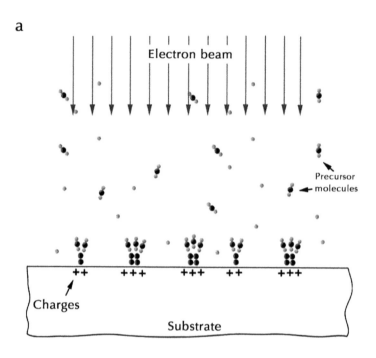

b

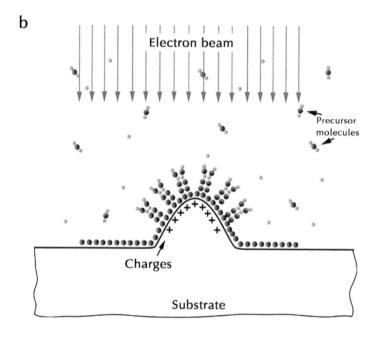

FIGURE 7   *(Continued)*

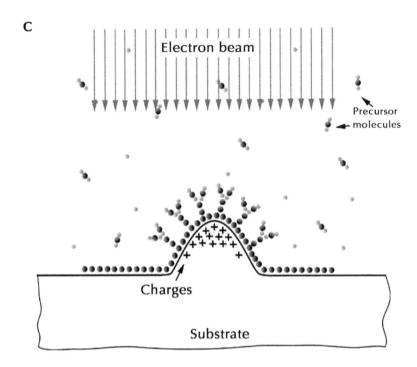

FIGURE 7 Schematic drawings showing the growth process of the nanostructures with different morphologies on insulator substrates by the EBID process. The dark dots represent the non-volatile fraction of the precursor and the deposit, while the brighter dots represent the volatile fraction of the precursor. (a) The growth of nanowhiskers (or nanowires) inside the area of irradiation with a weak electron beam; (b) the growth of nanodendrites on a convex surface under irradiation with a moderate electron beam; and (c) the growth of fractal-like nanotrees on a convex surface under irradiation with a strong electron beam [26].

13.4 EFFECT OF ACCELERATING VOLTAGE ON CRYSTALLIZATION OF THE NANOSTRUCTURES

The fabricated nanostructure morphologies at various electron beam accelerating voltages have been investigated with a conventional TEM. Figure 8 shows a set of TEM micrographs of dendrite-like structures grown on SiO_2 substrate at different electron beam accelerating voltages to an electron beam fluence of 6.0×10^{21} e cm^{-2} [27]. The electron beam irradiated the specimen from a direction perpendicular to the plane of the micrograph. The obvious difference in dendrite morphology and their composites for the deposits fabricated by various accelerating voltages is not observed. The length of the nanodendrite structures increases with an increase in electron beam irradiating fluence, and has not obvious effect with the increase of accelerating voltage, as shown in Figure 9.

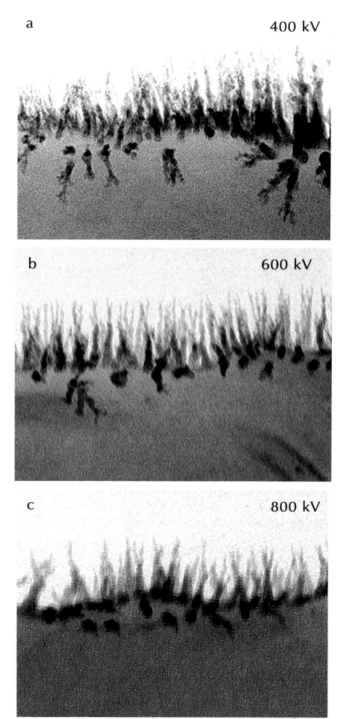

a   400 kV

b   600 kV

c   800 kV

FIGURE 8   *(Continued)*

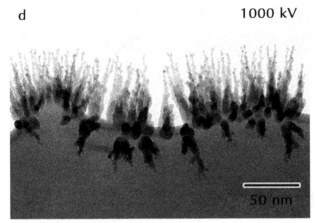

FIGURE 8 A set of bright-field TEM micrographs of the nanodendrite structures grown on SiO$_2$ substrate irradiated to an electron beam fluence of 6.0×10^{21} e cm^{-2} at different electron beam accelerating voltages. (a) 400 kV, (b) 600 kV, (c) 800 kV, and (d) 1000 kV. The current density of electron beam for irradiating the specimens was 1.1 A cm^{-2} for 400 kV and 1.6 A cm^{-2} for 600, 800, and 1000 kV, respectively [27].

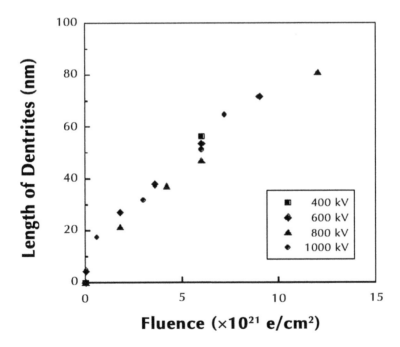

FIGURE 9 Dependence of the length of nanodendrites on electron beam irradiating fluence at different electron beam accelerating voltages [27].

The effect of electron beam accelerating voltage on crystallization of the nano-dendrite structures has been investigated with an HRTEM. Figure 10 shows a series of HRTEM micrographs of some branches of nanodendrite structures fabricated with various accelerating voltages to a fluence of 6.0×10^{21} e cm^{-2} [28]. Figure 10(a) is an HRTEM micrograph of some branches of nanodendrites grown with an accelerating voltage of 400 kV. Lattice fringes are observed at the most places. This indicates that they are crystal in several nanometers. The largest and also the most observed lattice spacing measured from micrographs is 0.22 nm. It is close to the lattice spacing, 0.224 nm, of {110} of bcc W crystals with a deviation smaller than 5%. Moreover, lattice fringes with spacing d = 0.22 nm have inter-fringe angle of 60° (grain A), as well as that of 90° (grain B), as indicated in Figure 10. They agree to zone axis of [111] and [001] of bcc W structure, respectively. Combined with nanometer-sized area diffraction pattern analyses and EDS analyses, it is clarified that these crystals are W grains in bcc structure. Furthermore, lattice fringes cannot be clearly observed in some places, indicating that a large part of them is in amorphous state, as shown by arrows in Figure 10. Figure 10(b) shows a micrograph of some branches of nanodendrites grown with an accelerating voltage of 600 kV. The fraction of amorphous state in the as-fabricated nanodendrites decreases obviously. For the dendrites fabricated with an accelerating voltage of 800 kV, as shown in Figure 10(c), the amorphous state is observed only in a few places. Figure 10(d) shows an HRTEM micrograph of some branches of nano-dendrites grown with an accelerating voltage of 1000 kV. Lattice fringes are observed clearly in almost all of the grains except in thick region, where small crystals in random orientations overlapped each other, so that their lattice fringes cannot be clearly observed. Therefore, it is indicated that the fraction of amorphous state in the as-fabricated nanodendrites decreases with an increase in electron beam accelerating voltage.

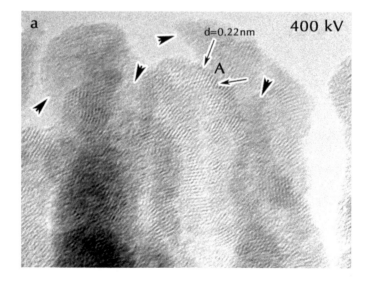

FIGURE 10   *(Continued)*

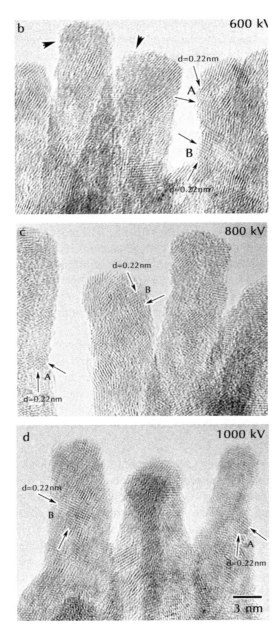

FIGURE 10   A series of HRTEM micrographs of some branches of nanodendrites grown on SiO$_2$ substrate irradiated to a fluence of 6.0×10^{21} e cm^{-2} with different electron beam accelerating voltages. (a) 400 kV, (b) 600 kV, (c) 800 kV, and (d) 1000 kV. Arrows indicate lattice fringes in grains A and B, which have inter-fringe angles of 60° (A) and 90° (B), respectively. The current density of electron beam for irradiating the specimens was 1.1 A cm^{-2} for 400 kV, and 1.6 A cm^{-2} for 600, 800, and 1000 kV, respectively [28].

It is known that the growth of a W-nanodendrite during the EBID process is controlled by a random accumulation of non-volatile elements with thermal energy. Therefore the element does not have enough energy to move and form crystalline grains. The electron beam during the EBID process may transfer some energy to the element, but it is not enough for total deposit to transform into crystalline state with a 400 keV electron beam. Therefore, some amorphous structures are remained in deposits (refer to Figure 10(a)). This also may explain the reason that an amorphous state has often been obtained in the W-deposits fabricated by a lower energy EBID [19, 26]. With an increase in electron beam accelerating voltage, the fraction transformed into crystalline state in the deposits increases, and the fraction of amorphous structure decreases. The results clearly indicate that high-energy electron irradiation enhances crystallization of an amorphous structure. Song et al. [37] has reported that the effect of 1 MeV electron beam irradiation on crystallization of nanometer-sized W-dendritic structure fabricated on Al_2O_3 substrates with the EBID process at an accelerating voltage of 200 kV. After irradiation at room temperature for 100 min at a current density of 6.4 A cm^{-2}, almost all the grains crystallized. The nanodendrite structure also changes morphology and shrinks its size. In general, there are two important factors in the occurrence of electron irradiation induced crystallization of an amorphous state:

(1) The promotion of atomic diffusion by electron irradiation in an amorphous state and

(2) The high stability of crystalline phase against electron irradiation [38].

When the two factors are satisfied simultaneously, electron irradiation induced crystallization of the amorphous structure. The change in morphology and shrinkage in size may be resulted from crystallization and sputtering during the irradiation.

13.5 COMPOSITE NANOSTRUCTURES OF NANOPARTICLES NANODENDRITES

Complex shape nanostructures have attracted great interest because advanced functional materials might be emerged if they can be formed with well-defined three-dimensional (3D) architectures [39-41]. Furthermore, the impact is greater for multi-element systems, as in the case of advanced nanostructured systems like Au/Cu, Pt/C composite materials [42, 43], and so on, which are used in catalysis, sensors, energy sources [44, 45] ,and in many other applications. It has been demonstrated that bimetallic bonding can induce significant changes in the properties of the surface, producing in many cases catalysts that have superior activity and/or selectivity [42, 43, 46, 47].

The nanostructures fabricated on insulator substrates by the EBID process are good base materials for the fabrication of complex-shape nanostructures because of their superior features. Pt nanoparticles or Au nanoparticles were deposited on W-nanodendrites fabricated on insulator Al_2O_3 or SiO_2 substrates, and a composite nanostructure consisting of Pt nanoparticles and W-nanodendrites (Pt-nanoparticle-decorated W-nanodendrites, or Pt nanoparticle/W nanodendrite), or Au nanoparticle and W-nanodendrites were fabricated by ion-sputtering [33]. Figure 11 shows the Pt nanoparticle/W-nanodendrite composite structures. Figure 11(a) is a bright-field TEM

image of the as-fabricated composite structure produced with an electron beam irradiation fluence of 5.0×10^{21} cm^{-2} and an ion sputtering time of 40 sec. The Pt-nanoparticles are nearly uniformly distributed on the W-nanodendrites. Figure 11(b) shows an HRTEM image of a tip of the composite nanostructure. Lattice fringes are observed in the image. By measuring the lattice spacing and the inter fringe angle from this image and other images, one can demonstrate that the as-fabricated composite nanostructures consist of nanocrystals of Pt and W. By using selected area diffraction (SAD) pattern (Figure 11(c)), these nanocrystals are identified to be equilibrium phases of fcc Pt and bcc W. Figure 11(d) is an EDS spectrum obtained at the tip corresponding to Figure 11(b). Pt and W peaks dominate in the spectrum although traces of C and O are also observed. Thus, it is confirmed that Pt has been effectively grown on the nanodendritic W structures. From the HRTEM micrographs, the average nanoparticle size at the present ion sputtering conditions was 2.3 nm. The particle size can be easily controlled by variation of the ion sputtering time.

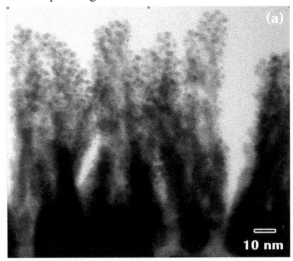

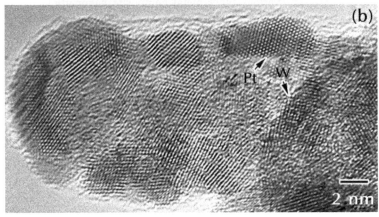

FIGURE 11   *(Continued)*

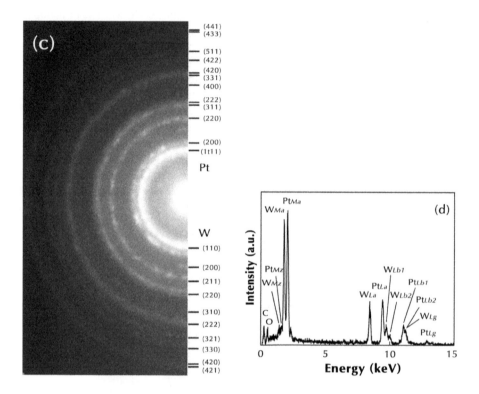

FIGURE 11 (a) Bright-field TEM image of the as-fabricated Pt-nanoparticle/W-nanodendrite composite structures on an Al_2O_3 substrate. (b) an HRTEM image, (c) SAD pattern, and (d) EDS spectrum taken from a tip of the composite nanostructures [33].

Using this process, various other metal composite nanoparticle/nanodendrite structures, such as Au/W, Mo/W, Au/Pt, and so on, can also be fabricated. Figure 12 shows an example of Au-nanoparticle/W-nanodendrite compound nanostructures grown on an insulator SiO_2 substrate. Therefore, the technique may be easily employed to fabricate metal nanoparticle/nanodendrite composite nanostructures of a wide range of materials. Because the nanodendrite structure possesses a comparatively large specific surface area, it has potential applications in catalysts, sensors, gas storages, and so on.

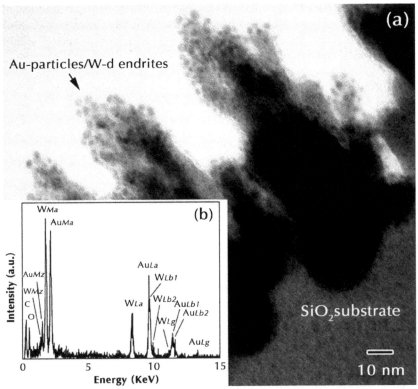

FIGURE 12 (a) Bright-field TEM image of the as-fabricated Au-nanoparticle/ W-nanodendrite composite structures on a SiO₂ substrate and (b) an EDS spectrum taken from a tip of the composite nanostructures. The W-nanodendrites were fabricated with an electron beam irradiation fluence of 4.4×10^{21} cm^{-2}. Au nanoparticles were deposited by a quick auto coater system (JEOL JFC-1500). The anodic voltage used during sputtering was 1 kV, and anodic current was 7 mA. The ion sputtering time was 7 sec. The average Au nanoparticle size was measured to be 2.1 nm [33].

13.6 CONCLUSION

Self-standing metal nano-sized branched structures including nanowire arrays, nano-dendrites, and nanofractal-like trees with controlled size and position were grown on insulator (SiO₂, Al₂O₃) substrates by the EBID process. The nanostructures of the different morphologies can be obtained by controlling the intensity of the electron beam during the EBID process. The nucleation and growth of the nano-sized branched structures are proposed to be related to a mechanism involving charge up produced on the surface of the substrate, and the movement of the charges to and charges accumulation at the convex surface of the substrate and the tips of the deposits. High-energy electron irradiation enhances diffusion of the metallic atoms in the nanostructures and hence promotes crystallization. More crystallized metal na-no-branched structures are achieved by the EBID process using high energy electron beams. The nano-branched structures can be easily decorated by metallic nanoparticles to form composite nano-structures such as the Pt nanoparticles/nanodendrites. Therefore, the nano-sized branched structures and their fabrication process may be applied in technology to realize various functional nanomaterials such as catalysts, sensor materials, and emitters.

KEYWORDS

- **Crystallization**
- **Electron-beam-induced deposition**
- **Insulator substrate**
- **Nanofabrication**
- **Nanostructures**

REFERENCES

1. Shi, J., Gider, S., Babcock, K., and Awschalom, D. D. *Science*, **271**, 937 (1996).
2. Favier, F., Walter, E., Zach, M., Benter, T., and Penner, R. M. *Science*, **293**, 2227 (2001).
3. Xia, Y. N., Yang, P. D., Sun, Y. G., Wu, Y. Y., Mayer, B., Gates, B., Yin, Y. D., Kim, F., and Yan, H. Q. *Adv. Mater.*, **15**, 353 (2003).
4. Lao, Y. L., Wen, J. G., and Ren, Z. F. *Nano Lett.*, **2**, 1287 (2002).
5. Manna, L., Milliron, D. J., Meisel, A., Scher, E. C., and Alivisatos, A. P. *Nature Mater.*, **2**, 382 (2003).
6. Yan, H., He, R., Pham, J., and Yang, P. *Adv. Mater.*, **15**, 402 (2003).
7. Dick, K. A., Deppert, K., Larsson, M. W., Martensson, T., Seifert, W., Wallenberg, L. R., and Samuelson, L. *Nature Mater.*, **3**, 380 (2004).
8. Zhang, H. F., Wang, C. M., Buck, E. C., and Wang, L. S. *Nano Lett.*, **3**, 577 (2003).
9. Shi, Z. J., Lian, Y. F., Liao, F. H., Zhou, X. H., Gu, Z. N., Zhang, T., Iijima, S., Li, H. D., Yue, K. T., and Zhang, S. L. *J. Phys. Chem. Solids*, **61**, 1031 (2000).
10. Dai, Z. R., Pan, Z. W., and Wang, Z. L. *Adv. Funct. Mater.*, **13**, 9 (2003).
11. Jin, Y., Tang, K. B., An, C. H., and Huang, L. Y. *J. Cryst. Growth*, **253**, 429 (2003).
12. Mitsuishi, K., Shimojo, M., Han, M., and Furuya, K. *Appl. Phys. Lett.*, **83**, 2064 (2003).
13. Dong, L. X., Arai, F., and Fukuda, T. *Appl. Phys. Lett.*, **81**, 1919 (2002).
14. Utke, I., Hoffmann, P., Dwir, B., Leifer, K., Kapon, E., and Doppelt, P., *J. Vac. Sci. Technol. B*, **18**, 3168 (2000).
15. Brückl, H., Kretz, J., Koops, H. W., and Reiss, G. *J. Vac. Sci. Technol. B*, **17**, 1350 (1999).
16. Koops, H. W. P., Kretz, J., Rudolph, M., Weber, M., Dahm, G., and Lee, K. L. *Jpn. J. Appl. Phys. Part 1*, **33**, 7099 (1994).
17. Koops, H. W. P., Weiel, R., Kern, D. P., and Baum, T. H., *J. Vac. Sci. Technol. B*, **6**, 477 (1988).
18. Hiroshima, H., Suzuki, N., Ogawa, N., and Komuro, M. *Jpn. J. Appl. Phys. Part 1*, **38**, 7135 (1999).
19. Hoyle, P. C., Cleaver, J. R. A., and Ahmed, H. *J. Vac. Sci. Technol. B*, **14**, 662 (1996).
20. Kohlmann-von Platen, K. T., Chlebek, J., Weiss, M. Reimer, K., Oertel, H., and Brünger, W. H. *J. Vac. Sci. Technol. B*, **11**, 2219 (1993).
21. Matsui, S., Kaito, T., Fujita, J., Komura, M., Kanda, K., and Haruyama, Y. *J. Vac. Sci. Technol. B*, **18**, 3181 (2000).

22. Liu, Z. Q., Mitsuishi, K., and Furuya, K. *Nanotechnology*, **15**, S414 (2004).
23. Banhart, F. *Phys. Rev. E*, **52**, 5156 (1995).
24. Zhang, J. Z., Ye, X. Y., Yang, X. J., and Liu, D. *Phys. Rev. E*, **55**, 5796 (1997).
25. Wang, H. Z., Liu, X. H., Yang, X. J., and Wang, X. *Mat. Sci. Eng. A*, **311**, 180 (2001).
26. Song, M., Mitsuishi, K., Tanaka, M., Takeguchi, M., Shimojo, M., and Furuya, K. *Appl. Phys. A*, **80**, 1431 (2004).
27. Xie, G. Q., Song, M., Mitsuishi, K., and Furuya, K. *J. Nanosci. Nanotechnol.*, **5**, 615 (2005).
28. Xie, G. Q., Song, M., Mitsuishi, K., and Furuya, K. *Physica E*, **29**, 564 (2005).
29. Xie, G. Q., Song, M., Mitsuishi, K., and Furuya, K. *Jpn. J. Appl. Phys.*, **44**, 5654 (2005).
30. Song, M., Mitsuishi, K., and Furuya, K. *Mater. Trans.*, **48**, 2551 (2007).
31. Song, M., Mitsuishi, K., Takeguchi, M., and Furuya, K. *Appl. Surf. Sci.*, **241**, 107 (2005).
32. Song, M., Mitsuishi, K., and Furuya, K. *Physica E*, **29**, 575 (2005).
33. Xie, G. Q., Song, M., Furuya, K., Louzguine, D. V., and Inoue, A. *Appl. Phys. Lett.*, **88**, 263120 (2006).
34. Xie, G. Q., Song, M., Mitsuishi, K., and Furuya, K. *J. Mater. Sci.*, **41**, 2567 (2006).
35. Xie, G. Q., Song, M., and Furuya, K. *J. Mater. Sci.*, **41**, 4537 (2006).
36. Furuya, K., Takeguchi, M. Song, M., Mitsuishi, K., and Tanaka, M. *J. Phys. Conf. Ser.*, **126**, 012024 (2008).
37. Song, M., Mitsuishi, K., and Furuya, K. *Mater. Sci. Forum*, **475–479**, 4035 (2005).
38. Nagase, T. and Umakoshi, Y. *Mater. Sci. Eng. A*, **352**, 251 (2003).
39. Gao, P. X., Ding, Y., Mai, W., Hughes, W. L., Lao, C., and Wang, Z. L. *Science*, **309**, 1700 (2005).
40. Dick, K. A., Deppert, K., Larsson, M. W., Martensson, T., Seifert, W., Wallenberg, L. R., and Samuelson, L. *Nature Mater.*, **3**, 380 (2004).
41. Li, M., Schnablegger, H., Mann, S. *Nature*, **402**, 393 (1999).
42. Pal, U., Ramirez, J. F. S., Liu, H. B., Medina, A., and Ascencio, J. A. *Appl. Phys. A*, **79**, 79 (2004).
43. Joo, S. H., Choi, S. J., Oh, I., Kwak, J., Liu, Z., Terasaki, O., and Ryoo, R. *Nature* **412**, 169 (2001).
44. Ruiz, A., Arbiol, J., Cirera, A., Cornet, A., and Morante, J. R. *Mater. Sci. Eng. C*, **19**, 105 (2005).
45. De Meijer, R. J., Stapel, C., Jones, D. G., Roberts, P. D., Rozendaal, A., Macdonald, W. G., Chen, K. Z., Zhang, Z. K., Cui, Z. L., Zuo, D. H., and Yang, D. Z. *Nanostruct. Mater.*, **8**, 205 (1997).
46. Wang, A., Liu, J., Lin, S., Lin, T., and Mou, C. *J. Catal.*, **233**, 1486 (2005).
47. Liu, P., Rodriguez, J. A., Muckerman, J. T., and Hrbek, J. *Surf. Sci.*, **530**, L313 (2003).

CHAPTER 14

PREPARATION, CHARACTERIZATION, AND APPLICATIONS OF MAGNESIUM STEARATE, COBALT STEARATE, AND COPPER STEARATE

MEHMET GÖNEN, THERESA O. EGBUCHUNAM, DEVRIM BALKÖSE, FIKRET INAL, and SEMRA ÜLKÜ

CONTENTS

14.1 INTRODUCTION

Metal soaps, such as zinc, calcium, copper, and magnesium are insoluble or sparingly soluble in water. Because of this property, they are commercially important compounds and find applications in industry, such as driers in paints or inks, components of greases, stabilizers for plastics, in fungicides, catalysts, waterproofing agents, fuel additives, components of creams and additive in drug formulation, and so on. Magnesium stearate ($MgSt_2$) is in widespread use as gelling, sanding and anti-sticking agents, stabilizer, lubricant, emulsifier and plasticizer for polymers, in the paint, food, rubber, paper, and pharmaceutical industries. Copper stearate ($CuSt_2$) is used mainly for rot proofing textiles, ropes, and so on. It is also used in paints since they are soluble in oils, white spirits, and so on. Quartz crystals coated with $CuSt_2$ was used in the detection of volatile organic compounds. Cobalt stearate ($CoSt_2$) has applications in producing Co nests and mesoporous silica as adhesion promoter.

Metal soaps are compounds of long-chain fatty acids with metals having different valences. Depending on the nature of cation and alkyl chain length, the physical properties of metal carboxylates may vary considerably. For instance, the general surface active materials, sodium and potassium carboxylates, are soluble in water, metal soaps, such as zinc, calcium, copper, magnesium are insoluble or sparingly soluble in water. Because of this property, they are commercially important compounds and find applications in industry, such as driers in paints or inks, components of greases, stabilizers for plastics, in fungicides, catalysts, waterproofing agents, fuel additives, components of creams and additive in drug formulation, and so on [1].

Metal soaps are produced in different forms such as fine powders, flakes, and granules. They are usually produced using precipitation or fusion techniques. Although, precipitation method produces very light, fine powders with a high surface area, and fusion technique produces flakes or pellets. Another issue relating to the product purity is that in precipitation process products with a high purity can be obtained at the expense of washing and filtering cost [2, 3]. In addition to mentioned applications, a number of other uses of polyvalent metal soaps have been suggested. Current interest in low dimensional compounds has led to a number of investigations on the potential application of metal soaps in this area, particularly as Langmuir–Blodgett (LB) multilayers [3].

The synthesis and characterization of metal stearates have commended considerable attention recently owing to their wide range of potential applications. Despite their wide application in industry, the fundamental characteristics of heavy metal soaps and their roles in various industrial preparations need to be investigated systematically. The characterization and structural elucidation of the soaps at room temperature are of considerable importance in elucidating the structure of greases, flatting agents, coatings, and other products made from these soaps. In all these fields, understanding of the phase state of the soaps, and the changes which they may undergo as a result of processing steps or of the action of solvents, may lead to greatly improved products or processes.

14.2 MAGNESIUM STEARATE

The Magnesium stearate $(Mg(C_{17}H_{35}COO)_2)$ is a fine white odorless bulky powder with a very high covering capacity [4]. The $MgSt_2$ is in widespread use as gelling, sanding and anti-sticking agents, stabilizer, lubricant, emulsifier and plasticizer for polymers, in the paint, food, rubber, paper, and pharmaceutical industries [5]. Magnesium soaps are also used as batting agents to reduce the gloss of paints and varnishes and also to thicken paints. Its production throughout the world is essentially based either on the reaction of stearic acid with a magnesium compound such as carbonate, oxide or on the reaction of magnesium chloride with sodium or ammonium stearate in aqueous solution leading to the precipitation of the dihydrate, $C_{36}H_{70}MgO_4 \cdot 2H_2O$. In the field of drug manufacturing, where it is mainly used as a solid lubricant, its lubricating capacity and overall activity in the various pharmaceutical forms in which it is incorporated may vary.

Lubricants are essential to the production of all tablet formulation. As with other classes of pharmaceutical excipients, lubricating agents aid in the manufacture of tablets and ensure that the finished products are of appropriate quality. The $MgSt_2$, with its low friction coefficient and large "covering potential", is an ideal lubricant widely used in tablet manufacturing [6]. Aerosol performance of micronized drug powders was increased when they were coated with $MgSt_2$. The agglomerate strength of the powders was decreased by the coating process [7].

The variability in the physical characteristics of $MgSt_2$ creates problems in its applications. Commercial $MgSt_2$ is a mixture of magnesium salts of different fatty acids, mainly stearic and palmitic, and of others in lower proportions. The magnesium weight fraction in the dried substance is 4% at the least and 5% at the most. The fatty acid fraction contains at least 40% of stearic acid and 90% of stearic and palmitic acids altogether [5]. the $MgSt_2$ exists in different hydration levels, such as anhydrous form, monohydrate, dehydrate, and trihydrate. The endotherm observed at 120°C in DSC curve of anhydrous $MgSt_2$ corresponds to destruction of the lamellar (LAM) mesophase, and melting of the ordered arrangements of the alkyl chains. Up to 190°C, an ordered hexagonal phase with molten alkyl tails exists. At higher temperatures a disordered phase is present [8].

The peak seen in DSC curve around 100°C is due to removal of hydrate water in DSC curve of magnesium stearate monohydrate. Melting endotherm was observed at higher temperatures. Magnesium stearate monohydrate, dehydrate, and trihydrate adsorbed moisture at 96% relative humidity. When the different samples were out gassed at 105°C under vacuum, peaks related to water removal disappeared in DSC curves and melting peaks were observed at 120°C and 130°C for magnesium stearate mononohydrate and dihydrate, respectively. For trihydrate two stage melting endotherm starting at 115°C was observed. X-ray diffraction peaks at two theta values of 3, 5, and 9° were present [9]. Formation of stable semisolid lipogels prepared from $MgSt_2$ and water in liquid paraffin depends on the type of $MgSt_2$ used and preparation technique. The $MgSt_2$ was essentially in crystalline state in semisolid lipogels producing α-crystalline LAM phases [10].

14.3   COBALT STEARATE

Cobalt stearate is synthesized by double decomposition of cobalt acetate with sodium stearate according to the procedure reported in the literature and the thermal characterization and other physicochemical properties of $CoSt_2$ have been reported [11, 12]. The prepared $CoSt_2$ was characterized in terms of its solubility and thermal behavior among others. The solubility of $CoSt_2$ was determined in polar/non-polar and protic/non-protic solvents and the results revealed that $CoSt_2$ is water insoluble but soluble in all the organic solvents like THF, DMF, xylene, and toluene. The FTIR spectra of $CoSt_2$ exhibited absorbance at 1560 cm^{-1} due to asymmetric vibration stretching of the carboxylic group coordinated to the metal ion. The TG curve showed single step decomposition with the initial temperature of degradation at 291.3°C. The cobalt content in the stearate was found to be 6.24% and from the results of the elemental analysis and the molecular formula of $CoSt_2$ was $Co(OOCC_{17}H_{35})_3.2H_2O$ [12]. The $CoSt_2$ exists in three different crystalline phases (Cr_1, Cr_2, and Cr_3), one mesophase (M) and isotropic liquid phase (I). The transition temperatures between the phases are 308.1, 380.9, and 404.4K for Cr_2 to Cr_1, Cr_1 to M and M to I phases, respectively [13].

The $CoSt_2$ had asymmetric and symmetric vibrations of carboxylate groups at 1589 and 1440 cm^{-1} indicating it existed as bridging (polymeric) complexes [14].

The $CoSt_2$ is used as pro-oxidants for polyethylene. The last few decades have seen a tremendous increase in the use of polyethylene particularly in the agriculture and packaging sectors. This has resulted in its increased production and associated plastic litter problem as polyethylene in its pure form is extremely resistant to environmental degradation. An excellent way to render polyethylene degradable is to blend it with pro-oxidant additives, which can effectively enhance the degradability of these materials [11]. Common pro-oxidants include transition metal salts with higher fatty acids, $CoSt_2$ being a typical example. The pro-oxidant activity of cobalt has been attributed primarily to (a) its ability to generate free radicals on polyethylene and (b) decompose the resulting hydro peroxides. The incorporation of these additives is expected to decrease the lifetime of polyethylene in general.

The $CoSt_2$ has applications in producing Co nests [15], mesoporous silica [16], as adhesion promoter [14]. The $CoSt_2$ assembled to micelles acted as soft template for the formation of primary nanorods during solvothermal processing of cobalt acetate and stearic acid. The nanorods then assembled to hollow cobalt spheres with a dense shell. These Co spheres transformed to Co nests constructed by netlike frameworks. Co nests are effective catalyst in hydrogenation of glycerol [15]. Tuning of porous structure of silica containing cobalt was possible using pink $CoSt_2.2H_2O$ as co-template in the synthesis [16]. The $CoSt_2$ was used as adhesion promoter in curing of rubber [14].

14.4   COPPER STEARATE

Copper Stearate is prepared by the interaction of the corresponding soap with copper sulfate solution. It is used mainly for rot proofing textiles, ropes, and so on. It is also used in paints since they are soluble in oils, white spirits, and so on. Quartz crystals coated with $CuSt_2$ was used in the detection of volatile organic compounds [17]. The vibration frequency of the crystal changed as the organic compounds were adsorbed on $CuSt_2$. A super hydrophobic copper surface with 153° contact angle was obtained by coating with $CuSt_2$ by applying DC voltage to copper electrodes immersed in stearic acid solution [18].

The C-O antisymmetric stretching vibration of $CuSt_2$ was at 1588 cm^{-1} [19]. The C-O antisymmetric and symmetric vibrations were at 1583 and 1417 cm^{-1} for mid chain monomethyl branched C17 copper soap with distinct hexagonal columnar mesophase [20].

14.5 POLYVINYL CHLORIDE THERMAL STABILIZER

Metal soaps are the most used heat stabilizers for polyvinyl chloride (PVC). The carboxylate group of the metal salt substitutes the tertiary or allylic chlorine atoms and stops the initiation of dehydrochlorination. The $MgSt_2$ was used in PVC thermal stabilization [21]. Copper-containing layered double hydroxide affected the thermal and smoke behavior of PVC [22]. Metal dicarboxylates were effective in retarding the dehydrochlorination reaction of PVC [23].

14.6 CONCLUSION

The preparation, characterization, and applications of metal soaps of stearic acid prepared from second group element magnesium and transition metal elements cobalt and copper were reviewed in the present study. They were mostly prepared by double decomposition reactions. They were crystalline solids with LAM structure. While magnesium was used mainly for its lubricating property, the catalytic and surface properties of $CuSt_2$ and $MgSt_2$ allowed them to be used as templates for mesoporous compounds and additives to polymers as either pro-oxidants or fire retardants.

KEYWORDS

- Cobalt stearate
- Copper stearate
- Magnesium stearate
- Metal soaps
- PVC thermal stability

REFERENCES

1. Elvers, B, Hawkins S., and Schulz G. *Ullmann's Encyclopedia of Industrial Chemistry*, Wiley: Weinheim, Germany (1990).
2. Gonen, M., Balkose, D., Inal F., and Ulku, S. Zinc Stearate Production by Precipitation and fusion Processes. *Ind. Eng. Chem. Res.*,**44**(6), 1627–1633 (2005).
3. Gonen, M., Ozturk, S., Balkose, D., and Ulku, S. Preparation and Characterization of Calcium Stearate Powders and Films Prepared by Precipitation and Langmuir-Blodgett Techniques. *Ind. Eng. Chem. Res.*, **49**, 1732–1736 (2010).

4.   Lower, E. S. Magnesium stearate: A review of its uses in paints, plastics, adhesives, and related industries, *Pigment & Resin Techn.*, **10**(12), 7–25 (1993).
5.   Bracconi, P., Andres, C., and Ndiaye, A. Structural properties of magnesium stearate pseudopolymorphs: effect of temperature. *Int. J. Pharm.*, **262**(1–2), 109–124 (2005).
6.   Patel, S., Kaushal, A. M., and Bansal, A. K. Lubrication Potential of Magnesium Stearate studied on Instrumented rotary tablet press. *AAPS PharmSciTech.*, **8**(4), E1–E8 (2007).
7.   Zhou, Q. T., Qu, L., Larson, I., Stewart, P. J., and Morton, D. A. V. Improving aerosolization of drug powders by reducing powder intrinsic cohesion via a mechanical dry coating approach. *Int. J. Pharm.*, **394**(1–2), 50–59 (2010).
8.   Wakabayashi, K. and Register, R. A. Phase Behavior of Magnesium Stearate Blended with Polyethylene Ionomers. *Ind. Eng. Chem. Res.*, **49**(23), 11906–11913 (2010).
9.   Koivista, M., Jalonen, H., and Lehto, V. P. Effect of Temperature and humidity on vegtable grade Magnesium Stearate. *Powder Techn.*, **147**, 79–85 (2004).
10. Sheikh, K. A., Kang Y. B., Rouse, J. J., and Eccleston, G. M. Influence of hydration state and homologue composition of magnesium stearate on the physical chemical properties of liquid paraffin lipogels. *Int. J. Pharm.*, **411**(1–2), 121–127 (2011).
11. Roy, P. K., Surekha, P., Rajagopal, C., and Choudhary, V. Thermal degradation studies of LDPE containing cobalt stearate as pro-oxidant. *Polymer Letters.*, **1**(4), 208–216 (2007).
12. Roy, P. K., Surekha, P., Rajagopal, C., and Choudhary, V. Effect of cobalt carboxylates on the photo-oxidative degradation of low-density polyethylene, *Poly. Degrad. Stab.*, **91**, 1980–1988 (2006).
13. Van Hecke, G. R., Nakamoto, T., Clements, T. G., and Sora M. Adiabatic calorimetry of the metallomesogen purple cobalt stearate $Co(O_2CC_{17}H_{35})_2$. *Liq. Cryst.*, **30**(7), 831–837 (2003).
14. Jona, E., Ondrusova, D., Pajtasova, M., Simon, P., Michaler, J. A study of curative interactions in the presence of cobalt (II) stearate. *J. App. Polym. Sci.*, **81**(12), 2936–2943 (2001).
15. Liu, Q. Y., Guo, X. H., Li, Y., and Shen, W. Synthesis of Hollow Co Structures with Net-like Framework. *Langmuir.*, **25**(11), 6425–6430 (2009).
16. Grosshans-Vieles, S., Tihay-Schweyer, F., Rabu, P., Paillaud, J. L., Braunstein, P., Lebeau, B., Estournes, C., Guille, J. L., and Rueff J. M. Direct synthesis of mesoporous silica containing cobalt: A new strategy using a cobalt soap as a co-template. *Microporous Mesoporous Mat.*, **106**(1–3), 17–27 (2007).
17. Filippov, A. P., Strizhak P. E., and Il'in, V. G. Quartz crystal microbalance modified with Cu(II) stearate and octadecylamine co-ordination chemical compounds for detection of volatile organic compounds. *Sens. Actuators B.*, **126**(2), 375–381 (2007).
18. Huang, Y., Sarkar, D. K., and Chen, X. G. A one-step process to engineer superhydrophobic copper surfaces. *Mater. Lett.*, **64**(24), 2722–2724 (2010).
19. Satake, I. and Matura, R. Studies with copper soaps, *Colloid-Zeitshrift*, **176**, 31–38 (1960).
20. Corkery, W. R. A variation on Luzzati's soap phases. Room temperature thermotropic liquid crystals, *Phys. Chem. Chem. Phys.*, **6**, 1534–1546 (2004).
21. Simon, P., Oremusava, P., Valko, L., and Kovarik ,P. Influence of metal stearates on thermal stability of poly(vinyl chloride)II. *Magnesium stearate. Chem. Pap.*, **45**(1),127–134 (1991).
22. Zhu, H., Wang, W., and Liu, T. Effects of Copper-Containing Layered Double Hydroxide on Thermal and Smoke Behavior of Poly(vinyl chloride). *J. App. Polym. Sci.*, **122**(1), 273–281 (2011).
23. Liu, Y. B., Liu, W. Q., and Hou, M. H. Metal dicarboxylates as thermal stabilizers for PVC, *Poly. Degrad. St.*, **97**, 1565–1571 (2007).

CHAPTER 15

CONTROL OF THE PARTICLE SIZE AND PURITY OF NANO ZINC OXIDE

FILIZ OZMIHÇI OMURLU and DEVRIM BALKÖSE

CONTENTS

15.1 INTRODUCTION

The effects of template, mechanical mixing, and/or ultrasound mixing on the size of the zinc oxide (ZnO) crystals obtained by precipitation at 30°C from aqueous zinc chloride ($ZnCl_2$) and potassium hydroxide (KOH) solutions were investigated by 2^k factorial design. The precipitation method is employed to synthesize nano ZnO particles. The monodisperse nano ZnO having 29 nm particle size was produced by adding triethylamine (TEA) and applying simultaneously mechanical and ultrasound mixing. The surface area and the density of the powder were 21 m^2/g and 4.8 g/cm^3. It contains 5.2% impurities present as CO_3^{-2} and bound OH^- groups. The volumetric resistivity was found as 1.3×10^7 Ohm cm. The absorption spectrum of the powder showed absorption peak at 353 nm. The room temperature fluorescence spectrum of the powder revealed a strong and sharp ultraviolet (UV) emission band at 391 nm due to free exciton or bound exciton of ZnO and a weak and broad violet emission band at 405 nm due to zinc (Zn) vacancies.

The nano crystalline materials have found an increasing research area on the material science, chemical, and electronic engineering during the past years. The ZnO is composed of tetrahedrally coordinated O^{2-} and Zn^{2+} ions, stacked along the c-axis. It is a semiconducting material with a band gap of about 3.2 eV and a large exciton binding energy of 60 meV [1, 2]. It is an important material due to its unique properties of near-UV emission value and has applications as electrical conductive and optically transparent additive in a polymer matrix [3-6].

There are many different methods for the preparation of nano ZnO powder and precipitation method is a good choice in the industrial point of view because of the low growth temperature and its potential for scale-up [7-9]. This synthesis method has the advantage of preparing highly crystallized particles with narrow size distribution and high purity without further treatment at higher temperature. The size and morphology can be controlled by controlling reaction temperature, reaction time, and additives [10-15].

The precipitation method was used to prepare ZnO nano sheets by sono chemical method using $ZnCl_2$ and NaOH as a precursor under constant stirring and at pH of 13 [16]. Different shapes of ZnO powders were prepared with sonochemical synthesis, which were in nanorod, trigonal, and dentritic shapes. The XRD patterns of the synthesized powders were in good agreement with the hexagonal wurtzite structure of ZnO [16].

Particles with different morphologies and sizes were obtained by adjusting the templates [17]. Nano ZnO particles were synthesized by Wei and Chang at room temperature and at 50°C under ultrasonic condition by hydrothermal method by using cetyl trimethyl ammonium bromide and triethanolamine surfactants [16]. While the bulk ZnO obtained by using only ultrasonic water bath treatment at 50°C had 454 nm particle size, the size was reduced approximately to 28–60 nm when surfactants were used [18]. Flower like ZnO microstructures were obtained from aqueous zinc nitrate, sodium hydroxide (NaOH) and TEA at 180°C. The TEA played a dual role both as the complexing agent and the alkaline reagent [19]. The ZnO was also obtained from aqueous zinc acetate ($Zn(O_2CCH_3)_2$) and TEA solutions [20]. Lai et al.

also investigated the hydrothermal synthesis of ZnO powder with the assistance of ultrasonic treatment. At ambient conditions, the aqueous solution of precursors that contains $Zn(O_2CCH_3)_2$ and NaOH was very clear. However, after ultrasonic treatment the clear solutions become cloudy and a white precipitate was observed. Due to acoustic cavitation, H_2O decomposed into H^- and OH^- radicals. The radicals react with Zn^{+2} ions to form ZnO and water molecules. The ultrasonic energy also converts $Zn(OH)^{-2}$ to $Zn(OH)_2$ and $Zn(OH)_2$ to ZnO [21].

The ZnO can also be obtained by hydrothermal transformation of zinc hydroxide chloride (ZHC). For instance, Zhang and Yanagisawa studied metal hydroxide salts (MHS) with layered structures. The common synthesis methods of the MHS include coprecipitation method and the obtained products usually have the lamellar morphologies such as films, sheets, and plates. In their chapter, ZHC sheets were synthesized the by a simple hydrothermal method. After thermal treatments, the ZHC sheets were transformed to sheet like dense ZnO [22].

The ZnO can exhibit unique optical, photocatalytic, piezoelectric, and pyroelectric properties, produces an efficient blue-green luminescence, and displays excitonic UV laser action. The ZnO has a relatively high absorption band starting at 380 nm [23] and extending into the far-UV. In addition to its excellent UV absorption characteristics, ZnO has several other advantages as a UV and visible light emitting additive material, it does not migrate, it is not degraded by absorbed light and in many cases it may improve mechanical, optical, and electrical properties of polymers which they are added.

Material synthesis involves the control of the particle size and morphology since the electrical and optical properties of materials depend both on the size and the shape of the particles. Therefore, morphologically controllable synthesis of ZnO having nano or microstructures is crucially important to answer the demand for exploring the potentials of ZnO [24].

The present investigation was focused on the preparation of mono dispersed nano ZnO powder by a hydrothermal precipitation method at low temperature. The effects of mechanical mixing, sonication, and using a template on the particle size were investigated by the aid of statistically designed experiments. The particle size distribution, morphology, and crystal structure of the powders were determined. Pure nano ZnO powder having the smallest particle size was characterized in more in more detail by the measurement of the volumetric resistivity, the density, and the optical properties.

15.2 MATERIALS AND METHODS

15.2.1 MATERIALS

The analytical grade chemicals, $ZnCl_2$ (98%, Aldrich), KOH (Pancreac), and TEA $(((CH_3CH_2)_3N)$, Merck), were used for the preparation of ZnO powders throughout the experimental study. Millipore ultrapure fresh water (18 ohm cm) was used in all steps of the synthesis. The TEA was used as a template.

15.2.2 EXPERIMENTAL DESIGN

The temperature, the concentration of the precursors, and the template type were held constant in preparation of the nano ZnO powder. The addition of the template (0.02 moldm^{-3}concentration), the sonication (for 30 minutes period), and the mechanical mixing (at 500 rpm) were the main factors. The particle size of the powders was chosen as the response. The experimental factors and the categorical levels are as given in Table 1. The experiments were performed by considering a 2^3 full factorial design consisting of 8 experiments as shown in Table 2. The analysis of variance full factorial design was carried out using Design of Expert 8.0.1.0.

TABLE 1 Experimental factors and levels investigated for optimum particle size of ZnO powder

Factors Name	Factors Symbol	Low	High
Template	a	0	1
Mechanical Mixing	b	0	1
Sonication	c	0	1

TABLE 2 Full factorial experimental design

Template	Sonication	Mechanical Mixing	Experimental Mean Particle Size (nm)	Predicted Mean Particle Size (nm)
1	1	1	29	78
0	0	0	650	433
0	0	1	1312	392
1	0	1	77	54
1	1	0	738	225
0	1	1	148	695
0	1	0	137	603
1	0	0	373	304

15.2.3 TYPICAL ZNO SYNTHESIS METHOD

100 cm^3 solution having 0.2 mol dm^{-3} KOH and 0.02 mol dm^{-3} template TEA was added instantly to 100 cm^3 0.1 mol dm^{-3} ZnCl$_2$ solutions. Control experiments without

template TEA addition were also made. The ultrasonic treatment was applied by immersing the beaker containing the reactants in an ultrasonic bath (Elma, Transsonic 660/H) at 30°C for 30 min. Mechanical mixing at 500 rpm was made using IKA RW 20 mechanical mixer. The solid and liquid phases were separated by centrifuging using Hettich, Rotofix 32. The solid phase was then washed for three times with water and dried at 50°C for 15 hr.

15.2.4 CHARACTERIZATION OF ZNO POWDERS

The phase identification and the crystal size of ZnO powders were determined by X-Ray diffractometer (Philips X'Pert diffractometer, Cu-K$_\alpha$ radiation). The powder morphology was determined by Scanning electron microscope (SEM) with Philips XL-30S FEG. The particle size distribution of the powders dispersed in water was determined by Zeta Sizer (Malvern Instruments 3000 HSA).

The detailed characterization of nano ZnO powder obtained by the template addition, the mechanical, and the ultrasonic mixing.

The helium pycnometer (Quantachrome Co. Ultrapycnometer 1000) was used to determine the density of the powder. The N$_2$ adsorption/desorption analysis were performed to determine the surface area of the powder (ASAP Micromeritics 2000). The impurities in the monodisperse nano ZnO powder was determined by fourier transform infrared (FTIR) spectroscopy using Shimadzu FTIR-8201 by KBr disc method. The ZnO pellets having 2.5 cm diameter and 2 mm were prepared from the nano ZnO powder by pressing under 10 MPa pressure. The silver contacts were formed by thermal evaporation of silver on both surfaces of the ZnO pellet for the resistivity measurement. The volumetric resistivity of the pellet was determined by changing potential between – 50 V and + 50 V and recording I-V data with Keithley 2420. The absorption spectrum of a dilute suspension of ZnO powder was obtained by using the UV-Vis spectrometer Perkin Elmer Lambda 45. The Fluorescence spectrum was obtained by using the fluorescence spectrometer Varian Cary Eclipse by using a ZnO pellet. The emission data were recorded in the 390nm and 600 nm range after exciting the sample at 380 nm for 15 sec.

15.3 DISCUSSION AND RESULTS

15.3.1 FACTORIAL DESIGN FOR PARTICLE SIZE OF ZnO

Table 2 gives the list of kinetic results for full factorial experimental design. The effects of the sonication time, the mechanical mixing and the template addition on the particle size of the powders were found according to fitted regression model with 0.5 confidence interval to experimental particle size data. Equation 1 is the fitted model for the particle size, as a function of the presence of template, a, the sonication, b and the mechanical mixing, c.

$$D = 433 - 128.8a - 170.0b - 41.5xc + 249.3ab - 209.8ac - 133.0bc + 29.8abc(1)$$

A factor was designated by "1" or "0" if it was present or not in the system respectively. Therefore, the values either 1 or 0 should be used for the variables a, b, and c when predicting the particle size using Equation 6.

The model results for particle size indicate that, template addition, sonication, and mechanical mixing have a negative effect on the particle size. All the main effects and interaction effects on particle size of the powder are significant. However, sonication has the largest negative effect on the response and the template affects the particle size more than the mechanical mixing. The fitted model predicts that the interaction between the template and the sonication is the most important interaction parameter for increasing the particle size. Both "sonication and mechanical mixing" and "template addition and mechanical mixing" make a synergistic effect for minimizing the particle size. To obtain minimum particle size the template addition, the sonication, and the mechanical mixing should be applied simultaneously during precipitation of the powder as seen in Table 2. The predicted particle size of each sample from Equation 6 is as reported in Table 2. However, the model predicts the smallest particle size for the template addition and the mechanical mixing case.

15.3.2   PURITY OF ZNO PARTICLES

The factorial design method has shown that the simultaneous template addition, sonication, and mechanical mixing will result in the minimum sized nanoparticles. However, the purity of ZnO is another fulfillment that should be met. Thus the effects of these three variables on the purity of the product ZnO were also investigated.

15.3.3   EFFECT OF TEMPLATE

In the present study, to investigate the effect of template on particulate properties TEA was added to the reaction medium. Figure 1 gives the X-ray diffraction (XRD) patterns of the powders prepared with and without the template for the case without any mixing. The powder synthesized in the absence of template was a complex compound as depicted in Figure 1(a). The XRD pattern belongs to a MHS with layered structure [22]. The product was confirmed to be $Zn_5(OH)_8Cl_2H_2O$ (ZHC) (JCPDS Card No: 07-0155) and no other impurity phases were found. In the XRD diagram of the TEA added powder in Figure 1(b) there are diffraction peaks at 2θ values of 31.6°, 34.26°, 36.1°, 47.35°, 56.4°, 62.66°, 66.2°, 67.76°, 68.86°, 72.2°, and 76.78°. In the XRD pattern of ZnO powder reported in joint committee on powder diffraction standards (JCPDS) Card No: 79-0207 there are peaks at 2θ values of 31.7°, 34.4°, 36.3°, 47.5°, 56.6°, 62.3°, 66.5°, 67.9°, and 69.1°. Thus the powder synthesized with TEA addition was pure ZnO. The ratio of the intensity of the peak of 002 planes at 34.3° to the intensity of the peak of the 101planes at 36.1° is 0.44 for the bulk wurtzite [25]. The sample prepared with the template had hexagonal ZnO crystals preferentially oriented in 002 directions since this ratio is 0.65. The SEM images and particle size distributions of the template added and template free powders are given in Figure 2 and Figure 3 respectively. As shown in Figure 2(a) hexagonal shaped sheets were obtained when there was no template in the medium. The sheets are 1.5–2 μm in size and their thickness is around 10 nm. Template free sample had 61, 25, and 14 mass % of Zn, O, and

Cl respectively as determined by Energy dispersive X-ray (EDX) analysis. Using the EDX and XRD data it was concluded that ZHC sheets were synthesized when there was no template. The mean particle size of ZHC sheets was around 650 nm as seen in Figure 3(a).

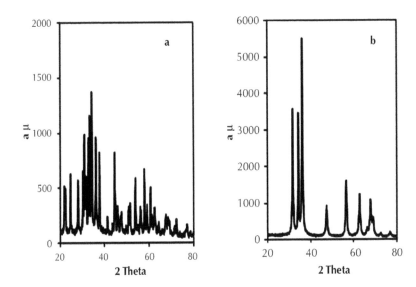

FIGURE 1   The XRD pattern of powders prepared without any mixing and (a) without TEA (b) with TEA.

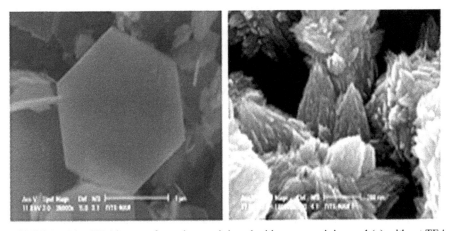

FIGURE 2   The SEM image of powder precipitated without any mixing and (a) without TEA (b) with TEA.

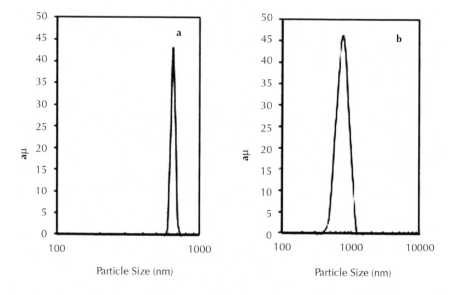

FIGURE 3 Particle size distribution of powder without any mixing and (a) without TEA (b) with TEA.

As depicted in Figure 2(b) the template TEA added powder was agglomerated to flower like particles similar to the ones made with TEA at 180°C by previous work-ers[19]. The flower like particles formation at 30°C in the present study indicated that the thermal treatments at high temperatures are not necessary for this purpose. The average particle size of the powder was found around 373 nm as shown in Figure 3(b) and as reported in Table 2.

Template addition had a direct influence on the type of the product obtained. In mixing $ZnCl_2$ and KOH solutions consecutive reactions shown in Equations 2–Equation 5 occur.

$$ZnCl_2(aq) + KOH(aq) \rightarrow ZnOHCl\downarrow + KCl \tag{2}$$

$$ZnOHCl\downarrow(s) + KOH(aq) \rightarrow ZnO\downarrow + H_2O + KCl \tag{3}$$

$$2Zncl_2(aq) + 4KOH(aq) \rightarrow Zn(OH)_4^{-2} Zn^{+2} + 4KCl \tag{4}$$

$$Zn(OH)_4^{-2} Zn^{+2} \rightarrow 2Zno \downarrow +2H_2O \tag{5}$$

The zinc hydroxychloride (ZnOHCl) sheets formed if the template was not used as shown in Equation 2. Large sheets of ZnOHCl were obtained due to faster growing of crystals than nucleation. The TEA molecules associate in one dimensional chains [19] and act as templates for small sized ZnO particles. The TEA also reduces the surface tension of water [26]. Template TEA creates nucleation centers by complexation with Zn ions and large numbers of $Zn(OH)_4^{-2}Zn^{+2}$ nuclei form and during slow crystal growth they are transformed to ZnO by reactions shown in Equation 4 and Equation 5.

Two experiments were made to understand the effect of template addition for the case of applying both the sonication and the mechanical mixing. The first powder was prepared without the template using sonication and mechanical mixing and the other one was prepared with the template using both sonication and mechanical mixing. The XRD patterns, SEM images, and the particle size distributions of the powders are as given in Figure 4, Figure 5, and Figure 6 respectively.

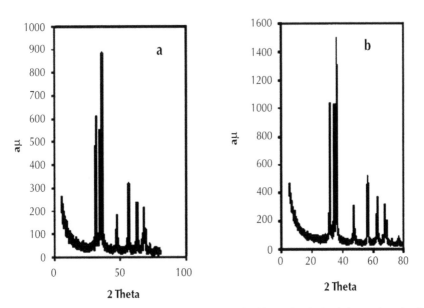

FIGURE 4 The XRD pattern of the powder prepared with mechanical mixing and sonication (a) without TEA (b) with TEA.

FIGURE 5 The SEM image of the powder prepared with mechanical mixing and sonication (a) without TEA (b) with TEA.

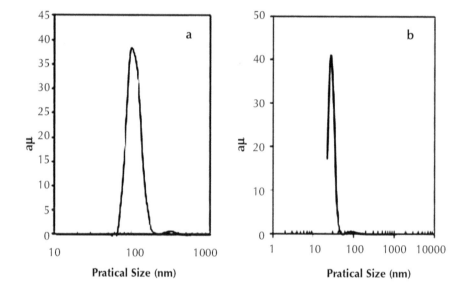

FIGURE 6 Particle size distribution of the powder prepared with mechanical mixing and sonication (a) without TEA (b) with TEA.

The XRD patterns of the powders in Figure 5 were found to be similar to each other and to that of ZnO. The ratio of the intensity of the peak of 002 planes at 34.3° to the intensity of the peak of the 101 planes at 36.1° is 0.61 and 0.7 for template free and template added samples. This indicated hexagonal crystals of template added ZnO were more oriented in 002 direction than the ZnO without template. However, the SEM image of TEA free sample shown in Figure 5(a) is polydisperse in particle size. Very small and very large particles were present in sheet like form. However, only a small fraction of larger particles was present when there was no template as seen in Figure 6(a). The average particle size was

148 nm when there was no template. On the other hand primary particles were ag-glomerated to form particles having the shape of a droplet are observed in the SEM image of the TEA added powders in Figure 5(b). The particle size distribution of the TEA added powder seen in Figure 6(b) confirms the monodispersity and the average particle size was found as 29 nm.

The application of sonication and mechanical mixing simultaneously reduced the particle size compared to unmixed case and the difference between particle sizes of the powders obtained with and without template addition was also reduced. With-out template ZnOHCl sheets were obtained when there was no mixing. When the reactants were mixed, due to faster growth of crystals than their nucleation, sheet like precipitates were formed. On the other hand TEA created nucleation centers and large numbers of nuclei formed and slower crystal growth occurred reducing the particle size.

15.3.4 MIXING EFFECT

The sonication and mechanical mixing were used to understand the mixing influence on particle size and morphology. The reaction temperature (30°C) and concentrations of the precursors and pH (10.5) were held constant and no template was used. Figure 7, Figure 8, and Figure 9 show the XRD patterns, SEM image, and particle size dis-tribution of precipitates obtained by applying only sonication and mechanical mixing respectively.

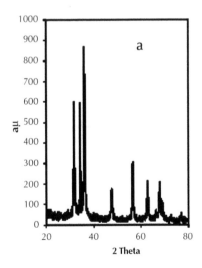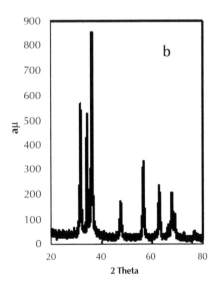

FIGURE 7 The XRD pattern of precipitates obtained without template and with (a) only sonication (b) only mechanical mixing.

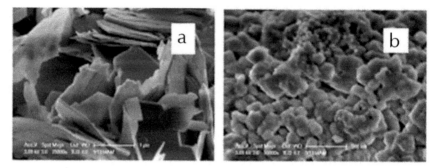

FIGURE 8 The SEM image of precipitates obtained without template and with (a) only sonication (b) only mechanical mixing.

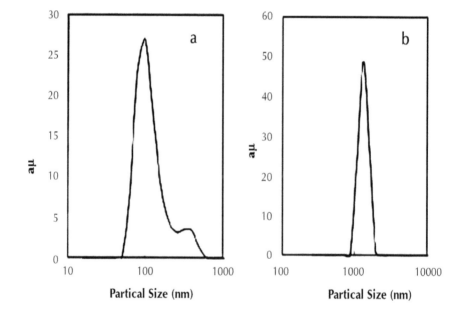

FIGURE 9 Particle size distribution of the precipitates obtained without template and with (a) only sonication (b) only mechanical mixing.

The XRD patterns of the precipitates had sharp peaks at 2θ values $31.6°$, $34.26°$, $36.1°$, $47.35°$, $56.4°$, $62.66°$, $66.2°$, $67.76°$, $68.86°$, $72.2°$, and $76.78°$. The peaks observed were identical with the characteristic XRD pattern of ZnO powders (JCPDS Card No: 79-0207). The ratios of the intensity of the peak of 002 planes at $34.3°$ to the intensity of the peak of the 101 planes at $36.1°$ are 0.68 and 0.5 for sonified and mechanically mixed samples. These indicated hexagonal crystals of ZnO obtained by sonication and mechanical mixing were oriented in 002 direction. Sheet like and polydisperse crystals are seen in the SEM image of the sonified precipitate in Figure 8(a). Mechanically mixed precipitate's SEM image in Figure 8(b) shows aggregated sphere like crystals. The particle size distribution of sonified and mechanically mixed samples seen in Figure 9(a) and Figure 9(b) indicated that mechanically mixed powder had larger crystal size. The sonified powder's size distribution in Figure 9(a) is bidisperse. A small fraction of the particles were larger in size. The monodisperse size distribution was obtained for only mechanically mixed particles as seen in Figure 9(b). The mean particle sizes for only sonified and only mechanically mixed samples were found as 137 nm and 1312 nm respectively. The results showed that applying only sonication and only mechanical mixing was not enough to have a monodisperse nano sized ZnO powder.

Experiments were also done to analyze the effect of mixing on template added precipitates. The XRD pattern, SEM image, and particle size distribution of the template TEA added precipitates are given in Figure 10, Figure 11, and Figure 12 respectively for sonication and mechanical mixing applied samples.

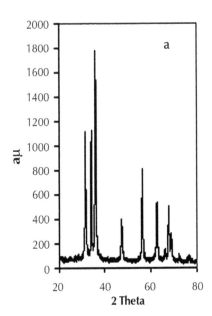 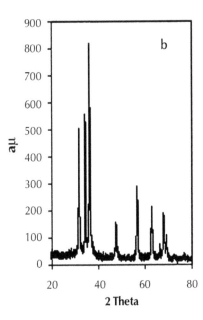

FIGURE 10   The XRD pattern of template added precipitates prepared with (a) only sonication (b) only mechanical mixing

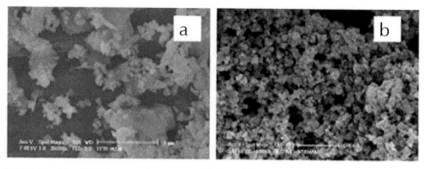

FIGURE 11   The SEM image of template added precipitates prepared with (a) only sonication (b) only mechanical mixing

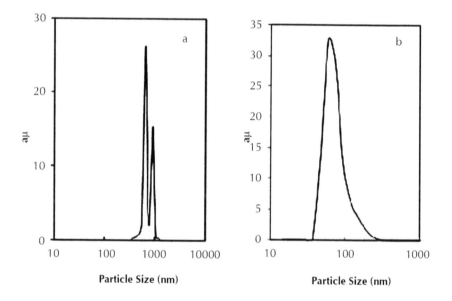

FIGURE 12   Particle size distribution of template added precipitates prepared with (a) only sonication (b) only mechanical mixing.

The template added precipitates XRD patterns give the pattern of typical ZnO as seen in Figure 10. In the SEM image of sonified powder in Figure 11 small and big shapeless particles and flake like structures are seen. The monodisperse particles smaller than 100 nm is seen in the SEM image of the mechanically mixed powder in Figure 11(b). The particle size distribution of sonified powders seen in Figure 12(a) was bidiperse and the mean particle size was determined to be 738 nm. This value is at the same order with the size (454 nm) of ZnO particles synthesized at 50°C under ultrasonic conditions Wei and Chang [16]. However, the particle size distribution range of the mechanically mixed powders found between 30–300 nm and the mean particle size of the mechanically mixed powder was 77 nm.

15.3.5 CHARACTERIZATION OF NANO ZNO POWDER SYNTHESIZED BY TEMPLATE ADDITION, MECHANICAL MIXING, AND SONICATION

The N_2 adsorption isotherm of nano ZnO powder is given in Figure 13. The Brunauer–Emmet–Teller (BET) surface area of the nano ZnO powder was determined to be 21 m^2/g using the data in Figure 13. If spherical particles are assumed and N_2 gas is adsorbed on the external surface of the particles, this corresponds to a particle size of 28 nm, confirming the average particle size, 29 nm determined by Zeta Sizer. The density of the powder was 4.8 g/cm^3 as determined by helium pycnometry. This value is lower than the density of pure ZnO, 5.1 g/cm^3. There were impurities in ZnO powder causing the density to be lower. The EDX showed that the surface composition of the powders was 81% Zn, 14 % O and 5 % C in mass. On the other hand pure ZnO should have 80% Zn and 20% O. Since there is no C-H stretching vibration peak at 2985 cm^{-1} in the FTIR spectrum of the powder shown in Figure 14, there was no TEA in the samples. Thus the presence of C in the powder could be due to the adsorbed CO_2 from atmosphere.

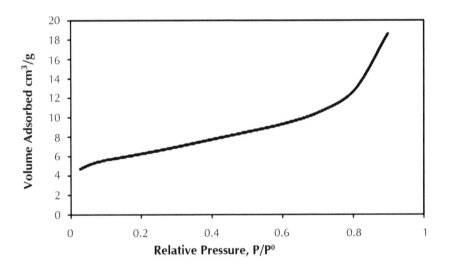

FIGURE 13   The N_2 adsorption isotherm of ZnO at 77K.

Peaks were present at 3400 cm^{-1} and 1660 cm^{-1} corresponding to hydrogen bonded OH stretching and bending vibration of H_2O respectively in the FTIR spectrum of the ZnO sample in Figure 14. The peaks at 908 cm^{-1}, 707 cm^{-1} belonged to OH group which may due to presence of $Zn(OH)_2$. The broad peak in the range 1517 and 1390 cm^{-1} could be attributed to v_3 stretching mode of carbonate ions. There were also peaks at 835 cm^{-1} (v_2 mode of carbonate), at 737 (sh) and 710 cm^{-1} (v_4 mode of carbonate) in the spectrum. The source of carbonate ions in nano ZnO could be the adsorbed CO_2

from atmosphere during preparation and drying of the particles. The basic pH of the precipitation medium caused absorption of CO_2 from air. The large surface area of the nano particles also allowed the adsorption of CO_2 from air during drying of the particles [23]. The CO_2 adsorption on ZnO was also reported by other workers [27, 28]. The Zn-O stretching vibrations at 473cm⁻¹ and 532 cm⁻¹ had the highest absorption value and indicated that the powder was mainly ZnO.

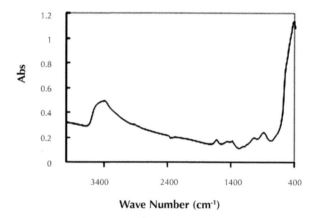

FIGURE 14   The FTIR spectrum of nano ZnO powder.

The thermogravimetric (TG) and differential thermal analysis (DTA) curves of the nano ZnO powder dried at 50°C are shown in Figure 15. The mass loss of the nano ZnO powder was 5.2% at 1000°C as seen in TG curve of the nano powder. This could be due to elimination of water and CO_2 from the sample on heating.

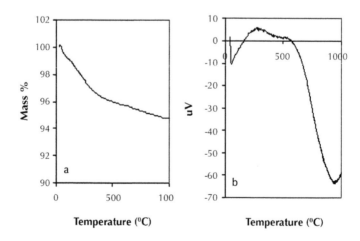

FIGURE 15   Thermal characterization of nano ZnO powder (a) TGA (b) DTA.

The DTA curve has three endothermic peak maxima at 59°C, ~430°C, and 940°C, which are related to the release of adsorbed water, dehydration of $Zn(OH)_2$ and decomposition of the other impurities such as carbonates in the powder and sintering of ZnO particles respectively. The presence of $Zn(OH)_2$ was also detected by the FTIR spectroscopy and DTA. Thus the mass loss in TGA was due to drying of ZnO and the dehydration of $Zn(OH)_2$ and evolution of adsorbed CO_2. The endotherm at 940°C could be due to the sintering of ZnO particles to each other. The melting point of ZnO is 2200°C, but the surface of the nano particles melts at a much lower temperature due to the high surface to volume ratio and sintering occurs at much lower temperatures.

RESISTIVITY OF THE POWDERS

Figure 16 shows the current versus sweeping voltage (I-V) for nano ZnO powder. The curve was linear with a very high "0.9984" correlation coefficient. The resistivity value was calculated according to Ohm's law using the inverse of the slope of the I-V line. The volumetric resistivity was found as 1.3×10^7 ohm cm. The resistivity of ZnO thin films was reported as 2.8×10^{-4} ohm cm [29] and for films prepared by spray pyrolysis and 1.4×10^{-4} ohm cm to 2×10^{-4} ohm cm independent of the preparation method [30]. The resistivity of the nano ZnO pellet was much higher than those of the thin films. However, the prepared ZnO was a semi conductive material that can be used in moderately conductive applications.

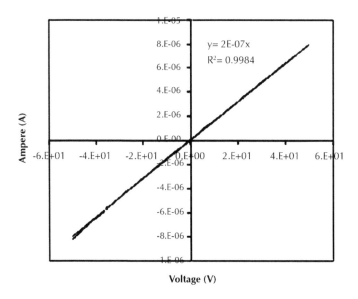

FIGURE 16   Sweeping voltage versus current values for nano ZnO powder.

LIGHT ABSORPTION BY THE POWDER

The ZnO has ability to absorb UV light. The peak maximum value is 353 nm in absorption spectrum of nano ZnO powder as seen in Figure 17(a). The UV-A light was strongly absorbed by the ZnO powder as observed by previous investigators [23].

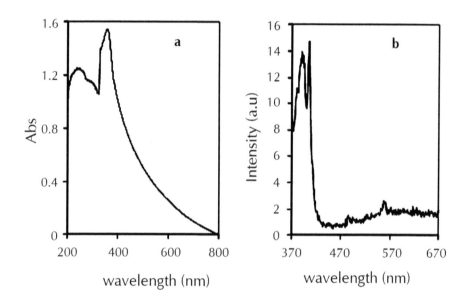

FIGURE 17   (a) Absorption and (b) fluorescence spectra of nano ZnO powders.

LIGHT EMISSION BY THE POWDER

Two peaks are observed in the fluorescence spectrum of the dry pressed pellet of ZnO powder as seen in Figure 17(b). The peak at 391 nm corresponds to free exciton or bound exciton of ZnO in the UV region. A violet luminescence which is attributed to the Zn vacancies is observed at 405 nm. This wavelength is higher than that of ZnO powders which has strong UV luminescence at 398 nm obtained by combustion technique [23]. However, the green and yellow luminescence which was mentioned for complex defects [28] was not observed for this powder. There was also no blue band emission attributed to singly ionized oxygen vacancies [31].

15.4   CONCLUSION

The experimental design was used to find out the most important variables affecting the size of the particles in ZnO preparation. It was found that minimum sized particles were obtained by TEA addition, sonication, and mechanical mixing. Template addition creates nucleation centers and a large number of nuclei forms and crystal growth stops at nano size level due to depletion of the ions in solution. Thus nano particles of ZnO were obtained. Mixing influenced the homogenous dispersion of the chemicals and nano ZnO crystals with a very narrow size distribution oriented in 002 directions were obtained.

The nano powder was synthesized using TEA under mechanical stirring and ultrasonic treatment simultaneously at 30°C. The crystals of the powder had 29 nm size. The XRD pattern gave the characteristic peaks of ZnO. However, there were some peaks related with $Zn(OH)_2$ and CO_3^{-2} in its FTIR spectrum. It was 95% ZnO.

Moderately conductive nano ZnO powder was obtained having 1.3×10^7 ohm cm electrical resistivity. The absorption spectrum of the powder showed absorption peak at UV-A region. The room temperature fluorescence spectrum of the powder revealed a strong and sharp UV emission band at 391 nm and a weak and broad violet emission band at 405 nm showing to free exciton or bound exciton of ZnO in the UV region and Zn vacancy, respectively.

The ZnO powder obtained by TEA addition, sonication, and mechanical mixing can be used as a polymer additive to produce statically dissipating composites with luminescence properties.

KEYWORDS

- **Electrical resistivity**
- **Luminescence**
- **Nano zinc oxide**
- **Precipitation**
- **Triethylamine**

REFERENCES

1. Wang, Z. L. *Journal of Physics: Condensed Matter.*, **16**, R829–R858 (2004).
2. Satyanarayana, V. N. T., Kuchibhatla, A. S., Karakoti, D., and Seal, S. One dimensional nanostructured materials. *Progress in Materials Scienc.*, **52**, 613–699 (2007).
3. Bangal, M., Ashtapure, S., Marathe, S., Ethiraj, A., Hebalkar, N., Gosavi, S. W., Urban, J., and Kulkarni, S. K. Semiconductor Nanoparticles. *Hyperfine Interactions*, **160**, 81–94 (2005).
4. Tang, E., Tian, B., Zheng, E., Fu, C., and Cheng, G. Preparation of Zinc Oxide Nanoparticle via Uniform Precipitation Method and Its Surface Modification by Methacryloxypropyltrimethoxysilane. *Chem. Eng. Comm.*, **195**,479–491 (2008).
5. Vergés, M., Martínez-Gallego, M., Lozano-Vila, A., and Díaz-Álvarez, J. A. Microscopy Techniques Applied to the Study of Zinc Oxide Microcrystalline Powders Synthesis. In: Current Issues on Multidisciplinary Microscopy Research and Education, A., Méndez-Vilas, and L., Labajos-Broncano (Eds.), FORMATEX Microscopy Book Series (N° 2), (2005).

6. Triboulet, R., Sanjosé, V. M., Zaera, R. T., Tomas, M. C. M., and Hassani, S. Zinc Oxide –
    A Material for Micro- and Optoelectronic Applications. In: The Scope of Zinc Oxide Bulk
    Growth. E.,Terukov (Ed.), VCH: Springer: Netherlands, (2005).

7. Karpinski, P. H. and Wey, J. S. Precipitation processes. In: Handbook of Industrial Crystal-
    lization (2nd ed.), A. S., Myerson (Ed.), Butterworth-Heinemann, Wolburn, (2002).

8. Zhang, W. and Yanagisawa, K. Hydrothermal Synthesis of Zinc Hydroxide Chloride
    Sheets and Their Conversion to ZnO. *Chemical Materials*, 2 19, 2329–2334(2007).

9. Yoshimura, M. and Byrappa, K. Hydrothermal processing of materials: past, present and
    future. *Journal of Materials Science*, **43**, 2085–2103 (2008).

10. Sue, K., Kimura, K., Murata, K., and Arai, K. Effect of Cations and Anions on Properties
    of Zinc Oxide Particles Synthesized in Supercritical Water. *J. of Supercritical Fluids*, **30**,
    325–331 (2004).

11. Loh, K. P. and Chua, S. J. Zinc Oxide Nanorod Arrays: Properties and Hydrothermal Syn-
    thesis, Springer, Berlin, (2007).

12. Ohara, S., Mousavanda, T., Umetsua, M., Takamia, S., Adschiria, T., Kurokib, Y. and
    Takatab, M. Hydrothermal Synthesis of Fine Zinc Oxide Particles Under Supercritical
    Conditions. *Solid State Ionics*, **172**, 261–264 (2004).

13. Deng, H. M., Ding, J., Shi, Y., Liu, X. Y., and Wang, J. Ultrafine Zinc Oxide Powders Pre-
    pared By Precipitation/Mechanical Milling. *Journal Of Materials Science*, **36**, 3273–3276
    (2001).

14. Xiao, Q., Huang, S., Zhang, J., Xiao, C. and Tan, X. Sonochemical synthesis of ZnO
    nanosheet. *Journal of Alloys and Compounds*, **59**, L18–L22 (2008).

15. Yu, J. and Yu, X. Hydrothermal Synthesis and Photocatalytic Activity of Zinc Oxide Hol-
    low Spheres. *Environmental Science and Technology*, **42**, 4902–4907 (2008).

16. Wei, Y. and Chang, P. Characteristics of Nano Zinc Oxide Synthesized Under Ultrasonic
    Condition. *Journal of Physics and Chemistry of Solids*, **69**, 688–692 (2008).

17. Chen, D., Jiao, X. and Cheng, G. Hydrothermal synthesis of zinc oxide powders with dif-
    ferent morphologies. *Solid State Communications*, **113**, 363–366 (2000).

18. Ammala, A., Hill, A. J., Meakin, P., Pas, S. J., and Turney, T. W. Degradation Studies of
    Polyolefins Incorporating Transparent Nanoparticulate Zinc Oxide UV Stabilizers. *Jour-
    nal of Nanoparticle Research*, **4**, 167–174 (2002).

19. Yi, R., Zhang, N., Zhou, H. F., Shi, R. R., Qiu, Z., and Liu, X. H. Selective synthesis and
    characterization of flower-like ZnO microstructures via a facile hydrothermal route. *Mate-
    rials Science and Engineering B-Advanced Functional Solid-State Materials*, **153**, 25–30
    (2008).

20. Colon, G., Hidalgo, M. C., Navio, J. A., Melian, E. P., Diaz, .O. G., Rodriguez, J. M. D.
    Photoluminescence and photocatalysis of the flower-like nano-ZnO photocatalysts pre-
    pared by a facile hydrothermal method with or without ultrasonic assistance. *Applied Ca-
    talysis B-Environmental*, **83**,30–38 (2008).

21. Lai, Y. L., Meng, M., Yu, Y. F., Wang, X. T., and Ding, T. Photoluminescence and pho-
    tocatalysis of the flower-like nano-ZnO photocatalysts prepared by a facile hydrother-
    mal method with or without ultrasonic assistance. *Applied Catalysis B-Environmental*,
    105,335–45 (2011).

22. Xie, J., Li, P., Li, Y., Wang, Y., and Wie, Y. Solvent-induced Growth of ZnO Particles at
    Low Temperature. *Materials Letters*, **62**, 2814–2816 (2008).

23. Tarwall, N. L., Jadhav, P. R., Vanalkar, S. A., Kalagi, S. S., Pawar, R. C., Skaikh, J. S.,
    Mali, S. S., Dalavi, D. S., Shinde, P. S., and Patil, P. S. Photolimunescence of zinc oxide
    nanopowder synthesized by a combustion method. *Powder Technology*, **208**, 183–188
    (2011).

24. Dijken, A. V., Meulenkamp, E. A.,Vanmaekelbergh, D., and Meijerink, A. The luminescence of nanocrystalline ZnO particles: the mechanism of the ultraviolet and visible emission. *Journal of Luminescence*, 87, 89, 454–456 (2000).

25. Jana, A., Bandyopadhyay, N. R., and Devi, P. S. Formation and assembly of blue emitting
water lily type ZnO flowers. *Solid State Sciences*, 13, 1633–1637 (2011).

26. Campbell, A. N. Activity coefficients, densities, dipole moments, and surface tensions of
the system triethylamine-methylethylketone-water. *Can. J. Chem.*, 51, 127–131 (1981).

27. Becheri, A., Durr, M., Nostra, P. L., and Baglioni, P. Synthesis and characterization of
zinc oxide nanoparticles: application to textiles as UV-absorbers. *J. Nanopart. Res.*, 10,
679–689 (2008).

28. Oo, W. M. H., McCluskey, M. D., Lalonde, A. D., and Norton, M. G. Infrared Spectroscopy of ZnO nanoparticles containing CO_2 impurities. *Applied Physics Letters*, 86, 07311-
1- 07311-3 (2005).

29. Sali, S., Boumaour, M., and Tala-Ighil, R. Preparation and characteristic of low resistive
zinc oxide thin films using chemical spray technique for solar cells application. The effect
of thickness and temperature substrate. Revue des Energies Renouvelables CICME'08
Sousse., 201–207 (2008).

30. Ellmer, K. Resistivity of polycrystalline zinc oxide films: current status and physical limit.
J. Phys. D: Appl. Phys., 34, 3097–3108 (2001).

31. Tian, Y., Ma, H., Shen, L., Wang, Z., Li, Y. Q. S. Novel and simple synthesis of ZnO nanospheres through decomposing zinc borate nanoplatelets. *Materials Letters*, 63, 1071–1073
(2009).

CHAPTER 16

WATER SORPTION OF POLYVINYL CHLORIDE–LUFFA CYLINDRICA COMPOSITES

HASAN DEMIR and DEVRIM BALKÖSE

CONTENTS

16.1 INTRODUCTION

Natural *Luffa* cylindrica fibers were modified with 0.1M sodium hydroxide (NaOH) for removing lignin and hemicellulose. Natural and modified *Luffa* fibers were characterized by using IR spectroscopy. Composites were produced with poly(vinyl chloride) (PVC) plastisol and natural *Luffa* fiber. Natural *Luffa* fiber is a highly hydrophilic substance. This feature increased the water sorption capacity of the composites. Flexible PVC-*Luffa* cylindrica composites had higher liquid water sorption capacity (0.3–0.6%) compared to that of flexible PVC (0.1%). There was no volume change of composites due to liquid water sorption.

Thermoplastics reinforced with special wood fillers are enjoying rapid growth due to their many advantages. Light weight, reasonable strength, and stiffness are some of these advantages. The composite is presenting flexible, economical, and ecological properties. Wood is polymeric composite consisting primarily of cellulose, hemicelluloses, and lignin. Lignin behaves a barrier and surrounds cellulose to hinder attack from enzymes and acids [1, 2]. Hemicellulose and lignin cause problems when wood is used as filler [3].

Luffa sponge products are readily available in the cosmetic and bath section of department stores, discount stores, pharmacies, and specialty shops. Many environmentally conscious consumers appreciate that *Luffa* products are biodegradable, natural, and renewable resources. The tough fibers can also be processed into industrial products such as filters, insulation, and packing materials [4]. *Luffa* fibers consist of 51.2% cellulose, 13.7% lignin, 11.2% hemicellulose, 1.8% ash, and 6% moisture at room temperature [5]. Siquear et al reported that *Luffa* cylindrica contained 60.0–63.0% cellulose, 19.4–22% hemicelluloses, and 10.6–11.2 % lignin [6]. Microcrystalline cellulose and cellulose nanocrystals were obtained from *Luffa* fibers [6].

Luffa fibers were used as filler in polypropylene and as nucleating agents in PVC foams [7, 8]. Composites having 0.3 volume fractions of *Luffa* fibers in polyester matrix absorbed 15% liquid water. The water diffusion coefficient in composites was found as $9.7 \times 10^{-6} mm^2/sec$ [9]. Microcrystalline cellulose PVC composites with 40 hr isononylphtalate were biodegradable since soil microorganisms could consume cellulose as a source of nutrient. The micropores formed by cellulose degradation allow water in the composites. The weight loss increased with time and reached to 10% after 8 weeks for 30 hr microcrystalline cellulose content [10].

Since *Luffa* cylindrica fibers had a network structure, it is expected that when composites are prepared from them two continuous phases, polymer and the interconnected cellulose phase will be obtained. The hydrohilic continuous network phase of the composites can transport water or water vapor at a controlled rate from high water content medium to low water content medium. Thus, this type of materials is controlled water release agents.

In this chapter, water sorption properties of *Luffa* fibers and its composite with PVC plastisol were aimed to be investigated. Samples were characterized by using infrared spectroscopy, optical micrography, scanning electron microscope (SEM), and differential scanning analysis. Water and water vapor sorption at 25°C were investigated.

16.2 MATERIALS AND THE METHODS

16.2.1 LUFFA FIBERS

Luffa cylindrica were obtained from local specialty shop. The *Luffa* fibers washed with water to remove the adhering dirt. They were dried in an oven at 70°C for 2 hr. After drying, they were cut with waring blendor for reducing dimensions to 2–3 mm. Some fibers were modified with 0.1M sodium hydroxide solution at boiling temperature for 10 min. Sodium hydroxide was obtained from Sigma Co. Modified fibers were washed with distilled water until all sodium hydroxide was removed. After washing, they were dried in an oven at 70°C for 2 hr. Natural and modified *Luffa* fibers were characterized by using KBr disc technique with Shimadzu IR-470 spectrophotometer. Differential scanning calorimetric curves of the samples in equilibrium with 75% relative humidity air at 25°C was obtained by Seteram DSC92 calorimeter. The samples were heated in 25–250°C range at 10°C/min heating rate.

COMPOSITE PREPARATION

Composites made from *Luffa* fiber as filler and a PVC plastisol as polymer matrix which contains 100 parts PVC, 60 parts dioctyl phthalate (DOP), 5 parts epoxidized soybean oil, and 5 parts zinc stearate. Composites were prepared in aluminum caps with 4 cm diameter. *Luffa* Fibers were cut into shape of the aluminum caps and pressed on PVC plastisol inside the caps. Two composites, composite I and II were prepared by using fiber network from inside and outside of the *Luffa* gourd. Inner fibres were thicker from outer fibres. Composites were put into an oven at 150°C for gelation of plastisols into a plastic mass. Plastic discs having 3.8 (composite I) and 4.0% w/w (composite II) *Luffa* fibers were obtained by this method. A control plastigel without any *Luffa* fibers was prepared in the same manner.

MICROSCOPY

Expansion of fiber diameter on wetting was also observed with time by optical microscopy. The micrographs of the fibers were taken using Orthomat Polarizing microscope in transmittance mode after wetting with a drop of water. Morpholgy of natural and modified *Luffa* was observed by using SEM with Philipps XL-305 FEG. Interface, between *Luffa* fiber and PVC plastisol matrix were also observed.

WATER ABSORPTION

Natural and modified fibers and composites were immersed into static distilled water bath for observing absorption of water. The samples were wiped with tissue paper to remove surface water before weighing. Water uptake of samples (x%) at time t was calculated from:

$$x\% = \frac{W_t - W_{t0}}{W_{t0}} x100 \tag{1}$$

where W_t—Weight of sample at time t
W_{t0}—Weight of sample at t = 0

WATER VAPOR ADSORPTION OF FIBERS

Water vapor adsorption isotherms of fibers at 25°C were obtained by using Omnisorp 100CX after out gassing the fibers at 110°C under 0.01 Pa pressure.

16.3   DISCUSSION AND RESULTS

16.3.1   *MORPHOLOGY OF NATURAL AND MODIFIED FIBERS*

Natural *Luffa* fibers were composed of cellulose, lignin, and hemicellulose. Water absorption into *Luffa* fibers became harder with lignin and hemicellulose structure. The lignin and hemicellulose could be removed with chemical processes. Natural *Luffa* fiber is processed with sodium hydroxide for dissolving lignin and hemicellulose. As seen in Figure 1 lignin layer was removed from surface of the fiber by NaOH treatment.

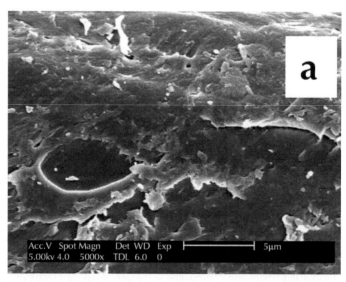

FIGURE 1   *(Continued)*

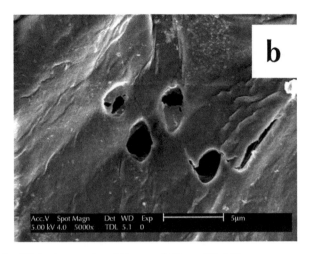

FIGURE 1   The SEM micrographs of (a) natural (b) modified fibers.

16.3.2 IR SPECTRA OF LUFFA FIBERS

Infrared spectra (IR) of modified and natural fibers are shown in Figure 2. The bands at 1070, 1115, and1165 cm^{-1} represent cellulose backbone vibrations of the polymer chain. Broad region of O-H vibration bond around 3450–3300 cm^{-1} is also characteristic peak for cellulose solids. The peak at 1740–1730cm^{-1} indicate the vibration of C=O stretching of carboxyl groups [1]. The IR spectrum of delignified fibers does not have the band at 1740–1730 cm^{-1} due to removal of lignin and had lower intensity band at 1640 cm^{-1}.

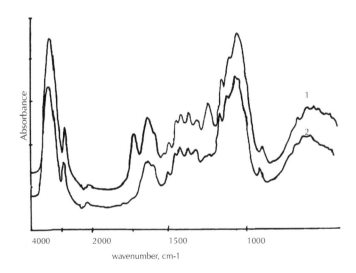

FIGURE 2   Fourier transform infrared (FTIR) spectra of natural and modified *Luffa* fiber.

16.3.3 DIFFERENTIAL SCANNING CALORIMETRY

The DSC curves for fibers equilibrated at 25°C at 75% relative humidity and heated from 25 to 250°C with 10°C/min heating rate are shown in Figure 3. Using the graphs, heat of vaporization of water, which was absorbs by fiber, was 2456.6 J/g and 2421 J/g for natural and modified *Luffa* fibers respectively. For free water that is 1714J/g at 25°C. Obviously heat of vaporization of adsorbed water is more than that for free water. During the heating, mass losses of samples are 15.4% and 14.5% for natural and modified *Luffa* fibers respectively. Heat of desorption of water from fibers was higher than heat of evaporation of free water.

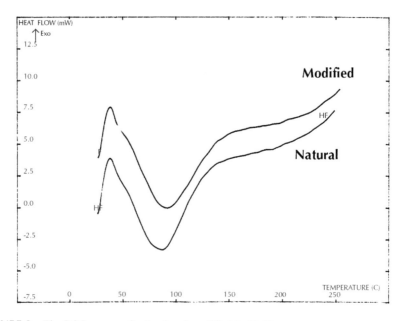

FIGURE 3   The DSC curves of natural and modified *Luffa* fiber.

16.3.4 LIQUID WATER SORPTION OF FIBER

Rate of water uptake versus time graph of fibers are shown in Figure 4. Modified *Luffa* fiber absorbs water much more than natural *Luffa* fiber. Removal of lignin made the fibers had more affinity to liquid water. Fibers dimensions increased with time due to water absorption. Figure 4 shows expansion of fiber diameter with respect to time. Diameter of modified *Luffa* fiber expands slower than natural *Luffa* fiber at the start of the process. But after 3.5 min, modified *Luffa* fiber diameter expands more than natural fiber.

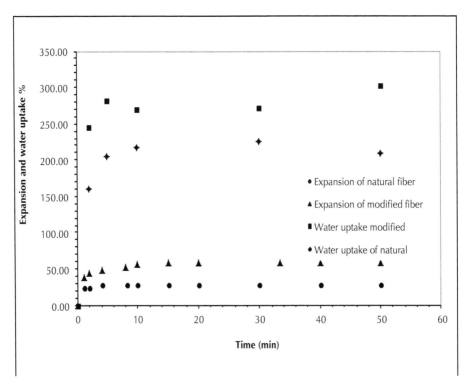

FIGURE 4   Expansion in water and water uptake of natural and modified fibers.

After that expansion of fiber diameter reaches equilibrium. Modified *Luffa* fiber absorbs water much more than natural *Luffa* fiber since it is more hydrophilic. While natural *Luffa* fiber absorbs 213% water, modified *Luffa* fiber takes up 281% water. At equilibrium 26.9% and 58.8% swelling occurred for natural and modified fibers. Liquid water files the pore spaces of the fibers and cause relaxation of the structure.

16.3.5   WATER VAPOR ADSORPTION OF FIBERS

The water vapor adsorption of the fibers shows a different behavior than liquid water adsorption. While the natural fibers adsorb 6.9% water vapor modified fibers adsorb less 4.9% at 95% relative humidity at 25°C as seen in Figure 5. The shape of the isotherm indicated cluster formation of water molecules in empty spaces of the fibers. Modified fibers adsorbed less water vapor than raw fibers.

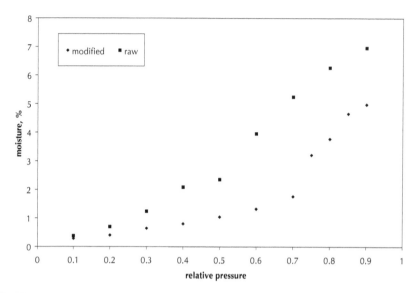

FIGURE 5   Water vapor adsorption isotherm of natural and modified fibers at 25°C.

16.3.6   FIBER PLASTIGEL INTERPHASE

There were a small space between the fiber and the matrix of the composites as seen in SEM micrographs in Figure 6. The surface of the fibers should be made more compatible with the matrix by silylation or malleation for enhancement of interphase.

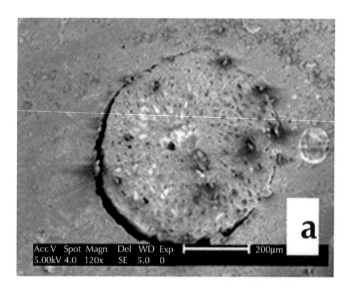

FIGURE 6   *(Continued)*

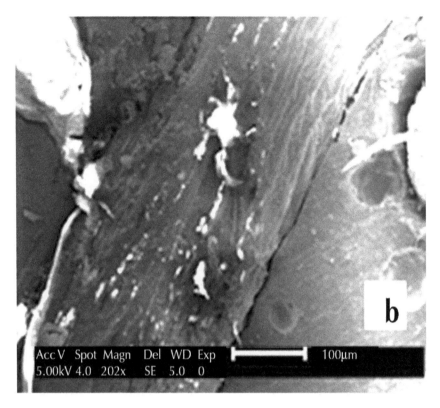

FIGURE 6   The SEM micrographs of composites cross sections: (a) cross section of fiber and (b) surface of fiber in composites.

16.3.7   LIQUID WATER ABSORPTION OF COMPOSITES

Water uptake ratios were calculated using Equation 1 for composites and plastigel. Figure 7 shows the water uptake percentages of pure plastisol and composite I and II. In the Figure, plastigel absorbs water rapidly and reaches equilibrium. Plastisol water uptake curve shows deviation due to time. After 10 min, plastigel weight was decreased, since some DOP was dissolved in water. Composite I and II's water uptake ratios were higher than that of pure plastisol. Consequently, sorption property of *Luffa* fiber affects structure of composites. Composite I show higher water uptake ratios than composite II. It could be depended fiber structure into the composites. However, composite I and II indicate similar water uptake path. Flexible PVC-*Luffa* cylindrica composites had higher liquid water sorption capacity (0.3–0.6%) compared to that of flexible PVC (0.1%). There was no volume change of composites due to liquid water sorption.

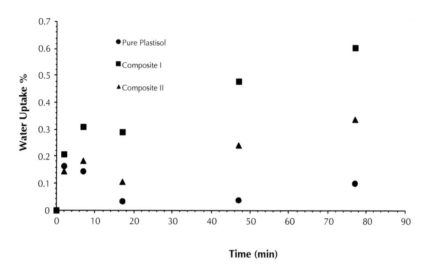

FIGURE 7   Liquid water uptake of pure plastisol and composites versus time.

16.3.7   LIQUID WATER DIFFUSIVITY IN FIBERS AND COMPOSITES

By assuming Fickian diffusion in fibers and composites the diffusivity of the liquid water was determined from the initial experimental rate data and Equation 2.

$$M_t/M_e = 4/L(Dt/\pi)^{1/2} \tag{2}$$

Where M_t and M_e are weight increase at time t and at equilibrium, L is the half thickness of slab or the radius of the fiber.

It was found as 1.5×10^{-10} m²/sec, 6.4×10^{-9} m²/sec, 2.9×10^{-10} m²/sec, 3.4×10^{-10} m²/sec, and 1.075×10^{-10} m²/sec for the natural fiber, modified fiber, plastigel, composite I, and composite II respectively.

16.4   CONCLUSION

The IR of fibers showed that lignin was removed with modification process. Successful modification is known to disrupt lignin barrier to increase the reactive sites of cellulose and increase pore volume as well as available surface area. The DSC curves predicted that natural and modified *Luffa* fibers had high water content. H eat of desorption of water, 2456.6 and 2421 J/g for natural and modified *Luffa* fibers respectively was higher than heat of evaporation of free water,1714 J/g at 25°C.

The results showed that the rate of water absorption of water was higher in the *Luffa* PVC composites than PVC plastigel. Flexible PVC-*Luffa* cylindrica composites also had higher liquid water sorption capacity (0.3–0.6%) compared to that of flexible PVC (0.1%). While *Luffa* fibers swell in water to a high extent, there was no volume change of composites due to liquid water sorption. Further studies are being made with modified fibers.

KEYWORDS

- **Luffa cylindrica fibers**
- **PVC plastisols**
- **Water sorption**
- **Water vapour adsorption**

REFERENCES

1. Cheng, W. *Pretreatment and enzymatic hydrolysis of lignocellulosic materials*, MS thesis, West Virginia University (2001).
2. Annadurai, G., Juang, R. S., and Lee, D. J. Use of cellulose-based wastes for adsorption of dyes from aqueous solutions. *J. Haz. Ma.*, **B92**, 263–274 (2002).
3. Avella, M., Bozzi, C., Erba, R., Focher, B., Morzetti, A., and Martuscelli, E. Steam-exploded wheat straw fibers as reinforcing material for polypropylene-based composites. Characterization and properties. *Angew. Macromol. Chem.*, **233**(4075), 149–166 (1995).
4. Davis, J. M. and DeCourley, C. D. *Luffa sponge gourds: A potential crop for small farms.* J. Janick and J. E. Simon (Eds.), New Crops. Wiley, New York, pp. 560–561 (1993).
5. Baltazar, A., Jimenez, A., and Bismarc, A. Wetting Behavior, moisture uptake, and electrokinetic parameters of lignocellulosic fibers. *Cellulose*, **14**, 115–127 (2007).
6. Siquera, G., Bras, J., and Dufresne, A. Luffa Cylindrica as a lignocellulosic source of fiber, microfibrillated cellulose, and cellulose nanocrystals. *Bioresources.*, **5**, 727–740 (2010).
7. **Demir H.** , Atikler U., Tihminlioglu F., and Balköse D. The effect of fiber surface treatments on the mechanical and water sorption properties of PP-Luffa composites. *Journal of Composite Part A.*, **37**, 447–456 (2006).
8. **Demir, H.,** Sipahioglu, M., Balköse, D., and Ülkü, S. Effect of additives on flexible PVC foam formation. *Journal of Materials Processing Technology,* **195**, 144–153 (2008).
9. Boynard, C. A. and D'Almedia, J. R. M. Water absorption by sponge guord (luffa cylindrica)-polyester composite materials. *Journal of materials Science Letters.*, **18**, 1789–1791 (1999).
10. Chuayhijit, S., Su-uthai, S., and Charachinda, S. Poly(vinyl chloride) film filled with microcrystalline cellulose prepared from cotton fabric waste: properties and biodegradability study. *Waste management research*, **28**, 109–117 (2010).

CHAPTER 17

KEY ENGINEERING PROPERTIES OF NANOFILLER PARTICLES

A. K. HAGHI and G. E. ZAIKOV

CONTENTS

17.1   INTRODUCTION

Improving material properties and creating more specific tailored properties have become more important over the last decades. Combining different materials to benefit from the usually very different properties creates better materials with the envisaged properties. Composite materials have been around for ages and have already proved their use.

In recent years an increasing interest is shown in nanocomposites. By choosing fillers with at least one dimension in the nanometre range, the surface to volume ratio increases tremendously. Thus, the interface between filler and matrix material increases, enabling a bigger impact of the filler properties on the overall properties, such as higher stiffness, higher melt strength, lower permittivity, and improved barrier properties.

Due to the positive influence of these nanofillers in the nanocomposites, an abundance of chapter on different methods to quantify the influence of the nanofiller can be found in literature. Many chapter on assessing the clay dispersion in a polymer matrix by morphological and rheological studies have been published. Due to the relatively easy sample preparation and sample loading, rheology is often used to screen or characterize the nanofiller dispersion, or more generally the influence of the nanofiller on the overall rheological behavior of (thermoplastic) nanocomposites.

Clearly industrial flows are complex, not only because the geometries are complex, but also because the constituents are not usually simple. Molecular weight distribution of a polymer is another level of complexity, as polymers are rarely synthesized in a sharp monodisperse population.

This brings in the need to study the behavior of polymeric liquid in simple flows and for simple systems, with the hope that the knowledge gained can be appropriately used in a complex flow pattern. The word rheology is defined as the science of deformation and flow. Rheology involves measurements in controlled flow, mainly the viscometric flow in which the velocity gradients are nearly uniform in space. In these simple flows, there is an applied force where the velocity (or the equivalent shear rate) is measured or *vice versa*.

Filled polymers exhibit a diverse range of rheological properties, varying from simple viscous fluids to highly elastic solids with increasing filler volume fraction. The effect of filling on rheology is well-known in the range of small volume fraction where the reinforcement could be attributed to hydrodynamic effects caused by the solid inclusions in the melt stream. For high volume fraction where direct particle contacts dominate the deformation, a straightforward solution of hydrodynamic equations is difficult and theoretical models based on realistic structural ideas are missing so far [1-8].

However, filled polymers usually show strong flow as well as strain and temperature history dependent rheological behaviors. It is always important to determine the dynamic viscoelastic properties at a strain that is low enough not to affect the material response.

Small strain amplitude frequency sweep is usually used to collect linear rheological data which are reproductive for repeated measurements within a certain experi-

mental error. Rheology is a way generally used to assess the state of dispersion of fillers in the melt. In nanocomposites, the terminal plateau can even be observed at considerably low.

Though the exact mechanism is not clear, the solid-like behavior is generally assigned to originations. In highly filled polymers, solid-like yielding can be observed even at temperatures above the quiescent melting temperature (Tm) or glass transition temperature (Tg) of the polymer [9-13].

It should be noted that dynamic rheology in the linearity regime is sensitive to filler dispersion in polymers. However, a straightforward description of how linear rheology varies with volume fraction is still missing so far [14-20].

Inorganic nanofiller of various types' usage for polymer nanocomposites production have been widely spread. However, the mentioned nanomaterials melt properties are not studied completely enough. As a rule, when nanofillers application is considered, then compromise between mechanical properties in solid state improvement, melt viscosity at processing enhancement, nanofillers dispersion problem and process economic characteristics is achieved. Proceeding from this, the relation between nanofiller concentration and geometry and nanocomposites melt properties is an important aspect of polymer nanocomposites study. Therefore, the purpose of this chapter is an investigation and theoretical description of the dependence of nanocomposite high density polyethylene/calcium carbonate melt viscosity on nanofiller concentration [20-29].

17.2 EXPERIMENTAL

High density polyethylene (HDPE) of industrial production was used as matrix polymer and nanodimensional calcium carbonate (CaCO$_3$) in the form of compound with particles size of 80 nm and mass contents 1–10 mass % was used as nanofiller.

Nanocomposites HDPE/CaCO$_3$ was prepared by components mixing in melt on twin-screw extruder. Mixing was performed at temperature 483–493K and screw speed of 15–25 rpm during 5 min. Testing samples were obtained by casting under pressure method on casting machine test samples molding apparate at temperature 473K and pressure 8 MPa.

Nanocomposites viscosity was characterized by melt flow index (MFI). The measurements of MFI were performed on extrusion type plastometer with capillary diameter of 2.095 ± 0.005, at temperature 513K and load of 2.16 kg. This sample was maintained at the indicated temperature during 4.5 ± 0.5 min.

Uniaxial tension mechanical tests were performed on samples in the form of two-sided spade with sizes. Tests were conducted on universal testing apparatus at temperature 293K and strain rate ~ 2×10^{-3} sec^{-1}.

17.3 DISCUSSION AND RESULTS

For polymer microcomposites (that is composites with filler of micron sizes) two simple relations between melt viscosity η, shear modulus G in solid-phase state and filling volume degree φ_n were obtained. The relationship between η and G has the following form:

$$\frac{\eta}{\eta_0} = \frac{G}{G_0}, \tag{1}$$

Where— η_0 and G_0 are melt viscosity and shear modulus of matrix polymer, accordingly.

Besides, microcomposite melt viscosity increase can be estimated as follows (for $\varphi_n < 0.40$):

$$\frac{\eta}{\eta_0} = 1 + \varphi_n. \tag{2}$$

In Figure 1 the dependences of ratios G_n/G_m and η_n/η_m, where G_n and η_n are shear modulus and melt viscosity of nanocomposite, G_m and η_m are the same characteristics for the initial matrix polymer, on $CaCO_3$ mass contents W_n for nanocomposites HDPE/$CaCO_3$. Shear modulus G was calculated according to the following general relationship:

$$G = \frac{E}{d_f}, \tag{3}$$

Where E is Young's modulus, d_f is nanocomposite structure fractal dimension, determined according to the equation:

$$d_f = (d-1)(1+v), \tag{4}$$

Where d is dimension of Euclidean space, in which a fractal was considered (it is obvious, that in case $d = 3$) and v is Poisson's ratio, estimated by mechanical tests results with the aid of the relationship:

$$\frac{\sigma_Y}{E} = \frac{1-2v}{6(1+v)}. \tag{5}$$

The MFI reciprocal value was accepted as melt viscosity η measure. The data of Figure 1 clearly demonstrate that in the case of the studied nanocomposites the relationship (Equation (1)) is not fulfilled either qualitatively or quantitatively: the ratio η_n/η_m decay at W_n growth corresponds to G_n/G_m enhancement and η_n/η_m absolute values are much larger than the corresponding G_n/G_m magnitudes.

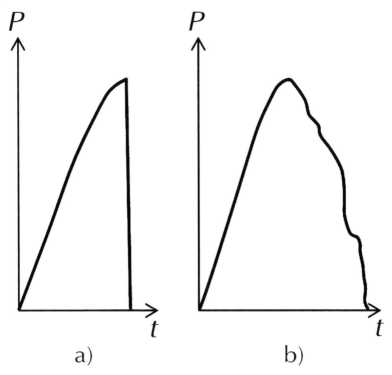

a)   b)

FIGURE 1 The dependences of shear moduli G_n/G_m (1) and melt viscosities η_n/η_m (Equation (2)) ratios of nanocomposite G_n, η_n, and matrix polymer G_m, η_m on nanofiller mass contents W_n for nanocomposites HDPE/CaCO$_3$.

In Figure 2 the comparison of parameters η_n/η_m and $(1 + \varphi_n)$ for nanocomposites HDPE/CaCO$_3$ is adduced. The discrepancy between the experimental data and the relationship (Equation (2)) is obtained again—the absolute values η_n/η_m and $(1 + \varphi_n)$ discrepancy is observed and $(1 + \varphi_n)$ enhancement corresponds to melt relative viscosity reduction. At the plot of Figure 2 graphing the nominal φ_n value was used, which does not take into consideration nanofiller particles aggregation and estimated according to the equation:

$$\varphi_n = \frac{W_n}{\rho_n},\qquad(6)$$

Where ρ_n is nanofiller particles density, which was determined according to the formula:

$$\rho_n = 0.188(D_p)^{1/3},\qquad(7)$$

Where D_p is $CaCO_3$ initial particles diameter.

Hence, the data of Figure 1 and Figure 2 show that the relationships (Equation (1)) and (Equation (2)) fulfilled in case of polymer microcomposites are incorrect for nanocomposites. In case of the relationship (Equation (1)) correctness and Kerner's equation application for G calculation viscosity η_n lower boundary can be obtained according to the equation:

$$\frac{\eta_n}{\eta_m} = 1 + \frac{2.5\varphi_n}{1-\varphi_n}.$$

(8)

Since η value is inversely proportional to MFI, then in such treatment the Equation (8) can be rewritten as follows:

$$\frac{MFI_m}{MFI_n} = 1 + \frac{2.5\varphi_n}{1-\varphi_n},$$

(9)

Where MFI_m and MFI_n are MFI values for matrix polymer and nanocomposite, accordingly.

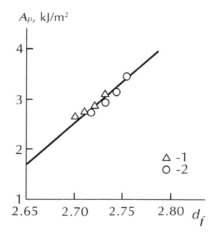

FIGURE 2   The dependence of nanocomposite and matrix polymer melt viscosities ratio η_n/η_m on nanofiller volume contents $(1 + \varphi_n)$ (1) for nanocomposites HDPE/$CaCO_3$. The straight line 2 shows relation 1:1.

Three methods can be used for the value φ_n estimation in the Equations (8) and (9). The first of them was described, which gives nominal value φ_n. The second method is usually applied for microcomposites, when massive filler density is used as ρ_n, that is $\rho_n = const \approx 2000$ kg/m³ in case of $CaCO_3$. And at last, the third method also uses the Equations (6) and (7), but it takes into consideration nanofiller particles aggregation and in this case in the Equation (7) initial nanofiller particles diameter D_p is replaced to such particles aggregate diameter D_{ag}. To estimate $CaCO_3$ nanoparticles aggregation degree and hence, D_{ag} value can be estimated within the frameworks of the strength dispersive theory, where yield stress at shear τ_n of nanocomposite is determined as follows:

$$\tau_n = \tau_m + \frac{G_n b_B}{\lambda},$$ (10)

Where τ_m is yield stress at shear of polymer matrix, b_B is Burgers vector, and λ is distance between nanofiller particles.

In case of nanofiller particles aggregation the Equation (10) assumes the look:

$$\tau_n = \tau_m + \frac{G_n b_B}{k(\rho)\lambda},$$ (11)

Where $k(\rho)$ is an aggregation parameter.

The included in the Equations (10) and (11) parameters are determined as follows. The general relation between normal stress σ and shear stress τ assumes the look:

$$\tau = \frac{\sigma}{\sqrt{3}}.$$ (12)

Burgers vector value b_B for polymer materials is determined from the relationship:

$$b_B = \left(\frac{60.5}{C_\infty}\right)^{1/2}, \text{ Å,}$$ (13)

Where C_∞ is characteristic ratio, connected with d_f by the equation:

$$C_\infty = \frac{2d_f}{d(d-1)(d-d_f)} + \frac{4}{3}.$$ (14)

And at last, the distance λ between nonaggregated nanofiller particles is determined according to the equation:

$$\lambda = \left[\left(\frac{4\pi}{3\varphi_n}\right)^{1/3} - 2\right]\frac{D_p}{2}.$$ (15)

From the Equations (11) and (15) $k(\rho)$ growth follows from 5.5 up to 11.8 in the range of $W_n = 1 - 10$ mass % for the studied nanocomposites. Let us consider, now such $k(\rho)$ growth is reflected on nanofiller particles aggregates diameter D_{ag}. The Equations (6), (7), and (15) combination gives the following expression:

$$k(\rho)\lambda = \left[\left(\frac{0.251\pi D_{ag}^{1/3}}{Wn}\right)^{1/3} - 2\right]\frac{D_{ag}}{2}, \tag{16}$$

Allowing at replacement of D_p on D_{ag} to determine real, that is with accounting of nanofiller particles aggregation, nanoparticles $CaCO_3$ aggregates diameter. Calculation according to the Equation (16) shows D_{ag} increase (corresponding to $k(\rho)$ growth) from 320 up to 580 nm in the indicated W_n range. Further the real value ρ_n for aggregated nanofiller can be calculated according to the Equation (7) and real filling degree φ_n—according to the Equation (6). In Figure 3 the comparison of the dependences $MFI_n(W_n)$, obtained experimentally and calculated according to the Equation (9) with the usage of the values φ_n, estimated by three indicated methods. As one can see, the obtained according to the Equation (9) theoretical results correspond to the experimental data neither qualitatively nor quantitatively.

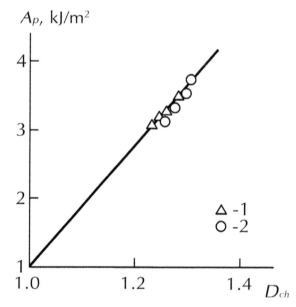

FIGURE 3 The dependences of MFI_n on nanofiller mass contents W_n for nanocomposites HDPE/$CaCO_3$. 1—Experimental data, 2–4—calculation according to the Equation (9) without appreciation Equation (2), with appreciation Equation (3) of nanofiller particles aggregation and at the condition ρ_n = const Equation (4), 5—calculation according to the relationship Equation (17).

The indicated discrepancy requires the application of principally differing approach at polymer nanocomposites melt viscosity description. Such approach can be fractal analysis, within the frameworks of which the authors were offered the following relationship for fractal liquid viscosity η estimation:

$$\eta(l) \sim \eta_0 l^{2-d_f},$$ (17)

Where l is characteristic linear scale of flow, η_0 is constant, and d_f is fractal dimension.

In the considered case the nanoparticles $CaCO_3$ aggregate radius $D_{ag}/2$ follows to the accepted as l. Since the indicated aggregate surface comes into contact with polymer, then its fractal dimension d_{surf} was chosen as d_f. The indicated dimension can be calculated as follows. The value of nanofiller particles aggregate specific surface S_u was estimated according to the equation:

$$S_u = \frac{6}{\rho_n D_{ag}},$$ (18)

and then the dimension d_{surf} was calculated with the aid of the equation:

$$S_u = 410 \left(\frac{D_{ag}}{2} \right)^{d_{surf}-d}.$$ (19)

As earlier, the value η was considered as reciprocal value of MFI_n and constant η_0 was accepted equal to $(MFI_m)^{-1}$. At these conditions and replacement of proportionality sign in the relationship (Equation (17)) on equality sign the theoretical values MFI_n can be calculated, if D_{ag} magnitude is expessed in microns. In Figure 3 the comparison of the received by the indicated mode MFI_n values with the experimental dependence $MFI_n(W_n)$, from which theory and experiment good correspondence follows.

The relationship (Equation (17)) allows making a number of conclusions. So, at the mentioned conditions conservation D_{ag} increase, that is initial nanoparticles aggregation intensification, results to nanocomposite melt viscosity reduction, whereas d_{surf} enhancement, that is nanoparticles surface roughness degree increasing, raises melt viscosity. At $d_{surf} = 2.0$, that is nanofiller particles smooth surface, melts viscosity for matrix polymer and nanocomposite will be equal. It is interesting that extrapolation of the MFI_n dependence, obtained experimentally, on the calculated according to the Equation (19) d_{surf} values gives the value $MFI_n = 0.602$ g/10 min at $d_{surf} = 2.0$, that is practically equal to the experimental magnitude $MFI_m = 0.622$ g/10 min. The indicated factors, critical ones for nanocomposites, are not taken into consideration in continuous treatment of melt viscosity for polymer composites (the Equation (8)).

17.4 CONCLUSION

Nanotechnology refers broadly to manipulating matter at the atomic or molecular scale and using materials and structures with nanosized dimension, usually ranging from 1 to 100 nm. Due to their nanoscale size, nanoparticles show unique physical and chemical properties such as large surface area to volume ratios or high interfacial reactivity. Till now increasing nanoparticles have been demonstrated to exhibit specific interaction with contaminants in waters, gases, and even soils, and such properties give hope for exciting novel and improved environmental technology.

However, the small particle size also brings issues involving mass transport and excessive pressure drops when applied in fixed bed or any other flow-through systems, as well as certain difficulties in separation and reuse, and even possible risk to ecosystems and human health caused by the potential release of nanoparticles into the environment. An effective approach to overcoming the technical bottlenecks is to fabricate hybrid nanocomposite by impregnating or coating the fine particles onto solid particles of larger size. The widely used host materials for nanocomposite fabrication include carbonaceous materials like granular activated carbon, silica, cellulose, sands, and polymers, and polymeric hosts are particularly an attractive option partly because of their controllable pore space and surface chemistry as well as their excellent mechanical strength for long-term use. The resultant polymer-based nanocomposite (PNC) retains the inherent properties of nanoparticles, while the polymer support materials provide higher stability, process ability and some interesting improvements caused by the nanoparticle–matrix interaction. The generally used nanoparticles include zero-valent metals, metallic oxides, biopolymers, and single-enzyme nanoparticles (SENs). These nanoparticles could be loaded onto porous resins, cellulose or carboxymethyl cellulose, chitosan, and so on. The choice of the polymeric supports is usually guided by their mechanical and thermal behavior. Other properties such as hydrophobic/hydrophilic balance, chemical stability, bio-compatibility, optical and/or electronic properties, and chemical functionalities (that is solvation, wettability, templating effect, and so on.) have to be considered to select the organic hosts.

Observed results in this chapter emphasizes that the microcomposites rheology description models do not give adequate treatment of melt viscosity for particulate-filled nanocomposites. The indicated nanocomposites rheological properties correct description can be obtained within the frameworks of viscous liquid flow fractal model. It is significant, that such approach differs principally from the used ones at microcomposites description. So, nanofiller particles aggregation reduces both melt viscosity and elasticity modulus of nanocomposites in solid-phase state. For microcomposites melt viscosity enhancement is accompanied by elasticity modulus increase.

APPENDIX

In Figure 1 the schematic diagrams load-time (P-t) are adduced for two cases of polymeric materials samples fracture in impact testing: by both instable (a) and stable cracks (b). As it is known, the area under P-t diagram, gives mechanical energy consumed with samples fracture. The polymeric materials macroscopic fracture process, defined by the magisterial crack propagation, begins at the greatest load P.

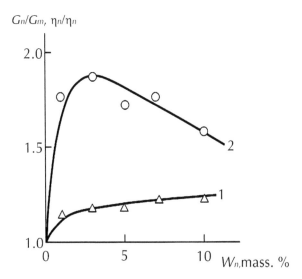

FIGURE 1 Schematic diagrams load-time (P-t) in instrumented impact tests. The fracture by instable crack (a) and stable one (b).

The fractal dimension d_f is the most general informant of an object structure (in case—polymeric material) and the true structural characteristic, describing structure elements distribution in space. The value d_f can be determined according to the following equation:

$$d_f = 3 - 6\left(\frac{\varphi_{cl}}{C_\infty S}\right)^{1/2},$$

(1)

Where φ_{cl} is a relative fraction of local order domains (clusters) in polymeric material structure, C_∞ is characteristic ratio, which is equal to 7 for polyethylenes, and S is macromolecule cross-sectional area, which is equal to 14.3 Å² for HDPE.

In its turn, the value φ_{cl} is determined according to the following percolation relationship:

$$\varphi_{cl} = 0.03(1 - K)(T_m - T)^{0.55},$$

(2)

Where K is cristallinity degree, equal to 0.48 and 0.55 for neat HDPE and nanocomposite HDPE/CaCO$_3$, respectively, T_m is melting temperature, equal to ~ 406 and 405K for the mentioned materials, respectively, and T is testing temperature.

Let us note, that d_f calculation according to the Equation (1) gives values, corresponding to other methods of this parameter estimation. So, the value d_f can be calculated alternatively according to the following equation:

$$d_f = (d-1)(1+v),$$ (3)

Where d is dimension of Euclidean space, in which fractal is considered (it is obvious, that in our case $d = 3$) and v is Poisson's ratio, estimated with the aid of the relationship:

$$\frac{\sigma_Y}{E} = \frac{1-2v}{6(1+v)} \, ,$$ (4)

Where σ_Y is yield stress and E is elasticity modulus.

The estimations according to the Equations (1) and (4) have given the following values d_f at testing temperature 293K for HDPE 2.73 and 2.68, for nanocomposite HDPE/CaCO$_3$—2.75 and 2.73, respectively. As one can see, the good enough correspondence is obtained (the discrepancy by d_f fractional part, which has the main information amount about structure, does not exceed 7%).

In Figure 2 the dependence $A_p(d_f)$ for the studied polymeric materials is adduced, which has turned out to be linear, common for the near HDPE and nanocomposite HDPE/CaCO$_3$ and is described by the following empirical correlation:

$$A_p = 13.5(d_f - 2.5)\,kj\,/\,m^2$$ (5)

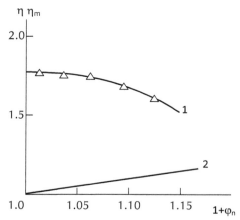

FIGURE 2   The dependence of impact toughness A_p on structure fractal dimension d_f for HDPE (Equation (1)) and nanocomposite HDPE/CaCO$_3$ (Equation (2)).

From the Equation (5) it follows, that at $d_f = 2.5$ the value $A_p = 0$. The mentioned fractal dimension corresponds to the ideally brittle fracture condition, that defines the condition $Ap = 0$. For real solids the greatest fractal dimension of their structure is equal to 2.95, that allows to determine the greatest value A_p according to the Equation (5), which is equal to ~ 6.1 kJ/m².

In general, energy dissipation at an impact grows at polymeric materials molecular mobility level increase. Within the frameworks of fractal analysis this level can be characterized with the aid of the fractal dimension D_{ch} of a polymer chain part between its fixation points (chemical cross-links, physical entanglements nodes, clusters, and so on). Such analysis method can be applied successfully for the value A_p description it case of particulate-filled nanocomposites phenylone/β-sialone. The value D_{ch} can be determined with the aid of the following equation:

$$D_{ch} = \frac{In\,N_{cl}}{In\left(4 - d_f\right) - In\left(3 - df\right)},$$ (6)

Where N_{cl} is a statistical segments number per chain part between clusters, which is determined as follows. Firstly the density of physical entanglements cluster network ν_{cl} is determined:

$$\nu_{cl} = \frac{\varphi_{cl}}{C_\infty l_0 S},$$ (7)

Where l_0 is the main chain skeletal bond length, which is equal to 1.54 Å for polyethylene.

Then the estimation of polymer chains total length per polymer volume unit L was carried out as follows:

$$L = S^{-1}.$$ (8)

The chain part length between clusters L_{cl} is determined according to the equation:

$$L_{cl} = \frac{2L}{\nu_{cl}}.$$ (9)

The statistical segment length l_{st} is determined as follows:

$$l_{st} = l_0 C_\infty.$$ (10)

And at last the value N_{cl} can be determined as ratio:

$$N_{cl} = \frac{L_{cl}}{l_{st}}.$$
(11)

In Figure 3 the dependence of impact toughness A_p on fractal dimension D_{ch} for the studied materials is adduced. As it should be expected, A_p growth at D_{ch} increase is observed, analytically described by the following relationship:

$$A_p = 6.75(D_{ch} - 1), kJ / m^2.$$
(12)

The Equation .(12) allows to determine the greatest value A_p for the studied materials at the condition $D_{ch} = 2.0$, this value is equal to 6.75 kJ/m² that is close to the cited estimation according to the Equation (5)—the average discrepancy makes less than 10%.

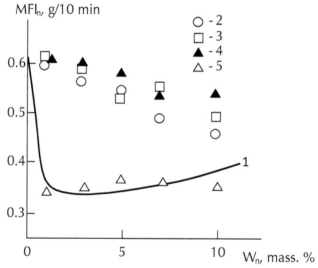

FIGURE 3   The dependence of impact toughness A_p on fractal dimension D_{ch} of chain part between clusters for HDPE (Equation (1)) and nanocomposite HDPE/CaCO₃ (Equation (2)).

KEYWORDS

- **High density polyethylene**
- **Inorganic nanofiller**
- **Nanotechnology**
- **Rheological properties**
- **Viscosity**

REFERENCES

1. Tan, S. H. and Inai, R. *Polymer*, **46**, 6128 (2005).
2. Deitzel, J. M. *Polymer*, **42**, 8163 (2001).
3. Zarkoob, S. *Polymer*, **45**, 3973 (2004).
4. Atheron, S. *Polymer*, **45**, 2017 (2004).
5. He, J. and Wan, Y.Q. *Polymer*, **45**, 19, 6731 (2004).
6. Qin, X. H., Wan, Y. Q., and He, J. H. *Polymer*, **45**, 18, 6409 (2004).
7. Therona, S., Zussmana, E., and Yarin, A. L. *Polymer*, **45**, 2017 (2004).
8. Demir, M., Yilgor, I., Yilgor, E., and Erman, B. *Polymer*, **43**, 3303 (2002).
9. Wan, Y., Guo, Q., and Pan, N. *International Journal of Nonlinear Sciences and Numerical Simulation*, **5**, 5 (2004).
10. He, J., Wan, Y., and Yu, J. *International Journal of Nonlinear Sciences and Numerical Simulation*, **5**, 3, 243 (2004).
11. He, J., Wan, Y. Q., and Yu, J. Y. *International Journal of Nonlinear Sciences and Numerical Simulation*, **5**, 3, 253 (2004).
12. Shin, Y. M., Hohman, M. M., Brenner, M. P., and Rutledge, G. C. *Applied Physics Letters*, **78**, 1149 (2001).
13. Reneker, D. H., Yarin, A. L., Fong, H., and Koombhongse, S. *Journal of Applied Physics*, **87**, 4531 (2000).
14. Sukigara, S., Gandhi, M., Ayutsede, J., Micklus, M., and Ko, F. *Polymer*, **44**, 5727 (2003).
15. Sukigara, S., Gandhi, M., Ayutsede, J., Micklus, M., and Ko, F. *Polymer*, **45**, 3708 (2004).
16. Park, K. E., Jung, S. Y. Lee, S. J., Min, B. M., and Park, W. H. *International Journal of Biological Macromolecules*, **38**, 165 (2006).
17. Yuan, X., Zhang, Y., Dong, C., and Sheng, J. *Polymer International*, **53**, 1704 (2004).
18. Ki, C. S., Baek, D. H., Gang, K. D., Lee, K. H., Um, I. C., and Park, Y. H. *Polymer*, **46**, 5094 (2005).
19. Deitzel, J. M., Kleinmeyer, J., Harris, D., and Beck Tan, N. C. *Polymer*, **42**, 261 (2001).
20. Buchko, C. J., Chen, L. C., Shen, Y., and Martin, D. C. *Polymer*, **40**, 7397 (1999).
21. Lee, J. S., Choi, K. H., Ghim, H. D., Kim, S. S., Chun, D. H., Kim, H. Y., and Lyool, W. S. *Journal of Applied Polymer Science*, **93**, 1638 (2004).
22. Fennessey, S. F. and Farris, R. J. *Polymer*, **45**, 4217 (2004).
23. Kidoaki, S., Kwon, I. K., and Matsuda, T. *Biomaterials*, **26**, 37 (2005).
24. Zong, X., Kim, K., Fang, D., Ran, S., Hsiao, B. S., and Chu, B. *Polymer*, **43**, 4403 (2002).
25. Li, D. and Xia, Y. *Nano Letters*, **3**, 4, 555 (2003).
26. Jin, W. Z., Duan, H. W., Zhang, Y. J., and Li, F. F. In the *Proceedings of the 1st IEEE International Conference on Nano/Micro Engineered and Molecular Systems*, Zhuhai, China, p.42 (2006).
27. Ryu, Y. J., Kim, H. Y., Lee, K. H., Park, H. C., and Lee, D. R. *European Polymer Journal*, **39**, 1883 (2003).
28. Mo, X. M., Xu, C.Y., Kotaki, M., and Ramakrishna, S. *Biomaterials*, **25**, 1883 (2004).
29. Zhao, S., Wu, X., Wang, L., and Huang, Y. *Journal of Applied Polymer Science*, **91**, 242 (2004).

CHAPTER 18

NANOPARTICLES IN POLYMERIC NANOCOMPOSITES

A. K. HAGHI and G. E. ZAIKOV

CONTENTS

18.1 INTRODUCTION

The nanoparticles structure is defined by chemical interactions between atoms nature forming them. The fundamental properties of nanoparticles, forming in strongly non-equilibrium conditions are their ability to:
- Structures self-organization by adaptation to external influence,
- An optimal structure self-choice in bifurcation points, corresponding to preceding structure stability threshold and new stable formation,
- A self-operating synthesis (self-assembly) of stable nanoparticles, which is ensured by information exchange about system structural state in the previous bifurcation point at a stable structure self-choice in the following beyond it bifurcation point.

These theoretical postulates were confirmed experimentally. In particular, it has been shown that nanoparticles sizes are not arbitrary ones, but change discretely and obey to synergetics laws. This postulate is important from the practical point of view, since nanoparticles size is the information parameter, defining surface energy critical level [1-15].

Let us consider these general definitions in respect to polymer particulate filled nanocomposites, for which there are certain distinctions with the considered the criterions. As it is well-known, nanofiller aggregation processes in either form are inherent in all types of polymer nanocomposites and influence essentially on their properties. In this case, although nanofiller initial particles have size (diameter) less than 100 nm, but these nanoparticles aggregates can exceed essentially the indicated the boundary value for nanoworld objects. Secondly, nanofiller particles aggregates are formed at the expense of physical interactions, but not chemical ones. Therefore, the present update reports the synergetics laws applicability for nanofiller aggregation processes and interfacial phenomena description in particulate-filled polymer nanocomposites on the example of nanocomposite polypropylene (PP)/calcium carbonate ($CaCO_3$) [16-37].

18.2 EXPERIMENTAL

The PP with average molecular weight M_w was used as matrix polymer and nanodimensional $CaCO_3$ in the form of compound was used as nanofiller.

The nanocomposites $PP/CaCO_3$ was prepared by components mixing in melt on twin-screw extruder. Mixing was performed at temperature 463–503 K and screw speed of 50 rpm during 5 min. Testing samples were obtained by casting under pressure method on casting machine at temperature of 483 K and pressure of 43 MPa.

The electron microscopy was used for nanocomposites $PP/CaCO_3$ structure study. The study objects were prepared in liquid nitrogen with the purpose of microscopic sections obtaining. The scanning electron microscope with autoemissive cathode of high resolution was used for microscopic sections surface images obtaining. The images were obtained in the mode of low-energetic secondary electrons, since this mode ensures the highest resolution.

Uniaxial tension mechanical tests have been performed on the samples. The tests have been conducted on universal testing apparatus at temperature of 293 K and strain rate $\sim 2 \times 10^{-3}$ s^{-1}.

18.3 DISCUSSION AND RESULTS

A particulate nanofiller particles aggregate size (diameter) D_{agr} estimation can be performed according to the following formula:

$$k(r)\lambda = \left[\left(\frac{25.11\pi D_{agr}^{1/3}}{W_n}\right)^{1/3} - 2\right]\frac{D_{agr}}{2} \tag{1}$$

where $k(r)$ is an aggregation parameter, λ is distance between nanofiller particles, and W_n is nanofiller mass contents in mass%.

In its turn, the value $k(r)\lambda$ is determined within the frameworks of strength dispersion theory with the help of the relationship:

$$\tau_n = \tau_m + \frac{Gb_B}{k(r)\lambda} \tag{2}$$

where τ_n and τ_m are yield stress in compression testing of nanocomposite and matrix polymer, accordingly, G is shear modulus and b_B is Burders vector.

The general relationship between normal stress σ and shear stress τ has the following look:

$$\tau = \frac{\sigma}{\sqrt{3}}. \tag{3}$$

Young's modulus E and shear modulus G are connected between themselves by the simple relationship:

$$G = \frac{E}{d_f} \tag{4}$$

where d_f is nanocomposite structure fractal dimension, which is determined according to the equation:

$$d_f = (d-1)(1+v), \tag{5}$$

where d is the dimension of Euclidean space, in which fractal is considered and v is Poisson's ratio, which is estimated by the mechanical testing results with the help of the relationship:

$$\frac{\sigma_Y}{E_n} = \frac{1-2v}{6(1-v)}, \tag{6}$$

where σ_Y and E_n are yield stress and elasticity modulus of nanocomposite, respectively.

Burgers vector value b_B for polymeric materials is determined according to the equation [11]:

$$b_B = \left(\frac{60.5}{C_\infty}\right)^{1/2} \tag{7}$$

where C is characteristic ratio, connected with d_f by the equation:

$$C_\infty = \frac{2d_f}{d(d-1)(d-d_f)} + \frac{4}{3} \tag{8}$$

The calculation according to the Equations (1–8) showed CaCO$_3$ nanoparticles aggregates mean diameter growth from 85 nm up to 190 nm within the range of W_n = 1–7 mass% for the considered nanocomposites PP/CaCO$_3$. These calculations can be confirmed experimentally by the electron microscopy methods. In Figure 1 the nanocomposites PP/CaCO$_3$ with nanofiller contents W_n = 1 mass% and 4 mass% sections electron micrographs are adduced. As one can see, if at W_n = 1 mass% nanofiller particles are not aggregated practically, that is, their diameter is close to CaCO$_3$ initial nanoparticles diameter ($\sim$ 80 nm), then at W_n = 4 mass% the initial nanoparticles aggregation is observed even visually and these particles aggregates sizes are varied within the limits of 80–360 nm.

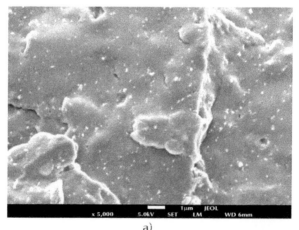

a)

FIGURE 1   *(Continued)*

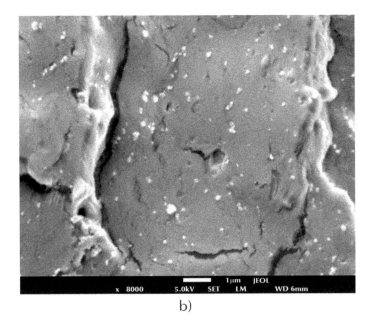

b)

FIGURE 1   Electron micrographs of sections of nanocomposites PP/CaCO3 with nanofiller mass contents Wn = 1 (a) and 4 (b) mass%.

The adduced estimations correspond to the results of calculation according to the Equation (1) at the indicated $CaCO_3$ contents, 85 nm and 142 nm, accordingly. Hence, the considered technique gives reliable enough estimations of nanofiller particles aggregates diameter.

It has been shown earlier on the example of different physical-chemical processes, that the self-similarity function has an iteration type function look, connecting structural bifurcation points by the relationship:

$$A_m = \frac{\lambda_n}{\lambda_{n+1}} = \Delta_i^{1/m} \qquad (9)$$

where A_m is the measure of aggregate structure adaptability to external influence, λ_n and λ_{n+1} are preceding and subsequent critical values of operating parameter at the transition from preceding to subsequent bifurcation point, Δ_i is the structure stability measure, remaining constant at its reorganization up to symmetry violation, and m is an exponent of feedback type, the value $m = 1$ corresponds to linear feedback, at which transitions on other spatial levels are realized by multiplicative structure reproduction mechanism and at $m \geq 2$ (nonlinear feedback) – replicative (with structure improvement) one.

Selecting as the operating parameter critical value nanoparticles aggregates diameter Dagr at successive Wn change, the dependence of adaptability measure Am on Wn can be plotted, which is shown in Figure 2. As one can see, for the considered nanocomposites the condition is fulfilled with precision of 2%.

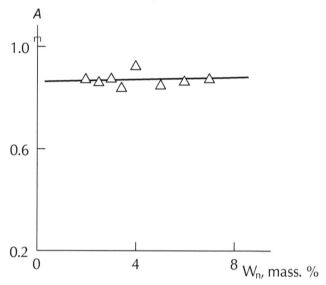

FIGURE 2   The dependence of adaptability measure Am on nanofiller mass contents Wn for nanocomposites PP/CaCO$_3$.

$$A_m = \frac{D_{agri}}{D_{agri+1}} = \text{const} = 0.899 \qquad (10)$$

This means, that aggregation processes in the considered nanocomposites obey to the synergetics laws, although their aggregates size exceeds the boundary value of 100 nm for nanoworld [3]. Let us note the important aspect of the dependence $A_m(W_n)$, adduced in Figure 2. The condition A_m = const is kept irrespective of gradation, with which W_n changes –0.5 mass% or 1.0 mass%.

As it is known, an interfacial layer in polymer nanocomposites can be considered as a result of two fractal objects (polymer matrix and nanofiller particles surface) interaction, for which there is the only linear scale l, defining these objects interpenetration distance. Since the filler elasticity modulus is, as a rule, considerably higher than the corresponding parameter for polymer matrix, then the indicated interaction comes to filler indentation in polymer matrix and then $l = l_{if}$, where l_{if} is interfacial layer thickness. In this case it can be written as:

$$l_{if} = a\left(\frac{D_{agr}}{2a}\right)^{2(d-d_{surf})/d} \qquad (11)$$

where a is a lower linear scale of fractal behavior, which for polymeric materials is accepted equal to statistical segment length l_{st}, d_{surf} is nanofiller particles (aggregates of particles) surface fractal dimension.

l_{st} is determined according to the equation:

$$l_{st} = l_0 C_\infty,$$

(12)

where l_0 is the main chain skeletal bond length, equal to 0.154 nm for PP.

The dimension d_{surf} is calculated in the following succession. First the nanofiller particles aggregate density ρ_n is estimated according to the formula:

$$\rho_n = 188 \left(D_{agr} \right)^{1/3} \ kg/m^3,$$

(13)

then the indicated aggregate specific surface S_u is determined:

$$S_u = \frac{6}{\rho_n D_{agr}}$$

(14)

And at last, the value d_{surf} calculation can be fulfilled with the help of the equation:

$$S_u = \frac{6}{\rho_n D_{agr}}$$

(15)

where S_u is given in m²/g, D_{agr} – in nm.

The calculation according to the offered technique has shown l_{if} increase from 1.78 nm up to 5.23 nm at W_n enhancement within the range of 1–7 mass%. The estimations according to the Equation (9), where as λ_n and λ_{n+1} the values l_{if_n} and $l_{if_{n+1}}$ were accepted, showed that the following condition was fulfilled:

$$S_u = 410 \left(\frac{D_{agr}}{2} \right)^{d_{surf} - d}$$

(16)

With the precision of 7%. Hence, an interfacial layers formation in polymer nanocomposites, characterizing interfacial phenomena in these nanomaterials, obeys to the synergetics laws with the same adaptability measure, as nanofiller particles aggregation. Nevertheless, it should be noted, that this analogy is not complete for the considered nanocomposites within the range of W_n = 1–7 mass% the value D_{agr} increases in 2.24 times, whereas the value l_{if} does almost in three times.

Let us consider further the exponent m in the Equation (9) choice, characterizing feedback type in aggregation process. As it has been noted, this exponent is equal to 2 in case of aggregates structure improvement, which can be characterized by their fractal dimension d_f^{agr}. This dimension can be calculated with the help of the equation:

$$\rho_n = \rho_{dens}\left(\frac{D_{agr}}{2a}\right)^{d_f^{agr}-d},\qquad(17)$$

where ρ_{dens} is massive material density, which is equal to 2000 kg/m³ for CaCO$_3$, a is a lower linear scale of fractal behavior, accepted equal to 10 nm.

In Figure 3 the dependence of d_f^{agr} on CaCO$_3$ mass contents W_n for the considered nanocomposites is adduced. As one can see, within the studied range of W_n the essential d_f^{agr} growth (from 2.34 up to 2.73 at general d_f^{agr} variation within the limit of 2.0 to 2.95) is observed, that can be classified as nanofiller particles aggregates structure improvement, as a minimum, by two reasons—their disaggregation level reduction and critical structural defect sizes decreasing. Therefore, proceeding from the said, it should be accepted that $m = 2$, which according to the Equation (9) gives $\Delta_I = 0.808$. Let us note, that Δ_I value defines very stable nanostructures. So, for self-operating nano-solid solutions synthesis the values $\Delta_I = 0.255–0.465$ at $m = 2$ were obtained and in addition it has been shown that an optimal technological regime indicator is $\Delta_I = 0.465$ attainment at nonlinear feedback ($m = 2$) realization.

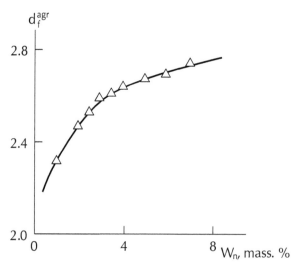

FIGURE 3 The dependence of nanofiller particles aggregates structure fractal dimension d_f^{agr} on its mass contents W_n for nanocomposites PP/CaCO$_3$.

Let us consider in conclusion the possibility of $CaCO_3$ nanoparticles aggregation process prediction within the frameworks of synergetics treatment. In Figure 4 the comparison of $CaCO_3$ aggregates diameter values, calculated according to the Equation (1) D_{agr} and Equation (9) D_{agr}^{syn} at A_m = const = 0.899. As one can see, this comparison demonstrates very good conformity of nanofiller particles aggregates diameter, calculated by both indicated methods (the average discrepancy of D_{agr} and D_{agr}^{syn} makes up 2%).

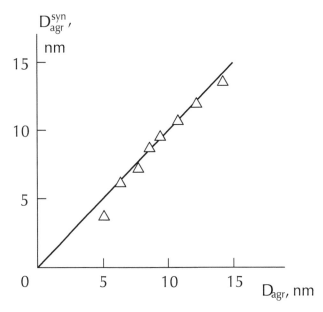

FIGURE 4 The comparison of nanofiller particles aggregates diameter, calculated according to the Equation (1) D_{agr} and Equation (9) D_{agr}^{syn}, for nanocomposites PP/CaCO_3.

18.4 CONCLUSION

Hence, the present chapter results have demonstrated that disperse nanoparticles aggregates formation in polymer nanocomposites obeys to synergetics laws even at these aggregates size larger than upper dimensional boundary for nanoparticles, equal to 100 nm. These aggregates are formed by structure reproduction replicative mechanism with nonlinear feedback. In this case nanofiller contents discrete change points are bifurcation points. The synergetics methods can be used for prediction of forming nanoparticles aggregates size. The parameters, characterizing interfacial phenomena in polymer nanocomposites (for example, interfacial layer thickness) obey also to synergetics laws.

KEYWORDS

- **Calcium carbonate**
- **Electron microscopy**
- **Nanoparticles**
- **Polypropylene**
- **Young's modulus**

REFERENCES

1. Ziabari, M., Mottaghitalab, V., McGovern, S. T., and Haghi, A. K. *Chemical Physics Letters*, **25**, 3071 (2008).
2. Ziabari, M., Mottaghitalab, V., McGovern, S. T., and Haghi, A. K. *Nanoscale Research Letter*, **2**, 297 (2007).
3. Ziabari, M., Mottaghitalab, V., and Haghi, A. K. *Korean Journal of Chemical Engineering*, **25**, 905 (2008).
4. Ziabari, M., Mottaghitalab, V., and Haghi, A. K. *Korean Journal of Chemical Engineering*, **25**, 919 (2008).
5. Haghi, A. K. and Akbari, M. *Physica Status Solidi*, **204**, 1830 (2007).
6. Kanafchian, M., Valizadeh, M., and Haghi, A. K. *Korean Journal of Chemical Engineering*, **28**, 428 (2011).
7. Kanafchian, M., Valizadeh, M., and Haghi, A. K. *Korean Journal of Chemical Engineering*, **28**, 763 (2011).
8. Kanafchian, M., Valizadeh, M., and Haghi, A. K. *Korean Journal of Chemical Engineering*, **28**, 751 (2011).
9. Kanafchian, M., Valizadeh, M., and Haghi, A. K. *Korean Journal of Chemical Engineering*, **28**, 445 (2011).
10. Afzali, A., Mottaghitalab, V., Motlagh, M., and Haghi, A. K. *Korean Journal of Chemical Engineering*, **27**, 1145 (27).
11. Wan, Y. Q., Guo, Q., and Pan, N. *International Journal of Nonlinear Sciences and Numerical Simulation*, **5**, 5 (2004).
12. Feng, J. J. *Journal of Non-Newtonian Fluid Mechanics*, **116**, 55 (2003).
13. He, J., Wan, Y., and Yu, J. Y. *Polymer*, **46**, 2799 (2005).
14. Zussman, E., Theron, A., and Yarin, A. L. *Applied Physics Letters*, **82**, 73 (2003).
15. Reneker, D. H., Yarin, A. L., Fong, H., and Koombhongse, S. *Journal of Applied Physics*, **87**, 4531 (2000).
16. Theron, S. A., Yarin, A. L., Zussman, E., and Kroll, E. *Polymer*, **46**, 2889 (2005).
17. Huang, Z. M., Zhang, Y. Z., Kotak, M., and Ramakrishna, S. *Composites Science and Technology*, **63**, 2223 (2003).
18. Schreuder-Gibson, H. L., Gibson, P., Senecal, K., Sennett, M., Walker, J., Yeomans, W., and Ziegler, D. *Journal of Advanced Materials*, **34**(3), 44 (2002).
19. Ma, Z., Kotaki, M., Inai, R., and Ramakrishna, S. *Tissue Engineering*, **11**, 101 (2005).
20. Ma, Z., Kotaki, M., Yong, T., He, W., and Ramakrishna, S. *Biomaterials*, **26**, 2527 (2005).
21. Jin, H. J., Fridrikh, S., Rutledge, G. C., and Kaplan, D. *Abstracts of Papers American Chemical Society*, **224**(1–2), 408 (2002).

22. Luu, Y. K., Kim, K., Hsiao, B. S., Chu, B., and Hadjiargyrou, M. *Journal of Controlled Release*, **89**, 341 (2003).
23. Senecal, K. J., Samuelson, L., Sennett, M., and Schreuder-Gibson, H. L. Inventors; US 0045547, (2001).
24. Sawicka, K., Goum, P., and Simon, S. *Sensors and Actuators B*, **108**, 585 (2005).
25. Fujihara, K., Kotak, M., and Ramakrishn, S. *Biomaterials*, **26**, 4139 (2005).
26. Fang, X. and Reneker, D. H. *Journal of Macromolecular Science, Part B: Physics*, **36**, 169 (1997).
27. Taylor, G. I. *Proceedings of the Royal Society Series A*, **313**, 453 (1969).
28. Kenawy, E. R., Bowlin, G. L., Mansfield, K., Layman, J., Simpson, D. G., Sanders, E. H., and Wnek, G. E. *Journal of Controlled Release*, **81**, 57 (2002).
29. Fennessey, S. F. and Farris, J. R. *Polymer*, **45**, 4217 (2004).
30. Zussman, E., Theron, A., and Yarin, A. L. *Applied Physics Letters*, **82**, 973 (2003).
31. Deitzel, J. M., Kleinmeyer, J., Harris, D., and Beck, T. N. *Polymer*, **42**, 261 (2001).
32. Zhang, C. H., Yuan, X., Wu, L., Han, Y., and Sheng, J. *European Polymer Journal*, **41**, 423 (2005).
33. Spivak, A. F. and Dzenis, Y. A. *Applied Physics Letters*, **73**, 3067 (1998).
34. Hohman, M. M., Shin, M., Rutledge, G., and Brenner, M. P. *Physics of Fluids*, **13**, 2201 (2001).
35. Hohman, M. M., Shin, M., Rutledge, G., and Brenner, M. P. *Physics of Fluids*, **13**, 2221 (2001).
36. Reneker, D. H., Yarin, A. L., Fong, H., and Koombhongse, S. *Journal of Applied Physics*, **87**, 4531 (2000).
37. Yarin, A. L., Koombhongse, S., and Reneker, D. H. *Journal of Applied Physics*, **89**, 47 (2001).

CHAPTER 19

BIODEGRADATION MECHANISM OF SOME POLYMERS

ELENA L. PEKHTASHEVA and GENNADY E. ZAIKOV

CONTENTS

19.1 INTRODUCTION

This chapter is including information about biodamages and protection of textile materials and fibers, cotton fiber biodamaging, bast fiber biodegradation, biodamaging of artificial fibers and wool fiber, and biodegradation of synthetic fibers. The chapter also is including information about methods of textile material protection against damaging by microorganisms.

19.2 BIODAMAGES AND PROTECTION OF TEXTILE MATERIALS AND FIBERS

Materials produced from fibers and threads are classified as textile—fabrics, nonwoven materials, fur fabric, carpets and rugs, and so on [1].

Textile fibers are the main raw material of the textile industry. According to their origin, these fibers are divided into natural and chemical ones.

Natural fibers are plant (cotton and bast fibers), animal (wool, silk), and mineral (asbestos) origin ones.

Chemical fibers are produced from modified natural or synthetic high-molecular substances and are classified as artificial ones obtained by chemical processing of natural raw material, commonly cellulose (viscose, acetate), and synthetic ones obtained from synthetic polymers (nylon-6, polyester, acryl, PVC fibers, and so on.).

Textile materials are damaged by microorganisms, insects, rodents, and other biodamaging agents. Fibers and fabrics resistance to biodamages primarily depends upon chemical nature of the fibers from which they are made. Most frequently, we have to put up with microbiological damages of textile materials based on natural fibers—cotton, linen, and so on, utilized by saprophyte microflora. Chemical fibers and fabrics, especially synthetic ones, are higher biologically resistant but microorganisms-biodegraders can also adapt to them.

Textile material degradation by microorganisms depends on their wear rate, kind and origin, organic composition, temperature and humidity conditions, degree of aeration, and so on.

With increased humidity and temperature, and restricted air exchange microorganisms damage fibers and fabrics at different stages of their manufacture and application, starting from the primary processing of fibers including spinning, weaving, finishing and storage, transportation and operation of textile materials and articles from them. Fiber and fabric biodamage intensity sharply increases when contacting with the soil and water, specifically in the regions with warm and humid climate.

Textile materials are damaged by bacteria and microscopic fungi. Bacterial degradation of textile materials is more intensive than the fungal one. The damaging bacterium genus are *Cytophaga, Micrococcus, Bacterium, Bacillus, Cellulobacillus, Pseudomonas,* and *Sarcina.* Among fungi damaging textile materials in the air and in the soil the following are detected: *Aspergillus, Penicillium, Alternaria, Cladosporium, Fusarium, Trichoderma,* and so on.

Annual losses due to microbiological damaging of fabrics reach hundreds of millions of dollars [2-10].

Fiber and fabric biodamaging by microorganisms is usually accompanied by the mass loss and mechanical strength of the material as a result, for example, of fiber degradation by microorganism metabolites enzymes, organic acids, and so on.

CURIOUS FACTS

The immediate future of the textile industry belongs to biotechnology. Even today suggestions on the synthesis of various polysaccharides using microbiological methods are present. These methods may be applied to synthesis of fiber-forming monomers and polymers.

Scientists have demonstrated possibilities of microbiological synthesis of some monomers to produce dicarboxylic acids, caprolactam, and so on. Some kinds of fiber-forming polymers, polyethers, in particular, can also be obtained by microbiological synthesis.

Some kinds of fiber-forming polypeptides have already been obtained by the microbiological synthesis. In some cases, concentration of these products may reach 40% of the biomass weight and they can be used as a perspective raw material for synthetic fibers. Studies in this direction are widely performed in many countries all over the world.

The impact of microorganisms on textile materials that causes their degradation is performed, at least, by two main ways (direct and indirect):
1. Fungi and bacteria use textile materials as the nutrient source (assimilation)
2. Textile materials are damaged by microorganism metabolism (degradation)

Biodamages of textile materials induced by microorganisms and their metabolites manifest in coloring (occurrence of spots on textile materials or their coatings), defects (formation of bubbles on colored surfaces of textile materials), bond breaks in fibrous materials, penetration deep inside (penetration of microorganisms into the cavity of the natural fiber), deterioration of mechanical properties (for example strength at break reduction), mass loss, a change of chemical properties (cellulose degradation by microorganisms), liberation of volatile substances, and changes of other properties.

It is known that when microorganisms completely consume one part of the substrate they then are able to liberate enzymes degrading other components of the culture medium. It is found that, degrading fiber components each group of microorganisms due to their physiological features decomposes some definite part of the fiber, damages it differently and in a different degree. It is found that along with the enzymes, textile materials are also degraded by organic acids produced by microorganisms—actic, gluconic, acetic, succinic, fumaric, malic, citric, oxalic, and so on. It is also found that enzymes and organic acids liberated by microorganisms continue degrading textile materials even after microorganisms die. As noted, the content of cellulose, proteins, pectins, and alcohol soluble waxes increases with the fiber damage degree in it; pH increases, and concentration of water-soluble substances increases that, probably, is explained by increased accumulation of metabolites and consumption of nutrients by microorganisms for their vital activity.

`The typical feature of textile material damaging by microorganisms is occurrence of honey dew, red violet or olive spots with respect to a pigment produced by microorganisms, and fabric color. As microorganism's pigment interacts with the fabric dye, spots of different hue and tints not removed by laundering or by hydrogen peroxide oxidation. They may sometimes be removed by hot treatment in blankly solution. Spot occurrence on textile materials is usually accompanied by a strong musty odor.

CURIOUS FACTS

Textile materials are also furnished by biotechnological methods based on the use of enzymes performing various physicochemical processes. Biotechnologies are mostly full for preparation operations (cloth softening, boil-off and bleaching of cotton fabrics, and wool washing) and bleaching effects of jeans and other fabrics, bio-polishing, and making articles softer.

Enzymes are used to improve sorption properties of cellulose fibers, to increase specific area and volume of fibers, to remove pectin "companions" of cotton and linen cellulose. Enzymes also hydrolyze ether bonds on the surface of polyether fibers.

Cotton fabric dyeing technologies in the presence of enzymes improving coloristic parameters of prepared textile articles under softer conditions, which reduce pollutants in the sewage. To complete treatment of the fabric surface, that is removal of surface fiber fibrilla, enzymes are applied. The enzymatic processing also decreases and even eliminates the adverse effect of long wool fiber puncturing. The application of enzymes to treatment of dyed tissues is industrially proven. This processing cause's irregular bleaching that adhere the articles fashionable "worn" style.

Temperature and humidity are conditions promoting biodamaging of fibrous materials. The comparison of requirements to biological resistance of textile materials shall be based on the features of every type of fibers, among which mineral fiber are most biologically resistant.

Different microorganism damaging rate for fabrics is due to their different structure. Thinner fabrics with the lower surface density and higher through porosity are subject to the greatest biodamaging, because these properties of the material provide large contact area for microorganisms and allow their easy penetration deep into it. The thread bioresistance increases with the yarning rate.

19.3   COTTON FIBER BIODAMAGING

Cotton fiber is a valuable raw material for the textile industry. Its technological value is due to a complex of properties that should be retained during harvesting, storage, primary, and further processing to provide high quality of products. One of the factors providing retention of the primary fiber properties is its resistance to bacteria and fungi impacts. This is tightly associated with the chemical and physical features of cotton fiber structures [11].

Ripe cotton fiber represents a unit extended plant cell shaped as a flattened tube with a corkscrew waviness. The upper fiber end is cone-shaped and dead-ended. The lower end attached to the seed is a torn open channel.

The cotton fiber structure is formed during its maturation when cellulose is biosynthesized and its macromolecules are regularly disposed.

The basic elements of cotton fiber morphological structure are known to be the cuticle, primary wall, convoluted layer, secondary wall, and tertiary wall with the central channel [11-14]. The fiber surface is covered by a thin layer of wax substances—the primary wall (cuticle). This layer is a protective one and possesses rather high chemical resistance.

There is a primary wall under the cuticle consisting of a cellulose framework and fatty-wax-pectin substances. The upper layer of the primary wall is less densely packed as compared with the inner one, due to fiber surface expansion during its growth. Cellulose fibrils in the primary wall are not regularly oriented.

The primary layer covers the convoluted layer having structure different from the secondary wall layers. It is more resistant to dissolution, as compared with the main cellulose mass.

The secondary cotton fabric wall is more homogeneous and contains the greatest amount of cellulose. It consists of densely packed and concurrently oriented cellulose fibrils composed in thin layers. The fibril layers are spiral shaped twisted around the fiber axis. There are just few micropores in this layer.

The tertiary wall is an area adjacent to the fiber channel. Some authors think that the tertiary wall contains many pores and consists of weakly ordered cellulose fibrils and plenty of protein admixtures, protoplasm, and pectin substances.

The fiber channel is filled with protoplasm residues which are proteins, and contains various mineral salts and a complex of microelements. For the mature fiber, channel cross-section is 4–8% of total cross-section.

Cotton fiber is formed during maturation. This includes not only cellulose biosynthesis, but also ordering of the cellulose macromolecules shaped as chains and formed by repetitive units consisting two β-D-glucose residues bound by glycosidic bonds. Among all plant fibers, cotton contains the maximum amount of cellulose (95–96%).

The morphological structural unit of cellulose is a cluster of macromolecules—a fibril 1.0–1.5 µm long and 8–15 nm thick, rather than an individual molecule. Cellulose fibers consist of fibril clusters uniform oriented along the fiber or at some angle to it.

It is known that cellulose consists not only of crystalline areas—micelles where molecule chains are concurrently oriented and bound by intermolecular forces, but also of amorphous areas.

Amorphous areas in the cellulose fiber are responsible for the finest capillaries formation, that is "sub-microscopic" space inside the cellulose structure is formed. The presence of the submicroscopic system of capillaries in the cellulose fibers is of paramount significance, because it is the channel of chemical reactions by which water-soluble reagents penetrate deep in the cellulose structure. More active hydroxyl groups interacting with various substances are also disposed here [15–18].

It is known that hydrogen bonds are present between hydroxyl groups of cellulose molecules in the crystalline areas. Hydroxyl groups of the amorphous area may occur free of weakly bound and as a consequence, they are accessible for sorption. These hydroxyl groups represent active sorption centers able to attract water.

Among all plant fibers, cotton has the highest quantity of cellulose (95–96%). Along with cellulose, the fibers contain some fatty, wax, coloring mineral substances (4–5%). Cellulose concomitant substances are disposed between macromolecule clusters and fibrils. Raw cotton contains mineral substances (K, Na, Ca, and Mg) that promote mold growth and also contains microelements (Fe, Cu, and Zn) stimulating growth of microorganisms. Moreover, it contains sulfates, phosphorus, glucose, glycidols, and nitrogenous substances, which also stimulate growth of microbes. Differences in their concentrations are one of the reasons for different aggressiveness of microorganisms in relation to the cotton fiber.

The presence of cellulose, pectin, nitrogen-containing, and other organic substances in the cotton fiber, as well as its hygroscopicity makes it a good culture medium for abundant microflora.

Cotton is infected by microorganisms during harvesting, transportation, and storage. When machine harvested, the raw cotton is clogged by various admixtures. It obtains multiple fractures of leaves and cottonseed hulls with humidity higher than of the fibers. Such admixtures create a humid macro zone around, where microorganisms intensively propagate[13]. Fiber humidity above 9% is the favorable condition for cotton fiber degradation by microorganisms.

It is found that cotton fiber damage rate directly in hulls may reach 42–59%, hence, the fiber damage rate depends on a number of factors, for example cultivation conditions, harvesting period, type of selection, and so on.

Cotton fiber maturity is characterized by filling with cellulose. As the fiber becomes more mature, its strength, elasticity, and coloring value increase.

Low quality cotton having higher humidity is damaged by microorganisms to a greater extent. The fifths fibers contain microorganisms 3–5 times more than the first quality fibers. When cotton hulls open, the quantity of microorganisms sharply increases in them, because along with dust wind brings fungus spores and bacteria to the fibers.

Cotton is most seriously damaged during storage in compartments with high humidity up to 24% of cotton is damaged. Cotton storage in compact bales covered by tarpaulin is of high danger, especially after rains. For instance, after one and half months of such storage the fiber is damaged by 50% or higher.

Cotton microflora remains active under conditions of the spinning industry. As a result, the initial damage degree of cotton significantly increases.

CURIOUS FACTS

On some textile enterprises of Ivanovskaya Oblast, sickness cases of spinning-preparatory workshops were observed. When processing biologically contaminated cotton, plenty of dust particles with microorganisms present on them are liberated to the air. This can affect the health status of the employees.

Under natural conditions, cotton products are widely used in contact with the soil (fabrics for tents) receiving damage both from the inside and the outside. The main role here is played by cellulose degrading bacteria and fungi.

It was considered over a number of years that the main role in cotton fiber damage is played by cellulose degrading microorganisms [13]. Not denying participation of cellulose degrading bacteria and fungi in damaging of the cotton fiber, it is noted that a group of bacteria with yellow pigmented mucoid colonies representing epiphytic microflora always present on the cotton plant dominate in the process of the fiber degradation. Nonspore-forming epiphytic bacteria inhabiting in the cotton plants penetrate from their seeds into the fiber channel and begin developing there. Using chemical substances of the channel, these microorganisms then permeate into the submicroscopic space of the tertiary wall primarily consuming pectins of the walls and proteins of the channels.

Enzymes and metabolites produced by microorganisms induce hydrolysis of cellulose macromolecules, increasing damage of internal areas of the fiber. Thus, the fiber delivered to processing factories may already be significantly damaged by microorganisms that inevitably affect production of raw yarn, fabric, and so on.

135 strains of fungi of different genus capable of damaging cotton fibers are currently determined [13]. It is found that the population of phytopathogenic fungi is much lower than that of cellulose degraders—*Chaetomium globosum, Aspergillus flavus, Aspergillus niger, Rhizopus nigricans, and Trichothecium roseum*. These species significantly deteriorate the raw cotton condition sharply reduce spinning properties of the fiber, in particular.

It is also found that the following species of fungi are usually present on cotton fibers—*Mucor* (consumes water-soluble substances), *Aspergillus* and *Penicillium* (consumes insoluble compounds), *Chaetomium, Trichoderma*, and so on. (degrade cellulose) [13]. This points to the fact that some species of mold fungi induce the real fiber decomposition that shall be distinguished from simple surface growth of microorganisms. For example, *Mucor* fungi incapable of inducing cellulose degradation may actively vegetate on the yarn finish [14]. Along with fungi, bacteria are always present on the raw cotton, most represented by *Bacillus* and *Pseudomonas* genus species.

Figures 1 and 2 present surface micrographs of the first and fifth quality grade primary cotton fibers. Figure 3, a micrograph, shows the cotton fiber surface after the impact of spontaneous microflora during 7 days. It is observed that bacterial cells are accumulated in places of fiber damage, at clearly noticeable cracks. Figure 4 shows the first quality cotton fiber surface after impact of *Aspergillus niger* culture during 14 days. On the fiber surface mycelium is observed. Figures 5 and 6 show photos of cotton fiber surfaces infected by *Bac. subtilis* (14 day exposure). Bacterial cells on first quality fibers are separated by the surface, forming no conglomerates, whereas on the fifth quality fibers conglomerates are observed that indicates their activity (Figure 7).

Academician A. A. Imshenetsky has demonstrated that aerobic cellulose bacteria are able to propagate under increased humidity, whereas fungi propagate at lower humidity. Textile products are destroyed by fungi at their humidity about 10%, whereas bacteria destroy them at humidity level of, at least, 20%. As a consequence, the main attention at cotton processing to yarn should be paid to struggle against fungi, and at wet spinning and finishing fabrics and knitwear not only fungi, but mostly bacteria should be struggled against.

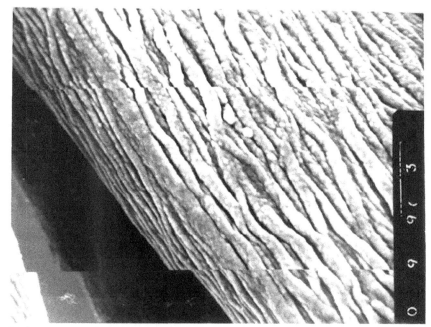

FIGURE 1   The surface of initial first quality cotton specimen (x 4500).

FIGURE 2   The surface of initial fifth quality cotton specimen (x 4500).

FIGURE 3   The first quality cotton, 7 days exposure (spontaneous microflora) (x 10000).

FIGURE 4   The first quality cotton, *Asp. niger* contaminated, 14 days exposure (x 3000).

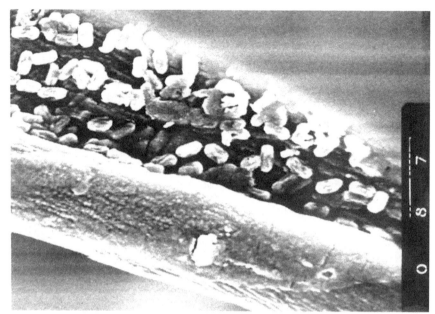

FIGURE 5   The first quality cotton, *Bacillus subtilis* contaminated, 14 day exposure (x 4500).

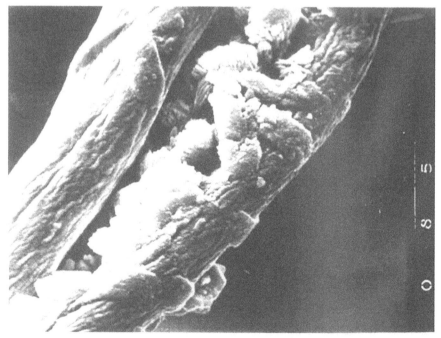

FIGURE 6   The fifth quality cotton, *Bacillus subtilis* contaminated, 14 day exposure (x 4500).

Cotton damage leads to:
- Significant decrease of strength of the fibers and articles from them,
- Disturbance of technological process (the smallest particles of sticky mucus excreted by some species of bacteria and fungi become the reason for sticking executive parts of machines),
- Abruptness increase,
- Waste volume increase.

Damaging of cotton fibers, fabrics, and textile products by microorganisms is primarily accompanied by occurrence of colored yellow, orange, red, violet, and so on spots and then by putrefactive odor, and, finally, the product loses strength and degrades [18-23].

The effect of microorganisms results in noticeable changes in chemical composition and physical structure of cotton fibers.

As found by electron microscopy, cotton fiber degradation by enzymes is most intensive in the zones of lower fibril structure density [24].

In the cotton fiber damaged by microorganisms, cellulose concentration decreases by 7.5%, pectin substances by 60.7%, hemicellulase by 20%, and non-cellulose polysaccharides content also decreases. Cellulose biostability increases with its crystalinity degree and macromolecule orientation, as well as with hydroxyl group replacement by other functional groups. Microscopic fungi and bacteria are able to degrade cellulose and as a result glucose is accumulated in the medium, used as a source of nutrition by microorganisms. However, some part of cellulose is not destroyed and completely preserves its primary structure.

Cellulose of undamaged cotton fiber has 76.5% of well-ordered area, 7.8% of weakly ordered area, and 15.7% of disordered area. Microbiological degradation reduces the part of disordered area to 12.7%, whereas the part of well-ordered area increases to 80.4%. The ratio of weakly ordered area changes insignificantly. This goes to prove that the order degree of cotton cellulose increases due to destruction of disordered areas.

A definite type of fiber degradation corresponds to each stage of the cotton fiber damage. The initial degree of damage is manifested in streakiness, when the fiber surface obtains cracks of different length and width due to its wall break.

Swellings are formed resulting abundant accumulation of microorganisms and their metabolites in a definite part of the fiber. They may be accompanied by fiber wall break induced by biomass pressure. In this case, microorganisms and their metabolites splay out that causes blobs formation from the fiber and breaks in the yarn, as well as irregular fineness and strength.

The external microflora induces the wall damage. The highest degradation stage is fiber decomposition and breakdown into separate fibrils. Hence, perfect fiber structure is absent in this case [13, 25].

In all cases of damage, a high amount of fungal mycelium may be present on the fiber surface, which hyphae penetrate through the fiber or wrap about it thus preventing spinning and coloring of textile materials.

Enzymatic activity of fungi is manifested in strictly defined places of cellulose microfibrils, and the strength loss rate depends on both external climate conditions and contamination conditions. Cotton fabrics inoculated by the microscopic fungus *Aspergillus niger* under laboratory conditions at a temperature +29°C lose 66% of the initial strength 2–3 weeks after contamination [26], whereas inoculation by *Chaetomium globosum* induces 98.7% loss of strength, that is completely destroys the material.

The same fabric exposed to soil at +29°C during 6 days loses 92% of the initial strength.

And cotton fabric exposed to sea water for 65 days loses up to 90% of strength.

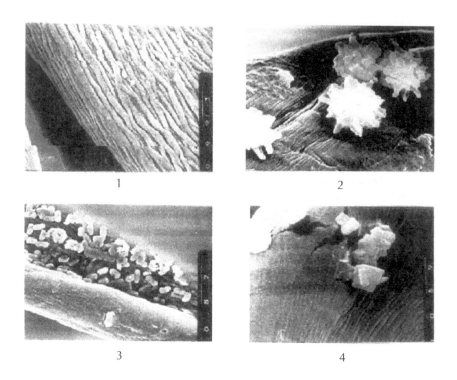

1

2

3

4

FIGURE 7 Cotton fiber micrographs—(1) initial fiber (× 4500), (2–4) fibers damages by different microorganisms—(2) *Aspergillus niger* (× 3000), (3) *Bacillus subtilis* (× 4500), and (4) *Pseudomonas fluorescens* (× 10000).

19.4 BAST FIBER BIODAMADING

Fibers produced from stalks, leaves or fruit covers of plants are called bast fibers. Hemp stalks give strong, coarse fibers—the hemp used for packing cloth and ropes. Coarse technical fibers: jute, ambary, ramie, and so on are produced from stalks of cognominal plants. Among all bast fibers, linen ones are most widely used.

The linen complex fiber, from which yarn and fabrics are manufactured, represents a batch of agglutinated filaments (plant cells) stretched and arrow-headed. The linen filament represents a plant cell with thick walls, narrow channel, and knee-shaped nodes called shifts. Shifts are traces of fractures or bends of the fiber occurred during growth, and especially during mechanical treatment. Fiber ends are arrow-shaped, and the channel is closed. The cross-section represents an irregular polygon with five or six edges and a channel in the center. Coarser fibers have oval cross-section with wider and slightly flattened channel.

Complex fibers consist of filament batches (15–30 pieces in a batch) linked by middle lamellae. Middle lamellae consist of various substances pectins, lignin, hemocellulase, and so on.

Bast fibers contain a bit lower amount of cellulose (about 70%) than cotton ones. Moreover, they contain such components as lignin (10%), wax and trace amounts of antibiotics, some of which increase biostability of the fiber. The presence of lignins induces coarsening (lignifications) of plant cells that promotes the loss of softness, flexibility, elasticity, and increased friability of fibers.

The main method for fiber separation from the flax is microbiological one, in which vital activity of pectin degrading microorganisms degrade pectins linking bast batches to the stalk tissues. After that the fiber can be easily detached by mechanical processing.

Microorganisms affect straw either at its spreading directly at the farm that lasts 20–30 days or at its retting at a flax processing plant where retting lasts 2–4 days.

In the case of spreading and retting of spread straw by atmospheric fallouts and dew under anaerobic conditions, the main role is played by microscopic fungi. According to data by foreign investigators, the following fungi are the most widespread at straw spreading—*Pullularia* (spires in the stalk bark), *Cladosporium* (forms a velvet taint of olive to dark green color), and *Alternaria* (grows through the bark by a flexible colorless chain and unambiguously plays an important role at dew spreading).

The studies indicate that *Cladosporium* fungus is the most active degrader of flax straw pectins. When retting linen at flax processing plants, conditions different from spreading are created for microflora. Here flax is submerged to the liquid with low oxygen content due to its displacement from straws by the liquid and consumption by aerobic bacteria, which propagate on easily accessible nutrients extracted from the straw.

These conditions are favorable for multiplication of anaerobic, pectin degrading clostridia related to the group of soil spore bacteria which includes just few species. Most of them are thermophiles and, therefore, the process takes 2–4 days in the warmed up water, however, at lower temperature (+15–20°C) it takes 10–15 days.

CURIOUS FACTS

In Russia and check republic, spreading is the most popular way of processing flax. In Poland, Romania, and Hungary, the flax is processed at flax processing plants by retting, and I Netherlands by retting and partly by spreading.

The linen fiber obtained by different methods (spread or retted straw) has different spinning properties. The spread straw is now considered to be the best, where the main role in degradation of stalk pectins is played by mold fungi. In production of retted fiber, this role is played by pectin degrading bacteria, some strains of which being able to form an enzyme (cellulase) that degrades cellulose itself. Such impact may be one of the damaging factors in the processes of linen retting. Thus, biostability of the flax depends on the method of fiber production.

This chapter shows that all kinds of biological treatment increase the quantity of various microbial damages of the fiber. Meanwhile, the spread fiber had lower total number of microscopic damages compared with any other industrial method.

There are other methods for flax production, steaming, for example, that gives steamed fiber. It has been found that steamed flax is the most biostable fiber. Possible reasons for so high biostability are high structure ordering of this fiber and high content of modified lignin in it. Moreover, during retting and spreading the fiber is enriched with microorganisms able to degrade cellulose under favorable conditions, whereas steaming sterilizes the fiber.

When exposed to microorganisms, pectins content in the linen fiber decreases by 38%, whereas cellulose content by 1.2% only. The quantity of wax and ash content of the fiber exposed to microorganisms do not virtually change.

The ordered area share in the linen cellulose is 83.6%, the weakly ordered area –5.1%, and disordered area –15.7%. During microbiological degradation the share of disordered areas in the linen cellulose decreases to 7.8%, and the share of ordered areas increases to 86.9%. The share of weakly ordered areas varies insignificantly.

Microbiological damages of linen, jute and other bast fibers and fabrics are manifested by separate staining (occurrence of splotches of color or fiber darkening) and putrefactive odor. On damaged bast fibers, microscopic cross fractures and chips, and microholes and scabs in the fiber walls are observed.

The studies of relative biostability of bast fibers demonstrate that Manilla hemp and jute are most stable, whereas linen and cannabis fibers have the lowest stability.

Natural biostability of bast fibers is generally low and in high humidity and temperature conditions, when exposed to microorganisms, physicochemical, and strength indices of both fibers and articles from them rapidly deteriorate. Generally, bast fibers are considered to have virtually the same biostability, as cotton fibers do.

Biostability of cellulose fivers is highly affected by further treatment with finishing solutions (sizing and finishing) containing starch, powder, resins and other substances which confer wearing capacity, wrinkle resistance, fire endurance, and so on to textile materials. Many of these substances represent a good culture medium for microorganisms. Therefore, at the stage of yarn and fabric sizing and finishing, the main attention is paid to strict compliance with sanitary and technological measures which are to prevent fabric infection by microorganisms and further biodamaging.

19.5 BIODAMAGING OF ARTIFICIAL FIBERS

Artificial fibers and fabrics are produced by chemical treatment of natural cellulose obtained from spruce, pine tree, and fir. It is based on cellulose are viscose, acetate, and so on ones. These fibers obtained from natural raw material have higher amorphous structure as compared with high-molecular natural material and, therefore, have lower stability, higher moisture, and swelling capacity.

By chemical structure and microbiological stability viscose fibers are similar to common cotton fibers. Biostability of these fibers is low. Many cellulosolytic microorganisms are capable of degrading them. Under laboratory conditions, some species of mold fungi shortly (within a month) induces complete degradation of viscose fibers, whereas wool fibers under the same conditions preserve up to 50% of initial stability. For viscose fabrics, the loss of stability induced by soil microorganisms during 12–14 days gives 54–76%. These parameters of artificial fibers and fabrics are somewhat higher than for cotton.

Acetate fibers are produced from acetyl cellulose—the product of cellulose etherification by acetic anhydride. Their properties significantly differ from those of viscose fibers and more resemble artificial fibers. For instance, they possess lower moisture retaining property, lesser swelling, and loss of strength under wet condition. They are more stable to damaging effect of cellulosolytic enzymes of bacteria and microscopic fungi, because contrary to common cellulose fibers possessing side hydroxyl groups in macromolecules, acetate fiber macromolecules have side acetate groups hindering interaction of macromolecules with enzymes.

Among artificial textile materials of the new generation, textile fibers from bamboo, primarily obtained by Japanese, are highlighted. Bamboo possesses the reference antimicrobial properties due to the presence of "bambocane" substance in the fiber. Bamboo fivers possess extremely porous structure that makes them much more hygroscopic that cotton. Clothes from bamboo fibers struggle against sweat secretion— moisture is immediately absorbed and evaporated by fabric due to presence of pores, and high antimicrobial properties of bamboo prevent perspiration odor.

Various modified viscose fibers, micromodal, and modal, for example, produced from beech were not studies for biostability. Information on biostability of artificial fibers produced from lactic casein, soybean protein, maize, peanut, and corn is absent.

19.6   WOOL FIBER BIODAMAGING

By wool the animal hair is called, widely used in textile and light industry. The structure and chemical composition of the wool fiber significantly differ it from other types of fibers and shows great variety and heterogeneity of properties. Sheep, camel, goat, and rabbit wool is used as the raw material.

After thorough cleaning, the wool fiber can be considered virtually consisting of a single protein—keratin. The wool contains the following elements (in %): carbon—50, hydrogen—6–7, nitrogen—15–21, oxygen—21–24, sulfur—2–5, and other elements.

The chemical feature of wool is high content of various amino acids. It is known that wool is a copolymer of, at least, 17 amino acids, whereas the most of synthetic fibers represent copolymers of two monomers.

Different content of amino acids in wool fibers promotes the features of their chemical properties. Of the great importance is the quantity of cystine containing virtually all sulfur and which is extremely important for the wool fiber properties. The higher sulfur content in the wool is, the better its processing properties are, the higher resistance to chemical and other impacts is and the higher physic-mechanical properties are.

Wool fiber layers, in turn, differ by the sulfur content: it is higher in the cortical layer that in the core. Among all textile fibers, wool has the most complex structure.

The fine merino wool fiber consists of two layers—external flaky layer or cuticle and internal cortical layer—the cortex. Coarser fibers have the third layer—the core.

The cuticle consists of flattened cells overlapping one another (the flakes) and tightly linked to one another and the cortical layer inside.

Cuticular cells have a membrane, the so-called epicuticle, right around. It is found that epicuticle gives about 2% of the fiber mass. Cuticle cells limited by walls quite tightly adjoin one another, but, nevertheless, there is a thin layer of intercellular protein substance between them, which mass is 3–4% of the fiber mass.

The cortical layer, the cortex, is located under the cuticle and forms the main mass of the fiber and, consequently, defines basic physico-mechanical and many other properties of the wool. Cortex is composed of spindle-shaped cells connivent to one another. Protein substance is also located between the cells.

The cortical layer cells are composed of densely located cylindrical, thread-like macrofibrils of about 0.05–0.2 μm in diameter. Macrofibrils of the cortical layer are composed of microfibrils with the average diameter of 7–7.5 nm [26, 27].

Microfibrils, sometimes call the secondary agents, are composed of primary aggregates—protofibrils. Protofibril represents two or three twisted α-spiral chains.

It is suggested [27, 28] that α-spirals are twisted due to periodic repetition of amino acid residues in the chain, hence, side radicals of the same spiral are disposed in the inner space of another α-spiral providing strong interaction, including for the account of hydrophobic bonds, because each seventh residue has a hydrophobic radical.

According to the data by English investigator J. D. Leeder, the wool fiber can be considered as a collection of flaky and cortical cells bound by a cell membrane complex (CMC) which thus forms a uniform continuous phase in the keratine substance of the fiber. This intercellular cement can easily be chemically and microbiologically degraded, that is a δ-layer about 15 nm thick (CMC or intercellular cement) is located between cells filling in all gaps [29].

The chapter show that the composition of intercellular material between flaky cells may differ from that of the material between cortical cells. In the cuticle-cuticle, cuticle-cortex, and cortex-cortex complexes the intercellular "cement" has different chemical compositions.

Although, the CMC gives only 6% of the wool fiber mass, there are proofs that it causes the main effect on many properties of the fiber and fabric [29, 30]. For instance, a suggestion was made that CMC components may affect such mechanical properties, as wear resistance and torsion fatigue, as well as such chemical properties, as resistance to acids, proteolytic enzymes, and chemical finishing agents.

The core layer is present in the fibers of coarser wool with the core cell content up to 15%. Disposition and shape of the core layer cells significantly vary with respect to the fiber type. This layer can be continuous (along the whole fiber) or may be separated in sections. The cell carcass of the core layer is composed of protein similar to microfibril cortex protein.

By its chemical composition, wool is a protein substance. The main substance forming wool is keratine—a complex protein containing much sulfur in contrast with other proteins. Keratine is produced during amino acid biosynthesis in the hair bag epidermis in the hide. Keratine structure represents a complex of high-molecular chain batches interacting both laterally and transversally [30-32]. Along with keratine, wool contains lower amounts of other substances.

Wool keratine reactivity is defined by its primary, secondary, and tertiary structures, that is the structure of the main polypeptide chains, the nature of side radicals and the presence of cross bonds.

Among all amino acids, only cystine forms cross bonds, their presence considerably defines wool insolubility in many reagents. Cystine bond decomposition simplifies wool damaging by sunlight, oxidants, and other agents. Cystine contains almost all sulfur present in the wool fibers. Sulfur is very important for the wool quality, because it improves chemical properties, strength, and elasticity of fibers.

Along with general regularities in the structure of high-molecular compounds, fibers differ from one another by chemical composition, monomer structure, polymerization degree, orientation, intermolecular bond strength and type, and so on that defines different physico-mechanical and chemical properties of the fibers.

The main chemical component of wool—keratine, is nutrition for microorganisms.

Microorganisms may not directly consume proteins. Therefore, they are only consumed by microbes having proteolytic enzymes—exoproteases that are excreted by cells to the environment.

Wool damage may start already before sheepshearing that is in the fleece, where favorable nutritive (sebaceous matters, wax, and epithelium), temperature, aeration, and humidity conditions are formed.

Contrary to microorganisms damaging plant fibers, the wool microflora is versatile, generally represented by species typical of the soil and degrading plant residues.

Initiated in the fleece, wool fiber damages are intensified during its storage, processing, and transportation under unfavorable conditions.

Specific epiphytic microflora typical of this particular fiber is always present on its surface. Representatives of this microflora excrete proteolyric enzymes (mostly pepsin), which induce hydrolytic keratine decay by polypeptide bonds to separate amino acids.

Wool is degraded in several stages—first, microorganisms destroy the flaky layer and then penetrate into the cortical layer of the fiber, although, the cortical layer itself is not destroyed, because intercellular substance located between the cells is the culture medium. As a result, the fiber structure is disturbed—flakes and cells are not bound yet, the fiber cracks and decays.

The mechanism of wool fiber hydrolysis by microorganisms suggested by American scientist E. Race represents a sequence of transformations—proteins—peptones—polypeptides—water + ammonia + carboxylic acids.

The most active bacteria—*Alkaligenes bookeri, Pseudomonas aeroginosa, Proteus vulgaris, Bacillus agri, B. mycoides, B. mesentericus, B. megatherium, B. subtilis, and microscopic fungi: Aspergillus, Alternaria, Cephalothecium, Dematium, Fusarium, Oospora, and Penicillium, Trichoderma*, were extracted from the wool fiber surface [33-39].

However, the dominant role in the wool degradation is played by bacteria. Fungi are less active in degrading wool. Consuming fat and dermal excretion, fungi create conditions for further vital activity of bacteria degraders. The role of microscopic fungi may also be reduced to splitting the ends of fibers resulting mechanical efforts of growing hyphae. Such splitting allows bacteria to penetrate into the fiber. Fungi weakly use wool as the source of carbon.

In 1960s, the data on the effect of fat and dirt present on the surface of unclean fibers on the wool biodamaging were published. It is found that unclean wool is damaged much faster than clean one. The presence of fats on unclean wool promotes fungal microflora development.

The activity of microbiological processes developing on the wool depends on mechanical damages of the fiber and preliminary processing of the wool.

It is found that microorganism penetration may happen through fiber cuts or microcracks in the flaky layer. Cracks may be of different origins—mechanical, chemical, and so on. It is also found that wool subject to intensive mechanical or chemical treatment is easier degraded by microorganisms than untreated one [39].

For instance, high activity of microorganisms during wool bleaching by hydrogen peroxide in the presence of alkaline agents and on wool washed in the alkaline medium was observed. When wool is treated in a weak acid medium, the activity of microorganisms is abruptly suppressed. This also takes place on the wool colored by chrome and metal-containing dyes. The middle activity of microorganisms is observed on the wool colored by acid dyes.

When impacted by microorganisms, structural changes in the wool are observed—flaky layer damages, its complete exfoliation, and lamination of the cortical layer.

Wool fiber damages can be reduced to several generalized types provided by their structural features:

- Channeling and overgrowth—accumulation of bacteria or fungal hyphae and their metabolites on the fiber surface,
- FLaky layer damage, local, and spread,
- Cortical layer lamination to spindle-shaped cells,
- Spindle-shaped cell destruction.

Along with the fiber structure damage, some bacteria and fungi decrease its quality by making wool dirty blue or green that may not be removed by water or detergents. Splotches of color also occur on wool, for example, due to the impact of *Pseudomonas aeruginosa bacteria, in this case, color depends on medium pH—green splotches occurred in a weakly alkaline medium, and in weak acid medium they are red. Green splotches may also be caused by development of Dermatophilus congolensis* fungi. Black color of wool is provided by *Pyronellaea glomerata* fungi.

Thus, wool damage reduces its strength, increases waste quantity at combing and imparts undesirable blue, green or dirty color and putrefactive odor.

However, wool is degraded by microorganisms slower than plant fibers.

19.6.1   CHANGES OF STRUCTURE AND PROPERTIES OF WOOL FIBERS BY MICROORGANISMS

To evaluate bacterial contamination of wool fibers, it is suggested to use an index suggested by A. I. Sapozhnikova, which characterizes discoloration rate of resazurin solution, a weak organic dye and currently hydrogen acceptor. It is also indicator of both presence and activity of reductase enzyme [40].

The method is based on resazurin ability to lose color in the presence of reductase, which is microorganisms' metabolite, due to redox reaction proceeding. This enables judging about quantity of active microorganisms present in the studied objects by solution discoloration degree.

Discoloration of the dye solution was evaluated both visually and spectrophoto-metrically by optical density value.

Table 1 shows results of visual observations of color transitions and optical density measurements of incubation solutions after posing the reductase test.

As follows from the data obtained, coloration of water extracts smoothly changed from blue-purple for control sterile physiological solution (D = 0.889) to purple for initial wool samples (D_{thin} = 0.821 and D_{coarse} = 0.779), crimson (D_{thin} = 0.657 and D_{coarse} = 0.651), and light crimson at high bacterial contamination (D_{thin} = 0.548 and D_{coarse} = 0.449 and 0.328) depending on bacterial content of the fibers.

TABLE 1   Visual coloration and optical density of incubation solutions with wool fibers at the wavelength λ = 600 nm and different stages of spontaneous microflora development

Time of microor-ganism develop-ment, days	Thin merino wool		Coarse caracul wool	
	Optical density	Color (visual assessment)	Optical density	Color (visual assessment)
Control, physi-ological saline	0.889	Blue purple	0.889	Blue purple
0 (init.)	0.821	Purple	0.779	Purple
7	0.712	Purple	0.657	Crimson
14	0.651	Crimson	0.449	Light crimson
28	0.548	Light crimson	0.328	Light crimson

This dependence can be used for evaluation of bacterial contamination degree for wool samples applying color standard scale [39].

It is found that the impact of microorganisms usually reduces fiber strength, especially for coarse caracul wool after 28 days of impact strength decreased by 57–65%. The average rate of strength reduction is about 2% per day (Figure 8).

It is found that after 28 days of exposure, the highest reduction of breaking load of wool fibers is induced by Bac. subtilis bacteria [39].

Figure 9 clearly shows that 14 days after exposure to microorganisms the surface of coarse wool fiber is almost completely covered by bacterial cells. Meanwhile, note also (Figure 10) that cuticular cells themselves are not damaged, but their bonding is disturbed that grants access to cortical cells for microorganisms. Figure 11 shows the wool fiber decay to separate fibrils caused by microorganisms.

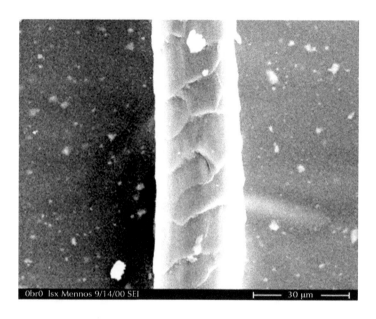

FIGURE 8   Micrographs of initial wool fibers (×1000)—(a) thin merino wool and (b) coarse caracul wool.

FIGURE 9   Micrographs of thin merino wool (a) and coarse caracul wool and (b) after 14 days of exposure to *Bac. subtilis* (×1000).

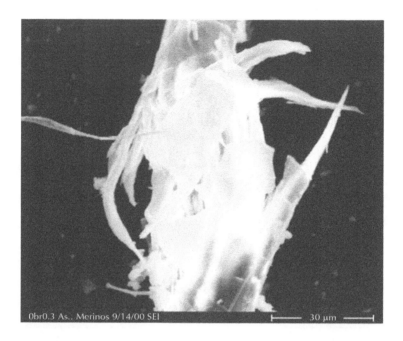

FIGURE 10 Micrographs of flaky layer destruction of thin (a) and coarse (b) wool fibers after 14 days of exposure to spontaneous microflora (×1000).

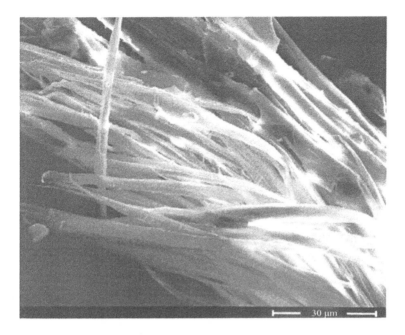

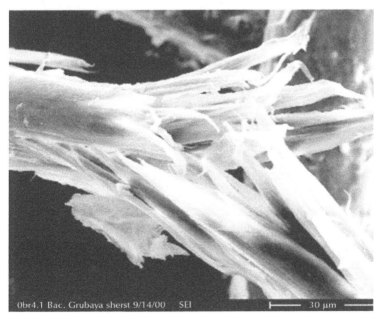

FIGURE 11 Micrographs of thin (a) and coarse and (b) wool fiber fibrillation after 28 days of exposure to microorganisms (×1000).

Wool fiber biodegradation changes the important quality indices, such as whiteness and yellowness. This process can be characterized as "yellowing" of wool fibers.

Table 2 shows data obtained on yellowness of thin merino and coarse caracul wool fibers exposed to spontaneous microflora, *Bac. subtilis* bacteria, and *Asp. niger* fungi during 7, 14, and 28 days.

TABLE 2   Yellowness of thin merino and coarse caracul wool fibers after different times of exposure to different microorganisms

| Time of exposure to microorganisms and days | | Yellowness, % | | |
| Type of fibers | | Impacting microorganisms | | |
		Bac. subtilis	Asp. niger	Spontaneous microflora
Thin,	0	27.7	27.7	27.7
Thin,	7	36.4	37.3	29.1
Thin,	14	45.4	42.5	34.8
Thin,	28	53.9	48.1	39.9
Coarse	0	39.3	39.3	39.3
Coarse	7	46.8	44.2	41.2
Coarse	14	51.8	46.4	42.1
Coarse	28	57.1	49.5	43.3

In terms of detection of biodegradation mechanism and changes of material properties, determination of relations between changes in properties and structure of fibers affected by microorganisms is of the highest importance. Yellowness increase of wool fibers affected by microorganisms testifies occurrence of additional coloring centers.

For the purpose of elucidating the mechanism of microorganisms' action on the wool fibers that leads to significant changes of properties and structure of the material, the changes of amino acid composition of wool keratine proteins being the nutrition source of microorganisms damaging the material are studied.

Exposure to microorganisms during 28 days leads to wool fiber degradation and, consequently, to noticeable mass reduction of all amino acids in the fiber composition. To the greatest extent, these changes are observed for coarse wool, where total quantity of amino acids is reduced by 10–12 rel.%, and for merino wool slightly lower reduction (4.7 rel.%) is observed. It should be noted that at comparatively low reduction of the average amount of amino acids in the system (not more than 12 rel.%) all types of wool fibers demonstrate a significant reduction of the quantity of some amino acids, such as serine, cystine, methionine, and so on (up to 25–33 rel.%).

Analysis of the data obtained testifies that in all types of wool fibers (but to different extent) reduction of quantity of amino acids with disulfide bonds (cystine, methionine) and ones related to polar (hydrophilic) amino acids, including serine, glycine, threonine, and tyrosine, is observed. These very amino acids provide hydrogen bonds imparting stability to keratine structure. In the primary structure of keratine, serine is the N-end group, and Tyrosine is C-end group. In this connection, reduction of the quantity of these amino acids testifies degradation of the primary structure of the protein. Changes observed in the amino acid composition of wool fiber proteins exposed to spontaneous microflora may testify that microorganisms degrade peptide and disulfide bonds, which provide stability of the primary structure of proteins, and break hydrogen bonds, which play the main role in stabilization of spatial structure of proteins (secondary, tertiary, and quaternary).

Very important data were obtained in the study of the wool fiber structure by IR-spectroscopy method. It is found that when microorganisms affect wool fibers, their surface layers demonstrate increasing quantity of hydroxyl groups that indicates accumulation of functional COO-groups, nitrogen concentration in keratine molecule decreases, and protein chain configuration partly changes—β-configuration (stretched chains) transits to α-configuration (a spiral). This transition depends on α- and β-forms ratio in the initial fiber and is more significant for thin fibers, which mostly have -configuration of chains in the initial fibers. Thus, it is found that microorganisms generally affect the CMC and degrade amino acids, such as cystine, methionine, serine, glycine, threonine, and tyrosine. Microorganisms reduce breaking load and causes "yellowing" of the wool fibers (Figure 12).

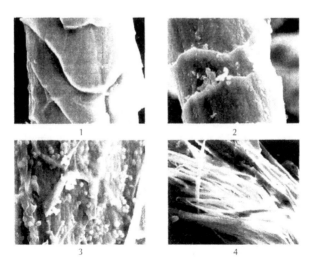

FIGURE 12 Micrographs of wool fibers—(1) initial undamaged fiber ($\times$ 3000), (2) and (3) bacterial cells on the fiber surface ($\times$ 3000), and (4) fiber fibrillation after exposure to microorganisms during 4 weeks ($\times$ 1000).

Synthetic fibers are principally different from natural and artificial ones by structure and, being an alien substrate for microorganisms, are harder damaged by them. Since occurrence of synthetic fabrics in 1950s, it is suggested that they are "everlasting" and are not utilized by microorganisms. However, it has been found with time that, firstly, microorganisms although slower, but yet are capable of colonizing synthetic fabrics and utilizing their carbon in the course of development (that is causing biodamage), and secondly, there are both more and less microorganism resistant fabrics among synthetic ones [41,42].

Among microorganisms damaging synthetic fibers, *Trichoderma* genus fungi are identified, at the initial stages developing due to lubricants and finishing agents without fiber damage and then wrap them with mycelium, loosen threads and, hence, reduce fabric strength.

When studying fabrics from nitrone, lavsan, and caprone, it has been found that soil fungi and bacteria cause roughly the same effect on characteristics of these fabrics increasing the fiber swelling degree by 20–25%, reducing strength by 10–15% and elongation at break by 15–20%.

Synthetic fibers represent potential source of energy and nutrition for microorganisms. The ability of microorganisms to attach to surfaces of insoluble solids, then using them as the nutritive substrate, is well-known. Living cells of microorganisms have complex structure, just on the surface of bacterial cells complexes of proteins, lipids, and polysaccharides were found, it contains hydrophilic and hydrophobic areas, various functional groups and mosaic electric charge (at total negative charge of the cells).

The first stage of microorganism interaction with synthetic fibers can be rightfully considered in terms of the adhesion theory with provision for the features of structure and properties of microorganisms as a biological system.

The entire process of microorganism impact of the fiber can conditionally be divided into several stages—attachment to the fiber, growth and multiplication on it and consumption of it, as the nutrition and energy source [43, 44].

Enzymes excreted by bacteria act just in the vicinity of bacterial membrane. Been adsorbed onto the fiber, living cells attach to the surface and adapt to new living conditions. The ability to be adsorbed onto the surface of synthetic fibers is caused by:

- The features of chemical structure of the fibers. For instance, fibers adsorbing microorganisms are polyamide and polyvinyl alcohol ones, the fiber not adsorbing microorganisms is, for example, ftorin,
- Physical structure of the fiber. For example, fibers with smaller linear density, with a lubricant on the surface absorb greater amount of microorganisms,
- The presence of electric charge on the surface, its value and sign. Positively charged chemical fibers adsorb virtually all bacteria, fibers having no electric charge adsorb the majority of bacteria, and negatively charged fibers do not adsorb bacteria.

Supermolecular structure also stipulates the possibility for microorganisms and their metabolites to diffuse inside the internal areas of the fiber. Microorganism assimilation of the fiber starts from the surface, and further degradation processes and their rate are determined by microphysical state of the fiber. Microorganism metabolite penetration into inner areas of the fiber and deep layers of a crystalline material is only possible in the presence of capillaries.

Chemical fiber damaging and degradation starting from the surface are, in many instances, promoted by defects like cracks, chips or hollows which may occur in the course of fiber production and finishing.

Along with physical inhomogeneity and chemical inhomogeneity may promote biodegradation of synthetic fibers. Chemical inhomogeneity occurs during polymer synthesis and its thermal treatments, manifesting itself in different content of monomers and various end groups. The possibility for microorganism metabolites to penetrate inside the structure of synthetic fibers depends on the quantity and accessibility of functional end groups in the polymer, which are abundant in oligomers.

The ability to synthetic fibers to swell also makes penetration of biological agents inside low ordered areas of fibers and weakens intermolecular interactions, off-orientation of macromolecules, and degradation in the amorphous and crystalline zones. Structural changes result in reduction of strength properties of fibers.

Theoretical statements that synthetic fibers with the lower ordered structure and higher content of oligomers possess lower stability to microorganism impact than fibers with highly organized structure and lower content of low-molecular compounds.

Thus, the most rapid occurrence and biodegradation of synthetic fibers are promoted by low ordering and low orientation of macromolecules in the fibers, their low density, low crystallinity, and the presence of defects in macro- and microstructure of the fibers, pores, and cavities in their internal zones.

Carbochain polymer based fibers are higher resistant to microbiological damages. These polymers are polyolefins, polyvinyl chloride, polyvinyl fluoride, polyacrylonitrile, and polyvinyl alcohol. Fibers based on heterochain polymers like polyamide, polyether, polyurethane, and so on are less bioresistant.

Comparative soil tests for biostability of artificial and synthetic fibers demonstrate that viscose fiber is completely destroyed on the 17th day of tests, bacterium and fungus colonies occur on lavsan on the 20th day, and caprone is overgrown by fungus mycelium on the 30th day. Chlorin and ftorlon have the highest biostability. The initial signs of their biodamage are only observed 3 months after the test initiation.

The studies of nitron, lavsan and capron fabric biostability have found that soil fungi and bacteria cause nearly equal influence on parameters of these fabrics, increasing swelling degree of the fibers by 20–25%, reducing strength by 10–15%, and elongation at break by 15–20%. Meanwhile, nitron demonstrated higher biostability, as compared with lavsan and capron.

19.7.1 CHANGES IN STRUCTURE AND PROPERTIES OF POLYAMIDE FIBERS INDUCED BY MICROORGANISMS

In contrast with natural fibers, chemical fibers have no permanent and particular microflora. Therefore, the most widespread species of microorganisms possessing increased adaptability are the main biodegraders of these materials.

Occurrence and progression of biological degradation of polyamide fibers is, in many instances, induced by their properties and properties of affecting microorganisms and their species composition. Generally, the species of microorganisms degrading polyamide and other chemical fibers are determined by their operation conditions, which form microflora, and its adaptive abilities.

Polyamide fibers are most frequently used in mixtures with natural fibers. Natural fibers contain specific microflora on the surface and inside. Therefore, capron fibers mixed with cotton, wool or linen are affected by their microflora. It is found [43-45] that capron fiber degradation by microorganisms obtained from wool is characterized as deep fiber decay, microorganisms extracted from natural silk cause streakiness of capron fibers, microorganisms extracted from cotton cause fading and decomposition, and microorganisms extracted from linen cause fading, streakiness and decomposition.

The microorganism interaction with polyamide fibers is most fully studied in the works by I. A. Ermilova [43-45]. For the purpose of detecting bacteria degraders of polyamide fibers, microorganisms were extracted from fibers damaged in the medium of active sewage silt, soil, microflora of natural fibers, and test-bacteria complex selected as degraders of polyamide materials. Capron fibers were inoculated by extracted cultures of microorganisms and types of damages were reproduced.

Polyamide fiber materials were natural nutritive and energy source for these microorganisms. Therefore, bacterial strains extracted from damaged fibers were different from the initial strains. It is proven that the existence on a new substrate has stipulated changes of intensity and direction of physiological-biochemical processes of bacterial cells, the change of their morphological and culture properties [43-45].

Extraction, cultivation and use of such adaptive strains are of both scientific and practical interest. Using bacterial strains adaptive to capron, polyamide production waste, and warn products, toxic substances may be utilized. This allows obtaining of secondary raw materials and solving the problem of environmental protection.

It is proved experimentally [43-45] that polycaproamide (PCA) fibers possess high adsorbability, which value depends on the properties of impacting bacteria. For instance, gram-positive, especially spore-forming bacteria *Bacillus subtilis, Bac. mesentericus* (from 84.5 to 99.3% of living cells), are most highly adsorbed, and adsorption of gram-positive bacteria varies significantly.

The extensive research of test-bacterium complex impact was performed [43-45]. These bacteria were chosen as degraders of polyamide fibers, along with microflora of active sewage silts, linen and jute microorganisms as degraders of a complex capron thread. It is found that the test-bacterium complex injected by the author, after 7 months of exposure, increases biodegradation index to 1.27 that testifies about intensive degradation of the fiber microstructure. The highest degradation of complex PCA thread is caused by from active sewage silt microorganisms.

High activity of silt microorganisms and test-bacteria is explained by the fact that they include bacterium strains *Bacillus subtilis* and *Bac. mesentericus*, which according to the data by a number of authors [39] may induce full degradation of caprolactam to amino acids using it, as the source of carbon and nitrogen.

Thus, adaptive forms of microorganisms induce the highest degradation of polyamide fibers.

Polyamide fibers are characterized by physical structure inhomogeneity, which occurs during processing and is associated with differences in crystallinity and orientation of macromolecules determining fiber accessibility for microorganisms and their metabolites penetration. The surface layer is damaged during orientation stretching and, consequently, has lower molecular alignment. That is why the surface layer is most intensively changed by microorganisms. The study of polyamide fiber macrostructure after exposure to microorganisms shows that streakiness and cover damage are the main damages of these fibers [43].

The studies of supermolecular structure of capron fiber surface show that after exposure to microorganisms the capron fiber cover becomes loose and uneven [39, 43]. Surface supermolecular structure degradation increases with the microbial impact—fibrils and their yarns become split and disaligned both laterally and transversely, multiple defects in the form of pores and cavities, and cracks of various depths are formed.

Chemical inhomogeneity of polyamide fibers also promotes changes in the fiber structure, when exposed to microorganisms [46-50]. It is found that the polyamide fiber degradation increases with low-molecular compound (LMC) content in them; meanwhile, at the same content of low-molecular compounds, thermally treated fibers were higher biostable, as compared with untreated specimens [43, 44].

Along with the morphological characters which characterize biodamaging of the fibers, functional features, such as strength decrease and increase of deformation properties of the fiber, were detected. The greatest strength decrease (by 46.4%) was observed for thermally untreated capron fiber containing 3.4% LMC, and the smallest decrease (by 5%) was observed for the fibers with 3.2% LMC, thermally treated at the optimum time of 5 sec [43, 44].

The IR-spectroscopy method was applied to detect PCA fiber damage by microorganisms [39]. It is found that carboxyl and amide groups are accumulated during their biodegradation.

The change of various property indices of polyamide fibers also results from macro-, micro,- and chemical changes.

Of interest are studies of the microorganism effect on polyamide fabric quality [43, 44]. Fabrics (both bleached and colored) from capron monofilament were exposed to a set of test cultures—*Bac.subtilis, Ps.fluorecsens, Ps.herbicola,* and *Bac.mesentericus.* After 3–9 month exposure, yellowness and dark spots occurred, coloring intensity decreased, and an odor appeared. The optical microscopy studies indicated that all fibers exposed to microorganisms, had damages typical of synthetic fibers—overgrowing, streakiness, bubbles, and wall damages. The increasing quantity of biodamages with time results in tensile strength reduction by 6–8% for capron fibers after 9 months, at inconsiderable change of relative elongation.

It all goes to show that development of microorganisms on polyamide fibrous materials results in changes of fibers morphology, their molecular and supermolecular structure and, as a consequence, reduction of strength properties, color change, and odor.

To clear up the mechanism of PCA fiber degradation by polarographic investigation, a possibility of ε-amino caproic acid (ACA) accumulation by *Bacillus subtilis k1* culture during degradation of PCA fibers 0.3 and 0.7 tex fineness was studied (Figure 13) [39].

As a source of carbon, ACA (10 mg/l) or PCA (0.5 g/l) was injected into the mineral medium.

It is found that at PCA fibrous materials exposure to *Bacillus subtilis k1* strain, the maximum quantity of ACA is liberated on the 5th day 32 mg/l, 66 mg/l. The ACA accumulation, mg\ L.

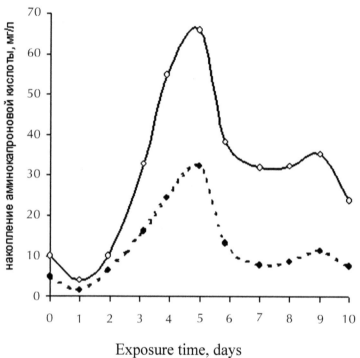

Exposure time, days

FIGURE 13   The change of ACA concentration during *Bacillus subtilis k1* development on PCA material fibers –0.7 tex and –0.3 tex.

It is known that if the medium includes several substrates metabolized by a particular strain of microorganisms, the substrate providing the maximum culture propagation rate is consumed first. As this substrate is going to exhaust, bacteria subsequently consume other substrates, which provide lower rates of cell multiplication.

In the mineral medium containing chemically pure ACA (with the initial concentration of 10 mg/L) as the source of carbon, its concentration decreased gradually, and 5 days after it was not detected in the solution.

Thus, basing on the data obtained, one may conclude that, firstly, PCA fibers of 0.3 tex fineness are more accessible for microorganisms, secondly, *Bacillus subtilis k1* strain can be used to utilize PCA fibrous materials, and thirdly, polarographic analysis has proven the mechanism of PCA fibrous material degradation with ACA liberation. As a consequence, suggested *Bacillus subtilis k1* strain VKM No.V-1676D degrades

PCA fibrous materials at the both macro and microstructure levels, with ACA formation.

19.8   METHODS OF TEXTILE MATERIAL PROTECTION AGAINST DAMAGING BY MICROORGANISMS

Imparting antimicrobial properties to textile materials pursues two main aims: protection of the objects contacting with textile materials against actions of microorganisms and pathogenic microflora.

In the first case, we speak about imparting biostability to materials and, as a consequence, about passive protection. The second case concerns creation of conditions for preventive attack of a textile material on pathogenic bacteria and fungi to prevent their impact on the protected object [51].

The basic method of increasing biostability of textile materials is application of antimicrobial agents (biocides). The requirements to the "ideal" biocide are the following:

- The efficacy against the most widespread microorganisms at minimal concentration and maximal action time,
- Non-toxicity of applied concentrations for people,
- The absence of color and odor,
- Low price and ease of application,
- Retaining of physico-mechanical, hygienic, and other properties of the product,
- Compatibility with other finishing agents and textile auxiliaries,
- Light stability and weather ability.

At any time, nearly every class of chemical compounds was applied to impart textile materials antibacterial or antifungal activity. Today, application of nanotechnologies, specifically injection of silver and iodine nanoparticles, to impart textile antimicrobial properties is of the greatest prospect.

At all times, copper, silver, tin, mercury, and so on salts were used to protect fibrous materials against biodamages. Among these biocides, the most widespread are copper salts due to their low cost and comparatively toxicity. The use of zinc salts is limited by their low biocide action, whereas mercury, tin and arsenic salts are highly toxic for the man [52-57]. However, there are organomercury preparations applied to synthetic and natural fibrous materials used as linings and shoe plates, widely advertised for antibacterial and antifungal finishing.

The data are cited [52-57] that impregnation of textile materials by a mixture of neomycin with tartaric, propionic, stearic, phthalic, and some other acids imparts them the bacteriostatic effect. Acids were dissolved in water, methyl alcohol or butyl alcohol and were sprayed on the material.

The methods of imparting textile materials biostability can be divided into the following groups:

- Impregnation by biocides, chemical and physical modification of fibers, and threads, which then form a textile material,
- Cloth impregnation by antimicrobial agent solutions of emulsions, its chemical modification,

- Injection of antimicrobial agents into the binder (at nonwoven material manufacture by the chemical method),
- Imparting antimicrobial properties to textile materials during their coloring and finishing,
- Application of disinfectants during chemical cleaning or laundry of textile products.

However, impregnation of fibers and cloth does not provide firm attachment of reagents. As a result, the antimicrobial action of such materials is nondurable. The most effective methods of imparting biocide properties to textile materials are those providing chemical bond formation that is chemical modification methods. Chemical modification methods for fibrous materials represent processing that leads to clathrate formation, for example injection of biological active agents into spinning melts or solutions.

At the stage of capron polymerization, an antibacterial organotin compound (tributyltin oxide or hydroxide) is added that retains the antibacterial effect after multiple laundries. Methods of imparting antimicrobial properties to textile materials by injection of nitrofuran compounds into spinning melts with further fixing them at molding in the fine structure of fibers similar to clathrates were designed.

There are data on imparting antimicrobial properties to synthetic materials during oiling. Prior to drafting, fibers are treated by compounds based on oxyquinoline derivatives, by aromatic amines or nitrofuran derivatives. Such fibers possess durable antimicrobial effect [58-60].

Nanotechnologies are actively intruded in the light industry, allowing obtaining of materials with antimicrobial properties. The following directions using nanotechnologies, which are now investigated, should be outlined, including creation of new textile materials on the account of:

- Primarily, the use of textile nonfibers and threads in the materials and
- Secondly, the use of nanodispersions and nanoemulsions for textile finishing.

It may be said that nanotechnologies allow a significant decrease of expenses at the main production stage, where consumption of raw materials and semi products is considerable. For nanoparticles imparting antimicrobial properties, silver, copper, palladium, and so on particles are widely used. Silver is the natural antimicrobial agent, which properties are intensified by the nanoscale of particles (the surface area sharply increases), so that such textile is able to kill multiple microorganisms and viruses. In this form, silver also reduces necessity in fabric cleaning, eliminates sweat odor as a result of microorganism development on the human body during wearing.

Properties of materials designed with the use of silver nanoparticles, which prevent multiplication of various microorganisms, may be useful in medicine, for example. The examples are surgical retention sutures, bandages, plasters, surgical boots, medical masks, whites, skull-caps, towels, and so on. Customers know well sports clothes, prophylactic socks with antimicrobial properties.

Along with chemical fibers and threads, natural ones are treated by nanoparticles. For example, silver and palladium nanoparticles (5–20 nm in diameter) were synthesized in citric acid, which prevented their agglutination, and then natural fibers were dipped in the solution with these negatively charged particles. Nanoparticles imparted antibacterial properties and even ability to purify air from pollutants and allergens to clothes and underwear.

When these products appeared at the world market, disputes about ecological properties and the influence of these technologies on the human organism have arisen. There are no accurate data yet how these developments may affect the human organism. However, it should be noted that some specialists do not recommend everyday use of antibacterial socks, because these antibacterial properties affect the natural skin microflora.

Nanomaterials are primarily hazardous due to their microscopic size. Firstly, owing to small size they are chemically more active because of a great total area of the nanosubstance. As a result, low toxic substance may become extremely toxic. Secondly, chemical properties of the nanosubstance may significantly change due to manifestations of quantum effects that, finally, may make a safe substance extremely hazardous. Thirdly, due to small size, nanoparticles freely permeate through cellular membranes damaging bioplasts and disturbing the cell operation.

Physical modification of fibers or threads is the direct change of their composition (without new chemical formations and transformations), structure (supermolecular and textile), properties, production technology, and processing. Modernization of the structure and increase of the fiber crystallinity degree induces biostability increase. However, in contrast with chemical modification, physical modification does not impart antimicrobial properties to the fibers, but may increase biostability.

By no means always textile materials produced completely from antimicrobial fibers are required. Even a small fracture of highly active antimicrobial fiber (for example 1/3 or even 1/4) is able to provide sufficient biostability to the entire material. This chapter show that antimicrobial fibers were found not only protected themselves against microorganism damage, but also capable of shielding plant fibers from their impact.

Manufacture of antimicrobial nonwoven materials by injection of active microcapsule ingredients into it is of interest. Microcapsules can contain solid particles of microdrops of antimicrobial substances liberated under particular conditions (for example by friction, pressure, dissolution of capsule coatings or their biodegradation).

Biostability of fibrous materials may be significantly affected by the dye selection. Dyes possessing antimicrobial activity on the fiber are known as salicylic acid derivatives capable of bonding copper, triphenylmethane, acridic, thiazonic, and so on ones. For instance, chromium-containing dyes possess antibacterial action, but resistance to mold fungi is not imparted.

It is known that synthetic fibers dyed by dispersed pigments are more intensively degraded by microorganisms. It is suggested that these pigments make the fiber surface more accessible for bacteria and fungi.

Single bath coloring and bioprotective finishing of textile materials are also applied. A combination of these processes is not only of theoretical interest, but is also perspective in terms of technology and economy.

Processing of textile materials by silicones also imparts antimicrobial properties to these clothes. Some authors state that textile material sizing by water repellents imparts them sufficient antimicrobial activity. Water repellency of materials may reduce the adverse impact of microorganisms, because the quantity of adsorbed moisture is reduced. However, Hydrophobic finishing itself may not fully eliminate the adverse effect of microorganisms. Therefore, antimicrobial properties imparted to some textile materials during silicone finishing may be related to application of metal salts as catalysts, such as copper, chromium, and aluminum.

Disinfectants, for example at laundry, may be applied by the customer himself. The method of sanitizing substance application for carpets, which is spraying or dispensing of a disinfectant on the surface of floor covers during operation is known. The acceptable disinfection level may be obtained during laundry of textile products by such detergents, which may create residual fungal and bacteriostatic activity [58-60].

KEYWORDS

- **Artificial fibers**
- **Bast fibers**
- **Biodegradation mechanism**
- **Cotton fiber**
- **Microorganism**
- **Wool keratine reactivity**

REFERENCES

1. Pekhtasheva, E. L. *Biodamages and protections of non-food materials* (in Russian). Masterstvo Publishing House, Moscow, pp. 224 (2002).
2. Emanuel, N. M. and Buchachenko, A. L. *Chemical physics of degradation and stabilization of polymers*, VSP International Science Publ., Utrecht, pp. 354 (1982).
3. Zaikov, G. E., Buchachenko, A. L., and Ivanov, V. B. *Aging of polymers, polymer blends, and polymer composites*. Nova Science Publ., New York, **1**, 258 (2002).
4. Zaikov, G. E., Buchachenko, A. L., and Ivanov, V. B. Aging of polymers, polymer blends and polymer composites. Nova Science Publ., New York, **2**, 253 (2002).
5. Zaikov, G. E., Buchachenko, A. L., and Ivanov, V. B. *Polymer aging at the cutting adge*. Nova Science Publ., New York, pp. 176 (2002).
6. Gumargalieva, K. Z. and Zaikov, G. E. *Biodegradation and biodeterioration of polymers*: *Kinetical aspects*. Nova Science Publ., New York, pp. 210 (1998).
7. Semenov, S. A., Gumargalieva, K. Z., and Zaikov, G. E. *Biodegradation and durability of materials under the effect of microorganisms*. VSP International Science Publ., Utrecht, pp. 199 (2003).
8. Polishchuk, A. Ya. and Zaikov, G. E. *Multicomponent transport in polymer systems*. Gordon & Breach, New York, pp. 231 (1996).
9. Moiseev, Yu. V. and Zaikov, G. E. Chemical resistance of polymers in reactive media. Plenum Press, New York, pp. 586 (1987).
10. Jimenez, A. and Zaikov, G. E. Polymer analysis and degradation. Nova Science Publ., New York, pp. 287 (2000).
11. Babaev, D. The quality of production. *Khlopkovodstvo (Cotton Production) journal* (Russian), (6), pp. 14–18 (1985).

12. Ermilova, I. A. and Semenova, D. I. Investigation of bioresistant properties of cotton fibres. *Tekstil'naya Promyshlennost' (Textile Industry) journal* (Russian), (4), pp. 13–14 (1999).
13. Ipatko, L. I. *Effect of microorganizmes on the structure and properties of cotton fibers*, Thesis, Leningrad Institute of Building Technology, Leningrad, pp. 143 (1988).
14. Aminov, Kh. A. Measurement of fibers quality of cotton, *Khlopkovaya Promyshlennost' (Cotton Idustry) journal* (Russian), (6), pp. 3–4 (1988).
15. Guban, I. N., Voropaeva, N. L., and Rashidova, S. Sh. Paisting of cotton fibers. Doklady Academy of Science Yuzbek. *USSR (Reports of Yuzbek. Academy of Sciences)* journal (Russian), (12), 48–50 (1988).
16. Chun, David T.W. High moisture storage effects on cotton stickiness. *Text. Res. J.*, **68**(9), 642–648 (1998).
17. Evans, Elaine, and Brain, Mc Carthy. Biodeterioration of natural fiber. *J. Soc. Dyers Colour.*, **114**(4), 114–116 (1998).
18. Mangialardi, Y. J., Lalor, W. F., Bassett, D., and Miravalle, R. J. Influence of Yrowth Period on neps in Cotton. *Text. Res. J.*, **57**(7), 421–427 (1987).
19. Perkins Henry, H. Spin Finishes for Cotton. *Text. Res. J.*, **58**(3) 173–179 (1988).
20. Rakhimov, A. Investigation of durability and degradation of cotton fibers (Russian), Donish Publishing House, Dushanbe, pp. 247 (1971).
21. Xu, B., Fang, C., and Watson, M. D. Investigation new factors in cotton color grading. *Text. Res. J.*, **68**(11) 779–787 (1998).
22. Bose, R. G. and Ghose, S. N. Detection of Mildew growth on jute and cotton textiles by ultraviolet light. *Text. Res. J.*, **39**(10) 982–983 (1969).
23. Kaplan, A. M., Mandels, M., and Greenberger, N. Mode of action of regins in preventing microbial degradation of cellulosic textiles. *In: Biodeterioration of materials. L.*, **2** 268–278 (1972).
24. Abu-Zeid, A. and Abou-Zeid. A technique for measuring microbial damage of cellulosic sources by microorganisms. *Pakistan J. Sci*, **23**(½) 21–25 (1971).
25. Piven', T. V. and Khodyrev, V. I. Biodegradation of flax and cotton. *Khimiya Drevesiny (Chemestry of wood) journal* (Russian), (1), pp. 100–105 (1988).
26. Fucumura, T. Hydrolysis of cyclic and liner oligomers of 6-aminocaproic acid by a bacterial cell extract. *J. of Biochemistry*, **59**, 531–536 (1966).
27. Alexander, P. and Hudson, R. F. *Physics and Chemistry of wool.* Chimiya (Chemistry) Publishing House (Russian), Moscow, p. 58 (1985).
28. Novorodovskaya, T. E. and Sadov, S. F. Chemistry and chemical technology of wool. *Lesprombytizdat* (Forest-industry Publishing House) (Russian), Moscow, pp. 245 (1986).
29. Wlochowicz, A. and Pielesz, A. Struktura wlokien welnianych w swetla aktualnych badan. *Prz. Wlok.*, (4) pp.4–8 (1997).
30. Leeder, J. D. *The cell membrane complex and its influence on the properties of the wool fibre.* Wool science review. International Wool Secretariat. Development Center, (63) pp. 3–35 (1986).
31. Lewis, J. *Microbial biodeterioration. Economic Microbiology.* A. H. Rose (Ed.), Academic Press, London, pp. 81–130 (1981).
32. Brian, J. Mc Carthy. Biodeterioration in wool textile processing. *International Dyer*, (164), pp. 59–62 (1980).
33. Onions, W. J. Wool an introduction to its properties, varieties, uses, and production. *Interscience*, p. 41 (1962).
34. Brian, J. Mc Carthy and Phil, H. Greavest. Mildew – causes, detection methods and prevention. *Wool sci.Rev.*, (65), pp. 27–48 (1988).

35. Lewis, J. Mildew proofing of wool in relation to modern finishing techniques. *Wool sci. Rev.*, **1**(46) 17–29 (1973).

36. Lewis, J. Mildew proofing of wool in relation to modern finishing techniques. *Wool sci. Rev.*, **2**(47), 17–23 (1973).

37. Jain, P. C. and Agrawal, S. C. A not on the keratin decomposing capability of some fungi. *Transactions of the Mycology Society of Japan*, (21), pp. 513–517 (1980).

38. Espie, S. A. and Manderson, G. J. Correlation of microbial spoilage of woolskins with curing treatments. *Journal of Applied Bacteriology*, (47), pp.113–119 (1979).

39. Evans, E. and Braian, Mc. Carthy. Biodeterioration of natural fibers. *J. Soc. Dyers Colour.*, **114**(4), 114–116 (1998).

40. Pekhtasheva, E. L., Sapozhnikova, A. I., Neverov, A. N., and Sinitsin, N.M. Estimation of amount of microbes in wool fibers. *Izvestiya Vuzov. Tekhnologiya Tekstil'noi Promyshlennosti (Herald of High School. Technology of Textile Industry) journal* (Russian), (2), (271) (2003).

41. Kato, K. and Fukumura, T. Bacterial breakdown of ε–caprolactam. *Chem.and Industr.*, (23), p. 1146 (1962).

42. Huang, S. J., Bell, J. P., and Knox, J. R. Desing, Synthesis, and Degradation of Polymers Susceptible to Hydrolysis by Proteolytic Enzymes. Proceeding of Third International Biodegradation Symposium (Kingston, USA), London Appl. Sci. Publ. LTD, pp. 731–741 (1975).

43. Ermilova, I. A. Theoretical and practical foundation of microbiological degradation of textile fibers and ways of defence of fibers against the action of microorganizmes. Phesis (Russian), S. M. Kirov Leningrad Institute of Textile and Light Industry, Leningrad, p. 470 (1982).

44. Ermilova, I. A. Theoretical and practical foundation of microbiological degradation of chemical fibers (Russian). Nauka (Science) Publishing House, Moscow, p. 248 (1991).

45. Ermilova, I. A., Alekseeva, L. N., Shamolina, I. I., and Khokhlova, V. A. Effect of microorganisms on the structure of synthetic fibers. *Tekstil'naya Promyshlennost' (Textile Industry) journal* (Russian), (9), pp. 55–57 (1981)

46. Watanabe, T. and Miyazaki, K. Morphological deterioration of acetate, acrylic, poliamide and polyester textiles by micro-organisms (Aspergillus spp., Penicillium spp.). *Sen.-1 Gakkaishi*, (36), pp. 409–415 (1980).

47. Perepelkin, K. E. Structure and properties of fibers (Russian), Khimiya (Chemistry) Publishing House, Moscow, pp. 208 (1985).

48. Mankriff, R. U. Chemical fibers (Russian), A. B. Pakshver (Ed.), Legkayz Industriya (Light Industry) Publishing House, Moscow, pp. 606 (1964).

49. Kudryavtsev, G. I., Nosov, M. P., and Volokhina, A. V. *Polyimide fibers* (Russian). Khimiya (Chemistry) Publishing House, Moscow, pp. 264 (1976).

50. Tager, A. A. *Physico-chemistry of polymers* (Russian), Khimiya (Chemistry) Publishing House, Moscow, pp. 544 (1978).

51. *Fibers with special properties*, (Russian). L. A. Wolf (Ed.), Moscow, Khimiya (Chemistry) Publishing House, pp. 240 (1980).

52. Hamlyn, P. F. Microbiological deterioration of textiles. *Textiles*, **12**(3), 73–76 (1983).

53. Hofman, H. P. Die antimikrobielle Ausrustung der Kleidung. *Textiltechnik*, **36**(1), S30–S32 (1986).

54. Vigo, T. L. and Benjaminson, M. A. Антибактериальная обработка волокон и дезинфекция. *Textile Research Journal*, **51**(7), 454–465 (1981).

55. Kozinda, Z. Yu., Gorbacheva, I. N., Suvorova, E. G., and Sukhova, L. M. *Obtaining methods of textile materials with specific (antimicrobes and flame retardants) properties* (Russian), Legprombytizdat (Light Industry) Publishing House, Moscow, pp. 112 (1988).
56. McCarthy, B. J. Rapid methods for the detecbion of biodeterioration in textiles. *International Biodeterioration*, (23), pp. 357–364 (1987).
57. Mc Carthy, B. J. Preservatives for use in the Wool Textile Industry. *Preservatives in the Food, Pharmaceutical, and Environmental Industries*. R. G. Board, M. C. Allwood, and J. G. Bauks (Eds.), Blackwell Scientific Publications, 75–98 (1987).
58. Anon. Preservative treatments for textiles. *Part I. Specification for treatments*. British Standard 2087. British Standards Institution, London (1981).
59. Kalontarov, I. Ya. *Properties and methods of application of active pigments* (Russian). Donish Publishing House, Dushanbe, pp. 126 (1970).
60. Emanuel, N. M., Zaikov, G. E., and Maizus, Z. K. *Oxidation of organic compounds. Medium effects in radical reactions*. Pergamon Press, Oxford, pp. 628 (1984).

CHAPTER 20

THE SOURCES OF BIODAMAGES

ELENA L. PEKHTASHEVA and GENNADY E. ZAIKOV

CONTENTS

20.1   INTRODUCTION

This chapter includes information about morphology, internal organization and chemicals composition of microorganisms (bacteria and fungi), aggressive metabolites of microorganisms, and degradation of materials by the action of bacteria and fungi. Some parts dedicate the problems of factors affecting biodamages processes (chemical, physical, and biological).

The problem of biodamages comprises a wide range of scientific and practical tasks associated with protection of raw materials, intermediates, and products against damaging by bacteria, fungi, insects, and rodents during long-term storage, production, transportation, and operation.

The biodamage treats glass, plastics, rubbers, radio equipment, textile, leather, wood, paper, valuable monuments, transportation, and facilities that may significantly change their properties, cause quality reduction and, in some cases, completely destroy them.

As a result of microorganisms, insects and rodents impact, economic merit of products is reduced and operation of articles is disturbed. Such impact is commonly called "biodamage".

The "biodegradation" term is used when useful action of the organisms aimed at degradation and utilization of worn-out materials and articles is studied. These investigations are urgent for ecology, the struggle against the environment contamination.

The biodamage problem is both scientifically complex and practically diversified. Scientifically, it is based on the knowledge of materials science, biology, and chemistry.

Protection of textile materials, leather and footwear, wood and paper, plastics and metals, optics and concrete, oil and fuel against biodamage becomes the problem equal to their saving and prudent management.

The biodamaging of materials by microorganisms are known a great while ago. Long since various protection measures against biodamages have been applied. When the Hanging Gardens of Babylon were constructed, for example, measures preventing wood damage were provided, that is liquid resin, lead sheet, and asphalt was applied. However, investigation of the role of microorganisms in damaging various industrial materials was generally initiated after the end of the World War II. The earlier period demonstrated just a few works, which not discovered the microbial damage specificity.

A lot of military accoutrements of the countries at war were lost during the war time in tropical regions. In New Guinea, for example, microorganisms have completely destroyed all the equipment of Australian Army—tarpaulins, footwear, rubber articles, electrical appliances, optics, radio transceivers, and so on.

In moderate climate, microorganisms cause damages of industrial materials during their manufacture at increased temperatures and humidity, in violation of storage and operation conditions, and at transportation. In tropical and subtropical climate, microorganisms develop much more intensively and, therefore, cause more significant damage.

The man was concerned with fur and wool depredator—moths and leather beetles, destructive activities of termites and rodents, and wood bores, a long while ago, since the first steps of husbandry, and struggled against their manifestations as far as possible.

Only in 1960s, the biodamage problem received the official status as a large international scientific and practical direction at the junction of sciences and practice of the mankind and pooling efforts of various specialists.

This monograph consists of three interrelated parts: bacteria and fungi—the sources of biodamages; insects and rodents—materials and products depredators; materials and products damaged by living organisms. The modern data on morphology and physiology of biodegrading microorganisms are presented. The microphotoes of the microorganisms damaging various materials and objects of microbiological impact are included as illustration. Some methods of assessing biological resistance of various materials are considered and analyzed.

A broad material on biological degradation on both natural and synthetic materials and products (plastics, textile fibers, leather and fur, wood, and so on) is systematized. Problems of utilization of polymeric wastes using microorganisms are discussed and the main protection methods and technology for raw materials and semi-products against biodamaging are shown, assuming examples of antimicrobial materials and products use.

The questions of biodamaging of materials by insects (moths, leather beetles, wood-fretters, cockroaches, and termites) and rodents (mice and rats) are given in the available form. Practical recommendations for struggle against these depredators are given.

20.2 MORPHOLOGY, INTERNAL ORGANIZATION, AND CHEMICALS COMPOSITION OF MICROORGANISMS

The tiny organisms sized in micrometers (1 μm = 10^{-6} m) or parts of micrometers, nanometers (1 nm = 10^{-9} m) are called microorganisms. They are the ancient organisms occurred billions of years before the man [1]. They feature high resistance, rapid propagation, and adaptability. No other living organisms in the nature, except for microbes, can stand pressure up to 800 atmospheres. They are known capable of existing at + 100°C, + 120°C, and – 250°C. Some species of microorganisms were found which stand the impact of hydrochloric and sulfuric acids inhabit in kerosene and formalin.

The microorganisms feature high diversity of metabolic processes, nutrition demands and sources of energy are different. The microbes have extremely intensive metabolism. A single cell, under favorable conditions, processes a substance mass 30–40 times greater than its own weight per day. As a consequence, the rate of microorganism biomass buildup is high. The main part of nutrition is consumed in the energy metabolism, when multiple metabolic products are excreted to the environment, acids, alcohols, carbon dioxide, hydrogen, and so on.

The prime chemical elements of the microorganism cells are carbon, oxygen, hydrogen and nitrogen. These are basic elements forming organic substance and therefore are called organogenic elements (90–97% of dry substance). Other elements giving 3–10% are called ashy or mineral. The greater part is given by phosphorus. Microorganisms have extremely low quantities of trace elements, copper, zinc, manganese, molybdenum, and so on. The elements are present in microbial cells in forms of various compounds, among which water gives 75–85% of the cell mass and is vitally important for them. All substances are delivered to the cell with water and metabolic products are removed with it.

Some part of water is bound in the cell with proteins, carbohydrates, and other substances and is a part of cell structures. The rest of it is free being the dispersion phase for colloids and a solvent for various organic and inorganic compounds formed in the cell by metabolic processes.

The organic substances comprise dry matter of the microorganism cells (15–25% of the cell mass), mostly (up to 85–95%) represented by proteins, nucleic acids, carbohydrates, lipids, and so on.

Proteins are the basic cell components, which content reaches 80% of the dry matter in bacteria, 60% in yeasts, and 40% in fungi. By the amino acid composition, microorganism proteins are similar to proteins of macroorganisms.

The carbohydrates are part of various cellular membranes of microorganisms, are used for synthesis of substances in the cell and as the source of energy. They can also deposit in the cell as reserved nutrition. In the most of bacteria, carbohydrates comprise 10–30% of the dry matter mass and 40–60% in fungi. In the microorganisms carbohydrates are mostly presented by polysaccharides.

Lipids in the cells of microorganisms comprise 3–10% of the dry matter mass. They are part of the cell wall, cytoplasmic and other cellular membranes, and deposit as reserve granules. Some part of lipids is bound to other substances in the cell forming complex aggregates.

Pigments and vitamins are also observed in microorganisms. Pigments, or staining agents, are responsible for dyeing of microorganisms and sometimes they are emitted into the environment.

Mineral substances comprise, at most, 5–15% of the dry matter in the cell represented by sulfates, carbonates, chlorides, and so on. Mineral compounds play an important role in regulation of intracellular osmotic pressure and the colloidal state of cytoplasm [2].

20.2.1 BACTERIA—GENERAL DESCRIPTION

SHAPE AND SIZE OF BACTERIA

Bacteria have extremely multiple shapes. Spherical (globular), cylindrical (rod-like) and flexibacterium shapes are most abundant [3].

Globular bacteria are distinguished by the combination of cells (Figure 1(a))— *cocci* or *micrococci*, *diplococcic*, *streptococci*, and *Sarcina*. *Cocci* are individual globular cells. Globular cells bound in pairs are called *diplococcic*; longer chains of cells are called *streptococci*; a bunch of grapes shaped aggregates are *staphylococci*; and a combination of four *cocci* formed by cell division in two orthogonally related directions gives *tetracocci*. If division happens in three orthogonally related directions, properly formed packs of eight or more cells, the so-called *Sarcina*, are formed. The *coccus* is 0.5–1.2 μm in diameter.

Rod-like bacteria (Figure 1(b)) are distinguished by sizes. Those forming spores under certain conditions are called bacilli, which the largest of the rod-like forms and those not forming spores are called bacteria proper. The rod-like bacteria may be single or coupled–diplobacteria, linked in chains of three-four or more cells–streptobacteria. The ratio between rod length and thickness may be extremely different.

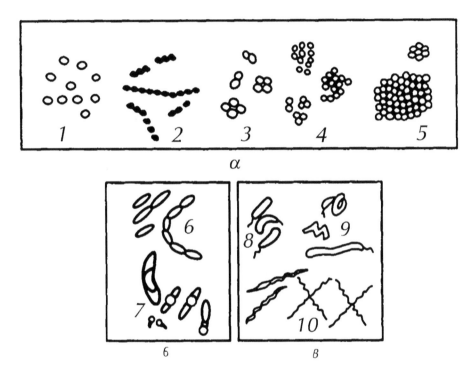

FIGURE 1   Bacterium shapes—*(a)* globular ((*1*) micrococci, (2) streptococci, (3) diplococcic and tetracocci, (4) staphylococci, (5) Sarcina), (*b*) rod-like, (*c*) flexibacteria ((*8*) virions, (*9*) spirillas, and (*10*) spirochetes).

The cell length of the rod-like bacteria varies from tenth parts of micrometer to 10–15 μm or more; the mean length of the rod-like bacteria is 2–5 μm; the cell diameter is 0.5 μm to 1 μm.

Flexibacteria or curved bacteria (Figure 1(c)) are distinguished by length, thickness, and curvature. The rods slightly curved to a comma shape are called *vibrions*; rods with one or several curls, corkscrew-shaped, are called *spirillas*; and thin rods with multiple curls are called spirochetes. Bacterium shape, as well as their sizes, may vary with respect to many reasons, for example, to growth conditions. Bacterial cell has extremely low mass of about $4 \cdot 10^{-13}$ g.

BACTERIAL CELL STRUCTURE

Bacterial cell consists of the cell wall, the cell membrane, the nucleus, and cytoplasm with inclusions (Figure 2).

The cell wall 0.01–0.04 μm thick preserves bacterium shape and is important for vital processes. Together with subjacent cell membrane, it regulates nutrition delivery to the cell and metabolic product excretion. The cell wall gives 5% to 20% of the cell dry matter mass. It is elastic mechanical barrier between the cell body and the environment. The chemical composition of the cell wall is differential for different species of bacteria. This has been found on the basis of different ability of the cell walls to retain triphenylmethane series dyes with iodine. With this respect, all bacteria are divided into two groups. One group includes bacteria, in which the complex with iodine is not discolored by alcohol. Another group comprises bacteria unable to retain dyes. This method has been suggested by Dutch physicist H. Gram. Bacteria dyed according to Gram are called gram-positive, others being gram-negative. In the preparation, gram-positive bacteria are colored purple-violet, and gram-negative–pink-crimson. Gram-positive bacteria have thick amorphous cell walls. Gram-negative bacteria have thinner laminated cell walls, with abundant lipids.

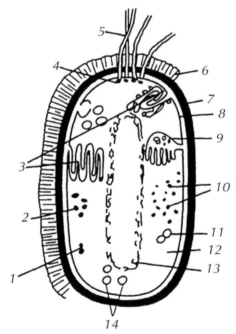

FIGURE 2 Bacterial cell structure sketch—*(1)* fat droplets, *(2)* polyphosphate granules, (3) intracellular membrane formations, (4) basal body, (5) flagellae, (6) capsule,(7) cell wall, (8) cell membrane, (9) mesosomes, (10) ribosomes, (11) polysaccharide granules, (12) cytoplasm, (13) nucleoid, and (14) sulfur inclusions.

Some bacteria are covered by a special protective mucous capsule, mucus protects against adverse impacts of the environment, mechanical damages and drying, creates additional osmotic barrier, makes an obstacle for phages and antibodies permeation, and sometimes, is the source of reserved nutrition. Sometimes, cell wall sliming is so intensive that capsules of separate cells conjugate in a mass with embedded bacterial cells. For instance, rawhide is subject to sliming.

The cell membrane 7–10 mm thick separates the cell content from the cell wall. When the wholeness of the cell membrane is disturbed, the cell devitalizes. The cell membrane gives 8–15% of the cell dry matter mass. The membrane contains up to 70–90% of cell lipids and various enzymes. The cell membrane is semipermeable, playing an important role in the substance exchange between the cell and the environment.

Cytoplasm of the bacterial cell represents a semi-liquid viscous colloid system. It is penetrated by mesosomes in places, which are membrane structures performing various functions. Enzymes supplying the cell with energy are located in mesosomes and associated cell membrane.

Cytoplasm contains ribosomes, the nucleus apparatus and various inclusions.

Ribosomes are spread in cytoplasm as granules sized 20–30 nm and consists approximately by 60% of ribonucleic acid (RNA) and by 40% of protein. Ribosomes are responsible for synthesis of the cell protein.

As found by the electron microscopy method, genetic information of the bacterial cell is carried by deoxyribonucleic acid molecule (DNA). It is shaped as double helical flagella in a closed ring; it is also called bacterial chromosome. It is located in a certain area of cytoplasm, not separated from it by the self membrane. The nuclear apparatus of bacterial cells is called nucleoid.

Bacterial cells have various cytoplasmic inclusions which are mainly nutrition reserved in cells, when they are developing in the nutrition excess in the medium, and consumed under starving conditions. Polysaccharides are settled in the bacterium cells, glycogen and an amyloid substance, which are carbon and energy sources; lipids shaped as granules and droplets which are the source of energy are also detected.

Under unfavorable conditions of development (nutrient shortage, medium temperature and pH variations, and metabolite accumulation), this ability is generally featured by rod-like bacteria. In each cell, one endospore located in the center or at the end of the cell is formed. Spores are commonly roundish or oval shaped. After spore maturation the mother cell dies, its membrane is destroyed and the spore releases. The spore is formed within several hours. Viewed under the microscope, spores appear colorless, brightly lighting bodies. In a commonly dyed smear spores are not dyed due to low permeability of multilayer membranes.

BACTERIUM MOBILITY

Globular bacteria are usually immovable. The rod-like bacteria are both mobile and immovable. Curved and spiral-form bacteria are mobile. Most of bacteria move with the help of flagella. Some bacteria displace by sliding [4].

Flagella are thin spiral shaped protein threads able to rotate. Their lengths are different and the thickness is so small (10–20 nm) that they only become visible under the optical microscope after special cell treatment. Monotrichate microorganisms are bacteria with one flagella at the cell end, lophotrichate microorganisms are lophotrichous bacteria (with a cluster of flagella at one end of the body), amphitrichate microorganisms have clusters of flagella at both ends of the body, and bacteria with flagella located by the entire body surface are called peritrichate microorganisms. The motion speed of such cells is high: the distance passed by a cell with flagella may reach 20–50 lengths of its body per second.

MULTIPLICATION OF BACTERIA

Simple division is typical of bacterial cells. The rod-like bacteria divide crosswise and globular forms divide by different equators. Various forms occur with respect to the equator orientation and the number of divisions: single, couples, chains, packs, and clusters. The multiplication of bacteria features fastness of the process. The division rate depends on bacterium species and cultivation conditions, some species divide every 15–20 min, other every 5–10 hr. At such division, the cell population becomes tremendously high within a day. For instance, just flayed raw hide is immediately cured, because two hours after it loses much of its commercial quality under the impact of putrefactive bacteria. If no nutrients are injected to the medium and metabolites are not removed, then bacteria appeared (inoculated) on a nutritive substrate develop in time by a particular dependence. Several stages (phases) alternating in a particular sequence, during which the multiplication rate and morphological, physiological and biochemical properties of microorganisms change, are recognized.

At the initial development stage (Figure 3(a)), the lag-phase, bacteria occurred on a new medium do not multiply for some time, as if they adapt to it. In this regard, the cells increase in size and protein and RNA content in them increases, too [5].

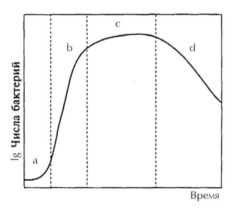

FIGURE 3   Bacterium growth curve, (*a*) lag-phase, (*b*) growth phase, (*c*) stationary phase, and (*d*) death phase.

As bacteria adapt to the medium, they start multiplying at an increasing rate. Then the cell division rate is maximal constant typical of each cell species and the medium for a definite time. This period (Figure 3(b)) is called the growth phase. The young active cells form great amounts of metabolites in the medium. The rate index of bacterium multiplication at this stage of development is the generation time (g) that is, time of the cell quantity doubling.

At the growth phase end, the number of cells becomes maximal and the stationary development phase begins (Figure 3(c)), when the number of living cells remains more or less constant. The numbers of forming and dying cells are approximately the same. Finally, the death phase begins (Figure 3(d)), when more and more cells devitalize and die. This happens due to nutritive medium exhaustion and accumulation of metabolites in it. Duration of separate development phases for different species and for bacteria of the same species may significantly vary with respect to the growth conditions.

FEATURES OF BACTERIA-DEGRADERS

Manifold possibilities of bacteria in biodamaging are due to their ability to use virtually all sources of energy and nutrition, both organic and inorganic, containing nitrogen and carbon.

For biodamages of inorganic industrial materials of importance is the ability of many bacteria to exist using no environmental organic substances.

Bacteria capable of using inorganic substances are called lithotrophic ones. Among lithotrophic bacteria, sulfate-reducing, thionic, nitrate-producing, and iron bacteria are the most active biodamaging agents. These bacteria induce colossal corrosion of metals, destruction of concrete, stone, bricks, and other inorganic materials.

CURIOUS FACTS

Pseudomonas aeruginosa bacterium strains capable of forming brightly dyed and fluorescent blue, turquoise and dark-green pigments were extracted from aviation kerosene and diesel samples in-situ tested in tropics [1].

This ability of the strains can be used as a diagnostic characteristic of biodegradation of fuels and other technical materials in laboratory.

The organotrophic bacteria obtain energy by oxidizing organic substances and, therefore, many of them degrade organic industrial materials. However, some of them also promote corrosion of metals, forming aggressive metabolites (organic acids, ammonia, and hydrogen sulfide). Virtually all organic materials, as well as raw materials and products are subject to biodegradation–leather, textile fibers, fur, plastics, cosmetic emulsions, and so on.

The destruction of solid materials by bacteria is, to a certain extent, associated with the expressed ability to be adsorbed on solid surfaces. This is frequently the first stage of bacterial degradation of solids. For example, the ability of bacteria to fix on the glass surface is commonly known. This is clearly observed at glass submerging to the culture fluid or water. Bacterium fixture to glass proceeds in two stages. The first stage is performed at a short-term glass contact with bacteria, when fixture is not yet strong enough and adsorbed bacteria are easily washed off the surface. At the second stage, bacteria are stronger fixed to the glass surface with the help of capsular substance or by other methods. In a number of cases, after bacterium fixture glass degradation begins. Using the electron microscopy, hollows in places of fixture of some slime-forming bacteria were observed.

Cellulose-degrading microorganisms are fixed to cellulose fibers along them. Cells of such microorganisms feature extremely high adsorbing capacity and are strongly adsorbed by cellulose fibers.

Experiments on polyvinyl chloride (PVC) film degradation in the soil show that definite forms of microorganisms are adsorbed and multiply on the films, and the strongest film change happens specifically under microcolonies of these microorganisms.

An active bacterial degradation of various materials is also promoted by the absence of specificity of some bacterial exoenzymes. For instance, beside protein hydrolysis, Bac. Subtilis enzymes are able to catalyze hydrolysis of amides, amino acid ethers and their derivatives, lower fatty acid ethers, and even some triglycerides. Of high importance is also ability of some bacterium species to stay alive under extreme environmental conditions, high temperature (up to + 80°C or even higher) and pressure, high acidity or alkalinity, intensive radiation, high salt concentration, and so on. As a consequence, the presence of such conditions at any manufacture or technological operation may not guarantee elimination of biocorrosion.

Putrefactive bacteria are the most active degraders of nonfood organic goods, raw materials, and products. They get energy by oxidizing organic substances. They comprise spore-forming, sporeless, aerobic and anaerobic bacteria. Many of them are mesophilic bacteria, but cold tolerant and heat-resistant ones are present, too. The most of them are medium pH and salinity sensitive.

The most widespread putrefactive bacteria are as follows:

Bacillus genus bacteria—rod-like, aerobic, mobile, gram-positive, and spore-forming bacteria (Figure 4(a)). Their spores feature high heat resistance. The optimal temperature for their development is + 35°C to + 45°C, and the maximal growth is observed at + 55°C to + 60°C; at temperatures below + 5°C they do not multiply [4].

Pseudomonas genus bacteria are aerobic mobile rods with polar flagella, forming no spores, and gram-negative.

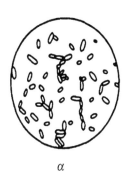

α 6

FIGURE 4   Putrefactive bacteria–(a) *Bacillus* (rods and oval spores) and (b) *Pseudomonas*.

Some species synthesizing pigments are called fluorescent *Pseudomonas*. There are cold tolerant species with the minimal growth temperature of –5°C to –2°C. Along with proteolytic activity, many *Pseudomonas* feature lipolytic activity, too, they are able to oxidize carbohydrates forming acids and mucify (Figure 4(b)).

Bacterium genus bacteria are gram-negative rods forming no spores [5].

20.2.2   MICROSCOPIC FUNGI—GENERAL SPECIFICATION

The microscopic fungi are widespread in the nature occurring in every area all over the Globe in various vegetable substrates and to a lesser extent in animal substrates. They take active part in degradation of organic residues and the soil-forming process [6]. Multiple groups of fungi cause a great economic damage degrading various industrial materials.

FUNGUS BODY STRUCTURE

The vegetative bodies of the most of fungi represent a spawn or a mycelium from branching threads–hyphae, with thickness varying from 2 μm to 3 μm. Such fungi are called filamentous (they are also called molds).

Some species of microscopic fungi have no mycelium. These are several representatives of lower fungi and yeasts, which represent singular roundish or elongated cells. Some fungi have cellular mycelium–hyphae are separated by septa, and the cells are frequently multinuclear; mycelium of other fungi is non-cellular, hyphae have no septa, and the whole mycelium represents somewhat like a giant cell with many nuclei. The so-called kames consist of compactly interlaced hyphae and contain reproductive organs.

Mycelium starts developing from spores growing out at particular temperature and humidity. The spore swells first absorbing water from the environment, then its cover breaks and one or several growth tubes appear initiating the new mycelium. Primarily, hyphae develop consuming reserved spore nutrition, and further on adsorbing nutrition from the substrate.

In accordance with the growth status, substrate, and aerial mycelium is distinguished. Mycelium can develop partly in the substrate (substrate mycelium) penetrating through it and absorbing water and nutrition and partly on the substrate surface (aerial mycelium) in the form of villous, spider-web or thin taints and films.

The growth status of the same fungus on the substrate may change with respect to the medium conditions (nutrient composition, humidity, and so on). However, for some species this feature is constant. For instance, lush aerial mycelium formation is typical of many fungi–the wood depredators. The aerial and substrate mycelia have different chemical composition and biochemical activity. The hyphae placed in the substrate contain more reserve nutrients (glycogen, proteins, and fats), than aerial ones.

In case of poor aeration at culture fluid, sometimes, fungi form films consisting of hyphae spread in different directions on the substrate surface or in cracks in wood. They look like a buck several millimeters thick. Further on, rhizomorphs or simple mycelium yield from this mycelial mfilm, and sometimes fruit bodies develop on it.

Some mucoral fungi form arched aerial hyphae–stolons. They help the fungus to spread rapidly by the substrate. The stolons are fixed by rhizoids developing as a response to contact with any solid substrate, with glass, for example (Figure 5).

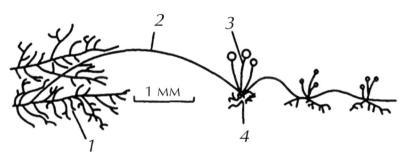

FIGURE 5   Mucoral fungus *Phizopus stolonifer*—(*1*) mycelium, (*2*) stolon, (*3*) conidium carrier, and (*4*) rhizoids.

FUNGUS CELL STRUCTURE

Cells of the most of fungi are covered with a rigid cover consisting of the cell wall and various extracellular excretions. The cell wall is the main structural component of the cover. It provides the cell with a typical stable shape and mechanically protects against osmotic pressure.

The cell wall consists of polysaccharides by 80–90%, it also includes few proteins, lipids, and polyphosphates. Quinine is the main polysaccharide of the cell wall of the most of fungi, and some have cellulose.

The cell wall is usually about 0.2 μm thick. It gives 10% to 50% of the dry mass of the organism. The cell wall material quantity varies during the life cycle of fungi, usually, it increases with age.

Under the cell was a three-layer cell membrane is located, about 8 nm thick. It plays the role of osmotic barrier of the organism and controls selective substance delivery to the cell.

The inner content of the cell can be divided into membrane structures and cytoplasm. Cytoplasm represents a colloid solution. It comprises enzymes, proteins, amino acids, carbohydrates, nucleic acids, and granules of reserved substances. Fungus cytoplasm contains a highly developed system of internal membranes.

The membrane structures are—endoplasmic reticulum, Golgi apparatus, and mitochondria.

Endoplasmic reticulum is the membrane system of interlinked tubules (constricting or expanding in places), which penetrates cytoplasm and is linked to the cell membrane and the nucleus membrane. This organelle synthesizes many substances (lipids, carbohydrates, and so on).

Mitochondria are formations of lipoprotein membranes, in which power processes precede and adenosine triphosphate (ATP), the high-energy substance, is synthesized.

Golgi apparatus is the membrane system linked to the nucleus membrane and endoplasmic reticulum. Its multiple functions also include transportation of substances synthesized in the endoplasmic reticulum and metabolic product excretion from the cell.

Ribosomes are very small roundish, highly abundant formations. A part of them is in a free state and the rest is fixed on membranes. Protein is synthesized in ribosomes.

Lisosomes are small roundish bodies covered by a membrane. They contain enzymes digesting (cleaving) proteins, carbohydrates, and lipids delivered from the outside.

Nucleus (or several nuclei) is encased in the double membrane. The nucleus comprises and nucleolus and chromosomes containing DNA. The nuclear cover has pores that provide substance transport from the nucleus to cytoplasm.

Vacuoles are cavities surrounded by the membrane and filled with the cellular fluid and inclusions of reserved nutrients.

FUNGUS PROPAGATION METHODS

Fungi feature manifold methods and organs for propagation. The same fungus often has several propagation forms [7].

Fungi propagate by vegetative, asexual, and sexual ways.

Vegetative propagation proceeds without special organs formation, any part of mycelium originates a new organism. It is usual for the culture of artificial nutrient media.

CURIOUS FACTS

*Fungi are very important for the leather industry, where hair removal from the treated hide of any animal was the dirtiest and labor consuming process for centuries, because enzymes contained in the dog turd and pigeon dung were used for this purpose. It took weeks and would not be considered complete. Now the enzyme proteinase both from animal's pancreas and As-*pergillus *genus fungi easily performs this work and abates leather simultaneously, within 24 hr or less. As a result, general quality of leather and its color are improved.*

Production experience in leather abating with the use of Penicillium chrysogenum *fungus enzyme has given extremely good result, whereas mycelium of this fungus has been used, preliminarily frozen for better preservation capacity. As it turned out, this significantly increases its enzymatic activity.*

At asexual and sexual propagation, specialized cells (spores) are formed, with the help of which propagation is performed.

At asexual propagation, spores are formed in special hyphae of aerial mycelium, which are exteriorly different from other hyphae.

In some fungi spores are formed exogenously (outright), externally, at the top of hyphae. Such spores are called conidia, and hyphae on which they are formed–conidium carriers (Figure 6).

Conidia are formed directly on the conidium carrier or special cells located on its top. These cells are usually bottle-shaped and are called sterigmata. The conidia are located on conidium carriers (or sterigmata) one by one, in groups, in chains, and so on.

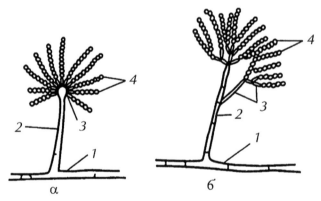

FIGURE 6 Conidium carriers for fungi of *(a)* Aspergillus *and (b)* Penicillium *genuses—(1)* vegetative *(2)* conidiophores, (3) sterigmata, and (4) conidia.

CURIOUS FACTS

When processing yarns in the textile manufacture, sizing is performed, that is sizing of the base with a substance from starch products. However, further on, this sizing compound shall be removed, otherwise no high-quality bleaching and further coloring of crude fabric may be performed.

In previous time, the sizing compound was removed by a complicated complex of physico-chemical procedures which, however, did not provide high quality of the fabric and appropriate hygienic and sanitary conditions of work. Therefore, desizing is now performed by enzymatic agents (malt extracts). Hence, "saccharifying mold" occurred more effective by time and product quality, which removes the sizing compound during 5 min and + 66°C to + 100°C. The enzyme producer is Penicillium chrysogenum fungus, which forms penicillin. Its mycelium also contains amylaze enzyme.

In other fungi spores are formed endogenously, inside special cells developed at the ends of hyphae. These cells, the spore cases, are called sporangia, spores present in them are called sporangiospores and hyphae carrying sporangia with spores–sporangia carriers (Figure 7).

In some fungi, mobile spores having flagellae, the zoospores, are formed in sporangia [8].

Sporangiospores and conidia have different shapes, sizes, and colors. That is why, at the stage of fruiting fungi look like colored blooms. Mature conidia shed. When sporangiospores become mature, sporangia break and spores precipitate from them. Conidia and sporangiospores are passively transported by air at the long distances. When appeared in favorable conditions, spores grow out to hyphae.

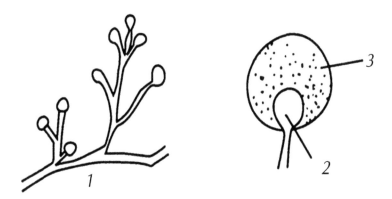

FIGURE 7 Sporangia and sporangia carriers of *Mucor* genus fungi—(*1*) fruiting mycelium,(2) sporangia carrier, and (3) sporangium with spores.

Spores of fungi damaging materials are usually distributed by wind, rain, insects, animals, and people. Asexually propagating spores serve for rapid substrate colonization. Industrial materials are generally polluted by asexual propagation spores.

In the case of sexual propagation of fungi, the spore formation is preceded by the sexual process–sexual cells fuse with further aggregation of nuclei. In this regard, specialized reproductive organs are formed. The development of these organs and sexual process forms of fungi are manifold.

The most of fungi are able to propagate both in asexual and sexual ways, and they are called perfect fungi. Some fungi are incapable of sexual propagation, and they are called imperfect fungi.

To distinguish fungi, the features of propagation methods and reproductive organs structure are used. These features form the basis for their classification.

FEATURES OF FUNGI CAUSING BIODAMAGES

Fungi have some morphological, physiological, and genetic features which make them predominant among organisms causing biodamages [1].

Fungi are extremely widely spread all over the Globe. They are present in the soil, water, and air. The most of fungi which induce material damage are of high propagation power. For example, dry spore forms (*Aspergillus*, *Penicillium*, *Trichoderma*, and so on) form spores amounted to hundred thousand or millions. Spores are so small and have so low mass that they raise very high and are transferred to long distances at any slightest motion of the air. Their microscopic size makes them capable of penetrating into invisible cracks and pores multiple even in such dense materials, as granite or metal.

Sometimes fungi are observed in polymeric materials, at the interface of high-molecular compound and the components in the material structure. Diffusing water may carry spores away from the surface deep in materials, especially porous ones. These examples testify that fungi can be observed everywhere, even there, where no other organisms can penetrate.

A great role in materials colonization by fungi is played by the ability of their spores to be adsorbed on a smooth surface. Adhesion is the first stage of biodamaging of solid insoluble substrates.

When fixed on the material surface and under favorable conditions, spores grow up and form mycelium. The mycelial structure of fungi is one of the most important biological features that define of their interrelation with the medium. Mycelium rapidly spreads by the substrate and captures large areas.

Among microorganisms, the dominant role of fungi in biodamage processes is caused by their metabolic features, which are due to rich enzymatic apparatus. They are therefore able to induce untimely failure of any industrial product.

The ability of microscopic fungi to grow under biologically extreme conditions is extremely important for biodamaging of materials by them. Fungus spores are resistant to drying. The instances are known, when they were dry during 20 years or more. A considerable part of fungal spores endures low temperatures without losing biochemical activity.

Fungi that cause biodamages enter the group of saprophytes. They are tightly bound to the substrate, have a large absorption surface and cause high impact on the environment *via* metabolic products. In relation to substrate, they are divided into two groups—nonspecific and specific saprophytes.

Nonspecific saprophytes are polyphage fungi observed on various substrates. Among them, *Aspergillus, Penicillium, Trichoderma, Alternaria and Fusarium* genus species most frequently develop on industrial materials.

Specific saprophytes observed on damaged materials consist of more or less specialized organisms. They were formed during adaptive evolution to various substrates. For instance, house-fungus *Serpula lacrymans* developing exclusively on stock timber correspond to this kind. Another example is *Cladosporium resinae*, which grows on petroleum derivatives preferring gasoline and kerosene [8].

Extraction from the centers of damage of nonfood raw materials, half-finished materials and products usually gives the following fungus genuses (Figure 8–10).

Aspergillus—fungi of this genus have unicellular unbranched conidium carriers. The tops of conidium carriers are ventricose to a greater or lesser extent and carry one or two layers of sterigmata with conidium chains located on their surface (see Figure 8). Conidia are most frequently roundish and differently colored (green, yellow or brown). The view of a conidium carrier is similar to mature dandelion.

Penicillium—fungi of this genus have multicellular branched conidium carriers. Sterigmata with conidium chains are located at the ends of conidium carrier branches. Conidia can be green, light blue, green-grey or colorless. The upper part of the conidium carrier (Figure 9) has a brash shape of various complexity what causes the name of this fungus—*Penicillium* (common blue-green mold).

FIGURE 8   *Aspergillus niger* micrograph (× 600).

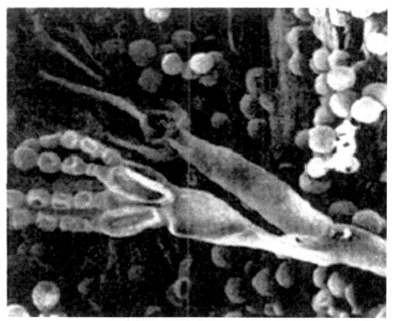

FIGURE 9   *Penicillium* micrograph (× 3000).

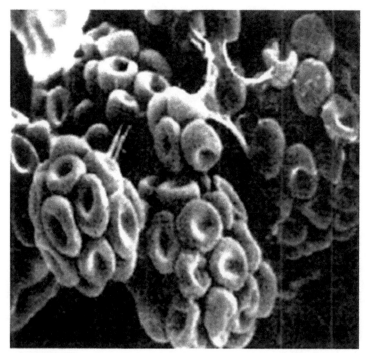

FIGURE 10   *Trichoderma* micrograph (× 3000).

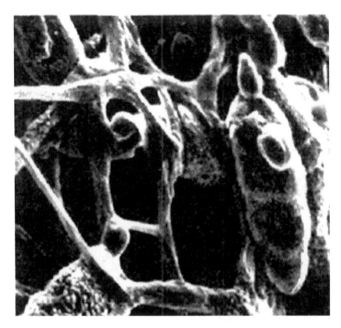

FIGURE 11   *Alternaria* micrograph (× 600).

Trichoderma—Condium carriers are much-branched; conidia are light-green or green, egg-shaped (sometimes elliptic). They are observed on polymeric materials (Figure 10).

Alternaria is characterized by the presence of multicellular, dark colored conidia of stretched claviform located in chains or in singles on poorly developed conidium carriers (Figure 11). Various *Alternaria* species are widespread in the soil and on vegetable residues. These fungi damage a broad range of polymeric materials of various chemical compositions covering them with black spots. Some species actively degrade cellulose.

Cladosporium has poorly branched conidium carriers with chains of conidia at their ends. Conidia have various shapes (roundish, oval, cylindrical, and so on) and sizes. Mycelium, conidium carriers, and conidia are colored olive green. These fungi feature excretion of a dark pigment to the medium.

Stemphylium—Conidium carriers are dark olive or body colored with single conidia, spiny or scabrous, having various sizes and shapes.

20.2.3   YEASTS—GENERAL SPECIFICATION

Yeasts are unicellular immovable microorganisms widely spread in the nature, they are met in the soil, on leaves, footstalks, and fruits of plants, in various food substrates of both vegetable and animal origin.

The wide use of yeasts in industry is based on their ability to induce alcohol fermentation [1, 2].

SHAPE AND STRUCTURE OF YEAST CELL

The yeast cells are most frequently roundish, oval-oviform or elliptic, less frequently cylindrical or lemon-shaped (Figure 12). Yeasts of special shapes: sickle-shaped, needle-shaped, halbert-shaped and triangular, are also observed. Yeast cell size is, at most, 10–15 μm.

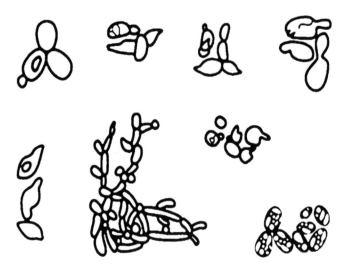

FIGURE 12   Yeasts.

The shape and size of yeasts may considerably change with respect to development conditions and the cell age. Internal structure of the yeast cell is similar to the fungus cell structure.

YEAST MULTIPLICATION

Budding is the most typical and widespread vegetative method of yeast multiplication, just a few yeasts divide.

Budding is as follows—a hummock (or several hummocks) occurs on the cell and gradually increases. This hummock is called bud. As the bud grows, a constriction at its connection with the mother cell is formed, bounding the young daughter cell which then either pinches (detaches) from the mother cell or remains on it. Under favorable conditions, this process lasts about 2 hr.

Along with budding, many yeasts multiply by spores, which may be formed by both asexual and sexual way.

20.3 AGGRESSIVE METABOLITES OF MICROORGANISMS

Many biochemical processes proceeding with the participation of microorganisms are used in food and light industry. The role of microorganisms in the cycle of matter in the nature is high. At present, a peculiar nutrition for microorganisms is represented by various industrial materials (textile, metal, concrete, plastics, rubber, leather, fuels, lacquers, paints, paper, and so on), as a result of "colonization" of which by saprophytes a specific group of bacteria and fungi-technophiles is formed.

Materials and products are biodamaged by mold fungi at the expense of mechanical destruction by proliferating mycelium, biological pollution, and especially, by the impact of enzymes and organic acids [3-5].

Biodamaging by bacteria generally happens due to impact of enzymes and organic acids.

20.3.1 DEGRADATION OF INDUSTRIAL MATERIAL BY ENZYMES— GENERAL CHARACTERIZATION OF ENZYMES

The various and multiple biochemical reactions proceeding in the living organism due to metabolism involve enzymes, which are biological catalysts produced by cells of the organism [9].

Enzymes are biological catalysts consisting of proteins and proteids and non-protein component–the active group. For instance, enzymes may have one and two components.

The active group of two-component enzymes comprises vitamins or their derivatives, various metals (Fe, Co, and Cu), nitrogenous bases, and so on.

The active group of enzymes is responsible for their catalytic ability and the protein part provides specific features, the selective ability to impact a particular substrate.

Enzymes are extremely active. A minute quantity of an enzyme is enough for involving a considerable mass of reagent (substrate) into reaction.

Enzymes act extremely rapidly. An enzyme molecule can induce conversion of tens and hundreds of thousand molecules of the appropriate substrate in 1 min.

The features of enzymes are their substrate and action specificity–every enzyme only interacts with a single particular substance and catalyzes only one of transformations this substance may be subject to.

The specificity of enzymes is stipulated by structural features of their molecules and the substrate. The substrate and the enzyme fit together as a key and a hole.

Each microorganism has a complex of various enzymes, which diversity and activity define its biochemical activity, selectivity in relation to nutrients, the role in the cycle of matter in the nature, and in biodegradation processes.

Enzymes typical of a particular microorganism and being among the components of its cell are called constitutive. There are inducible (adaptive) enzymes, produced exclusively in the presence of a substance (inducer) stimulating its synthesis in the medium. For example, a microorganism not assimilating polymers may be taught to use them by cultivating in the medium with this polymer, as the unique source of carbon. In these conditions, the microorganism synthesizes an enzyme, not yet produced by it.

Some enzymes which microorganism secrete to the environment are called exoenzymes. They play an important role in preparation of nutrients for delivery to the cell. Extracellular "digestion" of nutrients happens, that is cleavage of complex substances in the substrate (starch, proteins, and so on) to simpler components able to penetrate into the cell.

Cytoplasmic enzymes not secreted to the environment by the living cell are called endoenzymes (intracellular ones). These enzymes take part in intracellular metabolic processes. Enzymes secreted to the environment are of greater importance, rather than those only acting inside the cell.

At present, over 200 enzymes are known. In accordance with assumed classification, all enzymes are divided into six classes—1) oxidoreductases, 2) transferases, 3) hydrolases, 4) lyases, 5) isomerases, and 6) ligases.

The degradation of materials induced by enzymes happens by various reactions—oxidation, restoration, decarboxylation, etherification, hydrolysis, and so on Oxidoreductases, hydrolases, and lyases cause particularly active effect on the most materials.

Oxidoreductases are redox enzymes. This class comprises multiple enzymes, which catalyze redox reactions of the energy exchange (respiration and fermentation) in microorganisms.

Among oxidoreductases, oxygenases play the specific role in degradation of many industrial materials. These processes, first of all, are biodgradation of hydrophobic, non-polar substances of hydrocarbons type and materials formed by cyclic compounds. Oxygenases comprise enzymes which catalyze direct oxygen addition to the oxidized substrate. Such reactions usually represent the first stage of degradation of many alien substances in a living cell. For instance, *Pseudomonas genus bacteria catalyze the indole ring break.*

Of high importance for biodegradation processes are other enzyme of the oxidoreductase class, dehydrogenases and oxidases, too. Dehydrogenases catalyze hydrogen transfer from one compound to another. The "oxidase" term marks enzymes participating in reactions, where oxygen is direct acceptor of hydrogen. Dehydrogenases catalyze hydroxyl groups oxidation to aldehydes and then to carboxyls, as well as formation of unsaturated compounds from saturated ones.

In the class of oxidoreductases, peroxidase and catalase feature specific action. The peroxidase catalyzes oxidation of various organic compounds–phenols, amines, heterocyclic compounds–by hydrogen peroxide. Among mycelia fungi, representatives of *Penicillium, Aspergillus, Fusarium, Alternaria, Cladosporium* and so on genus manifest a considerable peroxidase activity.

Catalase intensifies hydrogen peroxide decay into water and molecular oxygen and also catalyzes oxidation by peroxides of various alcohols and other compounds. Active producers of catalase are some fungus species of *Penicillium* genus.

Hydrolases catalyze cleavage and synthesis of complex organic compounds by type of hydrolytic reactions with participation of water.

Of special importance in biodamaging of industrial materials are hydrolase class enzymes, because being exoenzymes, many of them participate in the primary preparation of nutritive substrate for cleavage and metabolism. Hydrolases catalyze cleavage reactions of complex compounds to simpler ones with simultaneous addition of water. In the context of biodamage problem, of special interest is the subclass of esterases, which catalyze hydrolytic break of ether bonds in various compounds.

Enzymes of the glycosidase group take active part in degradation of industrial materials containing cellulose and other carbohydrates, and their derivatives. Cellulose cleavage is catalyzed by a complex of enzymes–cellulase. It is based on enzymes hydrolyzing bonds between glucose residues in the cellulose molecule with formation of cellobiose and glucose.

The highest expressed ability to synthesize cellulolytic enzymes is manifested by microscopic fungi of *Alternaria, Trichoderma, Chaetomium, Aspergillus, Penicillium, Cladosporium* and so on genus. Various species of cellulose degrading fungi are only capable of growing and producing cellulase in a definite temperature range, mostly at + 28–+ 30°C.

Among hydrolases, of special interest are proteinase group enzymes. Their basic action is to cleave proteins by amide that is peptide bonds (proteolytic enzymes). However, some proteinases manifest quite broad specificity that is low selectivity to substrates. This gives grounds to suggest a possibility of any participation of proteinases synthesized by microorganisms in degradation of polymeric materials, first of all, these containing amide and ether bonds–urea formaldehyde polymers, acrylamides, polyamides (kapron, nylon), and polyurethanes (polyurethane foam).

Lyases catalyze nonhydrolytic cleavage of organic substances accompanied by detachment of one or another chemical group from the substrate—CO_2, H_2O, NH_3, and bonds -C-C, -C-O or -C-N to be broken simultaneously. The detached groups are—water, carbon dioxide, ammonia, and so on. The enzymes of this class are characterized by high substrate specificity and selectivity. Nevertheless, general direction of their action also provides a suggestion that these enzymes may participate in degradation of synthetic materials.

Transferases are transfer enzymes catalyzing transfer of parts of molecules or atomic groups from some compounds to others. Many of such enzymes are known, distinguished by the groups, which transfer they catalyze.

Isomerases transform organic compounds to isomers, which process is associated with intracmolecular transfer of radicals, atoms, and groups of atoms.

Ligases (synthetases) catalyze synthesis of complex organic compounds from simpler ones.

There is a clear correspondence between the type of damaged material and enzymatic properties of degradation originators, in which enzymes cleaving the main type of bonds in the particular material are specifically active. However, in general, many enzymes participate in degradation of the material. Several species of fungi and bacteria form a biocenosis on a material, and biochemistry of degradation is complex. For instance, biodegradation of polymers having amide and ester bonds in the molecule structure (polyamides, polyurethanes, urea formaldehyde resins, and polyesters) may be a result of microorganism attack, which produces active esterases and proteolytic enzymes.

20.3.2   DEGRADATION OF INDUSTRIAL MATERIALS BY ORGANIC ACIDS

The most aggressive metabolites of microorganisms are organic acids. They induce fast and deep degradation of industrial materials, both organic and inorganic, including metals [9].

Over 40 organic acids were successfully extracted from mold fungi cultures as follows—*Penicillium* genus fungi mostly produce citric and gluconic acids; *Aspergillus* genus produces citric, gluconic, and oxalic acids; and *Mucor* genus – succinic, fumaric, and oxalic acids.

The same fungus species is usually able to synthesize various allied acids. All fungi can be divided into three groups with respect to quantity of synthesized acids:

- Relatively high amount of organic acids secreted (*Aspergillus niger*, for example),
- Low amounts of acids secreted (the majority of fungus species, for example, *Penicillium* and *Trichoderma)*,
- Minute quantities of acids secreted (for example, *Mucor sp.* and *Alternaria tenuis*).

Most often, mold fungi secrete in high amounts the following acids—citric, gluconic, oxalic, lactic, fumaric, succinic, and malic acids.

In the most of cases, particular degradation mechanisms for some industrial materials by organic acids are studied insufficiently yet. The types of polymeric materials differ by resistance to organic acids—polyethylene, polypropylene, polystyrene, and phenoplasts are the most resistant ones; polyvinylchloride, poly(methyl methacrylate), and polyamide resins show lesser resistance. The presence of organic fillers in plastics, which are usually good nutrients for mold fungi, promotes active synthesis of organic acids, and therefore, increases the material degradation.

Organic acids play the leading role in degradation of varnishes. The damaging action of citric, tartaric and fumaric acids is observed already at rather low concentrations (0.09–0.4%). Strong degraders of varnishes are pyruvic, gluconic, acetic, and oxalic acids. All organic acids cause intensive degradation of various cellulose-containing materials.

Of special attention is corrosion of metals induced by organic acids. In some cases, they corrode metals even more intensively, than inorganic acids. Corrosion of tanks for petroleum products is sometimes caused by the action of organic acids formed at petroleum product decomposition by microorganisms on aluminum alloys.

The organic acids, enzymes, pigments, and some other metabolites of microorganisms induce considerable changes in physicomechanical, dielectric and other properties of materials, and abruptly decrease their technological parameters [8].

20.4   FACTORS AFFECTING BIODAMAGE PROCESSES

The development and vital activity of microorganisms is tightly associated with the medium conditions, where they inhabit. The environment may stimulate or suppress growth of biodegraders [1-11].

For solving basic problems associated with microbiological damages of materials, the knowledge of microorganisms' physiology is of high importance. For physiology, the processes of substance exchange between an organism and the environment, organism's growth and development, response to environmental impacts, and adaptation to the environment are assumed.

Metabolism represents a complex process of various chemical transformations of nutrients delivered to the organism from the environment (the substrate).

Processing of nutrients delivered to the cell is called anabolism. The process of getting energy by cell for implementing its vital functions is called catabolism. These processes are closely interacting and stipulate growth, development, and reproduction of the organism.

Microorgamisns grow and develop in the presence of accessible primary substances used for nutrition and energy supply. Some microorganisms have an exclusive ability to adapt for using various substrates in their vital activity, thus damaging raw materials, intermediates and products of various chemical and physical structures.

At the same time, beside dependence on the nutrition and energy sources, development of microorganisms is considerably stipulated by the influence of environmental factors, which may cause either positive or negative effect of their vital activity. Using unfavorable effect of some physical or chemical factors on microorganisms, one can affect the processes of biodamaging, protect raw materials and products against destruction.

The growth of microorganisms on industrial materials is tightly associated with conditions of the medium, where they are present. Note that the influence of some environmental factors on development of even widely spread microorganisms is poorly studied.

The knowledge of microorganisms response to the environmental factors are applied in practice to struggle against biodegraders. For example, death of microorganisms at high temperature and definite radiation or UV doses are used for sterilization. The increase of wood protection by waterproofing is based on sensitivity of wood-destroying and wood-coloring fungi to the lack of oxygen.

The environmental conditions, when microorganisms irreversible lose ability to propagate (total air humidity within 50–70%, temperature within + 12– + 18°C) are used in operation manuals for materials and products storage in warehouses.

20.4.1 CHEMICAL FACTORS

NUTRITION SOURCES

To make microorganisms growing and multiplying, the medium shall have accessible sources of energy and initial materials for biosynthesis. The metabolism consists of two main processes—biosynthesis of cell substances (constructive metabolism) and energy production (energy metabolism). Both processes proceed in the organism in the form of conjugated chemical reactions, hence, sometimes the same compound is used [1, 2].

The metabolism includes substance delivery to the cells and metabolites excretion from the organism to the environment. The chemical composition of the cell shows what substances are primarily needed for microorganisms to grow. As already mentioned, water gives 80–90% of total mass of cells, that is why, its presence in the environment in accessible form is mandatory. The dry substance of cells mostly consists of six elements called organogens—C, N, P, S, H, and O. Virtually all organisms also need Fe, Mn, Mg, Cu, Cl, K, Ca, Zn, Na, and so on. The quantities of these elements necessary for growth are extremely low – 0.3–1%; that is why, they are called micro-elements or trace elements.

CARBON SOURCES

With respect to carbon source used in the constructive metabolism, microorganisms are divided into two groups—autotrophs and heterotrophs.

Autotrophs use carbon dioxide (CO_2) as the unique or the main source of carbon for synthesis of organic substances. Biosynthesis of organic substances from CO_2 proceeds with energy consumption. For this purpose, some autotrophs use the energy of light and others apply energy of inorganic compound oxidation.

Heterotrophs mainly use organic compounds as the source of carbon. The required energy they produce by oxidizing organic compounds. The specific feature of heterotrophs is that frequently the same compound is used both energy and constructive metabolism, whereas, for other groups of organisms, there is a difference between sources of energy and carbon.

Heterotrophs represent a large complex group of microorganisms. They are subdivided into saprophytes (organisms growing on dead plant and animal residues) and parasites (fed from living organisms).

Microbes-parasites inhabit in a body of another organism–the host–and feed by its substances. Among these are human, animal, and plant pathogens.

Saprophytes decay various organic substances in the nature. Microorganisms that damage industrial materials are mostly saprophytes, although there are species among them, which simultaneously cause plant diseases. For example, colonies of *Alternaria genus species identical to tomato* Alternaria *spot pathogens were observed on glass ceramics specimens under cool humid summer conditions.*

Many saprophytes are pantophagous that is, are capable of using various organic compounds, some of them demonstrate expressed specificity (selectivity) in relation to carbon sources. There are also those demanding some particular compounds and are called substrate-specific microorganisms.

Among microscopic fungi able to use various carbon-containing substrates are— *Aspergillus niger, A. flavus, A. versicolor, Trichoderma viride, Penicillium chrysogenum, Alternaria,* and so on. They are able to oxidize any natural organic compound. *A. flavus* is the example of polytrophic fungus. This species fungi have been extracted from various food stuffs (cereals, bread, dried fruit, meat, nuts, and vegetables), as well as from leather products, paper mass, metal parts, varnishes, textile, insect dung, and human intestine.

There are also more specialized forms adapting to oxidation of a small number of compounds, for example, those using cellulose as the main nutritive source. Therefore, the damaging action of such microorganisms mostly affects industrial products with the cellulose base (paper, wood, fabrics, and so on).

The use of petroleum hydrocarbons and products—gasoline, kerosene, oils, asphalt, and so on, is most typical of *Cladosporium resinae*, which in popular-scientific literature is called the "kerosene" fungus.

Some fungi (*A. flavus* and *A. niger*) are able to use stable compounds, for example, waxes ad paraffins. It is common knowledge that paraffined paper, widely used in many technological cycles, in food industry is easily polluted by *A. flavus* fungus and lose water repellency. Fungi using wax destroy artwork that includes this component, for example, some kinds of paintings.

Microorganisms with high enzymatic activity also use well other hardly accessible carbon sources, including esters (synthetic and natural), polyolefins (polyethylene), and other carbochain polymers (PVA and PVC).

The organic compounds, which are nutritive sources for microorganisms, are part of many industrial materials that is frequently the reason of their colonization by microorganisms. Moreover, contamination of various kinds occurring on materials, which are not nutritive substrates due to their chemical composition (metals, glass, some polymers, concrete, and so on), may be the source of organic substances. In some cases, a small amount of organic substance is enough to start growth of biodegraders [3-5].

Along with organic compounds, saprophytes use small amounts of carbon dioxide, involving it in metabolism. Carbon dioxide is an additional source of carbon for biosynthesis of cell substances. If carbon dioxide is completely removed from the medium, the culture growth decelerates or is completely suppressed. Microorganisms usually satisfy their demands at the expense of amounts of carbon dioxide formed during consumption of organic substrates.

SOURCES OF NITROGEN

Nitrogen is a part of vitally important components of the microbial cell. In the vital activity of microorganisms, nutritive nitrogen volume is smaller than that of carbon. First of all, the reason is that nitrogen quantity in the cell is 5–6 times lower as compared with carbon, at second, carbon compounds are consumed in larger amounts, because they are used in both constructive and energy metabolism simultaneously. So far as concerns nitrogen value for the microorganism metabolites, it is equal to carbon.

All autotrophic microorganisms adopt nitrogen from its inorganic compounds.

Heterotrophs demonstrate selectivity in relation to sources of nitrogen. Parasites use organic nitrogen-containing substances of the host cells. For saprophytes, the source of nitrogen may be both organic and inorganic nitrogen-containing compounds. Some of these microorganisms are only able to grow on substrates containing complex nitrogen-containing substances (nitrogenous bases, peptides, and a broad selection of amino acids), because they themselves are unable to synthesize them from simpler compounds. Others can develop at limited number of nitrogen organic compounds, for example, in substrates only containing some amino acids or even one or two of them. The rest compounds necessary for protein synthesis they produce themselves.

Nitrogen organic compounds (such substrates as natural leather, fur, and wool or silk) are consumed by microorganisms capable of decomposing them with ammonia formation.

MINERAL CONSTITUENT SOURCES

The synthesis of cell substances also requires various mineral constituents—sulfur, phosphorus, potassium, calcium, magnesium, and iron. Though demand in them is low, shortage of any of these elements in the nutritive medium will hinder development of microorganisms and may cause their death.

For the most of microorganisms the sources of mineral constituents are mineral salts, although some of them adopt sulfur and phosphorus from organic compounds.

Small amounts of microelements are required for microorganism growth and can also be got from mineral salts.

VITAMIN DEMAND OF MICROORGANISMS

Microbial cell has various vitamins in their composition, which are required for its normal vital activity. Some microorganisms need to receive ready vitamins, and the absence of one vitamin or another seriously disturbs their metabolism. The addition of deficient vitamin to the nutritive medium eliminates growth lag. Therefore, vitamins are frequently called the growth factors and growth stimulators. Other microorganisms are able to synthesize vitamins themselves from substances of the nutritive culture. Some microorganisms synthesize vitamins in amounts much higher than their own demand.

In industry, many vitamins are produced microbiologically, using yeasts, bacteria, and fungi. Vitamins produced by microorganisms are widely used in agriculture, medicine, food industry, for cosmetic compounds and other purposes [6].

OXYGEN AND ENERGY METABOLISM IN MICROORGANISMS

Cell substance synthesis from delivered nutrients and many other processes of vital activity proceed with energy consumption.

Autotrophs consume energy of either visible light or chemical reactions of inorganic compound oxidation (NH_3, H_2S, and so on).

Heterotrophs obtain energy from organic compound oxidation. They can use any natural organic compound and many synthetic ones.

Regarding the possibility of using molecular oxygen in the energy processes, microorganisms are divided into two groups:

- *Aerobes* which oxidize organic substances by molecular oxygen,
- *Anaerobes* which are not using oxygen in energy processes.

Similar to higher organisms (plants and animals), many aerobic microorganisms, which are fungi, some yeast species and many bacteria, oxidize organic substances completely to carbon dioxide and water. This process is called respiration.

Anaerobic microorganisms, which are many bacteria and some yeast species, receive energy necessary for vital activity from fermentation.

Anaerobic microorganisms are divided into obligate, or unconditional, anaerobes, for which oxygen is not only required, but is even harmful, and facultative, or conditional, anaerobes of two types. One type is better developed under anaerobic conditions, though it can live in the presence of oxygen, but is incapable to use it (for example, lactic-acid bacteria). Another type of facultative anaerobes (for example, yeasts) is able to turn from anaerobic to aerobic type of energy supply with respect to conditions of development. In case of oxygen deficiency, temporary suspension of the vital activity is typical of aerobic microorganisms. For example, as a result of oxygen deficiency wood-destroying and wood-coloring fungi die rather rapidly in the wood, when timber submerges during the raft. This principle gives grounds for wood protection against rotting by the sprinkling method.

Aerobes consume nearly a half of energy and the rest is lost for heating. This explains the phenomenon of spontaneous heating of insufficiently dried bales of cotton, wool, and other materials, when various microorganisms intensively develop in them due to increased humidity. This spontaneous heating of cotton or wool may cause their spontaneous ignition. Some microorganisms emit unused energy in the form of light. This is typical of some bacteria and fungi. Fluorescence of sea water, rotten wood, and fish is explained by the presence of fluorescent microorganisms in them [7].

MEDIUM ACIDITY

An acidity (pH), or alkalinity, is of great importance for vital activity of microorganisms. Many nonfood materials (paper, glass, drum skin, and so on) can be acidic or alkaline with respect to their composition and designation. For example, water film always present on the glass surface in humid atmosphere has pH 5.5–9.0 depending on chemical composition of glass.

pH value has an effect on:

- Ionic state of the medium and, consequently, accessibility of many metabolites and inorganic ions for the organism,
- Activity of enzymes that may change biochemical activity of microbes,

- Electric charge of the cell surface that stipulates variation of cell permeability for some ions,
- Morphology of mycelium, for example, for multiplication and synthesis of pigments by fungi.

All other conditions being favorable, vital activity of every species of microorganisms is only possible within more or less certain pH limits, beyond which it is suppressed. The most of bacteria grow better within pH 6.8–7.3 range, that is in neutral or weakly alkaline medium. With only a few exceptions, they do not develop at below pH 4.0 and above pH 9.0, however, many of them remain vital for a long time.

Putrefactive bacteria manifest the highest proteolytic activity above pH 7.0.

For some species of bacteria, acidic medium is more harmful than alkaline one. Vegetative cells are usually less resistant than spores. Acidic medium is specifically unfavorable for putrefactive bacteria and those causing food poisoning. Bacteria giving an acid as a vital activity product are higher resistant to pH decrease. Some microorganisms, for example, lactic-acid bacteria, gradually die with accumulation of a definite amount of acid in the medium. Others are able to regulate acidity by synthesizing appropriate substances under these conditions, which render the medium either acidic or alkaline, preventing pH shift towards values unfavorable for their development.

Mycelial fungi can develop in a wide range of pH 1.2–11.0. Fungus spores grow out in a narrower pH range, as compared with mycelium.

For the majority of mycelia fungi and yeasts, weakly acidic medium (pH 5.0–6.0) is the most favorable. For example, the optimal growth of wood-destroying fungi is observed at pH 3.0, whereas, the upper limit of the vital activity falls within the range of pH 7.0–7.5. As a rule, fungi are suppressed at pH shift towards one side or another from the optimal value. For the most of fungi, extremely high acidity or alkalinity is toxic.

In the most of cases, microorganisms change pH of the medium, where they develop. In some cases, this is associated with consumption of particular components of the medium, and in other cases, it is associated for metabolic product formation (organic acids and ammonia). Intensive consumption of NO_3^- anion from KNO_3 inevitably leads to rendering medium alkaline. As a result of ammonia formation, growth of fungi on protein-containing substrates (leather, wool, and so on) is accompanied pH shift towards alkalinity. When fungi grow on substrates containing glucose, organic acids are accumulated which increase hydrogen ion concentration. This pH shift towards acidity is observed for *Chaethomium globosum* and *Trichoderma viride growth on the medium containing paper.*

One of the reasons for substitution of one form of microorganisms by another under natural conditions is unequal relation of microorganisms to acidity. If we know the response of microorganisms to pH and regulate the latter, it is possible to suppress or stimulate their development that is of high practical importance.

Up to now, just few data on microorganisms inhabiting both at high and low pH are recorded; some few is also known about the mechanisms providing survivability under these conditions. However, increasing environmental pollution due to industrial sewage, acid rains becoming more frequent may lead to formation of new habitats of organisms capable of enduring extreme conditions of this kind in future. This will also result in increasing danger of microbial damage of various objects.

CONCENTRATION OF SUBSTANCES DISSOLVED IN THE MEDIUM

In nature, microorganisms inhabit on substrates with various content of dissolved substances and consequently, with various osmotic pressures. Thus, intracellular osmotic pressure in microorganisms changes accordingly with the habitat conditions [8].

The microorganism normally develops, when its intracellular osmotic pressure is somewhat higher than pressure in the nutritive substrate. In this case, outside water is delivered to the cell and cytoplasm fits the wall, slightly stretching it. This state is called turgor.

When a microorganism occurs in a substrate with the trace concentrations of substances (for example, distilled water), cell plasmoptysis happens, cytoplasm is rapidly overfilled with water and the cell wall breaks.

Multiple microorganisms are extremely sensitive even to a small increase of substance concentration in the medium. Substrate osmotic overpressure compared with intracellular one causes dehydration–plasmolysis of cells, and delivery of nutrients into them intermits. In the case of plasmolysis, some microorganisms can remain alive for a long time, whereas others die more or less rapidly.

The most of bacteria are insensitive to NaCl concentration within the range of 0.5–2%; however, 3% content of this substance in the medium has an unfavorable effect on many microorganisms.

When sodium salt concentration is about 3–4%, multiplication of many putrefactive bacteria is suppressed and at 7–10% stops. The rod-like putrefactive bacteria are less resistant than cocci.

The suppressive impact of salt on microorganisms' growth is stipulated not only by osmotic pressure increase, but also by the fact that high concentrations of sodium salt in the substrate is toxic for microorganisms, suppressing respiration processes, disturbing cellular membrane functions, and so on

CHEMICAL SUBSTANCES

The effect of chemical substances on microorganisms can be different. Among chemical agents are such that are able to suppress development of microorganisms and even cause their death [9-11].

Substances harmful for microorganisms are call antiseptics. Their actions are manifold. Some of them suppress vital activity or decelerate multiplication of microbes sensitive to them; this action is called bacteriostatic (as regards bacteria) and fungistatic (as regards mycelia fungi). Other substances cause death of microorganisms, causing bactericide and fungicide action. Low doses of many chemical poisons cause even favorable effect stimulating multiplication or biochemical activity of microbes.

Sensitivity of various microorganisms to the same antiseptic is different. Spores are more resistant than vegetative cells.

Among inorganic compounds, the most powerful are heavy metal salts. Several heavy metal ions: copper, gold, and especially silver present in the solution even in trace concentrations, which cannot be directly detected, nevertheless, are harmful for microorganisms.

Many oxidants (chlorine, iodine, hydrogen peroxides, and potassium permanganate) and mineral acids (sulfurous, boric, and fluoric) manifest the bactericide action.

Hydrogen sulfide, carbon oxide, and sulfur dioxide also affect microorganisms.

Many organic compounds are poisonous for microbes. The impact of phenols, aldehydes, especially formaldehyde, alcohols, and some organic acids (salicylic, acetic, benzoic, and sorbic) is poisonous to a different extent. The impact of these acids is generally associated with penetration of non-dissociated molecules of these acids rather than medium pH decrease. Essential oils, resins, tanning agents, and multiple dyes (brilliant green, magenta) are bactericides.

Antiseptics act differently. Many of them damage cell walls and disturb permeability of the cell membrane. When penetrating into the cell, they begin interacting with one component or another that results in significant disturbance of metabolism. Heavy metal salts, formalin, and phenols affect cytoplasm proteins and are poisonous for enzymes. Alcohols and ethers dissolve lipids of cell membranes.

Antiseptics are used for protection of textile, wood, paper, articles from them and other materials, and objects against microbial damages.

20.4.2 PHYSICAL FACTORS

Among physical factors of the environment defining vital activity of microorganisms, the most important ones are humidity, temperature, illumination, and some others. The effect of physical factors on fungi is defined by many reasons, including climatic conditions, storage and operation conditions of materials.

The task is the following—basing on the knowledge of physiological features of microorganisms-biodegraders to determine conditions preventing their growth and reduce to a minimum the unfavorable process of substrate degradation.

MEDIUM HUMIDITY

The medium humidity makes a massive impact on the microorganism development. The cells of the most of microorganisms contain up to 75–85% of water, with which the cell receives nutrients and excretes metabolites.

The moisture requirement for various microorganisms varies in a broad range. By the minimal moisture requirement for growth, microorganisms are subdivided into the following groups—hydrophytes–moisture-loving, mesophytes–medium moisture-loving, and xerophytes–xerocolous. A number of mycelia fungi and yeasts are mesophytes, but hydrophytes and xenophytes are met, too.

For most fungi, the minimum relative air humidity is 70%, and for bacteria–95%.

The relative air humidity decreases with temperature and *vice versa*. Therefore, if temperature decreases during storage of materials, the existing quantity of water vapor in the air may appear above the saturation level that leads to wetting of the material surface and promotes development of microorganisms on it [1, 2].

Any substrate (material) capable of absorbing moisture is in equilibrium with the air humidity. If the latter increase, the material absorbs moisture, otherwise it extracts moisture.

At the present time, for most materials no accurate limits of their humidity, at which degrading activity of microorganisms is observed, are determined. This is explained by the fact that until now, the simplest method of measuring water content of the substrate was determination of total water content (moisture load) in it [10, 11].

Moisture load is total amount of water in the material expressed in percent of the mass of absolutely dry substrate. Such definition gives no idea of this water accessibility for microorganisms, because it includes both bound and free moisture of the substrate. The first form of water defines properties of the material, for example, it is strongly bound to fibers on paper and cannot be used by microorganisms. In this connection, materials of the same moisture load but different water accessibility, all other conditions being the same, will be differently damaged by fungi. Microorganisms may only use free or weakly bound water. The growth of microorganisms on a hygroscopic material starts at such water absorption, when free water occurs. When humidity is below the fiber saturation point (30%), wood is not damaged by fungi. Microscopic fungi are only grown on paper, when total moisture load of the material reaches 8–10%, because at this level capillary (free) water occurs. It is assumed that the minimal substrate humidity, at which bacterium development is possible, equals 20–30%; the same for fungi is 13–15%.

When fungi develop on a material, its humidity usually increases. Moistening happens due to water segregation as one of final metabolites. It is known, for example, that when destroying 1 m^3 of wood, *Serpula lacrymans* excretes 139% of water, and at *Coniophora puteana* fungus growth the initial humidity of wood equal 6.75% increases to 30–64%. Material moistening caused by fungi creates conditions for colonization by new, more moisture-loving species.

Selection of a definite level of relative air humidity, at which most of microorganisms stop growing, is one of the methods for struggling against material damaging. This is the reason for applying various waterproofing agents (sodium alkyl siliconates, polyalkyl hydroxyanes, and so on). Chemically bound covers on materials, formed in this case, create negative conditions for the growth of microorganisms.

MEDIUM TEMPERATURE

Medium temperature is one of the main factors defining possibility and intensity of microorganism development. Each group of microorganisms can only develop within definite temperature range, which is narrow for some of them and relatively broad, amounted to tens of degrees, for others.

Microorganisms can grow in a wide range of temperatures. *Serpula lacrymans fungus, which damages timber, is unable to grow at a temperature below* + 8°C or above + 27°C. Its optimal development is observed at + 23°C. Fungi growing at + 60–+ 62°C were extracted from mucus formed on the paper factory equipment. At the same time, some microorganisms inhabit in cold soils of tundra and in refrigerators at – 8– – 6°C.

In mountain springs at volcano slopes, bacteria able to grow at temperature of 100°C or even higher were detected.

Minimal and maximal temperatures define the range, beyond which microorganisms do not grow, independently of the incubation time. The optimal temperature is that, at which the growth rate is maximal.

Basing on the growth temperature range, microorganisms are subdivided into three large groups—psychrophiles, mesophiles, and thermophiles.

Psychrophiles (from Greek 'psychro'–cold) are cold-loving microorganisms, well multiplying and chemically active at relatively low temperatures. The minimum of – 12°C to – 10°C to 0°C, the optimum of – 15°C to – 10°C and the maximum of – 30°C are typical of them. They are, for example, microorganisms inhabiting in the Polar Region soils and in northern seas.

Thermophiles (from Greek 'thermo'–heat) are heat-loving microorganisms, growing most well at relatively high temperature. The minimum of, at least, + 30°C, the optimum of + 50°C to + 60°C and the maximum of about + 70°C to + 80°C are typical of them. In hot springs of Kamchatka, a rod-like non-sporing bacterium with the optimal temperature of + 70°C to + 80°C was found. It remains viable at water temperature of + 90°C.

Thermophiles are met in spontaneously heated accumulations of organic materials (wool and cotton bales, and wood chip heaps). Thermophilic fungi are the reason for spontaneous ignition of wood chips gathered in heaps. As a result, a considerable part of raw material for paper production is lost every year. Additional economic losses are caused by chip color change that makes processing more expensive and reduces the product quality.

Thermophiles can be used for utilization of cellulose waste and their conversion to microorganism protein, because many species of microorganisms feature high cellulase activity. The ability of thermophilic fungi to use various plastics as the unique source of carbon may help to solve the problem of domestic plastic waste utilization in future.

Mesophiles (from Greek 'mesos'–middle, intermediate) are microorganisms with the temperature minimum of about + 5°C to + 10°C, the optimum of + 25°C to + 35°C, and the maximum of + 45°C to + 50°C.

Most of microorganisms–material degraders, are mesophiles.

Temperature directly determines geographical regions of microorganisms' spreading.

Penicillium and *Aspergillus* genus fungi, most frequently extracted from damaged materials can be presented as an illustration. Although *Penicillium* genus comprises many species met at all latitudes, generally lower optimal temperatures are typical of it, as compared with *Aspergillus* genus. Most of *Aspergillus* species grow at optimal temperatures of + 30°C to + 35°C, whereas for *Penicillium* the range is + 25°C to + 30°C. This determines predominance of *Penicillium* in the North, where they are represented by a broad variety of species. In the South, *Aspergillus* dominate, which optimal growth temperature is higher. Among other fungi in the southern latitudes, *Penicillium* species are few in the soils [3-5].

In these examples the environment temperature effects on the qualitative composition of microorganisms-biodegraders detected in damaged materials. Those microorganisms are predominant, which optimal temperature is closest to the current environmental conditions.

MICROORGANISM RESPONSE TO HIGH TEMPERATURES

Environmental temperature rise above the optimal level has more unfavorable effect on microorganisms than its reduction. The response of various microorganisms to temperatures exceeding the maximum of their development characterizes their heat resistance, which is different for different species. Temperature rise above the maximum causes the heat shock. After a short-term increase of temperature the cells may reactivate, where longer impact causes their death. The most of sporeless bacteria die within 15–30 min, when heated up to + 60°C to + 70°C in a humid condition. At heating up to + 80°C to + 100°C, the time decreases to several seconds up to 1–2 min. Yeasts and mycelial fungi also die quite rapidly at a temperature of + 50°C to + 60°C.

The bacterial spores have the highest heat resistance. For many bacteria, they can endure the boiling water temperature during several hours. Bacterial spores die at + 120° to + 130°C during 20–30 min in the humid environment and at + 160°C to + 170°C during 1–2 hr, when dry. The heat resistance of spores of various bacteria is different—the spores of thermophilic bacteria are most resistant.

As compared with bacteria, spores of the most yeasts and molds are less heat resistant and die rather quickly at + 65°C to + 80°C. However, some mold species spores may endure boiling. Moreover, not all cells or spores of the same species die simultaneously: among them higher or lower resistant species are met.

When heated in a humid environment, microorganisms begin to die due to irreversible changes in the cell. The basic processes are protein and nucleic acid denaturation, as well as enzyme inactivation and possible damage of the cell membrane.

When impacted by dry heat (without moisture), cells die due to active oxidative processes and cell structure break.

Sterilization, one of the most important and widely used in microbiology and medicine methods, is based on the fatal action of high temperature.

MICROORGANISM RESPONSE TO LOW TEMPERATURES

The cold tolerance of various microorganisms varies in a wide range. The environmental temperature below the optimal level reduces the multiplication rate and metabolism intensity.

Many microorganisms are unable to develop at temperatures below zero. For example, some putrefactive bacteria do not usually multiply at temperatures below + 4°C to + 5°C, temperature minimum of many fungi also falls within the range of + 3°C to + 5°C. Microorganisms more sensitive to cold, which do not develop already at + 10°C are known. They are, for example, the most of pathogenic bacteria. Some microorganisms can temporarily endure extremely low temperatures. Coliform bacterium and typhoid bacillus survive at – 190°C to – 172°C for several days. Bacterium spores retain ability to grow out even after 10 hr exposure to –252°C (liquid hydrogen temperature). Some mycelial fungi and yeasts remain viable after exposure to – 190°C (liquid air temperature) during several days, under the same conditions, mycelial fungi spores remain viable for several months. In mammoth bodies preserved in the permafrost for tens of thousands years, viable bacteria, and their spores were detected.

Despite the fact that below the minimum temperature microbes do not multiply and active metabolism is suppressed, many of them remain viable for infinitely long time, transiting to the anabiotic state, that is the "suppressed life" state similar to the winter sleep of animals. As temperature increases, they return to active life. However, under such conditions some microorganisms die sooner or later. Die-away proceeds much slower than at high temperatures.

RADIATION

Microorganisms are exposed to various types of radiation which are, first of all: a complex spectrum of solar radiation, electromagnetic waves, UV-radiation, γ- and X-rays, high energy corpuscular particles (electrons, protons, neutrons, and so on) which ionize or excite atoms and molecules of the environment and substances forming microorganisms.

The solar radiation relates to environmental factors which cause a considerable impact on vital activities of microorganisms. The effect of different spectral regions of solar radiation on fungi is unequal—the long-wave radiation activates thermal receptors, ultraviolet radiation causes mutagenic and lethal effects, all photobiological processes (photosynthesis, photoprotective and photochemical processes) are due to visible spectrum.

Visible spectrum region of the Sun is necessary for photosynthesizing microbes, which use light energy during carbon dioxide assimilation. Microorganisms incapable of photosynthesis grow also well in darkness. Direct sunlight is harmful for microorganisms, even scattered light suppresses to some extent their growth. Pathogenic bacteria (with few exceptions) are less resistant to light than saprophytes.

The visible light may influence on chromogenesis. This explains the fact that brightly illuminated materials are in greater danger of undesirable pigment spot formation, than in shadow.

In the entire solar spectrum ultraviolet radiation is most dangerous for microorganisms. The effect of UV-radiation on microorganisms varies with respect to radiation dose and spectral region. Low doses cause stimulating effect. High UV-radiation doses cause mutagenic and lethal effect.

High-energy particles (electrons, neutrons, protons, and so on) and γ-rays are extremely chemically and biologically active.

Radioactive irradiation features the ability to ionize atoms and molecules that is accompanied by molecular structure disintegration.

Microorganisms are more radiation resistant, than higher organisms. Their lethal dose is hundreds and thousands times higher, than for animals and plants.

The efficiency of ionizing radiation for microorganisms depends on the absorbed radiation dose and many other factors. Extremely low doses activate some vital processes of microorganisms by impacting their enzymatic systems. It induces hereditary changes of microbe properties and, consequently, occurrence of mutations. As radiation dose increases, metabolism is disturbed to a greater extent and various pathological changes in cells (radiation disease), which may cause their death.

The study of microorganism response to the action of increased radiation level in the environment is of theoretical and practical importance. Lethal doses of microorganisms are applied to protect materials against microbiological degradation. For example, pharaoh Ramses II mummy, which back was completely covered with mold (about 60 fungus species) was sterilized by γ-rays. This method designed in the Center of Nuclear Research in Grenoble is now successfully applied to saving many artistic objects and archeological documents [1, 2, 9-11].

20.4.3 BIOLOGICAL FACTOR

Materials are usually degraded by an entire complex, which includes both bacteria and fungi rather than by any single group of microorganisms. One acting group of microorganisms prepares a substrate for another group. In this case, new connections between separate microorganisms occur and interconnected aggregations are permanently formed providing survival and adaptation of each species separately. This is extremely complex process determined by multiple factors, among which the substrate, on which new functionally interconnected units as microbial aggregations or biocenosis are formed, is of primary importance [6].

In the course of evolution, various types of interrelations between microorganisms adapted to coexistence–symbiosis, have been formed. The different kinds of symbiosis are observed: mutualism that is commensalism favorable for both symbionts, they develop even better together, than individually; synergism that is consensual action of two or several species, when some physiological functions, for example, synthesis of particular substances, are intensified during their cooperative development; metabiosis that is a phenomenon, when one of symbionts lives at the expense of metabolites of the other, making no harm to it, for example, proteolytic microorganisms create a medium for development of other microorganisms able to only use protein degradation products; parasitism that is a sort of microorganism commensalism, when only one of the partners benefits making its symbiont harm up to death. Agents of human and animal infections are parasites. Antagonism is a phenomenon, when one species of microorganisms suppresses or terminates development of the other species or causes its death.

In the world of microbes antagonistic interrelations are one of important factors determining microflora composition of the natural substrates. In many cases, antagonistic interrelations are defined by unfavorable impact of one species metabolites on the other species.

In many instances, the fatal action of microbes-antagonists is due to excretion of specific biologically active chemical agents to the medium by them. These substances are called antibiotics. Microorganisms that excrete antibiotics are widely spread in nature. This is the feature of many fungi and bacteria. Some microorganisms synthesize several antibiotics.

The chemical origins of antibiotics are manifold. Their typical feature is selective action, when each of antibiotics only impacts particular microorganisms', that is, it is characterized by specific antimicrobial spectrum of action. Some antibiotics actively affect fungi and others–bacteria. There are antibiotics impacting both fungi and bacteria; antiviral antibiotics do also exist.

Antibiotics activity of is extremely high; it is tens of thousands times higher, than the activity of strong antiseptics. Therefore, antimicrobial action is manifested at extremely low concentrations [9-11].

Microbial cell damaging mechanism is manifold and is not completely studied yet.

20.5   CONCLUSION

This chapter has general information about bacteria and microscopic fungi as the sources of biodamages (morphology, internal organization and chemicals composition of microorganisms, and factors effecting biodamages processes).

KEYWORDS

- **Adenosine triphosphate**
- **Biodamaging**
- **Deoxyribonucleic acid**
- **Pseudomonas**
- **Ribonucleic acid**

REFERENCES

1. Pekhtasheva, E. L. *Biodamages and protections of non-food materials* (in Russian). Masterstvo Publishing House, Moscow, 224 (2002).
2. Emanuel, N. M. and Buchachenko, A. L. *Chemical physics of degradation and stabilization of polymers*. VSP International Science Publ., Utrecht, 354 (1982).

3. Zaikov, G. E., Buchachenko, A. L., and Ivanov, V. B. *Aging of polymers, polymer blends and polymer composites*. Nova Science Publ., New York, **1**, 258 (2002).
4. Zaikov, G. E., Buchachenko, A. L., and Ivanov, V. B. *Aging of polymers, polymer blends and polymer composites*. Nova Science Publ., New York, **2**, 253 (2002).
5. Zaikov, G. E., Buchachenko, A. L., and Ivanov, V. B. *Polymer aging at the cutting adge*. Nova Science Publ., New York, 176 (2002).
6. Gumargalieva, K. Z. and Zaikov, G. E. *Biodegradation and biodeterioration of polymers: Kinetical aspects*. Nova Science Publ., New York, 210 (1998).
7. Semenov, S. A., Gumargalieva, K. Z., and Zaikov, G. E. *Biodegradation and durability of materials under the effect of microorganisms*. VSP International Science Publ., Utrecht, 199 (2003).
8. Polishchuk, A. Ya. and Zaikov, G. E. *Multicomponent transport in polymer systems*. Gordon & Breach, New York, 231 (1996).
9. Moiseev, Yu. V. and Zaikov, G. E. *Chemical resistance of polymers in reactive media*. Plenum Press, New York, 586 (1987).
10. Emanuel, N. M., Zaikov, G. E., and Maizus, Z. K. *Oxidation of organic compounds: Medium effects in radical reactions*. Pergamon Press, Oxford, 628 (1984).
11. Jimenez, A. and Zaikov, G. E. *Polymer analysis and degradation*. Nova Science Publ., New York, 287 (2000).

CHAPTER 21

ELASTOMERIC NANOCOMPOSITES AGGREGATION MODEL

G. V. KOZLOV, YU. G. YANOVSKII, and G. E. ZAIKOV

CONTENTS

21.1 INTRODUCTION

A nanofiller disperse particles aggregation process in elastomeric matrix has been studied. The modified model of irreversible aggregation particle-cluster was used for this process theoretical analysis. The modification necessity is defined by simultaneous formation of a large number of nanoparticles aggregates. The offered approach allows to predicting a nanoparticles aggregates final parameters as a function of the initial particles size, their contents and other factors number.

The aggregation of the initial nanofiller powder particles in more or less large particles aggregates always occurs in the course of technological process of making particulate-filled polymer composites in general [1] and elastomeric nanocomposites in particular [2]. The aggregation process tells on composites (nanocomposites) macroscopic properties [1, 3]. For nanocomposites nanofiller aggregation process gains special significance, since its intensity can be the one, that nanofiller particles aggregates size exceeds 100 nm – the value, which assumes (although and conditionally enough [4]) as an upper dimensional limit for nanoparticle. In other words, the aggregation process can result to the situation, when primordially supposed nanocomposite ceases to be the one. Therefore at present several methods exist, which allowed to suppress nanoparticles aggregation process [2, 5]. The proceeding from, in the present paper theoretical treatment of disperses nanofiller aggregation process in butadiene-styrene rubber matrix within the frameworks of irreversible aggregation models was carried out.

21.2 EXPERIMENTAL

The elastomeric particulate-filled nanocomposite on the basis of butadiene-styrene rubber was an object of the study. The mineral shungite nanodimensional and microdimensional particles and also industrially produced technical carbon with mass contents of 37 mass% were used as a filler. The analysis of the received milling process shungite particles were monitored with the aid of analytical disk centrifuge (CPS Instruments, Inc., USA), allowing to determine with the high precision the size and distribution by sizes within the range from 2 nm up to 50 mcm.

The nanostructure was studied on atomic-power microscopes Nano-DST (Pacific Nanotechnology, USA) and Easy Scan dynamic force mode (DFM) (Nanosurf, Switzerland) by semi-contact method in the force modulation regime. Atomic-power microscopy results were processed with the aid of specialized software package Scanning Probe Image Processor (SPIP), Denmark. The SPIP is a powerful programmes package for processing of images, obtained on scanning probe microscopy (SPM), atomic forced microscopy (AFM), scanning tunneling microscopy (STM), scanning electron microscopes, transmission electron microscopes, interferometers, confocal microscopes, profilometers, optical microscopes and so on. The given package possesses the whole functions number, which are necessary at images precise analysis, in the number of which the following are included:

1. The possibility of three-dimensional reflected objects obtaining, distortions automatized leveling, including Z-error mistakes removal for examination separate elements and so on;

2. Quantitative analysis of particles or grains, more than 40 parameters can be calculated for each found particle or pore: area, perimeter, average diameter, and the ratio of linear sizes of grain width to its height distance between grains, coordinates of grain center of mass a.a. can be presented in a diagram form or in a histogram form.

21.3 DISCUSSION AND RESULTS

For theoretical treatment of nanofiller particles aggregate growth processes and final sizes traditional irreversible aggregation models are inapplicable, since it is obvious, that in nanocomposites aggregates a large number of simultaneous growth takes place. Therefore the model of multiple growths, offered in paper [6], was used for nanofiller aggregation description.

In Figure 1 the images of the studied nanocomposites, obtained in the force modulation regime, and corresponding to them nanoparticles aggregates fractal dimension d_f distributions are adduced. As it follows from the adduced values d_f (d_f=2.40-2.48), nanofiller particles aggregates in the studied nanocomposites are formed by a mechanism particle-cluster (P-Cl), that is they are Witten-Sander clusters [7]. The variant A, was chosen which according to mobile particles are added to the lattice, consisting of a large number of "seeds" with density of c_0 at simulation beginning [6]. Such model generates structures, which have fractal geometry on length short scales with value $d_f \approx 2.5$ (see Figure 1) and homogeneous structure on length large scales. A relatively high particles concentration c is required in the model for uninterrupted network formation [6].

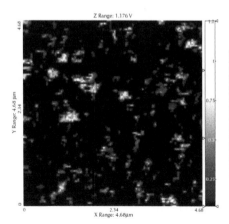

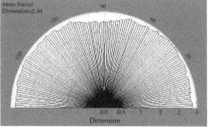

Mean fractal dimension d_f =2.40

FIGURE 1   *(Continued)*

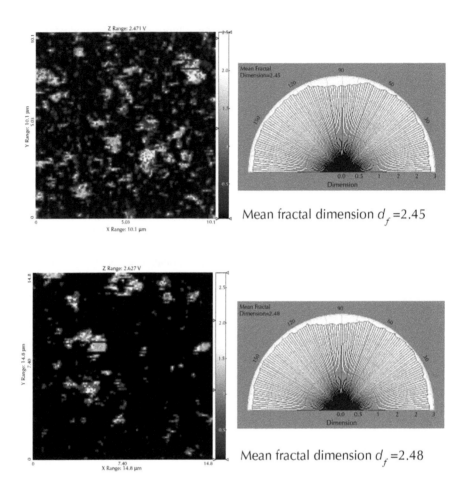

Mean fractal dimension d_f =2.45

Mean fractal dimension d_f =2.48

FIGURE 1 The images, obtained in the force modulation regime, for nanocomposites, filled with technical carbon (a), nanoshungite (b), microshungite (c) and corresponding to them fractal dimensions d_f.

In case of "seeds" high concentration c_0 for the variant A the following relationship was obtained [6]:

$$R_{\max}^{d_f} = N = c / c_0,$$ (1)

Where $R_{\max}$ is nanoparticles cluster (aggregate) greatest radius, N is nanoparticles number per one aggregate, c is nanoparticles concentration, c_0 is "seeds" number, which is equal to nanoparticles clusters (aggregates) number.

The value N can be estimated according to the following Equation [8]:

$$2R_{max} = \left(\frac{S_n N}{\pi \eta} \right)^{1/2}, \qquad (2)$$

Where S_n is cross-sectional area of nanoparticles, from which aggregate consists, η is packing coefficient, equal to 0.74.

The experimentally obtained nanoparticles aggregate diameter $2R_{agr}$ was accepted as $2R_{max}$ (Table 1) and the value S_n was also calculated according to the experimental values of nanoparticles radius r_n (Table 1). In Table 1 the values N for the studied nanofillers, obtained according to the indicated method, were adduced. It is significant that the value N is a maximum one for nanoshungite despite larger values r_n in comparison with technical carbon.

TABLE 1 The parameters of irreversible aggregation model of nanofiller particles aggregates growth

Filler	Experimental radius of nanofiller aggregate R_{agr}, nm	Radius of nanofiller particle r_n, nm	Number of particles in one aggregate N	Radius of nanofiller aggregate R_{max}^T, the Equation (1), nm	Radius of nanofiller aggregate R_{agr}^T, the equation (3), nm	Radius of nanofiller aggregate R_c, the Equation (8), nm
Technical carbon	34.6	10	35.4	34.7	34.7	33.9
Nanoshungite	83.6	20	51.8	45.0	90.0	71.0
Microshungite	117.1	100	4.1	15.8	158.0	255.0

Further the Equation (1) allows to estimating the greatest radius R_{max}^T of nanoparticles aggregate within the frameworks of the aggregation model [6]. These values R_{max}^T are adduced in Table 1, from which their reduction in a sequence of technical carbon-nanoshungite-microshungite, that fully contradicts to the experimental data, that is to R_{agr} change (Table 1). However, we must not neglect the fact that the Equation (1) was obtained within the frameworks of computer simulation, where the initial aggregating particles size are the same in all cases [6]. For real nanocomposites the values r_n can be distinguished essentially (Table 1). It is expected, that the value R_{agr} or R_{max}^T will be the higher, the larger is the radius of nanoparticles, forming aggregate, is r_n. Then theoretical value of nanofiller particles cluster (aggregate) radius R_{agr}^T can be determined as follows:

$$R_{agr}^T = k_n r_n N^{1/d_f},$$

(3)

Where k_n is proportionality coefficient, in the present work accepted empirically equal to 0.9.

The comparison of experimental R_{agr} and calculated according to the Equation (3) R_{agr}^T values of the studied nanofillers particles aggregates radius shows their good correspondence (the average discrepancy of R_{agr} and R_{agr}^T makes up 11.4%). Therefore, the theoretical model [6] gives a good correspondence to the experiment only in case of consideration of aggregating particles real characteristics and, in the first place, their size.

Let us consider two more important aspects of nanofiller particles aggregation within the frameworks of the model [6]. Some features of the indicated process are defined by nanoparticles diffusion at nanocomposites processing. Specifically, length scale, connected with diffusible nanoparticle, is correlation length ξ of diffusion. By definition, the growth phenomena in sites, remote more than ξ, are statistically independent. Such definition allows to connect the value ξ with the mean distance between nanofiller particles aggregates L_n. The value ξ can be calculated according to the equation [6]:

$$\xi^2 \approx \tilde{n}^{-1} R_{agr}^{d_f - d + 2},$$

(4)

Where c is nanoparticles concentration, d is dimension of Euclidean space, in which a fractal is considered (it is obvious, that in this case $d = 3$).

The value c should be accepted equal to nanofiller volume contents φ_n, which is calculated as follows [9]:

$$\varphi_n = \frac{W_n}{\rho_n},$$

(5)

Where W_n is nanofiller mass contents, ρ_n is its density, that determined according to the equation [3]:

$$\rho_n = 0.188(2r_n)^{1/3}.$$

(6)

The values r_n and R_{agr} were obtained experimentally (see histogram of Figure 2). In Figure 3 the relation between L_n and ξ is adduced, which, as it is expected, proves to be linear and passing through coordinates origin. This means, that the distance between nanofiller particles aggregates is limited by mean displacement of statistical walks, by which nanoparticles are simulated. The relationship between L_n and ξ can be expressed analytically as follows:

$$L_n = 9.6\xi, \text{ nm.} \tag{7}$$

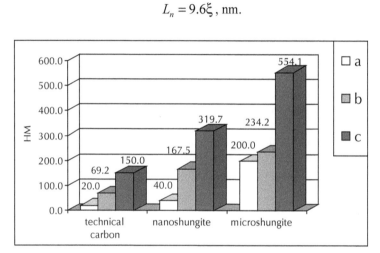

FIGURE 2 The initial particles diameter (a), their aggregates size in nanocomposite (b) and distance between nanoparticles aggregates (c) for nanocomposites, filled with technical carbon, nano- and microshungite.

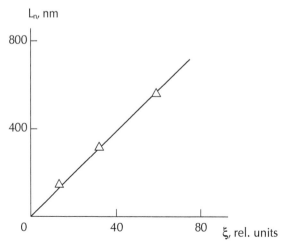

FIGURE 3 The relation between diffusion correlation length and distance between nanoparticles aggregates L_n for studied nanocomposites.

The second important aspect of the model [6] in reference to nanofiller particles aggregation simulation is a finite nonzero initial particles concentration c or φ_n effect, which takes place in any real systems. This effect is realized at the condition $\xi \approx R_{agr}$, that occurs at the critical value $R_{agr}(R_c)$, determined according to the relationship [6]:

$$c \sim R_c^{d_f - d}. \tag{8}$$

The Equation (8), right side represents cluster (particles aggregate) mean density. This equation establishes that fractal growth continues only, until cluster density reduces up to medium density, in which it grows. The calculated according to the Equation (8) values R_c for the considered nanoparticles are adduced in Table 1, from which it follows, that they give reasonable correspondence with this parameter experimental values R_{agr} (the average discrepancy of R_c and R_{agr} makes up 24%).

Since the treatment [6] was obtained within the frameworks of a more general model of diffusion-limited aggregation, then its correspondence to the experimental data indicated unequivocally, that aggregation processes in these systems were controlled by diffusion. Therefore let us consider briefly nanofiller particles diffusion. Statistical walkers diffusion constant ζ can be determined with the aid of the relationship [6]:

$$\xi \approx (\zeta t)^{1/2}, \tag{9}$$

Where t is walk duration.

The Equation (9) supposes (at $t = $ const) ζ increase in a number technical carbon-nanoshungite-microshungite as 196-1069-3434 relative units, that is diffusion intensification at diffusible particles size growth. At the same time diffusivity D for these particles can be described by the well-known Einstein's relationship [10]:

$$D = \frac{kT}{6\pi\eta r_n \alpha}, \tag{10}$$

Where k is Boltzmann constant, T is temperature, η is medium viscosity, and α is numerical coefficient, which further is accepted equal to 1.

In its turn, the value η can be estimated according to the equation [11]:

$$\frac{\eta}{\eta_0} = 1 + \frac{2.5\varphi_n}{1 - \varphi_n}, \tag{11}$$

Where η_0 and η are initial polymer and its mixture with nanofiller viscosity, accordingly, φ_n is nanofiller volume contents.

The calculation according to the Equations (10) and (11) shows, that within the indicated above nanofillers number the value D changes as 1.32-1.14-0.44 relative units, that is reduces in three times, that was expected. This apparent contradiction is due to the choice of the condition t = constant (where t is nanocomposite production duration) in the Equation (9). In real conditions the value t is restricted by nanoparticle contact with growing aggregate and then instead of t the value t/c_0 should be used, where c_0 is seeds concentration, determined according to the Equation (1). In this case the value ζ for the indicated nanofillers changes as 0.288-0.118-0.086, that is it reduces in 3.3 times, that corresponds fully to the calculation according to the Einstein's relationship (the Equation (10)). This means, that nanoparticles diffusion in polymer matrix obeys classical laws of Newtonian rheology [10].

21.4 CONCLUSIONS

Disperse nanofiller particles aggregation in elastomeric matrix can be described theoretically within the frameworks of a modified model of irreversible aggregation particle-cluster. The obligatory consideration of nanofiller initial particles size is a feature of the indicated model application to real systems description. The indicated particles diffusion in polymer matrix obeys classical laws of Newtonian liquids hydrodynamics. The offered approach allows to predict nanoparticles aggregates final parameters as a function of the initial particles size, their contents and other factors number.

KEYWORDS

- **Aggregation**
- **Diffusion**
- **Elastomer**
- **Nanocomposite**
- **Nanoparticle**

REFERENCES

1. Kozlov G. V., Yanovskii Yu. G., and Zaikov G. E. Structure and Properties of Particulate-Filled Polymer Composites: the Fractal Analysis. New York, Nova Science Publishers, Inc., p 282 (2010).
2. Edwards D. C. Polymer-filler interactions in rubber reinforcement. *J. Mater. Sci.*, **25** (12), 4175 (1990).
3. Mikitaev A. K., Kozlov G. V., and Zaikov G. E. Polymer Nanocomposites: the Variety of Structural Forms and Applications. New York, Nova Science Publishers, Inc., p 319 (2008).
4. Buchachenko A. L. The nanochemistry – direct way to high technologies of new century. *Uspekhi Khimii*, **72** (5), 419 (2003).
5. Kozlov G. V., Yanovskii Yu. G., Burya A. I., and Aphashagova Z. Kh. Structure and properties of particulate-filled nanocomposites phenylone/aerosol. *Mekhanika Kompozitsionnykh Materialov i Konstruktsii*, **13** (4), 479 (2007).

6. Witten T. A. and Meakin P. Diffusion-limited aggregation at multiple growth sites. *Phys. Rev. B*, **28** (10), 5632 (1983).

7. Witten T. A. and Sander L. M. Diffusion-limited aggregation. *Phys. Rev. B*, **27** (9), 5686 (1983).

8. Bobryshev A. N., Kozomazov V. N., Babin L. O., and Solomatov V. I. Synergetics of Composite Materials. Lipetsk, NPO ORIUS, p 154 (1994.).

9. Sheng N., Boyce M. C., Parks D. M., Rutledge G. C., Abes J. I., and Cohen R. E. Multi-scale micromechanical modeling of polymer/clay nanocomposites and the effective clay particle. *Polymer*, **45** (2), 487 (2004).

10. Happel J. and Brenner G. The Hydrodynamics at Small Reynolds Numbers. Moscow, Mir, p 418 (1976)..

11. Mills N. J. The rheology of filled polymers. *J. Appl. Polymer Sci.*, **15** (11), 2791 (1971).

CHAPTER 22

NANOFILLER IN ELASTOMERIC MATRIX—STRUCTURE AND PROPERTIES

YU. G. YANOVSKII, G. V. KOZLOV, and G. E. ZAIKOV

CONTENTS

22.1 INTRODUCTION

It has been shown that nanofiller particles (aggregates of particles) "chains" in elastomeric nanocomposites are physical fractal within the self-similarity (fractality) range ~500–1450 nm. The low dimensions of nanofiller particles (aggregates of particles) structure in elastomeric nanocomposites are due to high fractal dimension of nanofiller initial particles surface.

It is well-known [1, 2], that in particulate-filled elastomeric nanocomposites (rubbers) nanofiller particles form linear spatial structures ("chains"). At the same time in polymer composites, filled with disperse microparticles (microcomposites) particles (aggregates of particles) of filler form a fractal network, which defines polymer matrix structure (analog of fractal lattice in computer simulation) [3]. This results to different mechanisms of polymer matrix structure formation in micro and nanocomposites. If in the first filler particles (aggregates of particles) fractal network availability results to "disturbance" of polymer matrix structure, that is expressed in the increase of its fractal dimension d_f [3], then in case of polymer nanocomposites at nanofiller contents change the value d_f is not changed and equal to matrix polymer structure fractal dimension [4]. As it has to been expected, composites indicated classes structure formation mechanism change defines their properties change, in particular, reinforcement degree.

At present there are several methods of filler structure (distribution) determination in polymer matrix, both experimental [5, 6] and theoretical [3]. All the indicated methods describe this distribution by fractal dimension D_n of filler particles network. However, correct determination of any object fractal (Hausdorff) dimension includes three obligatory conditions. The first from them is the indicated above determination of fractal dimension numerical magnitude, which should not be equal to object topological dimension. As it is known [7], any real (physical) fractal possesses fractal properties within a certain scales range [8]. And at last, the third condition is the correct choice of measurement scales range itself. As it has been shown in this chapter [9, 10], the minimum range should exceed at any rate one self-similarity iteration.

The present chapter purpose is dimension D_n estimation, both experimentally and theoretically, and checking two indicated conditions fulfillment, that is obtaining of nanofiller particles (aggregates of particles) network ("chains") fractality strict proof in elastomeric nanocomposites on the example of particulate-filled butadiene-styrene rubber.

22.2 EXPERIMENTAL

The elastomeric particulate-filled nanocomposite on the basis of butadiene-styrene rubber (BSR) was an object of the study. The technical carbon of mark N9 220 (TC) of industrial production, nano and microshungite (the mean filler particles size makes up 20 nm, 40 nm and 200 nm, accordingly) were used as a filler. All fillers content makes up 37 mass%. Nano and microdimensional disperse shungite particles were obtained from industrially extractive material by processing according to the original technology. A size and polydispersity of the received in milling process shungite particles were monitored with the aid of analytical disk centrifuge (CPS Instruments, Inc., USA), allowing to determine with high precision the size and distribution by sizes within the range from 2 nm up to 50 mcm.

The nanostructure was studied on atomic-power microscopes Nano-DST (Pacific Nanotechnology, USA) and Easy Scan dynamic force microscope (DFM) (Nanosurf, Switzerland) by semi-contact method in the force modulation regime. Atomic-power microscopy results were processed with the aid of specialized software package scanning probe image processor (SPIP) (SPIP, Denmark). SPIP is a powerful programmes package for processing of images, obtained on scanning probe microscope (SPM), Atomic Force Microscopy (AFM), scanning tunneling microscopy (STM), scanning electron microscopes, transmission electron microscopes, interferometers, confocal microscopes, profilometers, optical microscopes, and so on. The given package possesses the whole functions number, which are necessary at images precise analysis, in the number of which the following are included:

- The possibility of three-dimensional reflecting objects obtaining, distortions automatized leveling, including Z-error mistakes removal for examination separate elements and so on,
- Quantitative analysis of particles or grains, more than 40 parameters can be calculated for each found particle or pore: area, perimeter, average diameter, the ratio of linear sizes of grain width to its height distance between grains, coordinates of grain center of mass a.a. can be presented in a diagram form or in a histogram form.

22.3 DISCUSSION AND RESULTS

The first method of dimension D_n experimental determination uses the following fractal Equation [11, 12]:

$$D_n = \frac{ln N}{ln \rho} \tag{1}$$

where N is a number of particles with size ρ.

Particles sizes were established on the basis of atomic-power microscopy data (see Figure 1). For each from the three studied nanocomposites no less than 200 particles were measured, the sizes of which were united into 10 groups and mean values N and ρ were obtained. The dependences $N(\rho)$ in double logarithmic coordinates were plotted, which proved to be linear and the values D_n were calculated according to their slope (see Figure 2). It is obvious, that at such approach fractal dimension D_n is determined in two-dimensional Euclidean space, whereas, real nanocomposite should be considered in three-dimensional Euclidean space. The following Equation can be used for D_n re-calculation for the case of three-dimensional space [13]:

$$D_3 = \frac{d + D_2 \pm \left[\left(d - D_2 \right)^2 - 2 \right]^{1/2}}{2} \tag{2}$$

where D_3 and D_2 are corresponding fractal dimensions in three- and two-dimensional Euclidean spaces, $d = 3$.

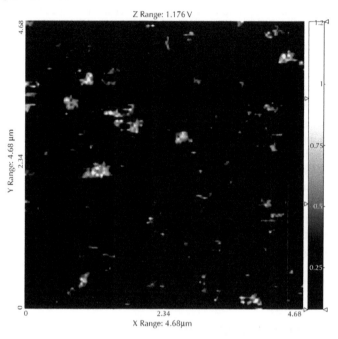

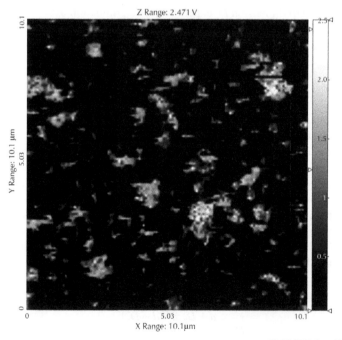

FIGURE 1 *(Continued)*

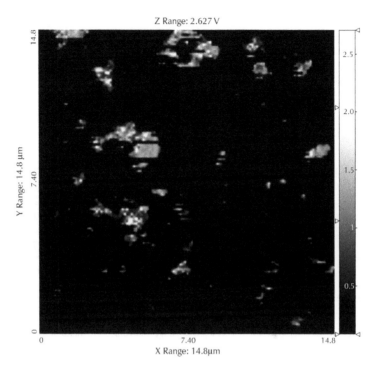

FIGURE 1 The electron micrographs of nanocomposites BSR/TC (a) BSR/nanoshungite, (b) and BSR/microshungite, and (c) obtained by atomic-power microscopy in the force modulation regime.

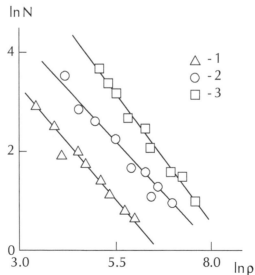

FIGURE 2 The dependence of nanofiller particles number N on their size ρ for nanocomposites BSR/TC (1) BSR/nanoshungite (2) and, BSR/microshungite (3).

The calculated according to the indicated method dimensions D_n are adduced in Table 1. As it follows from the data of this table, the values D_n for the studied nanocomposites are varied within the range of 1.10–1.36, that is they characterize more or less branched linear formations ("chains") of nanofiller particles (aggregates of particles) in elastomeric nanocomposite structure. Let us remind that for particulate-filled composites polyhydroxiether/graphite the value D_n changes within the range of ~ 2.30–2.80 [5], that is for these materials filler particles network is a bulk object, but not a linear one [7].

TABLE 1 The dimensions of nanofiller particles (aggregates of particles) structure in elastomeric nanocomposites

The nanocomposite	D_n, the Equation (1)	D_n, the Equation (3)	d_0	d_{surf}	φ_n	D_n, the Equation (7)
BSR/TC	1.19	1.17	2.86	2.64	0.48	1.11
BSR/nanoshungite	1.10	1.10	2.81	2.56	0.36	0.78
BSR/microshungite	1.36	1.39	2.41	2.39	0.32	1.47

Another method of D_n experimental determination uses the so-called "quadrates method" [14]. Its essence consists in the following. On the enlarged nanocomposite microphotograph (see Figure 1) a net of quadrates with quadrate side size α_i, changing from 4.5 up to 24 mm with constant ratio $\alpha_{i+1}/\alpha_i = 1.5$, is applied and then quadrates number N_i, in to which nanofiller particles hit (fully or partly), is calculated. Five arbitrary net positions concerning microphotograph were chosen for each measurement. If nanofiller particles network is fractal, then the following relationship should be fulfilled [14]:

$$N_i \sim S_i^{-D_n/2} \tag{3}$$

where S_i is quadrate area, which is equal to α_i^2.

In Figure 3 the dependences of N_i on S_i in double logarithmic coordinates for the three studied nanocomposites, corresponding to the Equation (3), is adduced. As one can see, these dependences are linear, that allows to determine the value D_n from their slope. The determined according to the Equation (3) values D_n are also adduced in Table 1, from which a good correspondence of dimensions D_n, obtained by the two described above methods, follows (their average discrepancy makes up 2.1% after these dimensions re-calculation for three dimensional space according to the Equation (2)).

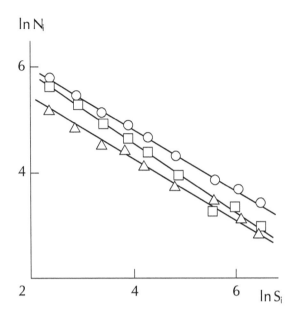

FIGURE 3 The dependences of covering quadrates number N_i on their area S_i, corresponding to the Equation (3), in double logarithmic coordinates for nanocomposites on the basis of BSR. The designations are the same, that in Figure 2.

As it has been shown in paper [15], at the Equation (3) the usage for self-similar fractal objects the condition should be fulfilled:

$$N_i - N_{i-1} \sim S_i^{-D_n} \tag{4}$$

In Figure 4 the dependence, corresponding to the Equation (4), for the three studied elastomeric nanocomposites is adduced. As one can see, this dependence is linear, passes through coordinates origin, that according to the Equation (4) is confirmed by nanofiller particles (aggregates of particles) "chains" self-similarity within the selected α_i range. It is obvious, that this self-similarity will be a statistical one [15]. Let us note, that the points, corresponding to $\alpha_i = 16$ mm for nanocomposites BSR/TC and BSR/microshungite, do not correspond to a common straight line. Accounting for electron microphotographs of Figure 1 enlargement this gives the self-similarity range for nanofiller "chains" of 464–1472 nm. For nanocomposite BSR/nanoshungite, which has no points deviating from a straight line of Figure 4, α_i range makes up 311–1510 nm, that corresponds well enough to the indicated above self-similarity range.

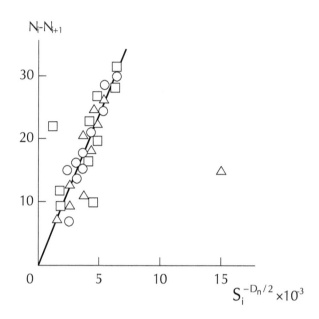

FIGURE 4 The dependences of $(N_i\text{-}N_{i+1})$ on the value $S_i^{-D_n/2}$, corresponding to the Equation (4), for nanocomposites on the basis of BSR. The designations are the same, that in Figure 2.

In chapters [9, 10] it has been shown, that measurement scales S_i minimum range should contained at least one self-similarity iteration. In this case the condition for ratio of maximum S_{max} and minimum S_{min} areas of covering quadrates should be fulfilled [10]:

$$\frac{S_{max}}{S_{min}} > 2^{2/D_n} \tag{5}$$

Hence, accounting for the defined above restriction let us obtain $S_{max}/S_{min} = 121/20.25 = 5.975$, that is larger than values $2^{2/D_n}$ for the studied nanocomposites, which are equal to 2.71–3.52. This means, that measurement scales range is chosen correctly.

The self-similarity iterations number μ can be estimated from the equation [10]:

$$\left(\frac{S_{max}}{S_{min}}\right)^{D_n/2} > 2^{\mu} \tag{6}$$

Using the indicated above values of the included in the Equation (6) parameters, $\mu = 1.42$–1.75 is obtained for the studied nanocomposites, that is in our experiment conditions self-similarity iterations number is larger than unity, that again is confirmed by the value D_n estimation correctness [6].

And let us consider in conclusion the physical grounds of smaller values D_n for elastomeric nanocomposites in comparison with polymer microcomposites, that is, the causes of nanofiller particles (aggregates of particles) "chains" formation in the first. The value D_n can be determined theoretically according to the equation [3]:

$$\varphi_{if} = \frac{Dn + 2.55 d_0 - 7.10}{4.18} \qquad (7)$$

where φ_{if} is interfacial regions relative fraction, d_0 is nanofiller initial particles surface dimension.

The dimension d_0 estimation can be carried out with the aid of the equation [4]:

$$S_u = 410 \left(\frac{D_p}{2} \right)^{d_0 - d} \qquad (8)$$

where S_u is nanofiller initial particles specific surface in m²/g, D_p is their diameter in nm, d is dimension of Euclidean space, in which a fractal is considered (it is obvious, in our case $d = 3$).

The value S_u can be calculated according to the equation [16]:

$$S_u = \frac{6}{\rho_n D_p} \qquad (9)$$

where ρ_n is nanofiller density, which is determined according to the empirical formula [4]:

$$\rho_n = 0.188 (D_p)^{1/3} \qquad (10)$$

The results of value d_0 theoretical estimation are adduced in Table 1. The value φ_{if} can be calculated according to the equation [4]:

$$\varphi_{if} = \varphi_n (d_{surf} - 2) \qquad (11)$$

where φ_n is nanofiller volume fraction, d_{surf} is fractal dimension of nanoparticles aggregate surface. The value φ_n is determined according to the equation [4]:

$$\varphi_n = \frac{W_n}{\rho_n} \qquad (12)$$

where W_n is nanofiller mass fraction and dimension d_{surf} is calculated according to the Equation (8)–(10) at diameter D_p replacement on nanoparticles aggregate diameter D_{agr}, which is determined experimentally (see Figure 5).

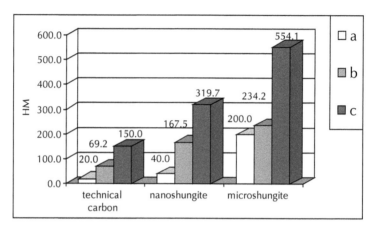

FIGURE 5   The initial particles diameter (a), their aggregates size in nanocomposite (b) and distance between nanoparticles aggregates (c) for nanocomposites on the basis of BSR, filled with technical carbon, nano- and microshungite.

The results of dimension D_n theoretical calculation according to the Equation (7)–(12) are adduced in Table 1, from which theory and experiment good correspondence follows. The Equation (7) indicates unequivocally the cause of filler in nano and microcomposites different behavior. The high (close to 3, see Table 1) values d_0 for nanoparticles and relatively small ($d_0 = 2.17$ for graphite) values d_0 for microparticles at comparable values φ_{if} for composites of the indicated classes [3, 4].

22.4   CONCLUSIONS

Therefore, the present chapter results have shown, that nanofiller particles (aggregates of particles) "chains" in elastomeric nanocomposites are physical fractal within self-similarity (and, hence, fractality [12]) range of ~ 500–1450 nm. In this range their dimension D_n can be estimated according to the Equation (1), Equation (3), and Equation (7). The cited examples demonstrate the necessity of the measurement scales range correct choice. As it has been noted earlier [17], linearity of the plots, corresponding to the Equation (1) and Equation (3), and D_n nonintegral value do not guarantee object self-similarity (and, hence, fractality). The nanofiller particles (aggregates of particles) structure low dimensions are due to the initial nanofiller particles surface high fractal dimension.

KEYWORDS

- Dynamic force microscope
- Elastomeric nanocomposites
- Microcomposites
- Nanofiller
- Scanning probe image processor

REFERENCES

1. Lipatov, Yu. S. *The Physical Chemistry of Filled Polymers.* Khimiya, Moscow, p. 304 (1977).
2. Bartenev, G. M. and Zelenev, Yu. V. *The Physics and Mechanics of Polymers.*, Vysshaya Shkola, Moscow, p. 391 (1983).
3. Kozlov, G. V., Yanovskii, Yu. G., and Zaikov, G. E. *Structure and Properties of Particulate-Filled Polymer Composites: the Fractal Analysis.* Nova Science Publishers Inc., New York, p. 282 (2010).
4. Mikitaev, A. K., Kozlov, G. V., and Zaikov, G. E. *Polymer Nanocomposites: the Variety of Structural Forms and Applications.* Nova Science Publishers, Inc., New York, p. 319 (2008).
5. Kozlov, G. V. and Mikitaev, A. K. *Mekhanika Kompozitsionnykh Materialov i Konstruktsii.* 2(3–4), 144–157 (1996).
6. Kozlov, G. V., Yanovskii, Yu. G., and Mikitaev, A. K. *Mekhanika Kompozitnykh Materialov.* 34(4), 539–544 (1998).
7. Balankin, A. S. *Synergetics of Deformable Body.* Publishers of Ministry Defence SSSR, Moscow, p. 404 (1991).
8. Hornbogen, E. *Intern. Mater. Rev.*, 34(6), 277–296 (1989).
9. Pfeifer, P. *Appl. Surf. Sci.*, 18(1), 146–164 (1984).
10. Avnir, D., Farin, D., and Pfeifer, P. *J. Colloid Interface Sci.*, 103(1), 112–123 (1985).
11. Ishikawa, K. *J. Mater. Sci. Lett.*, 9(4), 400–402 (1990).
12. Ivanova, V. S., Balankin, A. S., Bunin, I. Zh., and Oksogoev, A. A. *Synergetics and Fractals in Material Science.* Nauka, Moscow, p. 383 (1994).
13. Vstovskii, G. V., Kolmakov L. G., and Terent'ev, V. F. Metally, 4, pp. 164–178 (1993).
14. Hansen, J. P. and Skjeitorp, A. T. *Phys. Rev. B*, 38(4), 2635–2638 (1988).
15. Pfeifer, P., Avnir, D., and Farin, D. *J. Stat. Phys.*, 36(5–6), 699–716 (1984).
16. Bobryshev, A. N., Kozomazov, V. N., Babin, L. O., and Solomatov, V. I. *Synergetics of Composite Materials.* NPO ORIUS, Lipetsk, p. 154 (1994).
17. Farin, D., Peleg, S., Yavin, D., and Avnir, D. Langmuir, 1(4), 399–407 (1985).

CHAPTER 23

HYDRODYNAMICS OF GAS-LIQUID-SOLID FLUIDIZED BED REACTOR

A. K. HAGHI and G. E. ZAIKOV

CONTENTS

23.1   INTRODUCTION

In this chapter, the gained results of numerical simulation of gas velocity effect on the phases hold-up in a three phase fluidized bed column have been provided using computational fluid dynamics. Eulerian multiphase model and standard k–epsilon (K-ε) turbulence model are used for column simulation. The gained results of stimulation are compared with experimental data in one column with laboratory dimensions containing liquid phase with 100 cm height and 20 cm diameter and have good agreement with experimental data. The solid phase with 0.15 volume fraction dispersion into liquid phase and the gas phase entered the column through a sparger of 2 cm diameter with various velocities. The results show as the gas phase velocity increases from 0.02 to 0.08 m/sec, the solid hold-up will decrease and simultaneously the gas hold-up will increase. The radial examination of the gas hold-up also demonstrates that the level of hold-up in the center of the column, specially the gas phase hold-up, is more than the one by the walls.

The gas-liquid-solid fluidize bed has in emerged recent years as one of the most promising devices for three phase operation. Such a device is of considerable industrial importance as evident from its wide application in chemical, petrochemical, and biochemical processing [1].

In this type of reactor, gas and liquid are passed through a granular solid material at high enough velocities to suspend the solid in fluidized state .The solid particles in the fluidized bed are typically supported by porous plate known as distributor at the static condition. The fluid is then forced through the distributor up through the solid materials at lower fluid velocities, the solid remain in place as the fluid passes through the voids in the material as the fluid velocity is increased, the bed reaches a stage where the force of the fluid on the solids is enough to balance the weight of the solid material. This stage is known as incipient fluidization and the corresponding fluid velocity is called the minimum fluidization velocity in fluidized bed reactors, the density of are much higher than the density of the liquid and particles size is normally large (above 150 μm) and volume fraction of particles varies from 0.6(packed stage) to 0.2 as close to dilute transport stage (paneerselvam et al 2009) [2-11].

Even though a large number of experimental studies have been directed towards the quantification of flow structure and flow regime identification for different parameters and physical properties , the complex hydrodynamics of these reactors are not well understood due to the interaction of all the three phases simultaneously. It has been a very tedious task to analyze the hydrodynamic property experimental way of three phase fluidized bed reactor, so another advanced modeling approaches based on computational fluid dynamics (CFD) techniques have been applied for investigation of three phases fore accurate design and scale up. Basically two approaches namely, the Euler-Euler formulation based on the interpenetrating multi fluid model. And the Euler-Lagrangian approaches based on solving Newton equation motion for dispersed phase are used.

Bahary et al. (1994) have used Eulerian multi-fluid approach for three phase fluidized bed [2, 3]. Where gas phase treated as a particulate phase having 4 mm diameter and a kinetic theory granular flow model applied for solid phase. They have simulated both symmetric and axisymmetric model and verified the different model regimes in the fluidized bed by comparing with experimental data.

Schallenberg et al. (2005) have used 3-D multi-fluid Eulerian approach for three phase bubble column. Gas-liquid drag coefficient based on single bubble rise modified for the effect of solid phase [4]. Extended K-ε turbulence model to account for bubble-induced turbulence has been used and the interphase momentum between two dispersed phases include. Local gas and solid hold-up as well as liquid velocities have been validated with experimental data.

Panneerselvam et al. (2009) have work in 3-D Elerian multi-fluid approach for gas-liquid-solid fluidized bed [5]. The kinetic theory granular flow (KTGF) model for describing the particulate phase and a K-ε based turbulence model for liquid phase turbulence have been used. The interphase momentum between tow dispersed phases has been included .Radial distributions of axial and radial solid velocities, axial and radial solid turbulent velocities, shear stress axial bubble velocity, axial liquid velocity, and average gas hold-up and various energy flows have been studied.

23.2 THEORY AND HYDRODYNAMICS

Hydrodynamic equations used in this research are based on the following equations:

23.2.1 BASE MODEL EQUATION

In the Equation (1), which is continuity equation, ρ_k density and ε_k as volume fraction of gas, liquid and solid phases. Thus the sum of volume fraction of three phases presents $\varepsilon_l + \varepsilon_v + \varepsilon_g$ in the Equation (2), namely momentum equation, the phases volume

fraction r_α are correlated to interphase movement Mαi for Np in Naviar stocks equation. Hence, the right quotations of the Equation (2) describe those operating forces on fluidized element of phase X into the control volume. The total pressure gradient, the viscosity tensions, the gravity forces and interphase momentum forces have been mixed in M. The Equation (3) expresses the drag force among liquid and gas phases which has different amounts in different Reynolds. The Equation (4) shows the drag force among solid and gas phases that is known as Gidaspaow [12.13].

$$\frac{\partial\left(\varepsilon_k \rho_k\right)}{\partial t} + \nabla\left(\rho_k \varepsilon_k \mu_k\right) \tag{1}$$

23.2.2 MULTIPHASE MODELS

This research has employed the multiphase model Eulerian. The model considers the phases as come turbulence environments in with the possibility of each phase existing in calculating range will be determined by its volume fraction and sum of the volume fractions is equal to unity. The interphase momentum appears as a drag which is a function of stumbling velocity between the phases in this research, liquid acts as a continual phase and gas acts as a diffuse phase which get into the system from the bottom of column.

23.2.3 MODEL AND THE WALL BOUNDARY CONDITION

Boundary conditions on the wall occur as the boundary condition of inlet at the spurger, the boundary condition of outlet pressure at the top of column as well as the wall boundary condition at the walls. The model consists of a cylinder 20 cm in diameter and 100 cm in height. The gas flow lets in the bottom of column, passing through the static liquid and also the solid phase dispersed inside the liquid phase with 0.15 volume fraction , then lets out of the top. Air crosses over the fluidized column with various velocity .The research applies two dimensional simulations under condition of axis symmetry and inlet dry air.

23.2.4 EXPERIENCES

Image shown in Figure (1) illustrates the experience equipment. This equipment includes a column with 100 cm height, 20 cm diameter made of heat-resistant glass (Pyrex). At the top of the column, separator equipment and at the bottom of the column, a multi-rod gas distributer few rods of crusade type that is interchangeable with other types of sparger have been illustrated for surveying the effect of gas injection on hydrodynamics of the system. Sparger has 2 cm thickness and is made of copper. Moreover, this column is equipped with material drain valve at the bottom of the flow cylinder as well as manometer and thermometer to measure temperature and pressure. A compressor connected to a pressure regulator and outlet and a Rotameter for flow air direction. Exhaust air flow enter a check valve from the compressor through the Rotameter (with operation range from 1 to 200 liters per min) and then enter to Sparger *via* blowing from the bottom of the column. Total Column hold-up is done *via* seeing the changes in the column's height that is graded. To increase accuracy and reduce test error for each test, trial was repeated 3 or 4 times. Details of column and other equipment in Table (1) are reported below.

TABLE 1   Details of equipment's Installation

Column's diameter	8cm
Column's height	90cm
Number of distributer air faces	45
The total surface of sparger	0.005 square meter
Active efficient level and diameter of sparger	70% , 3cm
Internal cylinder diameter	5cm
Gas flow rate	5 to 50 liter per minute

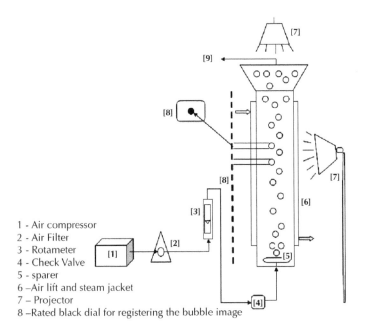

1 - Air compressor
2 - Air Filter
3 - Rotameter
4 - Check Valve
5 - sparer
6 – Air lift and steam jacket
7 – Projector
8 – Rated black dial for registering the bubble image

FIGURE 1   Schematic picture of experimental setup.

23.3 GEOMETRY AND MESH

The first step in CFD simulation of fluidized bed column is preprocessor, which has been done by GAMBIT tools, to design the problem in geometrical configuration and mesh the geometry. Before fluid flow problems can be solved, FLUENT needs the domain in which the flow takes place to evaluate the solution. The flow domains as well as the grid generation into the specific domain have been created in GAMBIT which is shown in Figure 1.

23.4 DISCUSSION AND CONCLUSION

The Figure (1) shows the bubble column containing liquid phase and solid particles of volume fraction 0.15 at T = 0 sec. Gas phase enters the column through a spurger 2 cm diameter with velocity of 2 cm/sec as the determined time pass, the solid phase hold-up will decrease.

The Figure (2a) looks over the solid phase hold-up changes in a radial line in the fluidized bed column. It shows that the solid phase hold-up in center is the most and it is decreasing besides the walls.

The Figure (2b) studies the effect of gas phase velocity on the solid phase hold-up on various velocity 2, 4, 6, 8 cm/sec. The result demonstrates as the gas velocity goes up, the solid hold-up goes down.

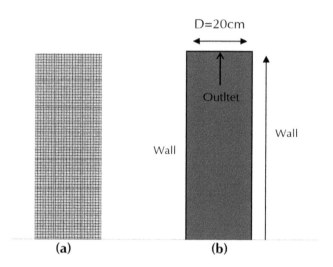

FIGURE 2   Computational grid (a) and column dimensions (b).

   The Figure (3a) studies the gas hold-up changes over a radial line in the fluidized bed column. It displays that the gas hold-up is in its maximum level at the center of column but in minimum besides the walls comparing the Figure (2a) and (3a), it can be concluded that the solid phase hold-up is more than the gas phase hold-up at the same conditions.

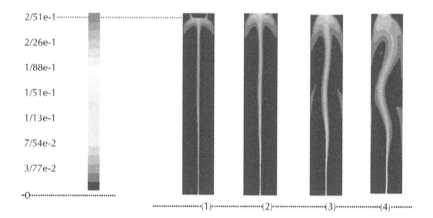

FIGURE 3A   Liquid volume fraction changes in different velocities 2,4,6,8 cm/s.

   The Figure (3b) shows the effect of increasing the gas phase velocity on the gas hold-up. The results represent that the increase of velocity from 2 to 8 cm/sec makes the gas phase hold-up also increase, demonstrating the inverse effect of the gas velocity increase on the gas and solid phase hold-up.

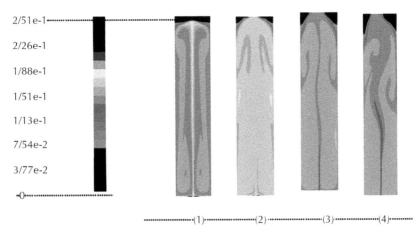

FIGURE 3B Solid volume fraction changes in different velocities 2,4,6,8 cm/s.

KEYWORDS

- **Computational fluid dynamics (CFD)**
- **Gas-liquid-solid fluidize bed**
- **K–epsilon (K-ε) turbulence model**
- **Kinetic theory granular flow (KTGF)**

REFERENCES

1. Panneerselvam, R., Savithri, S., and Surender, G. D. CFD simulation of hydrodynamics of gas-liquid-solid fluidized bed reactor. *Chemical engineering science.* **64**, 1119-1135 (2009).
2. Bahary, M., *experimental and computational studies of hydrodynamics in three phase and two phase fluidized bed*. Ph.D. thesis .I11inois Institute of technology, Chicago.1994
3. Pain, C. C, Mansourzadeh, S., and Oliveira, C. R. E. A study of bubbling and slugging fluidized beds using two flow granular temperature model. *International Journal multiphase flow*, **27**, 527-551 (2001).
4. Schallenberg, J., Enb, J. H., and Hempel, D. C. The important role of local dispersed phase hold-ups for the calculation of three phase bubble columns. *Chemical engineering science.* **60**, 6027-603 (2005).
5. Panneerselvam, R., Savithri, S., and Surender, G .D. CFD simulation of hydrodynamics of gas-liquid-solid fluidized bed reactor. *Chemical engineering science*, **64**, 1119-1135 (2009).
6. Kunii, D. and Levenspiel, O. Fluidization Engineering. 2nd edition Butterwo Washington (1991).

7. Howard, J .R. Fluidized Bed Technology:Principles and Applications. Adam Higler, New York (1989).
8. Park, H. CFD Advantages and Practical Applications Glumac Inc. (2009).
9. Trambouze, P. and Euzen, J. Chemical Reactors:From Design to Operation. Technip, Paris (2004).
10. Muroyama, K. and Fan, L .S. Fundamentals of gas-liquid-solidfluidization.A.I.Ch.E. Jounral **31**, 1-34 (1985).
11. Epstein, N. Three-Phase Fluidization: Some knowledge gaps. *Can. J. Chem. Engg.* **59**, 649-757 (1981).
12. Dudukovic, M. P., Larachi, F., and Mills, P .L. Multiphase reactors-revisited *Chemical Engineering Science* **54**, 1975–1995 (1999).
13. Patankar, S. K. "Numerical Heat Transfer and Flow," ISBN: 0-89116-522-3 (1980).

CHAPTER 24

RUBBER TECHNOLOGY COMPOUNDING AND TESTING FOR PERFORMANCE: PART I - THE BUTYL RUBBER COMPOUNDS. ADHESIVE PROPERTIES

MARIA RAJKIEWICZ, MARCIN ŚLĄCZKA, and JAKUB CZAKAJ

CONTENTS

24.1   INTRODUCTION

The butyl rubber isobutylene-isoprene rubber (IIR) plays an important role in the building products. Particular suitability of the butyl compounds in the building industry is a result of good resistance to action of ozone, low gas migration rate, good electric insulating properties, and vibration damping. For sealing products, the suitability of the butyl compounds is based on their adhesive properties. In this chapter a relation between the adhesive properties of the butyl rubber compounds and amount of the applied paraffinic oil and alkylphelolic resin has been described.

The IIR was blended with alkylphelolic resin, paraffinic oil, and whiting. Mixing was conducted in a Z-Blade Mixer. Adhesive, rheological, and mechanical properties were determined by standard methods. A good adhesion and loop tack of the blends to steel plates was achieved.

Butyl rubber compounds are widely used in the industry, in the form of vulcanizates as well as in the form of the unvulcanized rubber; they mainly play a sealing role in tyres, stem hoses, gaskets, putties, sealants, and roof coverings. Due to the small permeability of the moisture, the butyl compounds are commonly used as an insulating material in the electric installations [1, 2, 4, 7].

For the mastics, the adhesion of the butyl rubber to most surfaces is insufficient. In the production practice compatible resins and oil is added to improve the adhesion and the mineral filler is added to reduce the cost of the compound. The commercial compounds are available in the form of tapes, thermoplastic adhesives (hot-melt), and in the solvent form [1]. An advantage to the solvent-based compounds is the ease of dosing of high-molecular rubber-based sealants. The mostly used solvent is mineral spirit (which is a mixture of paraffin, cycloparaffins, and aromatic alkali, used as a cheap and odorless solvent). Due to the cost of solvent evaporation and harmful impact on the natural environment, solvent-based sealants are gradually withdrawn from use. In case of thermoplastic compounds the consistency is adjusted with content of the plasticizers – paraffinic or naphthenic oil is used for the butyl rubber.

24.2   SOLVENTLESS BUTYL SEALANTS

Thermoplastic sealing compounds usually contain much filler (100-800 parts per 100 parts of the basic elastomer). The filler is added mainly for cost efficiency reasons in order to reduce the cost of the final compound. Increasing the content of the filler results in increased hardness of the compound, worsened mechanical properties with high filling, and reduced adhesive properties.

Besides of the butyl rubber, the polyisobutylene (PIB) is used as the base elastomer. Molecular weights of IIR and PIB used in the compound have significant impact on the processing and usability properties of the product.

The PIB and IIR with average and high molecular weights (200 000-1100 000) are used as a base providing good mechanical properties of the compound. Despite of the relatively good adhesion of the base elastomer itself, addition of resins is necessary to improve adhesion. For IIR and PIB the alkali resins C_5, rosin esters, polyterpenic, and coumarone-indene resins are used most often for improved stickiness.

Another method of improving adhesion is by adding plasticizers. They adjust the compound hardness and adhesion, but such an addition significantly reduces cohesion of the material. The most commonly used plasticizers include paraffinic and naphthenic oils [4, 5, 6, 8].

For adhesive materials the mechanism of bonding to the surface and tearing off from the surface of the substrate is to be considered. Ensuring optimum dampness requires that the surface tension of the material and the substrate are similar. Tearing off consists in a deformation of the material layer (stretching) and subsequent separation from the substrate surface.

Influence of the mixture components on adhesion is as follows: Increasing the molecular weight of the elastomer base increases G' and G'', hence it reduces the ability to bond to the surface (it explains better adhesion of the low-molecular PIB and IIR to the substrate). Addition of the compatible resin results in the increased temperature of the glass transition (T_g) of the compound, it results in the increased G''. Next stage of addition of the resin is dilution of the base elastomer, what results in the decreased G'. Addition of the plasticizer dilutes the elastomer and decreases T_g of the compound, causing a decrease of G' and G''. It leads to higher adhesion, but lower cohesion of the material [3].

24.3 EXPERIMENTAL PART

24.3.1 RAW MATERIALS

In the technological tests of the compounds a butyl rubber with the Mooney viscosity: ml (1 + 8) 51 + - 5 mu and saturation level 1.85% (by. Lanxess) were used. A resin with softening point of 55-75°C, oil with kinematic viscosity in 40°C within the range 62-88 cSt and the chalk with average granulation of 2.5 um and humidity of 0.2% as a filler were added.

24.3.2 TEST METHOD

The sealing compounds have been prepared using butyl and polyisobutylene rubber according to its usage guidelines and technical requirements of users (low gas permeability, permanent adhesion, and flexibility in low temperature conditions). A series of sample compounds has been made (each 10 for each base elastomer) with variable number of components: oil and resin (component I and component II) according to Table 1. Rheological properties of the compounds (plasticity) are marked with the Mooney MV 2000E equipment.

TABLE 1   Change of two components in the rubber compound

Symbol	1	2	3	4	5	6	7	8	9	10
Component I	40	40	40	50	50	50	50	60	60	60
Component II	60	70	80	60	70	70	80	60	70	80

Properties of the compounds are marked according to the following standards: Melt Flow Index – ISO 1133:2002, Mooney plasticity – PN ISO 289-1: 2007, density – PN ISO 2781 + AC1:1996 method A, rigidity – method worked out in the EiTG branch, durability (T_{sb}) according to PN-ISO-37:98.

Cohesion 180°, tearing off from metal, initial "loop tack" adhesion marked according to the FINAT 1 and FINAT 9 standards.

For assessment of test results the ANALISA software was used, which employs the approximation method. The software provides possibility to draw outline charts showing curves connecting points with the same property values.

The ANALIZA software also enables determining:

- Content of the composite meeting the preset requirements,
- Content of the composite corresponding to the minimum and the maximum of the set property.

The software allows for analyzing any groups of input data.

In the chapter the following approximation curves for the evaluated properties of the compounds (flow index, Mooney plasticity, density, rigidity, tear off from metal, durability, and bonding strength 180°) have been achieved.

24.4   DISCUSSION OF THE RESULTS

The technological tests of the sealing compounds covered butyl rubber with average molecular weight.

A series of compounds with symbol 'SB' based on butyl rubber have been made. Rheological and mechanical compounds have been examined. Significant impact of the dosing of oil and resin on the adhesive properties has been shown.

TABLE 2   Mooney viscosity data for SB compounds

Properties	SB-1	SB-2	SB-3	SB-4	SB-5	SB-6	SB-7	SB-8	SB-9	SB-10
Mooney plasticity ML max. temp. 50°C	63,4	46,0	37,8	56,5	47,2	47,2	35,5	9,35	44,9	34,2
Mooney plasticity, ML min. temp. 50°C	52,2	39,9	32,6	45,1	39,4	39,4	30,1	44,6	36,0	28,7

TABLE 3   Properties of SB series compounds

Properties	SB-1	SB-2	SB-3	SB-4	SB-5	SB-6	SB-7	SB-8	SB-9	SB-10
Flow index, g/10 min	0,06	0,25	0,94	0,12	0,29	0,29	1,07	0,21	0,45	1,21

Density, g/cm³	1,62	1,59	1,54	1,59	1,57	1,57	1,54	1,52	1,52	1,51
Initial adhesion, N	8,75	9,53	8,02	7,39	8,81	8,81	8,86	6,58	11,38	14,23
Bonding strength (cohesion), N	37,16	35,63	28,59	43,99	39,72	39,72	30,60	37,79	38,76	33,49
Rupture force weight + metal, N	107	94	71	86	89	83	83	83	82	72
Durability, MPa	0,111	0,098	0,073	0,09	0,093	0,086	0,086	0,086	0,085	0,075
Rigidity of the plastic, N	79	55	35	63	47	37	37	37	48	32

The test results are shown on the outline charts, which depict the change of properties of the compounds depending on content of the Component I and content of the Component II. Some examined properties in relation to the Component I and Component II are shown in Figures (1-4).

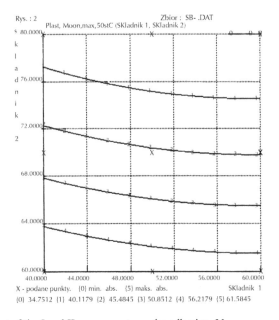

FIGURE 1 Effect of the I and II components on the adhesion, N

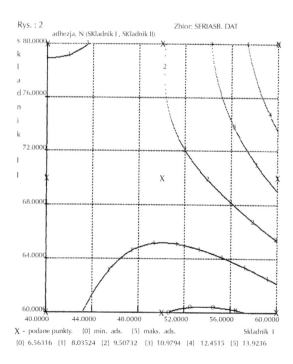

FIGURE 2   Effect of the I and II components on the Mooney plasticity, ML

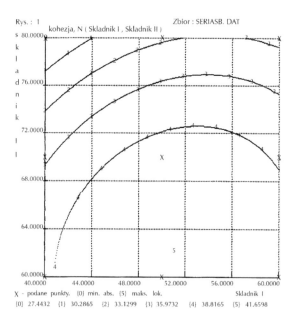

FIGURE 3   Effect of the I and II components on the index flow, g/10 min

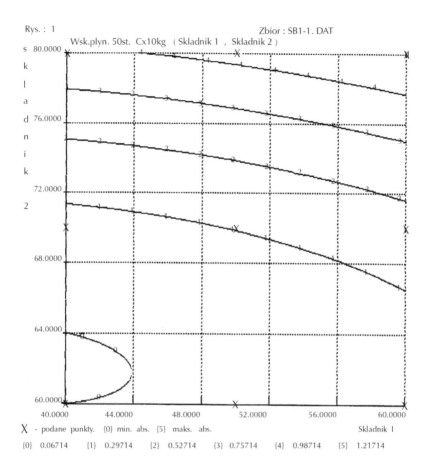

FIGURE 4 Effect of the I and II components on the cohesion, N

Then a series of tests of polyisobutylene-based compounds with symbol 'S' were performed.

TABLE 4 Mooney viscosity data for the S series compounds

Properties	S-1	S-2	S-3	S-4	S-5	S-6	S-7	S-8	S-9	S-10
Mooney plasticity, ML max. temp. 50°C	54,8	45,3	31,7	48,3	30,1	30,1	29,5	49,3	38,1	30,7
Mooney plasticity, ML min. temp. 50°C	45,8	38,4	27,7	38,5	23,7	23,7	25,1	38,1	30,8	25,5

TABLE 5 Properties of S series compounds

Properties	S-1	S-2	S-3	S-4	S-5	S-6	S-7	S-8	S-9	S-10
Index flow, g/10 min	0,7	0,26	1,29	0,31	1,86	1,86	1,77	0,30	0,82	1,76
Density, g/cm^3	1,55	1,56	1,55	1,58	1,42	1,42	1,55	1,55	1,53	1,51
Initial adhesion, N	4,93	6,99	3,1	5,75	11,78	11,78	14,35	5,47	10,2	14,29
Bonding strength (cohesion), N	40,77	32,01	22,38	45,19	30,70	30,70	25,82	46,93	39,58	28,34
Rupture force weight+metal, N	86	89	88	78	57	80	80	104	79	69
Durability, MPa	0,089	0,093	0,091	0,081	0,059	0,059	0,084	0,108	0,082	0,072
Rigidity of the plastic, N	74	51	38	64	36	42	42	85	55	44

The test results are shown on the outline charts, which depict the change of properties of the compounds depending on content of the Component I and content of the Component II. Some examined properties in relation to the Component I and Component II are shown in Figures (5–8).

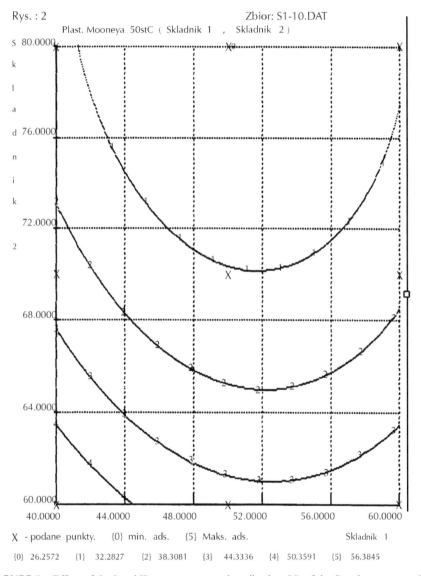

FIGURE 5   Effect of the I and II components on the adhesion (N) of the S series compounds

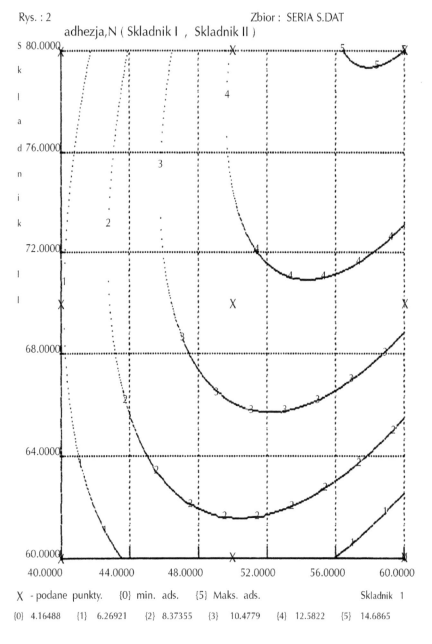

FIGURE 6 Effect of the I and II components on the Mooney plasticity, ML of the S series compounds

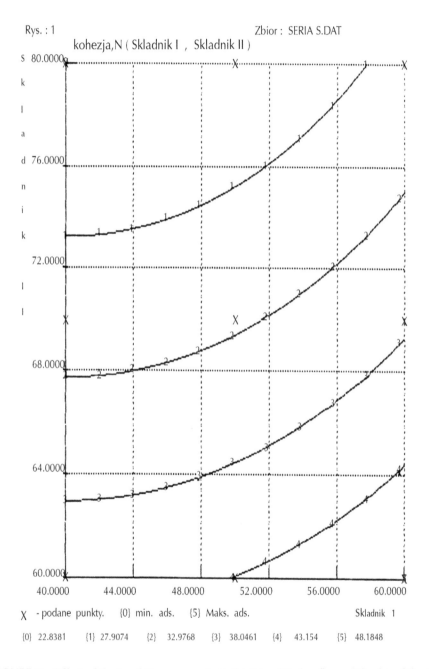

FIGURE 7   Effect of the I and II components on the tear off index flow, g/10 min of the S
series compounds

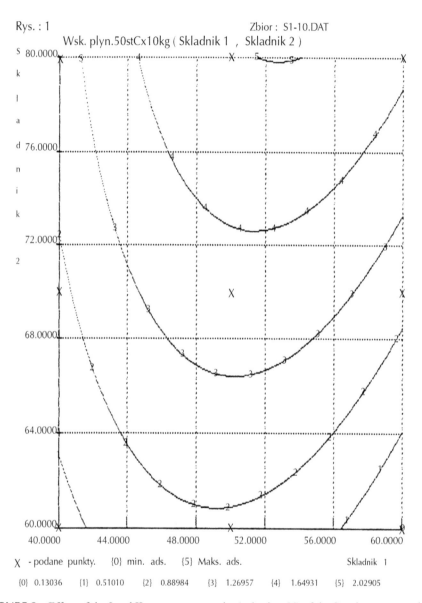

FIGURE 8   Effect of the I and II components on the (cohesion, N) of the S series compounds

24.5   CONCLUSIONS

The adhesive mastics for the building industry, based on butyl and isobutylene rubber, with law plasticity, high adhesion (20 N), and high cohesion of the adhesive mastic (40 N), have been achieved.

KEYWORDS

- **Butyl rubber**
- **Isobutylene-isoprene rubber (IIR)**
- **Mooney viscosity**
- **Plasticizers**
- **Polyisobutylene (PIB)**

REFERENCES

1. Petrie, E. Handbook of adhesives and sealants – 2nd ed., McGraw-Hill, New York (2007).
2. Dick, J .S. Rubber Technology Compounding and Testing for Performance 2nd Edition, Carl Hanser Verlag, Munich (2009).
3. Benedek, I. Handbook of Pressure-Sensitive Adhesives and Products Fundamentals of Pressure Sensitivity, Taylor & Francis Group, Boca Raton, **5** (2009).
4. Benedek, I. Handbook of Pressure-Sensitive Adhesives and Products Technology of Pressure-Sensitive Adhesives and Product , Taylor & Francis Group, Boca Raton, **4** (2009).
5. Benedek, I. Pressure-Sensitive Formulation, VSP BV, Zeist, 159-160 (2000).
6. Sealant Formulation Guide, Royal Elastomers.
7. White, J. R., S.K. De S.K., Rubber Technologist's Handbook, RAPRA Technology Ltd., 64-66, 499-510 (2001).
8. Pat. USA US 2005/0147813 (2005).

CHAPTER 25

RUBBER TECHNOLOGY COMPOUNDING AND TESTING FOR PERFORMANCE: PART II-COMPOSITES ON THE BASE OF POLYPROPYLENE AND ETHYLENE-OCTENE RUBBER

MARIA RAJKIEWICZ, MARCIN ŚLĄCZKA, and JAKUB CZAKAJ

CONTENTS

25.1 INTRODUCTION

The structure and physical properties of the thermoplastic vulcanisates (TPE-V) produced in the process of the reactive processing of polypropylene (PP) and ethylene-
octene elastomer (EOE) in the form of alloy, using the cross-linking system was analysed. With the DMTA, SEM and DSC it has been demonstrated that the dynamically
produced vulcanisates constitute a typical dispersoid, where semi-crystal PP produces
a continuous phase, and the dispersed phase consists of molecules of the cross-linked
ethylene-octene elastomer, which play a role of a modifier of the properties and a
stabiliser of the two-phase structure. It has been found that the mechanical as well as
the thermal properties depend on the content of the elastomer in the blends, exposed
to mechanical strain and temperature. The best results have been achieved for grafted/
cross-linked blends with the contents of iPP/EOE-55/45%.

Three com. ethene/n-octene copolymers (I) were blended with a com. isotactic
polypropylene (II). grafted/crosslinked with mixts. of unsatd silanes and $(PhCMe_2O)_2O$
under conditions of reactive extrusion at 170-190°C and studied for mech. Properties,
thermal stability and microstructure. The composite materials consisted of a semicryst.
II matrix and dispersed amall pertides of crosslinked I. The best mech. properties were
when the II / I mass   ratio was 55/45.

The thermoplastic elastomers (TPE) are a new class of the polymeric materials,
which combine the properties of the chemically cross-linked rubbers and easiness of
processing and recycling of the thermoplastics (1-8). The characteristics of the TPE
are phase micro-nonuniformity and specific domain morphology. Their properties are
intermediate and are in the range between those, which characterise the polymers,
which produce the rigid and elastic phase. These properties of TPE, regardless of its
type and structure, are a function of its type, structure and content of both phases, nature and value of inter-phase actions and manner the phases are linked in the system.

The progress in the area of TPE is connected with the research oriented to improve
thermal stability of the rigid phase (higher T_g) and to increase chemical resistance
as well as thermal and thermo-oxidative stability of the elastic phase (2). A specific
group among the TPE described in the literature and used in the technology are micro-
heterogeneous mixes of rubbers and plastomers, where the plastomer constitutes a
continuous phase and the molecules of rubber dispersed in it are cross-linked during
a dynamic vulcanising process. The dynamic vulcanisation is conducted during the
reactive mixing of rubber and thermoplast in the smelted state, in conditions of action
of variable coagulating and stretching stresses and of high coagulating speed, caused
by operating unit of the equipment. Manner of producing the mixtures and their properties as well as morphological traits made them be called thermoplastic vulcanisates
(TPE-V) (9,10). They are a group of "customisable materials" with configurable properties. To their advantage is that most of them can be produced in standard equipment
for processing synthetics and rubbers, using the already available generations of rubbers as well as generations of rubbers newly introduced to the market with improved
properties. A requirement of developing the system morphology (a dispersion of macromolecules of the cross-linked rubber with optimum size in the continuous phase of
plastomer) and achieving an appropriate thermoplasticity necessary for TPE-V are

the carefully selected conditions of preparing the mixture, (temperature, coagulation speed, type of equipment, type and amount of the cross-linking substance). When selecting type and content of elastomer, the properties of the newly created material can be adjusted toward the desirable direction. Presence of the cross-linked elastomer phase allows for avoiding glutinous flow under the load, what means better elasticity and less permanent distortion when squeezing and stretching the material produced in such a manner as compared to the traditional mixtures prepared from identical input materials, each of which produces its own continuous phase. With the dynamic vulcanisation process many new materials with configured properties have been achieved and introduced to the market. The most important group of TPE-V, which has commercial significance, are products of the dynamic vulcanisation of isotactic polypropylene (iPP) and ethylene-propylene-diene elastomer (EPDM). It is a result of properties of the PP and EPDM system, which, due to presence of the double bonds, may be cross-linked with conventional systems, which are of relatively low price, good contents miscibility and ability to be used within the temperature range 233 – 408 K (11-15).

Next level in the field of thermoplastic elastomers began with development of the metallocene catalysts and their use in stereoblock polymerisation of ethylene and propylene and co-polymerisation of these olefins with other monomers, leading to macromolecules with a "customised" structure with a microstructure and stereoregularity defined upfront. The catalysts enabled production of the homogeneous olefin copolymers, which have narrow distribution of molecular weights ($RCC=M_w/M_n<2.5$). according to the developed technology called Insite and using the on-place catalyst (16,17). The Dow Chemical Co company produces the olefin elastomers Engage™ which contain over 8% of octene. Co-polymers Engage are characterised by lack of relation between the traditional Mooney viscosity and the technological properties. Compared to other homogeneous polymers with the same flow index, they are characterised by higher dynamic viscosity at zero coagulation speed and decreasing viscosity at increasing coagulation speed. They have no fixed yield point. Saturated nature of elastomer, caused by absence of diene in the chain, results in some restrictions in choice of the cross-linking system. The ethylene-octene elastomers can be easily radiation cross-linked with peroxides or moisture, if they are formerly grafted with silanes. There is relatively not much description of the behavior of the ethylene-octene elastomer in the dynamic vulcanisation process in the literature. The specific physical properties of such elastomers and possibility to process them within a periodic process as well as within a continuous process, due to convenient form of the commercial product (granulate), encouraged us to start the recognition works on development of the technology of producing thermoplastic vulcanisates from the mixture of elastomer Engage and iPP.

Use a silane-based cross-linking system in the dynamic vulcanisation process seemed the most interesting. For the research works one of the known methods of cross-linking poliolefins with silanes was used, assuming that the cross-linking of EOE would proceed according to the analogous mechanism. In the seventies of the 20th century the Dow Corning Co. company developed two methods of the hydrolytic cross-linking of polyolefins grafted with vinylosilanes according to the radical mechanism (18,19). Nowadays three polyolefin cross-linking methods are widely used in the industrial production. The grounds for distinguishing them are technological equipment and procedure. It is one-phase and two-phase method and a "dry silane method", available only under licence (20). The mechanism of cross-linking PE with the cross-linking system: silane/peroxide/moisture is shown schematically in fig. 1.

The process of catalytic hydrolyse of the alco-xylene groups of the grafted silane to the silane groups and, then, the catalytic condensation of the silane groups leads to production of the cross-linked structure through the siloxane groups. The hydrolyse and the condensation take place in an increased temperature with presence of the catalyst and water. Dibutyl tin dilaurate (DBTL) is most often used as a catalyst of the reaction. The catalyst may be added either to the polymeric blend (it constitutes an increased risk of the premature cross-linking), or in the form of a premixed reagent during the processing. The mechanism of action of DBTL, as a cross-linking catalyst, is complex and has not been sufficiently explained.

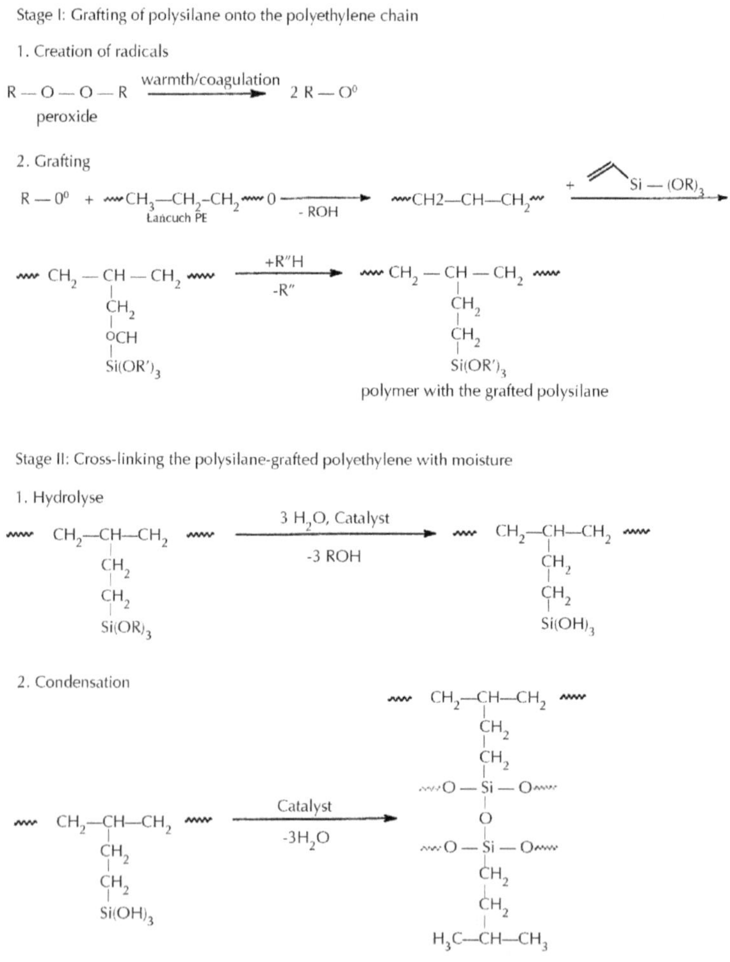

FIGURE 1   Crosslinking mechanism of polyolefins with a silane

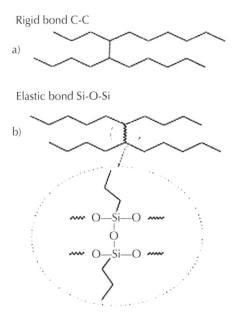

Rigid bond C-C

a)

Elastic bond Si-O-Si

b)

FIGURE 2   Structure of polyolefins crosslinked with a (a) peroxide or radiation, (b) with a silane

As a result of cross-linking the polyolefins with silanes the Si-O-Si bonds are produced, which are more elastic than the rigid bonds C-C created as a result of cross-linking of polymers induced by radiation and peroxides (fig. 2). Use of silanes gives more elastic products and the cross-linking process is more cost-effective.

25.2   EXPERIMENTAL PART

25.2.1   RAW MATERIALS

- Isotactic polypropylene Malen P-F401 iPP, for extrusion, made by Orlen SA, flow index 2.4–33.2 g/10 min, yield point in stretching 28.4 MPa, crystallinity level 95%,
- The ethylene-n-octene elastomers (EOE) type engage, synthesized according to the insite technological process, manufactured by DuPont Dow Chemical Elastomers (Table 1)
- Silanes——Silquest A-172 vinyl-tris (2-methoxyethoxysilane) and Silquest A-174 3-methacryloxypropyltrimethoxysilane, manufactured by Vitco SA,
- Dicumyl peroxide with 99% content of the neat peroxide, manufactured by ELF Atochem,
- Antioxidant tetra-kis (3,5-di-tetra-butyl-4-hydroxyphenyl) propionate, manufactured Ciba-Geigy.

25.2.2 TEST METHOD

Three types of EOE from the wide range offered by the manufacturer were selected (Table 1). The general purpose elastomers were selected with high content of octene and a defined characteristic.

The test were made aimed for determining a threshold value of content of elastomer in the iPP/EOE mixture, which was subject of the dynamic vulcanization process, considering the influence of these parameters on variable properties of iPP. A series of tests was made, in which the proportions of PP and elastomer Engage I were changed in the range 15–60%, with continuous addition of the cross-linking system (silane A-172/ dicumyl peroxide) 3/0.03% in relation to elastomer and antioxidant additive 0.2%.

For the tests of preparing dynamic vulcanizates in the continuous process of reactive extrusion a twin-screw mixer-extruder DSK 42/6D manufactured by Brabender-was used. The vulcanizates were produced dynamically in the process of one-stage or two-stage extrusion process, setting the favorable operating parameters for the device, which had been determined based on multiple tests: distribution of temperatures in each heating area of the extruder: 170/180/190°C, screw rotation—40/min. In the one stage process all the components provided in the formula (elastomer, iPP, antioxidant, and silane initiating system/peroxide) was initially mixed in a fast-rotating mixer type Stephan in temperature of 50°C, next a granulate was extruded. In the two stage process in the first stage the iPP, elastomer, and antioxidant mixture was extruded, next—after mixing the granulate with the cross-linking system—it was extruded again.

The profiles were formed from granulates with an injection moulding machine type ARBURG-420M1000-25 All rounder. For the tests the actual injection at speed of 10 cm³/sec was used with addition at speed of 15 cm³/sec, injection temperatures—–195/200/210/210°C and blend injection time was slightly lower than for iPP itself.

METHODS OF ANALYZING THE BLEND

Hardness was marked according to the Shore method, scale D according to PN-ISO 868 or according to the ball insertion method according to the PN-ISO 868 (MPa). Flow speed index (MFR) was determined according to PN-ISO 1033. Resistance properties of the blend with static stretching were tested according to ISO-527, using a digital tester Instron 4505 (tear off speed: 50 mm/min). The bending properties were determined according to PN-EN ISO 178. In addition to the regular tests, the selected blends were subject to specialist examination, such as the thermogravimetric analysis (TGA), scanning electron microscopy (SEM), differential scanning calorimetry (DSC), and dynamic thermal analysis of mechanical properties (DMTA).

The samples were heated in the ambient temperature in temperature range of 30–490°C with speed of 5 deg/min. The test was conducted with thermobalance TGA manufactured by "Perkin Elmer". Turning points were made after freezing the samples in the liquid nitrogen for about 3 min. The surfaces of the turning points were concocted with gold with vacuum powdering. The SEM JSM 6100 manufactured by JEOL was used to conduct the tests. The photographs have been made in magnification of 2000x.

TABLE 1   Properties of EOE Engage

Elastomer type Properties	Engage I	Engage II	Engage III
Co-monomer content, % of weight (^{13}C NMR/FTR)	42	40	38
Density, g/cm^3, ASTM	0.863	0.868	0.870
Mooney viscosity, ML (1 + 4) 121°C	35	35	8
MFR, deg/min, ASTM D-1238	0.5	0.5	5.0
Shore hardness A, ASTM D-2240	66	75	75

TABLE 2   Selected properties of the dynamically cross-linked blends in relations to PP/EOE ratio

Ratio PP / Engage I ,% of weight	100/0	85/15	70/30	55/45	40/60
Hardness, °ShD	80	63	57	50	36
MFR (190°C, 2, 16 kg), g/10 min	2.4	1.63	1.29	1.28	1.15
MFR (190°C, 5), g/10 min	-	5.06	5.89	5.80	4.90
T_{A120}, °C*	152	143	130	106	~60
Hardness HK, MPa**	24.7	16.1	12.2	11.8	8.7
Solubility of elastomer in cyclo-hexane, %	-	-	13.9	12.03	14.2
Solubility of elastomer<t4/> in boiling xylene, %	-	-	24.0	33.0	42.0

*T_{A120}—Vicat softening point
**HK—ball pan hardness method

25.2.3   RESULTS OF THE TESTING

Influence of the content of elastomer Engage I on physical properties of TPE with PP and EOE modified (grafting/cross-linking) with a silane/peroxide cross-linking system has been shown in Figure 3 and Table 2. The content of co-monomer had significant influence on such properties of the elastomer as elasticity, modulus, density, and hardness. Values of two last parameters decreased with the increase of content of *n*-octene in elastomer. It has been stated that properties of the dynamically vulcanized blends could be adjusted with content of the elastomer phase. With the increase of content of EOE in range 15–45% tensile strength increased (18–30 MPa), and, in the same time, relative elongation increased with tear off (300 –700%). With elastomer content over 50% a visible decrease of both properties occurred, which came to 15 MPa and 600% respectively. Whereas hardness expressed in Sh degrees or in MPa) systematically decreased with the increase of the content of EOE in the blend. The optimum content of EOE introduced to PP was 45% and therefore in most subsequent tests a blend was used, in which iPP/EOE ratio was 55/45. Such contents had also the blends listed in Table 3, made of three types of EOE and two types of silane, with constant content of the cross-linking system (silane A–174/dicumyl peroxide 3.0/0.01%, irganox 1010–0.2%. The blends containing elastomers Engage I and Engage II, with difference of content of octene by 2% Shore hardness A (66 and 75 respectively) and with very similar Mooney viscosities, showed comparable resistance and rheological characteristic. The blends containing elastomer Engage III, with the lowest octene content, were characterized by slightly lower variables of tensile strength (tension at the tear off), elongation and hardness, but by a much higher tension at the yield point and high flow index.

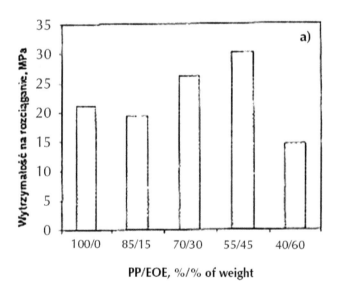

FIGURE 3   *(Continued)*

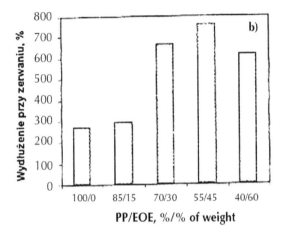

FIGURE 3 Mechanical properties of the dynamically cross-linked blends in relation to PP/EOE ratio, (a) tensile strength, (b) elongation at the tear off.

Such behavior of elastomer Engage III blended with PP resulted probably from its different rheological characteristic, including its four times lower viscosity and very high flow index as compared to Engage I, which was recognized as the most suitable for production of non-saturated blends using the dynamic vulcanization method.

TABLE 3 Effect of type of elastomer Engage modified with Silane A-174 on the properties of the dynamically cross-linked PP/EOE–55/45% blends

Elastomer	Engage I	Engage II	Engage III
Blend properties			
MFR, g/10 min (2.16 kg, 190°C)	1.86	1.80	4.07
Gardbess, Sh, D	42/39	42/40	39/38
ϵ_B, %	720	752	660
σ_M, MPa	25.2	29.5	21.9
$\sigma_{100\%}$, MPa	11.1	12.6	11.0
σ_y, MPa	11.1	12.6	11.1
ϵ_y, %	24.0	27.9	39.9

Symbols: $\epsilon_{100\%}$ —tension at 100% elongation, σ_M—maximum tension, ϵ_B–relative elongation at the tear off, σ_y—yield point, and ϵ_y—elongation on yield point.

Blend with the selected optimum contents iPP/Engage I 55/45% and the selected cross-linking system (silane/peroxide 3/0.03%) were characterized by high thermal stability, independent from type of the material employed to cross-linking silane. It has been confirmed with tests of TGA of blend containing silane A-172 and silane A-174 (samples PL-1 and PL-2, respectively), what is shown in Table 4. In temperature of 230°C the decrease of weight did not exceed 0.5%.

TABLE 4   Results of the TGA of iPP/EOE-55/45% blend (Engage I)

PL-1(Silane A-172)		PL-2 (Silane A-174)	
Temperature, °C	Decrease of weight, %	Temperature, °C	Decrease of weight, %
230	0.23	230	0.36
300	7.43	300	7.66
352	25.97	378	54.36
363	40.06	405	89.63
430	94.05	426	94.36

SEM micro-photographs of: (a) neat iPP

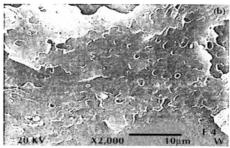

dynamically cross-linked PP/EOE blend
– 55/45%; magnification 2000 x

FIGURE 4   The SEM micro-photographs of (a) neat iPP, (b) dynamically cross-linked PP/EOE blend—55/45%, magnification 2000x.

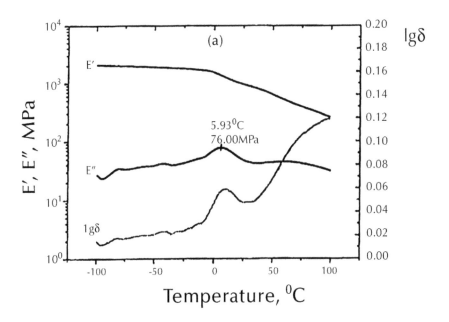

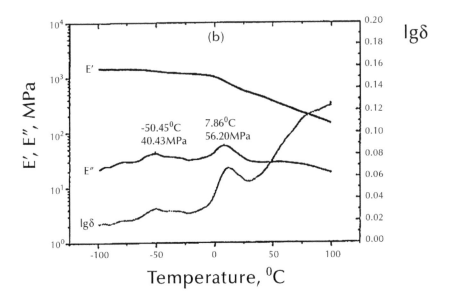

FIGURE 5   *(Continued)*

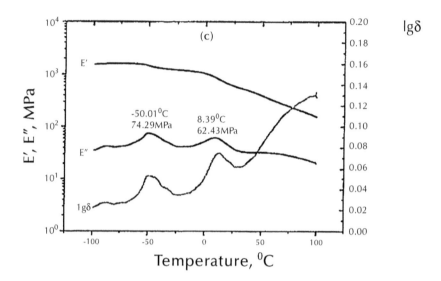

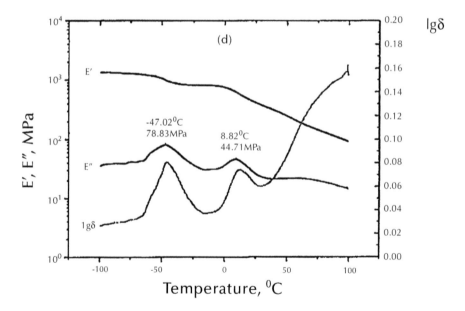

FIGURE 5 The dynamic mechanical properties of neat PP and dynamically cross-linked PP/ EOE blends in relation to temperature: storage modulus (E'), loss modulus (E"), loss tangent (tanδ), (a) PP (b), PP/EOE 85/15, (c)-PP/EOE 70/30, and (d)PP/EOE 55/45.

In temperature.300°C came to as much as 7.5%, and the further increase of temperature caused the progressive degradation process. Analysis of the morphological structure of grafted/cross-linked iPP/Engage I blend using the SEM, DSC, and DMTA methods showed that the blends produced with dynamic vulcanization had a special two-phase structure. With SEM photographs of surface of turning points of iPP samples and iPP/Engage I 55/45% blends have been made (Figure 4). The SEM analysis showed that the obtained blends were mixtures of two thermodynamically non-miscible structures. The continuous phase of iPP had a visible semi-crystal structure and the spherical and oval molecules of the dispersed phase of the cross-linked elastomer were not connected to the continuous phase. The viscoelastic properties, assessed with the DMTA Mk II equipment manufactured by Polymer Laboratories in the sinusoidaly variable load conditions at bending with frequency of 1 Hz, in the temperature range between –100 and +100°C also showed heterogeneous structure of the produced blends. In Figure 5 course of changes of the storage modulus E', loss modulus E" and vibration damping factor gδ for iPP and iPP/Engage I blends with content of 85/15, 70/30, and 55/45% in relation to temperature has been shown. For iPP/EOE blends two, clear relaxation transitions in the range of glass transition are visible, near glass points of iPP and EOE.

Addition of elastomer slightly moved the glass point of iPP toward higher temperatures. PP showed higher values of the E' modulus as compared to the analyzed composites, whereas in the chart E" one maximum appeared corresponding to T_g PP.

TABLE 5 Selected properties of PP and dynamically cross-linked PP/EOE blend PP/EOE (55/45%) in the reactive processing

Properties	PP	TE-1 (Silane A-174)	TE-2 (Silane A-172
Yield point, MPa	31.4	12.1	12.5
Relative elongation of yield point, %	9.4	25.1	33.0
Tensile strength, MPa	15.1	15.5	18.3
Elongation at the tear off, %	196	443	440
Tensile shear modulus, MPa	1455	476	430
Bending strength, MPa	39.6	13.0	10.4
Tensile bending modulus, MPa	1499	504	415
Young modulus, MPa	1520	515	502
HDT, load 1.8 MPa, °C	50.5	38	38
Izod notched impact strength, kJ/m²	3.16	46.7	43.0

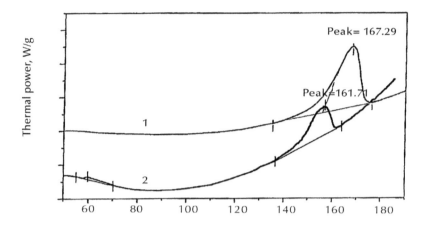

FIGURE 6   The DSC curves of [1] PP, [2] PP/EOE dynamically cross-linked 55/45%.

On the DSC thermal images made in positive temperatures (50–210°C) a visible maximum appeared, which was connected with thermal transition corresponding to the melting point of iPP. Systematic decrease of melting point of the thermoplastic phase of iPP in iPP/Engage I blends related to the original polymer was observed (Figure 6). Causes of these changes could not be unambiguously determined. It is supposed that here the phenomena such, as degradation of iPP in conditions of the high-temperature processing change of semi-crystal structure of iPP may have occurred. In order to compare the properties of iPP/Engage I blend with content of 55/45%, produced with periodical method and in the one-stage or two stage continuous process, a series of tests with use of the general formula was performed. Properties of the selected blends cross-linked with the silane A-174/dicumyl peroxide (TE-1) and silane A-172/dicumyl peroxide (TE-2) systems have been listed in Table 5.

25.3   CONCLUSIONS

The conducted tests leaded to developing grounds for the technology for dynamic vulcanization of materials with thermo-elastoplastic properties, in which a thermoplastic polymer constitutes a continuous phase, whereas the dispersed phase consists of cross-linked elastomer particles. Basic elastomers are polyisoprene with isotactivity level of 85% or higher and co-polymer EOE containing over 30% of n-octene.

The blend properties can be adjusted with content of the elastomeric phase and with cross-linking silane/peroxide system. The achieved material has a heterogeneous structure and favorable set of resistance properties, including higher Izod notched impact strength as compared to PP.

The developed technology allows for achieving new materials with preconfigured properties, competitive to the unmodified iPP and to physical PP/elastomer mixtures. They could be processed in the equipment for synthetics. These blends may be used as structural materials, characterized by higher thermal resistance as compared to the unmodified PP. They may also be a generation of modifiers for polyolefins and polymeric mixtures.

KEYWORDS

- Ethylene-octene elastomers
- Scanning electron microscopy
- Thermogravimetric analysis
- Thermoplastic elastomers
- Thermoplastic vulcaniates

REFERENCES

1. Rzymski, W. and Radusch, H. J. *Polimery*, **47**, 229 (2002).
2. Spontak Richard, J. and Patel Nikunj, P. *Current Opinion in Colloids and Interface Sci.*, **5**, 333 (2000).
3. Rzymski, W. and Radusch, H. J. *Polimery*, **50**, 247 (2005).
4. Radusch., H. J., Dosher, P., and Lohse, G. *Polimery*, **50**, 279 (2005).
5. Rzymski, W. M. *Stosowanie i przetwórstwo materiałów polimerowych*. Wyd. Polit. Częstochowskiej, Częstochowa, s17–28 (1998).
6. Holden, G. *Understanding Thermoplastic Elastomers*. Hanser Publishers, Munich, (2000).
7. Rader, C. P. *Modern Plastic Encyclopedia* (1991).
8. Rader, C. P. *Kuststoffe*, **83**, 777 (1993).
9. Rzymski, W. and Radusch, H. J. *Elastomery*, **5**(2), 19 (2001).
10. Rzymski, W. and Radusch, H. J. *Elastomery*, **5**(3), 3 (2001).
11. Winters, R. *Polimery*, **42**, 9745 (2001).
12. An Huy, T., Luepke, T., and Radusch, H. J. *App. Polym. Sci.*, **80**, 148 (2001).
13. Jain, A. K., Nagpal, A. K., Singhal, R., and Gupta, Neeraj K. *J. Appl. Polym.Sci.*, **78**, 2089 (2000).
14. Gupta, Neeraj K., Janil, Anil K., Singhal, R., and Nagpal, A. K. *J. Appl. Polym. Sci.*, **78**, 2104 (2000).
15. Kumar, Suresh Chandra, Alagar, M., and Prabu, A. A. *Eur. Polym. Journal*, **39**, 805 (2003).
16. Fanicher, L. and Clayfield, T. *Elastomery*, **1**(4) (1997).
17. ENGAGE—polyolefin elastomers. A product of Du Pont Dow Elastomers, product Information (2003).
18. Voight, H. U. *Kautsch. Gum. Kunstst.*, **34**, 197 (1981).
19. Toynbee, J. *Polymer*, **35**, 428 (1994).
20. Special Chem. Crosslinking Agent Center, Dane techniczne (2004).

CHAPTER 26

ANTIFUNGAL ACTIVITY OF AMINATED CHITOSAN AGAINST THREE DIFFERENT FUNGI SPECIES

T. M. TAMER, M. M. SABET, E. A. SOLIMAN, A. I. HASHEM, and M. S. MOHY ELDIN

CONTENTS

26.1   INTRODUCTION

The antifungal activity of aminated chitosan against three different fungal species *Aspergillus Niger, Alternaria Alternata, and Fusarium Moniliforme* was measured and evaluated. Aminated chitosan was produced by chemically amination of chitosan *via* introducing further amino groups to the back bone of chitin using para-benzoquinone (pBQ) as activation agent and ethylenediamine (EDA) as amino group source. The aminated chitin was further deacetylated to obtain finally chemically modified chitosan with higher content of amine groups. The success of grafting process has been confirmed using fourier transform infrared (FTIR), thermo gravimetric analysis (TGA), differential scanning calorimetry (DSC), and scanning electron microscope (SEM). It was found that the antifungal activity of the modified chitosan is better than the native one, and increases by increasing external amine groups against given fungal species. Modification improves solubility of polymer along different acidic pH but still unsoluble in neutral and alkaline pH.

26.2   CHITIN AND CHITOSAN

Chitin is the second most abnormal polysaccharide in nature, second only to cellulose, and is primary present in the exoskeletons of crustaceans such as crab, shrimp, lobster, and so on. In addition to crustaceans, it is also found in various insects, worms, fungi, and mushrooms, in varying proportions from species and from region Table 1.

TABLE 1   Approximate chitin content in various living species [1]

Species	Weight% chitin by dry weight body
Fungi	5–20
Worms	20–38
Squids octopus	3–20
Scorpions	30
Spiders	38
Cockroaches	35
Water Beetle	37
Silkworm	44
Hermit Crab	69
Edible Crab	70

Chitin has the same backbone as cellulose, but it has an acetamide group on the C-2 position instead of a hydroxyl group and its molecular weight (MW), purity, and crystal morphology are dependent on its source [2]. Chitosan is the N-deacetylated derivative of chitin, and so it is a linear polysaccharide consisting of β-(1-4) 2 amino-2-deoxy-D glucopyranose as shown in Figure 1.

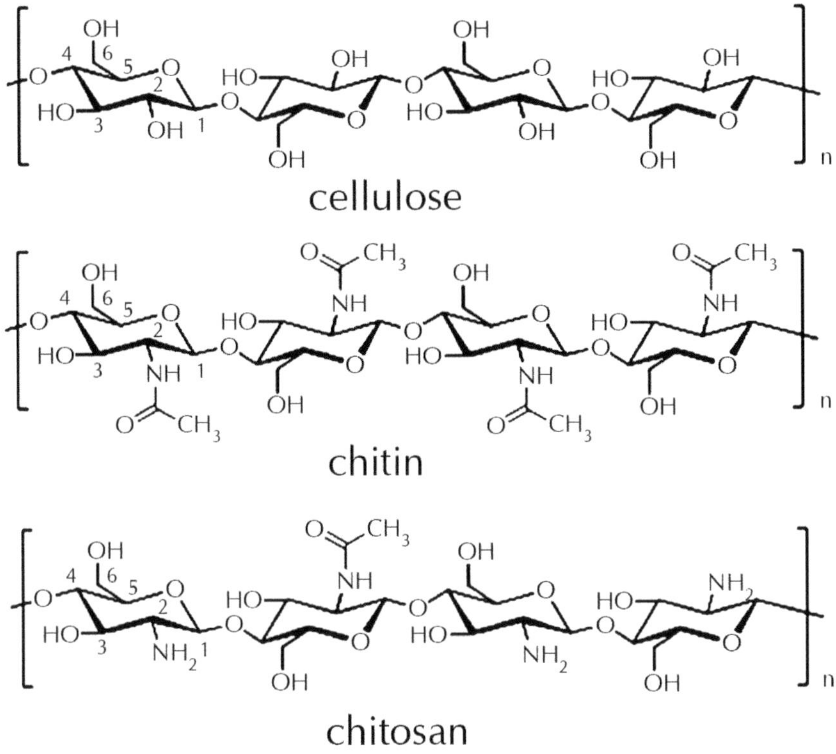

FIGURE 1   Chemical structures of cellulose, chitin, and chitosan.

26.2.1 ISOLATION OF CHITIN AND SYNTHESIS OF CHITOSAN

Within its natural resource of commercial interest, chitin exists not as a standalone biopolymer, but rather in conglomeration with other biomaterials, mainly proteins, lipids, and inorganic salts. The isolation process of chitin starts at sea-food industry (Figure 2), [3]. Shells from crab, shrimp, and so on are first crushed into fine powder to help make a greater surface area available for the heterogeneous processes to follow. An initial treatment of the shell with 5% sodium hydroxide dissolves various proteins, leaving behind chitin. Then treatment with 30% hydrochloric acid hydrolyzes lipids and calcium salts (mainly as $CaCO_3$) and other mineral inorganic constituents. Chitin thus obtained can be hydrolyzed using 50% sodium hydroxide at high temperature (100–150 °C) to provide chitosan, Figure 2.

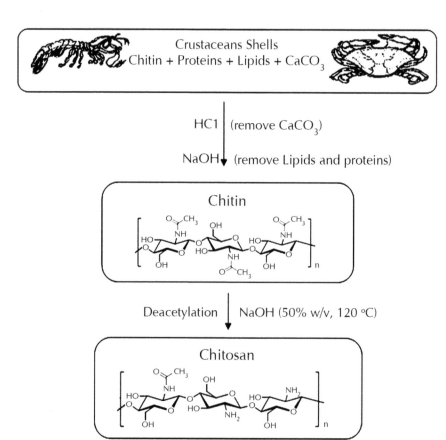

FIGURE 2   Schematic extraction of chitin and preparation of chitosan.

26.2.2 ANTIFUNGAL ACTIVITY OF CHITOSAN

Chitin and chitosan have been investigated as an antimicrobial material against a wide range of target organisms like algae, bacteria, yeasts, and fungi in experiments involving *in vivo* and *in vitro* interactions with chitosan in different forms (solutions, films, and composites). Early research describing the antimicrobial potential of chitin, chitosan, and their derivatives dated from the 1980–1990s [4–9].

26.3 MECHANISM OF THE ANTIFUNGAL ACTIVITY

Several different mechanisms for antifungal inhibition by chitosan have been proposed and recorded in the literature, but the exact mechanism is still unknown.

In general, it is known that the mode of chitosan action on phytopathogens fungi could development in an extra level (plasma membrane) and intracellular level (penetration of chitosan on fungal cell) [10, 11].

Several studies suggest that positively charged chitosan interact with the negatively charged residues at the cell surface of fungi, which causes extensive cell surface alterations and alters cell wall permeability, therefore, this interaction causes the leakage of intracellular electrolytes and proteinaceous material of the cell [10]. El Ghaouth et al demonstrated that chitosan provoked the leakage of amino acids and proteins of the *Rhizopus stolonifer* cell [12]. Similar results was obtained on three isolates on *R. stolonifer* grew in minimum medium, in that study there were an increased release of compounds at 260nm and 280 nm with chitosan of different MW [13]. In other studies, potassium ion leakage was demonstrated by effect of chitosan on fungal cell, being more pronounced for the first 5 min [14, 15]. In general, it is known that chitosan treatment causes changes in the membrane integrity of spores, modifications in pH media, and the proteins release. This effect was different depending on the isolate, kind of chitosan and used concentration [16].

On the other hand, the membrane integrity of *P. expansum* and *B. cinerea* spores was affected by chitosan. *P. expansum* was more sensible than *B. cinerea*; and the effect was related with the fungal species [17]. In other studies, chitosan affected the membrane integrity on *S. sapinea* allowing the out flow of cell components [14]. Besides, chitosan could be affecting the plasma membrane properties. It was demonstrated that this polymer caused a decrease in the H+-ATPase activity on plasma membrane of *R. stolonifer*; this effect could provoke the accumulation of protons inside the cell, which would result in the inhibition of the chemiosotic driven transport that allows the H^+/K^+ exchange [15].

Resent researches suggest that the plasma membrane forms a barrier to chitosan in chitosan-resistant but not chitosan-sensitive fungi. Additionally, it was reported that the plasma membranes of chitosan-sensitive fungi had more polyunsaturated fatty acids than chitosan-resistant fungi, suggesting that the permeabilization by chitosan may be dependent on membrane fluidity [18, 19].

For low MW and oligomer chitosan, the ability of molecules penetration will be increased that varying inhibition mechanism. Few reports demonstrated that chitosan could penetrate the fungal cell. Recent studies of chitosan-fungal cell interactions showed that the polymer penetrates the cell and cause intracellular affectations. Microscopic observation reported that chitosan oligomers diffuse inside hyphae interfering on the enzymes activity responsible for the fungus growth [20]. It was found that chitosan by an energy-dependent process quickly penetrated the conidia of *F. oxysporum* (less than 15 min) and caused ultra structural alterations (disorganized cytoplasm, retraction of the plasma membrane and loss of intracellular content) in the treated spores [11]. However, is evident that a chitosan tracer is needed to evaluate the capture and dissemination within the cell.

Earlier studies showed that oligochitosan penetrated the fungal cell and caused disruption on endomembrane system of *Phytophthora capsici*, such as, distortion and disruption of most vacuoles, thickening of plasmalemma and appearance of unique tubular materials [21]. Additionally, other studies in this plant pathogenic fungus with oligochitosan marked confirmed that, the polymer penetrated the membrane and binds to nucleic acids [22].

26.3.1 FACTORS AFFECT ON CHITOSAN ANTIFUNGAL ACTIVITY

The extent of the antifungal action of chitosan is influenced by intrinsic and extrinsic factors such as MW, pH, species of fungi, and so on. According to several authors, the antimicrobial activity of chitosan is directly proportional to the deacetylation degree (DD) of chitosan [23, 24]. The increase in DD means an increased number of amino groups on chitosan. As a result, chitosan has an increased number of protonated amino groups in an acidic condition and dissolves in solution completely, which leads to an increased chance of interaction between chitosan and negatively charged cell walls of microorganisms in solution.

The essential role of free amine groups in the antimicrobial mechanism of chitosan attract the attention of scientist to produces several derivatives of chitosan with higher amine contents. In this work, evaluation of antifungal activity of aminated chitosan was tested against three different fungal species *Aspergillus Niger, Alternaria Alternata, and Fusarium Moniliforme*.

CHITOSAN AMINATION

The aminated chitosan was prepared as our previous work [25], Briefly, Aminated chitosan derivatives were prepared through three steps. In the first step, 4 g of chitin was dispersed in 50 mL of pBQ–distilled water solution at a known pH and temperature and was stirred for 6 hr.

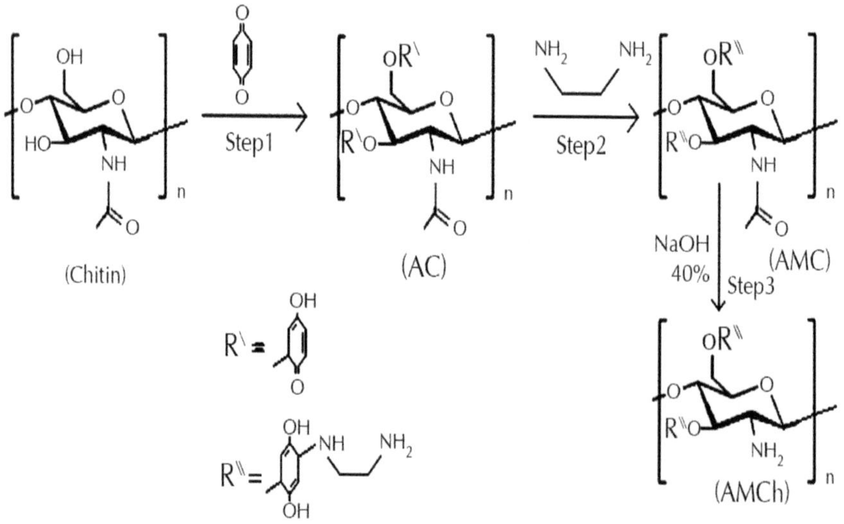

FIGURE 3   Schematically diagram for synthesis of aminated chitosan [25].

The pBQ-conjugated chitin was separated and washed with distilled water to remove unreacted pBQ. In the second step, pBQ-conjugated chitin was dispersed in 50 mL of EDA-distilled water solution of definite temperature and was stirred for 6 hr. The aminated modified chitin was separated and washed with distilled water to remove unreacted EDA. In the last step, aminated modified chitin was deacetylated according to the method of Rigby and Wolfarn [26]. The aminated modified chitin derivative was treated with 50% aqueous solution of NaOH at 120–150°C for 6 hr. The obtained aminated chitosan derivatives were separated and washed with distilled water to remove unreacted NaOH.

POLYMER SOLUBILITY

The chitosan solubility, biodegradability, and biological reactivity depend on the amount of protonated amino groups in the polymeric chain, therefore, on free amine groups along polymer backbone. The amino groups (pKa from 6.2 to 7.0) are completely protonated in acids with pKa smaller than 6.2 making chitosan soluble. Chitosan is insoluble in water, organic solvents and aqueous bases and it is soluble after stirring in acids such as acetic, nitric, hydrochloric, perchloric, and phosphoric [27–33]. It was demonstrated that intra and intermolecular hydrogen bonds play a significant role in forming chitosan's crystalline domains, and appear to provide the main factor limiting its aqueous solubility (it is soluble in water at pH< 6). Protonation of amine group in acidic environment form polycationic form that distorted crystal structure and provide solution stabilities. Solubility of chitosan polymer is one of very important factor for its biological applications. It increases the chance of polymer-microorganism interaction.

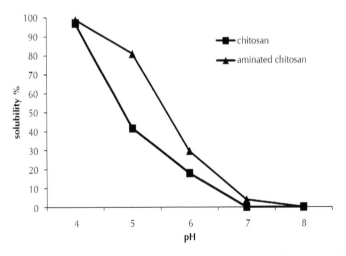

FIGURE 4   Solubility percent of chitosan and aminated chitosan at different pH. [25].

Solubility of aminated chitosan comparing to chitosan itself was studied and presented in Figure 4. A solubility test was performed by dissolving a weighted sample in 2% acetic acid and was stirred at room temperature for 1 hr, and then the sample was filtrated, dried, and weighed. The solubility was determined by using the following equation:

Solubility % = [1 – insoluble part/total weight sample] × 100.

26.4   AMINATED CHITOSAN CHARACTERIZATION

26.4.1   FOURIER TRANSFORM INFRARED (FTIR)

The FTIR of chitosan and aminated chitosan was done using FTIR-8400S SHIM-DZU, Japan. Figure 5 show stretching vibration band at 3430 cm^{-1} that attributed stretching vibration of NH$_2$ and OH groups. Aminated chitosan exhibit more fine at this region that may be attributed increase amine content in the modified polymer. Bands at 2970 cm^{-1} represent to (C-H stretching on methyl) and 2935 cm^{-1} for (C-H stretching in methylene). The bands at 1654 cm^{-1} correspond to stretching of carbonyl group (C=O) of primary amide (amide I). The band at 1633 cm^{-1} corresponds to deformation vibration of –NH$_2$ in plane. The band at 1568 cm^{-1} corresponds to deformation vibration of groups –NH– of amines. The bands at 1427, 1388, and 1159 cm^{-1} correspond to deformation vibration of C–N and the band at 1055 cm^{-1} corresponds to asymmetric stretching of C–N–C.

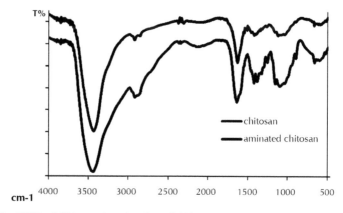

FIGURE 5   FTIR of Chitosan A and aminated Chitosan B.

26.4.2   THERMAL GRAVIMETRIC ANALYSIS (TGA)

The TGA of chitosan and aminated chitosan were carried out using TGA-50 SHI-MADZU Japan. Figure 6 illustrate the thermal degradation of chitosan and aminated chitosan under Nitrogen atmosphere. First depuration from ambient temperature to about 150°C was attributed to elevation of moisture content in polymer. Increasing of the amount of moisture from 7.2% to about 11.15% was attributed to increase hydrophilic nature of modified polymer by amination process. Table 2 illustrate the most important depuration in thermal degradation behavior of chitosan and aminated chitosan.

TABLE 2   Thermal gravimetric depuration steps of Chitosan and aminated chitosan

	Ambient-150°C	220-350°C	350–750	T50
Chitosan	7.2%	35.33%	12.16%	309.4°C
Aminated chitosan	11.15%	36.35%	49.23%	347.22°C

According to Pawlak and Mucha, the main depression of chitosan TGA curve ranged from 220 to 350 was a result of oxidative decomposition of the chitosan backbone. In this stage first depression was resulted from destruction of amine groups to form crosslinked fragment and the second decomposition step, which appears at high temperature, may result from the thermal degradation of a new cross-linked material formed by thermal cross-linking reactions occurring in the first stage of degradation process [34]. Results in Table 2 show decrease the thermal stability of chitosan as grafted with external amine groups. That may be attributed to role of amine group in enhancing thermal degradation process.

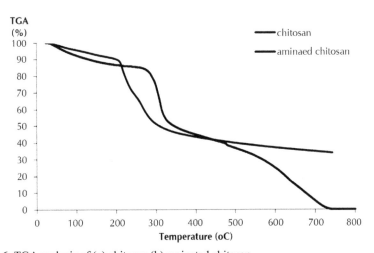

Figure 6. TGA analysis of (a) chitosan (b) aminated chitosan.

26.4.3   DIFFERENTIAL SCANNING CALORIMETRY (DSC)

The differential scanning calorimetric analyses of chitosan and aminated chitosan are illustrated by Figure 7. The first endothermic peak that starting from 50°C to 120°C can be ascribed to the loss of moisture content. Polysaccharides usually have a strong affinity for water, and in solid state these macromolecules may have disordered structures which can be easily hydrated. As is known, the hydration properties of these polysaccharides depend on the primary and supramolecular structures [35, 36].

The second thermal event may be related to the decomposition of Glucose amine (GlcN) units with correspondent exothermic peak at 295°C [37, 38]. Increase the intensity that attributed to increase decomposition process amine (GcN) unites by Increase of amine content.

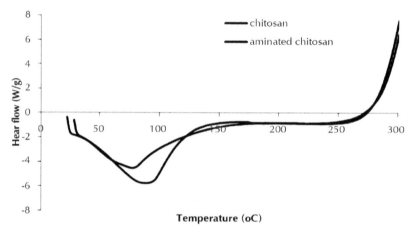

FIGURE 7   DSC analysis of chitosan aminated chitosan.

26.4.4   SCAN ELECTRON MICROSCOPE (SEM)

The surface morphological analysis of chitosan and their aminated derivatives were performed using SEM.

The SEM graphs, Figures 8 show that the surfaces of chitosan become rougher upon amination that attributed to modification process. The increase in surface roughness is usually accompanied with increase in the surface area leading to enhanced adhesion with the microorganism's cell walls.

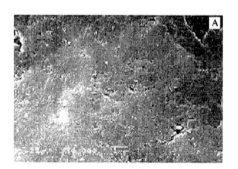
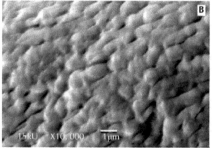

FIGURE 8   SEM graph of Chitosan A and aminated Chitosan B [25].

26.5   EVALUATION OF THE ANTIFUNGAL ACTIVITY OF AMINATED CHITOSAN

The antifungal activity of chitosan and aminated chitosan was tested against three different fungi species *Aspergillus Niger, Alternaria Alternata, and Fusarium Moniliforme.*

The fusarium species are frequently reported as the causative agent in opportunistic infections in human [39]. *A. niger* is the most common causative agent encountered in food contamination cases. Although it is not a common human pathogen, in high concentration, it may cause Aspergillosis [40]. In the other hand Alternaria produces a number of toxins as pathogenicity factors, among them alternariol and alternariol monomethylether are major ones, since these are produced by most *Alternaria* species in large quantities [41]. The toxins of *Alternaria* have been detected as natural contaminants of plants like tomato fruit and tomato products, apples [42], and olives [43].

In this chapter, The mycelial disks (7 mm in diameter) from two-week-old cultures of the fungi were placed in the centre of Petri dishes (90 mm in diameter) with 10 ml solid potato dextrose agar (PDA) or prostate-specific antigen (PSA) medium layered with 800µL of chitosan or aminated chitosan solution (2%), then incubated at 25°C. The mycelia growth was determined by measuring colony diameter daily and antifungal index was calculated as following equation:

$$\text{Antifungal index (\%)} = (1 - D_a/D_b) \times 100$$

Where D_a is the diameter of the growth zone in the experimental dish (cm) and D_b is the diameter of the growth zone in the control dish (cm).

Figures (9–11), show daily growth of different fungi species. Aminated chitosan show always lower growth than that of chitosan in all selected fungi species. This could be explained by the fact that the negatively charged plasma membrane is the main target site of polycation [14]. Therefore, the polycationic aminated chitosan will interact more effectively with the fungus compared with free form of chitosan itself and disrupt the membrane integrity [44].

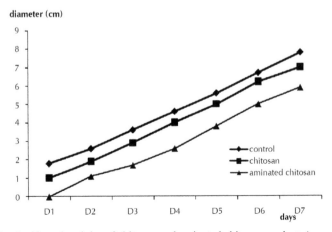

FIGURE 9   Antifungal activity of chitosan and aminated chitosan against *Aspergillus Niger.*

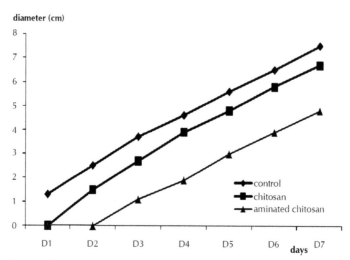

FIGURE 10   Antifungal activity of chitosan and aminated chitosan against *Alternaria Alternat*a.

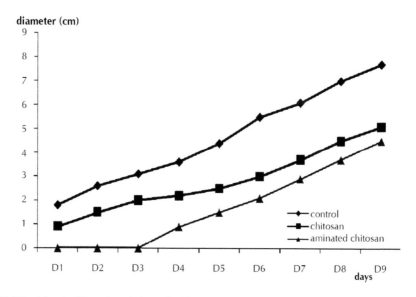

FIGURE 11   Antifungal activity of chitosan and aminated chitosan against *Fusarium Moniliforme.*

Figure 12 show antifungal index of chitosan and aminated chitosan against different fungi species. It is clear that promotion of antifungal activity by modifications, study show also increase the activity of polymer solutions against *Fusarium M* species rather than *Alternaria A* and *Aspergillus N* that may be attributed to its internal structure of cell wall membrane.

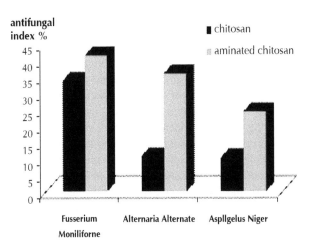

FIGURE 12   Antifungal index of chitosan and aminated chitosan against *Aspergillus Niger, Alternaria Alternata,* and *Fusarium Moniliforme.*

26.6   DISCUSSION

The mycotoxins can be characterized as secondary metabolites of various toxigenic fungi. Mycotoxins occur in a wide variety of foods and feeds and have been implicated in a range of human and animal diseases [45]. Exposure to mycotoxins can produce both acute and chronic toxicities ranging from death to deleterious effects upon, for example, the central nervous, cardiovascular, and pulmonary systems. Their general teratogenicity, cancerogenicity and their toxicological properties constitute a high human and animal health risk. The mycotoxins also attract attention because of the significant economic losses associated with their impact on human health and animal productivity.

Chitosan's fungal inhibition was observed on different development stages such as mycelial growth, sporulation, spore viability and germination, and the production of fungal virulence factors.

It has been commonly recognized that antifungal activity of chitosan depends on its MW, DD, pH of chitosan solution and, of course, the target organism. Mechanisms proposed for the antifungal activity of chitosan focused mainly on its effect on fungal cell wall [46] and cell membrane [47].

The antifungal activity of chitosan has been reported and developed in several studies both *in vitro* and *in vivo*, although chitosan activity against fungus has been shown to be less efficient as compared with its activity against bacteria [48]. The inhibitory efficiency of chitosan has been related to chitosan properties such as its MW, DD, and pH of chitosan solution and, of course, the target organism. In others works, researchers reported that the level of inhibition of fungi was also highly correlated with chitosan concentration, indicating that chitosan performance is related to the application of an appropriate rate. On the other hand, results from Bautista-Banos et al. [48] and Guo-Jane et al [49], showed important differences among them. Nevertheless, all these studies indicated that the polycationic nature of chitosan is the key to its antifungal properties and that the length of the polymer chain enhances that activity. An additional explanation includes the possible effect that chitosan might have on the synthesis of certain fungal enzymes [12]. Recent studies have shown that not only chitosan is effective in stopping the growth of the pathogen, but it also induces marked morphological changes, structural alterations, and molecular disorganization of the fungal cells [48]. The positive charge of the chitosan is due to the protonization of its functional amino group. This group reacts with the negatively charged cell walls of macromolecules, causing a dramatic increase in the level of the permeability of cell membrane, causing disruptions that lead to cell death [40].

In the presented results, it can show increase of anrifungal activity of *Fusarium Moniliforme* and *Alternaria Alternata* more than that in *Aspergillus Niger*. This variation of antifungal activity may be attribute to nature and consists of fungal cell wall. *A. niger* was found to be highly resistant to both chitosan and aminated chitosan. Fungi that have chitosan as one of the components in the cell wall are more resistant to externally amended chitosan. This fact could therefore explain the high resistance of *A. niger* as it contains 10% of chitin in its cell wall [50].

KEWWORDS

- **Antifungal**
- **Aminated chitosan**
- **Aspergillus niger**
- **Alternaria alternata**
- **Fusarium moniliforme**

REFERENCES

1. Mrunal, R. Thatte. Ph.D thesis. *Synthesis and antibacterial assessment of water-soluble hydrophobic chitosan derivatives bearing quaternary ammonium functionality*. Louisiana State University (2004).
2. Salmon. S. and Hudson, S. M. *J. Macomol. Sci. M. C. C*, **37**, 199–276 (1997).
3. Brine, C. J. In chitin, chitosan, and Related Enzymes. J. P.Zikakis (Ed.) Academic ress Inc Chpt., 17–23 (1984).
4. Chen, C. S., Liau, W. Y., and Tsai, G. J. *J. Food Prot.*, **61**, 1124–1128 (1998).
5. Hadwiger, L. A., Kendra, D. G., Fristensky, B. W., and Wagoner, W. Chitosan both activated genes in plants and inhibits RNA synthesis in fungi, in: "*Chitin in nature and technology*". R. A. A. Muzzarelli, C. Jeuniaux, and G. W. Gooday (Eds.), Plenum, New York (1981).
6. Papineau, A. M., Hoover, D. G., Knorr, D., and Farkas, D. F. *Food Biotechnol.*, **5**, 45–57 (1991).
7. Shahidi, F., Arachchi, J., and Jeon, Y. J. *Trends Food Sci. Technol.*, **10**, 37–51 (1999).
8. Sudarshan, N. R., Hoover, D. G., and Knorr, D. *Food Biotechnol.*, **6**, 257–272 (1992).
9. Young, D. H., Kohle, H., and Kauss, H. *Plant Physiol.*, **70**, 1449–1454 (1982).

10. Guo, Z., Xing, R., Liu, S., Zhong, Z., Ji, X., Wang, L., and Li, P. The influence of molecular weight of quaternized chitosan on antifungal activity. *Carbohydr. Polym.*, **71**(4), 694–697 (2008).

11. Palma-Guerrero, J., Jansson, H., Salinas, J., and Lopez-Llorca, J. V. Effect of chitosan on hyphal growth and spore germination of plant pathogenic and biocontrol fungi. *J. Appl. Microbiol.*, **104**(2), 541– 553 (2008).

12. El, G., Arul, J., Asselin, A., and Benhamou., N. Antifungal activity of chitosan on postharvest pathogens: induction of morphological and cytological alterations in Rhizopus stolonifer. *Mycology Research*, **96**, 769–779 (1992).

13. Guerra-Sánchez, M. G., Vega-Pérez, J., Velázquez-del Valle, M. G., and Hernández-Lauzardo, A. N. Antifungal activity and release of compounds on Rhizopus stolonifer (Ehrenb.:Fr.) Vuill. by effect of chitosan with different molecular weights. *Pestic. Biochem. Phys.*, **93**(1), 18–22 (2009).

14. Singh, T., Vesentini, D., Singh, A. P., and Daniel, G. Effect of chitosan on physiological, morphological, and ultrastructural characteristics of wood-degrading fungi. *Int. Biodet. Biodeg.*, **62**(2), 116–124 (2008).

15. García-Rincón, J., Vega-Pérez, J., Guerra-Sánchez, M. G., Hernández- Lauzardo, A. N., Peña-Díaz, A., and Velázquez-Del Valle, M. G. Effect of chitosan on growth and plasma membrane properties of Rhizopus stolonifer (Ehrenb.: Fr.) Vuill. *Pestic. Biochem. Phys.*, **97**(3), 275–278 (2010).

16. Hernández-Lauzardo, A. N., Guerra-Sánchez, M. G., Hernández-Rodríguez, A., Heydrich-Pérez, M., Vega-Pérez, J., and Velázquez-del Valle, M. G. Assessment of the effect of chitosan of different molecular weights in controlling Rhizopus rots in tomato fruits. *Arch Phytopathology Plant Protect*, **45**(1), 33–41 (2010a) *(2012)*.

17. Liu, J., Tian, S., Meng, X., and Xu, Y. Effects of chitosan on control of postharvest diseases and physiological responses of tomato fruit. *Postharvest Biol. Technol.*, **44**(3), 300–306 (2007).

18. Palma-Guerrero, J., Lopez-Jimenez, J. A., Pérez-Berná, A. J., Huang, I. C., Jansson, H. B., Salinas, J., Villalaín, J., Read, N. D., and Lopez-Llorca, L. V. Membrane fluidity determines sensitivity of filamentous fungi to chitosan. *Mol. Microbiol.*, **75**(4), 1021–1032 (2010).

19. Hernández-Lauzardo, M. G., Velázquez-del Valle, M. G., Guerra-Sánch. Current status of action mode and effect of chitosan against phytopathogens fungi. Afr. *J. Microbiol. Res.*, **5**(25), 4243–4247 (2011).

20. Eweis, M., Elkholy, S. S., and Elsabee, M. Z. *Int. J. Biological Macrom.*, 38, 1–8 (2006).

21. Xu, J., Zhao, X., Han, X., and Du, Y. Antifungal activity of oligochitosan against Phytophtora capsici and other pathogenic fungi in vitro. *Pestic. Biochem. Phys.*, **87**(3), 220–228 (2007a).

22. Xu, J., Zhao, X., Wang, X., Zhao, Z., and Du, Y. Oligochitosan inhibits Phytophtora capsici by penetrating the cell membrane and putative binding to intracellular target. *Pestic. Biochem. Phys.*, **88**(2), 167–175(2007b).

23. T. Tanigawa, Y. Tanaka, H. Sashiwa, H. Saimoto, Y., J. Shigeasa, P. A. Sandford, and J. P. Zikakis (Eds.)Elsevier science Publishers Ltd., London, New York, pp. 206–15 (1992).

24. Hirano, S. and Nagao, N. *Agric. Biol, Chem.* (53) pp. 3065–66 (1989).

25. Mohy Eldin, M. S., Soliman, E. A., Hashem, A. I., and Tamer, T. M. Antimicrobial Activity of Novel Aminated Chitosan Derivatives for Biomedical Applications. *Advances in Polymer Technology*, **31**(4), 414–428 (2012).

26. Rigby, G. patent. Substantially Undegraded Deacetylated Chitin and Processes for Producing the Same.USA 2,040,879 (May 19, 1936).

27. Guibal, E. Interactions of metal ions with chitosan-based sorbents: A review. *Separation and Purification Technology*, **38**, 43–74 (2004).

28. Klug, M., Sanches, M. N. M., Laranjeira, M. C. M., Favere, V. T., and Rodrigues, C. A. (1998).

29. Kubota, N., Tastumoto, N., Sano, T., and Toya, K. A simple preparation of half Nacetylated chitosan highly soluble in water and aqueous organic solvents. *Carbohydrate Research*, **324**, 268–274 (2000).

30. Kurita, K. Chitin and chitosan: Functional biopolymers from marine crustaceans. *Marine Biotechnology*, **8**, 203–226 (2006).

31. Anthonsen, M. W. and Smidsroed, O. Hydrogen ion titration of chitosans with varying degrees of N-acetylation by monitoring induced 1H NMR chemical shifts. *Carbohydrate Polymers*, **26**, 303–305 (1995).

32. Rinaudo, M. Chitin and chitosan: Properties and applications. *Progress in Polymer Science*, **31**, 603–632 (2006).

33. Sankararamakrishnan, N. and Sanghi, R. Preparation and characterization of a novel xanthated chitosan. *Carbohydrate Polymers*, **66**, 160–167 (2006).

34. Pawlak, A. and Mucha, M. *Thermochim. Acta*, **396** 153 (2003).

35. Kacurakova, M., Belton, P. S., Hirsch, J., and Ebringerova, A. *Physiology and Molecular plant Pathology*, **35**, 215–230 (1998).

36. Phillips, G. O., Takigami, S., and Takigami, M. *Food hydrocolloids*, **10**, 11–19 (1996).

37. Guinesi, L. S. and Cavalheiro, E. T. G. The use of DSC curves to determine the acetylation degree of chitin/chitosan samples. *Thermochimica Acta*, **444**(2006) 128–133 (1996).

38. Kittur, F. S., Prashanth, H., Sankar, K. U, and Tharanathan, R. N. characterization of chitin, chitosan, and their carboxymethyl derivatives by different scanning calorimetry. *Carbohydr. Polym.*, **49**(2002) 185 (1996).

39. Godoy, P., Nunes, F., Silva, V., Tomimori-Yamashita, J., Zaror, L., and Fischman, O. Onychomycosis caused by *Fusarium solani* and *Fusarium oxysporum* in S˜ao Paulo, Brazil. *Mycopathologia*, **157**(3), 287–290 (2004).

40. Sebti, I., Martial-Gros, A., Carnet-Pantiez, A., Grelier, S., and Coma, V. Chitosan polymer as bioactive coating and film against Aspergillus niger contamination. *Journal of Food Science*, **70**(2), 100–104 (2005).

41. Heisler, E. G., Siciliano, J., Stinson, E. E., and Osman, S. F. High performance liquid chromatographic determination of major mycotoxins produced by *Alternaria* molds. *J. Chromatogr.*, **194**, 89–94 (1980).

42. Stinson, E. E., Osman, S. F., Heisler, E. G., Siciliano, J., and Bills, D. D. Mycotoxin production in whole tomatoes, apples, oranges and lemons. *J. Agr. Food Chem.*, **29**, 790–792 (1981).

43. Visconti, A., Logrieco, A., and Bottalico, A. Natural occurrence of *Alternaria* mycotoxins in olives—Their production and possible transfer into the oil. *Food Additives and Contaminants*, **3**, 323–330 (1986).

44. Qi, L., Xu, Z., Jiang, X., Hu, C., and Zou, X. Preparation and antibacterial activity of chitosan nanoparticles. *Carbohydrate Research*, **339**(16) 2693–2700, (2004).

45. Coker, R. D. *Mycotoxins and their control: Constraints and opportunities*. NRI Bulletin 73. Natural Resources Institute, Central Avenue, Chatham Maritime, Chatham, Kent, ME4 4TB (1997).

46. Allan, C. R. and Hadwiger, L. A. The fungicidal effect of chitosan on fungi of varying cell wall composition. *Exp. Myco.*, **3**, 285–287 (1979).

47. Zakrzewska, A., Boorsma, A., Brul, S., Hellingwerf, K. J., and Klis, F. M. Transcriptional response of Saccharomyces cerevisiae to the plasma membrane-perturbing compound chitosan. *Eukaryot. Cell.*, **4**, 703–715 (2005).

48. Tsai, G. J., Wu, Z. Y., and Su, W. H. Antibacterial activity of a chitooligosaccharide mixture prepared by cellulase digestion of shrimp chitosan and its application to milk preservation. *Journal Food Protection*, **63**, 747–752 (2000).

49. Guo, J., Tsai, M. T., Lee, J. M., and Zhong, M. Z. Effects of chitosan and a low- Molecular-Weight chitosan on Bacillus cereus and application in the preservation of cooked rice. *Journal of Food Protection*, **69**(9), 2168–2175 (2006).

50. Klis, F. M., Ram, A. F. J., and De Groot, P. W. J. *A molecular and genomic view of the fungal cell wall, in Biology of the Fungal Cell.* R. J. Howard and N. A. R. Gow (Eds.), Springer, Berlin, Germany, pp. 97–112 (2007).

CHAPTER 27

BIODEGRADABLE BINARY AND TERNARY BLENDS

S. Z. ROGOVINA, K. V. ALEKSANYAN, S. M. LOMAKIN,
and E. V. PRUT

CONTENTS

27.1 INTRODUCTION

Blends of cellulose and ethyl cellulose with low density polyethylene (LDPE) are obtained under action of high-temperature shear deformations at different initial ratios of the components. It is shown that the composition of the powder fractions is identical to the original blend composition that testifies the homogeneity of the blends obtained. The films from these polymer blends demonstrate high mechanical characteristics. The X-ray diffraction analysis shows that mixing of cellulose and ethyl cellulose with LDPE under these conditions results in changes of the polymer structure and leads to a decrease of their crystallinity. The peculiarities of thermal degradation of the obtained blends are discussed in detail. The observed results are interpreted by the model kinetic analysis based on thermogravimetric analysis (TGA). The biodegradability of binary blends is determined and it is established that the addition of poly(ethylene oxide) (PEO) as well as polysaccharides chitin or chitosan as a third component promotes an increase in the biodegradability of compositions that, consequently, results in the expansion of the possible areas of application of the related materials.

The creation of partially or completely biodegradable polymer materials by addition of a biodegradable polymer to synthetic polymer allows one to solve the problem of polymer waste utilization. The presence of biodegradable components facilitates the destruction of polymer materials under the influence of the microorganisms and environmental action. At first stage the macromolecules of the natural origin are subjected to degradation with formation of microcracks on the material surface. Due to increase of degraded material surface the oxidation processes leading to polymer fragmentation proceed more intensively and finally cause the material decomposition.

Nowadays, a lot of methods of creation of the biodegradable blends, which are common for production of polymer compositions, are known [1]. One of the most efficient and economically profitable among these methods is their production *via* mixing of synthetic and natural polymers. The production of the polymer compositions allowing one to use properties of each component are successfully applied in numerous areas especially for production of packaging materials, films for food products, and articles for short-term usage.

Among various polymers of natural origin for production of biodegradable composite materials the polysaccharides (cellulose, starch, chitin, chitosan, and so on.) which are almost inexhaustible raw materials are widely used [2].

Although, cellulose is the most widely spread natural polysaccharide, there is not so many works in literature dedicated to investigation of its blends with various synthetic polymers. Since the cellulose has three hydroxyl groups, so it easily interacts with synthetic polymers, which can form hydrogen bonds. Although, the blends obtained from cellulose with other polymers must be compatible, there are some other factors which can hinder the cellulose–polymer interaction. For example, nylon-6 is incompatible with cellulose, and poly(ε-caprolactam) is partially compatible [3]. Cellulose–polyamide blends are heterogeneous two-phase systems [4], whereas a blend of cellulose with poly(4-vinylpyrrolidone) is homogeneous at any ratio [5]. The investigation of the films obtained from blends of cellulose with poly(ethylene terephthalate) showed that a specific polymer–polymer interaction takes place [5]. A number of works is dedicated to the study of cellulose–poly(vinyl alcohol) (PVA) blends. The optimal compatibility of PVA with cellulose is observed with 60% PVA in the blend [5]. Such compositions can have various applications, for example, the membranes on their base are used to remove the metal ions from aqueous solutions [6].

The properties of cellulose–polyethylene compositions were studied in [7-9]. It was shown that low biodegradability of a cellulose blend with low LDPE can be improved *via* preliminary UV irradiation of blend [7]. It was noted that the film degradation under the environmental conditions is initiated on the surface and further degradation proceeds more rapidly and efficiently. Such assumptions are confirmed by the study of the surface of the cellulose–PE films *via* scanning electron microscopy (SEM).

The blends of LDPE with cellulose treated by cellulase were investigated in [8]. First, the treatment of cellulose by cellulase leads to a decrease of the polysaccharide molecular weight, however, with increase of the treatment time, the rise of the molecular weight and crystallinity of cellulose is observed, that is the crystallization of the amorphous regions occurs. With addition of maleic anhydride to PE, the interfacial adhesion is improved that results in increase of dispersion and wetting of cellulose in the matrix. Due to an increase of the interfacial adhesion, the mechanical characteristics become higher. Another method to improve the interfacial adhesion of compositions by pretreatment by ozone was described in [9]. It was shown that the treatment of cellulose–LDPE blend by ozone leads to formation of carbonyl and hydroperoxide groups capable to interaction with the functional groups of cellulose that leads to improvement of adhesion between the blend components.

The properties of LDPE reinforced by cellulose fibers were investigated in [10-13]. It was established that the improvement of adhesion between the blend components can be achieved as a result of fibril formation by the cellulose fibers under the action of mechanical shaking [10]. In order to obtain a uniform distribution of the cellulose nanofibers in LDPE matrix the polymers were mixed in an extruder, in which water was supplied under high pressure and elevated temperature through the special window [12]. It was found that a higher dispersion degree of the forming composition can be obtained by increasing shear value and residence time of material in extruder. The values of the mechanical parameters rise with the cellulose content.

Cellulose ethers in contrast to cellulose are thermoplastic polymers and can be processed by common for polymers methods. As cellulose its derivatives are also biodegradable, therefore, they can be used as biodegradable components in the polymer compositions. Ethyl cellulose is one of the most well studied and widespread cellulose ether successfully used in different fields. Since ethyl cellulose possesses well film-forming properties it is used in production of strong coatings. In recent years ethyl cellulose is often used as coatings of drug pellets with controlled release [13-15]. In [15] it was shown that also copolymer of ethyl cellulose with graft poly(poly(ethylene glycol)methyl ether methacrylate) has a great potential for application as the controlled drug delivery system.

At the same time there is not so many works dedicated to study of thermal and mechanical properties of the blends of ethyl cellulose with other polymers [16-19]. So, the biodegradable blends of copolymer of ethylenevinyl acetate with ethyl cellulose were obtained in [17]. To improve the component compatibility the copolymer of ethylenevinyl alcohol promoting the rise in mechanical characteristics of composition was added. The TGA method showed that the temperature of the decomposition beginning decreases with introduction of ethyl cellulose.

The development of new production methods of blends based on cellulose and ethyl cellulose with various synthetic and natural polymers is necessary for creation of new biodegradable composite materials possessing wide range of properties depending on the blend composition and polymer nature. Detailed investigation of properties and determination of mechanical, thermal, and structural characteristics of the compositions obtained will allow one to find the possible application fields where their properties can be maximally used.

27.2   PREPARATION AND FRACTIONATION OF BLENDS

Among different mixing methods of polymers the solid-phase blending under conditions of high-temperature shear deformations developed at the Semenov Institute of Chemical Physics, Russian Academy of Sciences is of special interest, since allows one to obtain powder homogeneous blends even in the case when one of the components of the system grinded is an infusible polymer (polysaccharide).

To realize this method the mixing of polymers was performed in a rotor disperser where they undergo the joint action of pressure and shear deformations. The method is based on the physical principle that the energy accumulated in the sample after the application of pressure is consumed on forming of new surface under the influence of shear deformations with obtaining of highly dispersed powders [20].

Under these conditions the binary blends of cellulose and ethyl cellulose with LDPE as well as their ternary blends with LDPE and PEO of different molecular weight (low-molecular (PEO$_{Lm}$, $M = 3.5 \times 10^4$) and high molecular (PEO$_{Hm}$, $M = 5 \div 6 \times 10^6$)) and ternary blends of cellulose with chitin or chitosan and LDPE were obtained. The PEO is a synthetic polyether which was added to the system as the third component in order to increase the biodegradability of the compositions based on LDPE, cellulose, and ethyl cellulose. Thanks to low toxicity PEO is widely used in the medical items, and, consequently, expands the possible areas of application of the related compositions. Low-molecular PEO is an effective plasticizer and being added to composition provides the improvement of its service properties. The presence of two polysaccharides in system obviously also must increase its biodegradability. Among numerous natural polymers the special attention is given to chitin as a second abundant after cellulose polysaccharide in nature and its deacetylated derivative chitosan. In recent years chitin and chitosan are widely used in various fields as they have some advantages in comparison with other polysaccharides, especially chitosan possessing along with excellent biodegradability the high antibacterial activity and biocompatibility. Although, the cost of these polysaccharides is slightly higher than for cellulose, nevertheless thanks to their special properties they can be used in medicine, chemical industry, and so on. Since chitin and chitosan are rigid biodegradable polysaccharides their introduction leads to increase of the rigidity and biodegradation capacity of the final compositions.

Polysaccharide content both in the binary and ternary blends with two synthetic polymers was varied from 20 to 40 wt%, in the case of binary blend of ethyl cellulose its content was up to 50 wt%. The content of PEO in all blends was 20 wt%. There was only one composition for the ternary blends based on two polysaccharides: cellulose–LDPE–chitin (chitosan) 30:40:30 wt%.

While passing through a rotor disperser, the polymers are subjected to the joint action of pressure and shear deformations leading to changes in the polymer structure [21]. The resulting binary and ternary powders were fractionated. The histograms of cellulose–LDPE (30:70 wt%), cellulose–LDPE–PEO$_{Lm}$ (30:50:20 wt%), and cellulose–LDPE–chitin (30:40:30 wt%) blends obtained *via* fractionation of a finely dispersed powder formed at the output of the rotor disperser are presented in Figure 1 (a). As can be seen, there are two fractions of particles 0.071–0.09 and 0.09–0.315 mm in size are the most significant for cellulose–LDPE blend (Figure 1 (a)).

As compared to the binary blends, the addition of PEO leads to an appearance of the fractions with the particle size more than 1 mm and change of the ratio of the basic fractions. Thus, for a cellulose–LDPE–PEO$_{Lm}$ blend, the basic fraction (almost 50%) contains particles 0.09–0.315 mm in size, and the content of fraction with particles 0.315–0.63 mm in size is about 30% (Figure 1 (a)).

The results obtained for compositions containing two polysaccharides differ from systems based on two synthetic polymers (Figure 1 (a)). In this case the number of fine dispersed fractions is increased and for composition cellulose–LDPE–chitin a new fraction with size 0.071–0.09 mm is occurred again as and in the case of binary system. The same effect is observed for cellulose–LDPE–chitosan composition. These data can be explained by effect of cogrinding of polymers which was demonstrated in [20].

Figure 1 (b) shows the typical histogram of a powder obtained *via* blending of ethyl cellulose with LDPE at an initial component ratio of 30:70 wt% and their ternary blend with PEO. It is evident that the basic fraction (about 60%) is composed of particles 0.09–0.315 mm in size.

The fractionation results of compositions of LDPE with cellulose and ethyl cellulose at other component ratios are given in Table 1. It follows from table that, for LDPE blends with cellulose and ethyl cellulose, the basic fraction consists of particles 0.09–0.315 mm in size. The content of this fraction is, on average, 50%. (for the cellulose–LDPE blend at a ratio of 20:80 wt%, the percentage of this fraction is somewhat lower than that of the fraction of particles 0.315–0.63 mm in size).

The fractional composition of the ethyl cellulose–LDPE blend was determined through analysis of each fraction after ethyl cellulose was washed with an ethanol–toluene mixture (1:4 vol%), the residual LDPE was dried, and the fractional composition was determined from the weight difference. As can be seen from Table 2, the fractional composition is nearly consistent with the initial component ratio in the blend.

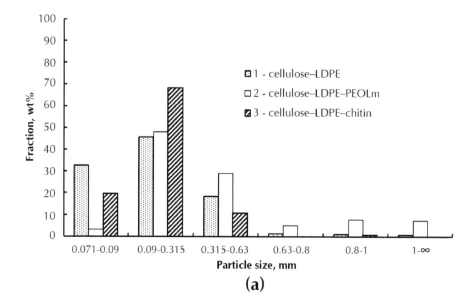

(a)

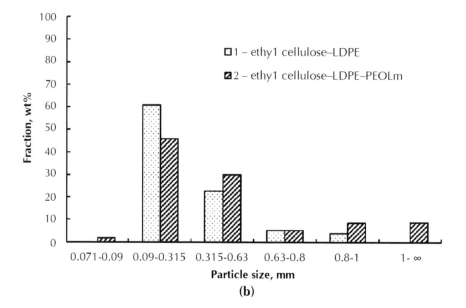

(b)

FIGURE 1 Histograms of the particle size distribution of blends—(a) cellulose–LDPE (30:70 wt%) (1), cellulose–LDPE–PEO$_{Lm}$ (30:50:20 wt%) (2), and cellulose–LDPE–chitin (30:40:30 wt%) (3) and (b) ethyl cellulose–LDPE (30:70 wt%) (1), ethyl cellulose–LDPE–PEO$_{Lm}$ and (30:50:20 wt%) (2).

TABLE 1 Fractional composition of blends of cellulose and ethyl cellulose with LDPE at different component ratios

Composition	Compo-nent ratio, %	Fraction with d (mm), %				
		0.071–0.09	0.09–0.315	0.315–0.63	0.63–0.8	0.8–1.0
Cellulose–LDPE-	20:80	1.6	43.0	49.2	2.1	3.5
	30:70	32.7	45.7	18.4	1.3	1.1
Ethyl cellu-lose–LDPE-	20:80	11.3	41.7	24.6	9.8	9.2
	50:50	3.2	53.2	21.6	6.4	10.0

TABLE 2 The LDPE content in fractions of ethyl cellulose–LDPE blends of different compositions determined *via* the chemical method

Composition, %	LDPE content in various fractions, %		
	0.071–0.09	0.09–0.315	0.315–0.63
20:80	77	84	83
30:70	71	72	70
50:50	51	54	56

27.3 FRACTIONAL COMPSITION OF BLENDS DETERMINED BY DIFERENTIAL SCANNING CALORIMETRIC (DSC) METHOD

The content of cellulose in different fractions of the obtained blends was determined by DSC method. As follows from the data presented in Figure 2 (a), curve 1 the LDPE melting curve is characterized by a single peak. At the same time there is no endo- or exo- peaks on the curve obtained *via* DSC for blend based on cellulose, that is on DSC curves of cellulose blends with LDPE the single peak is observed, which is characteristic for LDPE, and its T_m coincides with LDPE T_m (Figure 2 (a), curve 2). Since the change in the melting heat of LDPE unlike amorphization which is constant for each blend depends also on the content of the second component, so in this way the content of cellulose in composition can be calculated. The results of fractional composition of the blends obtained at different ratios of components are presented in Table 3. As it follows from these data the composition of fractions approximately corresponds to the initial ratio of the components in blend.

TABLE 3   Fractional composition of cellulose–LDPE blends determined by the DSC method

Polysaccha-ride–LDPE (wt%)	Fraction size (mm)	ΔH_{exp} (mJ)	Determined content of components (%)	
			LDPE	**Polysaccharide**
20:80	(0.09–0.315]	142	79.8	20.2
	(0.315–0.63]	148	83.2	16.8
30:70	(0.071–0.09]	128	72.05	27.95
	(0.09–0.315]	125	70.35	29.65

Under determination of the real ratio of components in cellulose–LDPE–PEO ternary blends two peaks on the melting curve corresponding to melting points of PEO and LDPE were detected by the DSC method (Figure 2 (b)). Changes in melting enthalpies of PEO also as of LDPE depend on amorphization and the content of components in blend. Thus, the polymer contents in the blends can be calculated by the difference in the melting enthalpies of pure PEO and LDPE and the enthalpies of the blends (Table 4). It is visible that the blend composition also approximately corresponds to the initial ratio of the components.

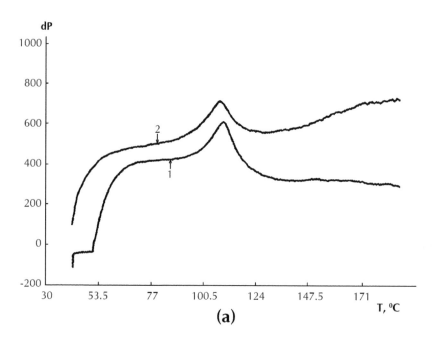

(a)

FIGURE 2   *(Continued)*

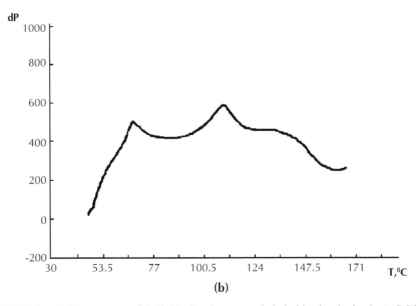

FIGURE 2 Melting curves of individual polymers and their blends obtained *via* DSC— (a) LDPE (1) and cellulose–LDPE blend (30:70 wt%) (2) and (b) cellulose–LDPE–PEO$_{Lm}$ (30:50:20 wt%).

TABLE 4 Polysaccharide–LDPE–PEO$_{Lm}$ blend composition determined by the DSC method

Composition	Polysaccha-ride–LDPE–PEO$_{Lm}$ (wt%)	ΔH (PEO$_{Lm}$) (J/g)	ΔH (LDPE) (J/g)	Determined content of com-ponents (%)		
				PEO$_{Lm}$	LDPE	Polysac charide
Cellulose–LDPE–PEO$_{Lm}$	20:60:20	18.42	34.3	14.01	61.3	24.69
	30:50:20	22.74	27.22	17.31	48.6	34.09
	40:40:20	19.44	22.89	14.8	40.9	55.7
Ethyl cellu-lose–LDPE–PEO$_{Lm}$	20:60:20	17.47	34.92	13.3	62.4	24.3
	30:50:20	19.79	29.01	15.1	51.8	33.1
	40:40:20	18.78	23.3	14.3	41.6	44.1

27.4 INVESTIGATION OF MECHANICAL PROPERTIES OF FILMS FROM BINARY AND TERNARY BLENDS

The mechanical characteristics of the compositions were investigated on films obtained by pressing of blends investigated at 160°C and 10 MPa on a tensile test machine. The results of mechanical tests are given in Table 5. Since elastic modulus E of polysaccharides is higher than that of LDPE, the elastic modulus of the films rises with the increase of polysaccharide content. As E is determined by the structural features of polysaccharide and grows with its rigidity, the highest E values were obtained for the cellulose–LDPE blends.

It is evident that the addition of polysaccharides to LDPE results in a decrease in the ultimate tensile strength, σ_b. For example, in the films containing 20% polysaccharides, σ_b decreases by a factor of 1.2 for cellulose–LDPE, 1.7 for ethyl cellulose–LDPE blend. However, σ_b is practically unaffected by a change in the blend composition.

The blending of LDPE with the polysaccharides leads to significant decrease of the elongation at break, ε_b. So, the addition of 20 wt% of cellulose results in the most pronounced decrease in ε_b (by a factor of 28) compared to that of the pure LDPE. The addition of 20 wt% of ethyl cellulose is accompanied by a 3.5-fold decrease in ε_b. With a further increase of the polysaccharide content in blends, ε_b decreases gradually, especially for ethyl cellulose–LDPE blends (for the component ratio of 50:50, $\varepsilon_b = 2.3\%$). Thus, the decrease of the mechanical parameters is connected with weak adhesion between polymers in the composition that is testified by the results obtained *via* SEM (see Figure 5).

TABLE 5   Mechanical characteristics of films obtained from polysaccharide–LDPE blends

Composition	Polysaccha-ride–LDPE (wt%)	E, MPa	σ_b, MPa	ε_b, %
LDPE	–	200 ± 5	13.3 ± 0.2	460 ± 10
	20:80	390 ± 10	9.2 ± 0.1	16.5 ± 0.5
Cellulose–LDPE-	30:70	660 ± 25	11.0 ± 0.2	5.5 ± 0.2
	40:60	720 ± 15	10.3 ± 0.1	4.6 ± 0.1
	20:80	240 ± 10	7.9 ± 0.2	130 ± 10
Ethyl cellulose–LDPE-	30:70	350 ± 10	6.1 ± 0.2	7.2 ± 0.5
	50:50	510 ± 10	6.4 ± 0.4	2.3 ± 0.2

Figure 3 (a, b, and c) shows the dependences of the mechanical parameters of polysaccharide-containing films obtained from binary and ternary blends on its content on the example of cellulose-based compositions. As was expected, the introduction of the third component leads to a drop of the mechanical characteristics.

From analysis of the curves of the elastic modulus E dependence on cellulose content it is evident that in comparison with two-component system cellulose–LDPE the introduction of PEO leads to a decrease of elastic modulus E, especially with addition of low-molecular PEO that is caused by influence of the molecular weight of polymer on the mechanical properties of the composite material (Figure 3 (a)). The ultimate tensile strength σ_b slightly depends on the cellulose content, but decreases with PEO introduction especially for samples containing low-molecular PEO (Figure 3(b)).

The greatest value of elongation at break ε_b was obtained for films from the blends containing 20% of cellulose with low-molecular PEO. With increase of cellulose content in the blend the values of the elongation at break become almost equal for all three compositions (Figure 3 (c)).

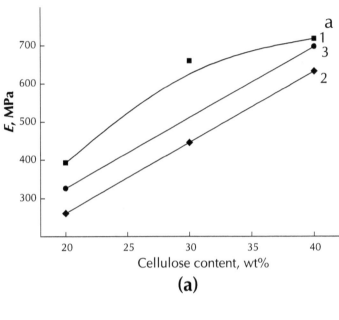

(a)

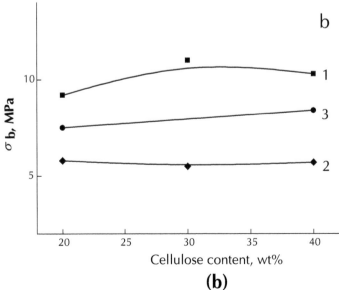

(b)

FIGURE 3 *(Continued)*

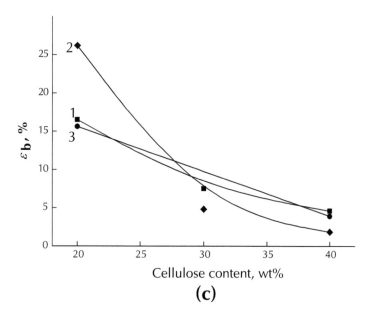

(c)

FIGURE 3 Dependence of elastic modulus E (a), ultimate tensile strength σ_b (b), and elongation at break ε_b (c) for cellulose–LDPE (1), cellulose–LDPE–PEO$_{Lm}$ (2), and cellulose–LDPE–PEO$_{Hm}$ (3) films on cellulose content.

The results of measurements of the mechanical characteristics of the films obtained from ternary blends based on two polysaccharides are presented in Table 6. In comparison with the data presented in Table 5 it is visible that the addition of rigid chitin and chitosan leads to increase of elastic modulus E and decrease of the elongation at break ε_b. As is seen from Table 6 the elastic modulus of cellulose–LDPE film containing 30 wt% of polysaccharide was 660, while E of cellulose–LDPE–chitin and cellulose–LDPE–chitosan at the same amount of cellulose comprised 1170 and 1190 MPa, respectively. Thus, the values of the mechanical characteristics of the blends depend on blend composition. The presented data clearly illustrate the influence of the molecular weight of PEO used and addition of rigid polysaccharides chitin or chitosan on the mechanical parameters of films.

TABLE 6 Mechanical characteristics of films obtained from cellulose–LDPE–chitin/chitosan blends

Composition	Cellulose–LDPE–chitin/chitosan (wt%)	E, MPa	σ_b, MPa	ε_b, %
Cellulose–LDPE–chitin	30:40:30	1170 ± 20	12.9 ± 0.3	1.9 ± 0.1
Cellulose–LDPE–chitosan	30:40:30	1190 ± 23	16.2 ± 0.2	2.9 ± 0.2

27.5 X-RAY ANALYSIS OF CELLULOSE AND ETHYL CELLULOSE BLENDS WITH LDPE

In order to establish the influence of shear deformations on the structure of studied polymers the initial polysaccharides, LDPE, and their blends obtained after passing through the rotor disperser were investigated by X-ray diffraction analysis. The X-ray diffraction patterns of cellulose, LDPE, and their blend prepared under high-temperature shear deformations at a component ratio of 30:70 wt% are shown in Figure 4. The X-ray pattern of the initial cellulose (Figure 4 (a), curve 1) displays two crystalline reflections with the angular positions $2\theta = 16.4°$ and $22.2°$. Using the Wolf–Bragg equation, $2d\sin\theta = \lambda$ ($\lambda = 1.54$ Å), the interplanar distances d were calculated: 5.4 and 4.0 Å, respectively. The quantity $d_1 = 5.4$ Å is approximately comparable to the average interchain distance, while $d_2 = 4.0$ Å is approximately comparable to the intrachain distance for cellulose. The decrease of the first halo of cellulose and characteristic peak of LDPE (Figure 4 (a), curve 2) after mixing under the action of high-temperature shear deformations (Figure 4 (a), curve 3) is related to the decrease of the crystallinity degree of polymers owing to their amorphization.

The X-ray pattern of ethyl cellulose (Figure 4 (b), curve 1) is characterized by two crystalline reflections at $2\theta = 7.8°$ (corresponds to an interchain distance of $d_1 = 11.3$ Å) and $2\theta = 20.3°$ (corresponds to an intrachain distance of $d_2 = 4.4$ Å). The ethyl cellulose–LDPE blend produced under high-temperature shear deformations is a partially ordered system. As can be seen from the X-ray pattern, the intensity of the first reflection decreases considerably during treatment in the rotor disperser. Thus, under shear deformations, the structure of ethyl cellulose changes significantly; as a result, the rearrangement of most chains takes place without retention of the interchain ordering typical for the original polymer.

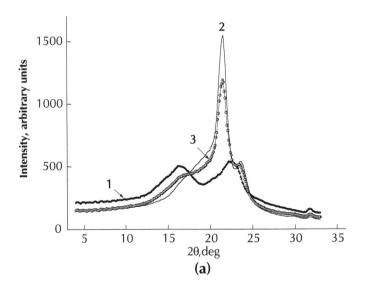

(a)

FIGURE 4   *(Continued)*

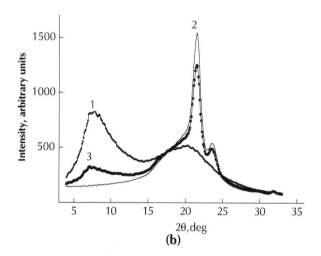

(b)

FIGURE 4 The X-ray diffraction patterns—(a) cellulose (1), LDPE (2), and their blend (3) (30: 70 wt%) and (b) ethyl cellulose (1), LDPE (2), and their blend (3) (30:70 wt%) prepared *via* mixing under high-temperature shear deformations.

27.6 INVESTIGATION OF BLENDS MORPHOLOGY BY SEM METHOD

Figure 5 shows the electron micrographs at various magnifications of the initial cellulose and cross-section of the film obtained from cellulose–LDPE blends (30:70 wt%) after passing through the rotor disperser. As is seen from Figure 5 (a) and (b), the initial cellulose represents stretched fibers with diameter d ≈ 20–80 μm and length 10–20 times higher, which can aggregate with formation of globular structures.

(a)

FIGURE 5 *(Continued)*

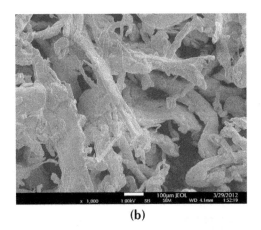

(b)

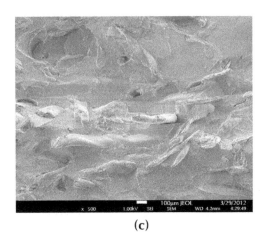

(c)

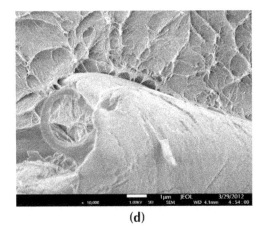

(d)

FIGURE 5 *(Continued)*

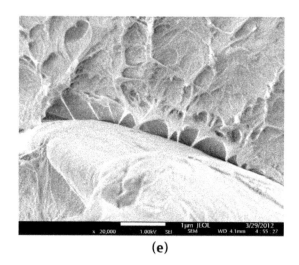

(e)

FIGURE 5   Electron micrographs of the initial cellulose (a, b) and cross-section of the film obtained by pressing of cellulose–LDPE blends (30:70 wt%) (c, d, e) at various magnifications— (a) 100, (b) 1000, (c) 500, (d) 10000, and (e) 20000

In electron micrographs of the film cross-section both phases of the composite are clearly seen that cellulose fibers are distributed in LDPE matrix (Figure 5 (c)). The cellulose fiber delaminations from LDPE matrix are observed in the composite structure that can suggest a weak adhesion interaction between polymers (Figure 5 (d)). At the same time, in Figure 5 (e) it is clearly seen that there are bands at the matrix–fiber interface connected with LDPE deformation and determined by the local high adhesion between the blend components.

Thus, the obtained results confirm the existence of a weak adhesion between the composite components investigated.

27.7   THERMAL STABILITY AND KINETIC ANALYSIS OF CELLULOSE–LDPE AND ETHYL CELLULOSE–LDPE BLENDS THERMAL DEGRADATION

27.7.1   FEATURES OF THERMAL DEGRADATION OF CELLULOSE, ETHYL CELLULOSE, LDPE, AND THEIR BLENDS

The found differences in the polymer structures connected with their amorphization allow one to assume that the character of the diffusion processes changes on the interphase boundary.

Thermal stability of LDPE and cellulose–LDPE and ethyl cellulose–LDPE blends for different application is necessary in determining of their temperature range of use and combustibility of compositions. Figures 6 and 7 present the TGA experimental data for cellulose, ethyl cellulose, LDPE, cellulose–LDPE (40:60 wt%), cellulose–LDPE (30:70 wt%), ethyl cellulose–LDPE (50:50 wt%), and ethyl cellulose–LDPE (20:80 wt%) blends obtained by TG method.

Onset temperatures (TG), the temperatures of the maximum rates of thermal degradation (T^1_{max} and T^2_{max}) and weight loss percentage are shown in Table 7.

TABLE 7   The TGA data for LDPE, cellulose, ethyl cellulose, and their blends

Sample	Onset T°C	T^1_{max} °C	T^2_{max} °C	Residue %
LDPE	408	-	460	-
Cellulose	309	360	-	11.9
Ethyl cellulose	316	363	-	3.3
Cellulose–LDPE (30:70 wt%)	312	343	464	2.5
Cellulose–LDPE (40:60 wt%)	311	360	476	4.5
Ethyl cellulose–LDPE (20:80 wt%)	335	365	480	0.1
Ethyl cellulose–LDPE (50:50 wt%)	318	350	458	0.3

The LDPE starts to degrade at 408°C while cellulose, ethyl cellulose, cellulose–LDPE (40:60 wt%), cellulose–LDPE (30:70 wt%), ethyl cellulose–LDPE (50:50 wt%), and ethyl cellulose–LDPE (20:80 wt%) start degrade at around 310°C due to independent degradation steps of cellulose and ethyl cellulose.

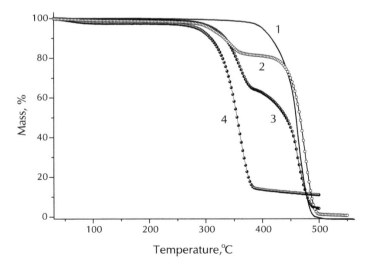

FIGURE 6   The TG curves of LDPE composites with cellulose—LDPE—1, cellulose–LDPE (30:70 wt%)—2, cellulose–LDPE (40:60 wt%)—3, and cellulose—4.

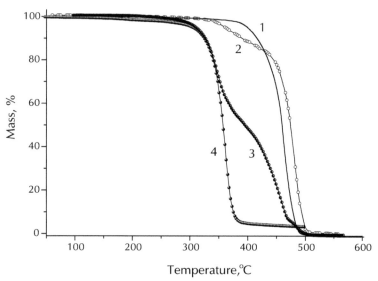

FIGURE 7   The TG curves of LDPE blends with ethyl cellulose—LDPE—1, ethyl cellulose–LDPE (20:80 wt%)—2, ethyl cellulose–LDPE (50:50 wt%)—3, and ethyl cellulose—4.

The radical mechanism of LDPE thermal degradation has been widely discussed as an example for random scission type reactions [22-30]. The products of decomposition include a wide range of alkanes, alkenes, and dienes. The kinetics of thermal degradation of polyethylene is frequently described by a first-order model of mass conversion of the sample [30]. A broad variation in Arrhenius parameters reported in literature, activation energy (E) ranging from 160 to 320 kJ/mol, pre-exponential factor (A) between 10^{11} and 10^{21} sec^{-1} [28-30]. The reasons for the broad range of E value are differences in molecular weight and various additives, and differences in experimental conditions [21-29, 31, 32]. Earlier Bockhorn et al. proposed a formal reaction mechanism of LDPE thermal degradation [28]. The following reaction mechanism considers only the main reactions in order to evaluate a simple kinetic model explaining the determined global kinetic data. The mechanism of LDPE thermal degradation is a radical chain mechanism (Scheme 1), initiated by random scission of the polymer chain into primary radicals R_p (1). β-scission of these radicals leads to ethylene (2). However at higher temperatures, the unzip reaction to ethylene is more evident [29]. At lower temperatures, intermolecular hydrogen transfer followed by β-scission occurs (3). This reaction leads to more stable secondary radicals R_s. The 1, 5 rearrangement reaction (3) in Scheme 1 stands for all preferred rearrangement reactions *via* cyclic intermediates such as 1, 9, 1, 13, 1, 17, and so on. Subsequent β-scission of the secondary radicals contributes to the radical chain mechanism because the primary radical is produced in each step (propagation). Two β-scission reactions (4, 4') are possible. Reaction (4) leads to alkenes, whereas reaction (4') leads to a short primary radical and a polymer with a terminated double bond. Important for the change in the reaction order is the intermolecular hydrogen transfer in reaction (5), which leads to the alkanes. In this case only the intermolecular hydrogen transfer of the primary radicals is considered because they are less stable than the secondary radicals. At high temperatures and at a high degree of conversion, alkane formation *via* reaction 5 is favored and, hence the reaction order is shifted from 0.5 to 1.5 [28].

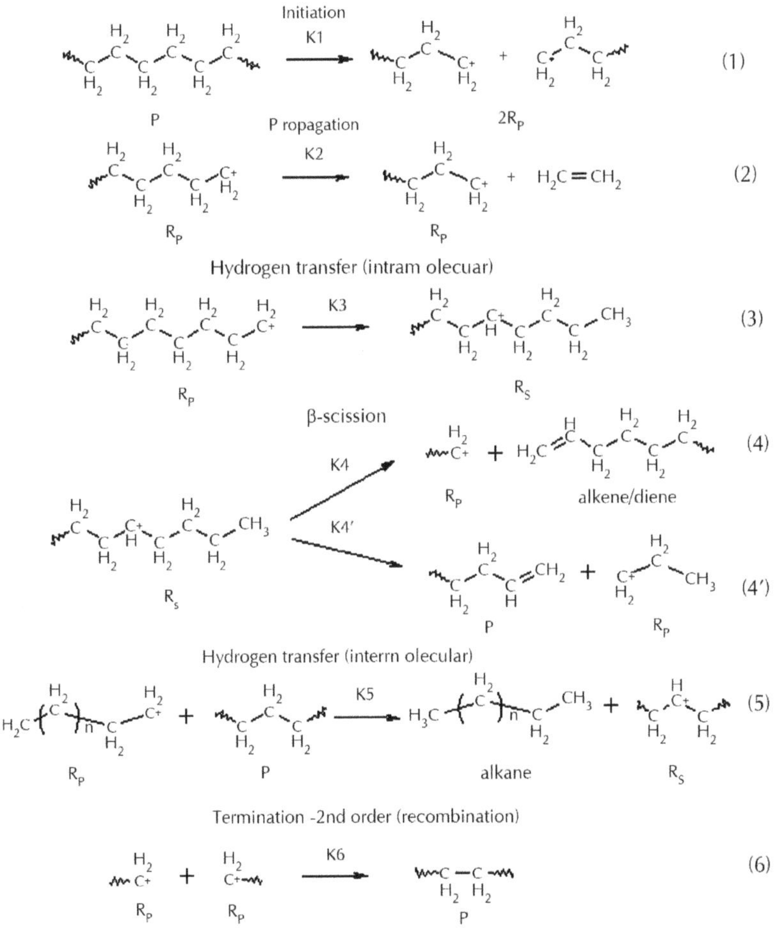

SCHEME 1   Mechanism of LDPE thermal degradation [28].

The thermal degradation of cellulose involves at least four processes in addition to simple desorption of physically bound water [33]. The first is the cross-linking of cellulose chains, with the evolution of water (dehydration). The second concurrent reaction is the unzipping of the cellulose chain (transglycosidation). Levoglucosan is formed from the monomer unit (Scheme 2).

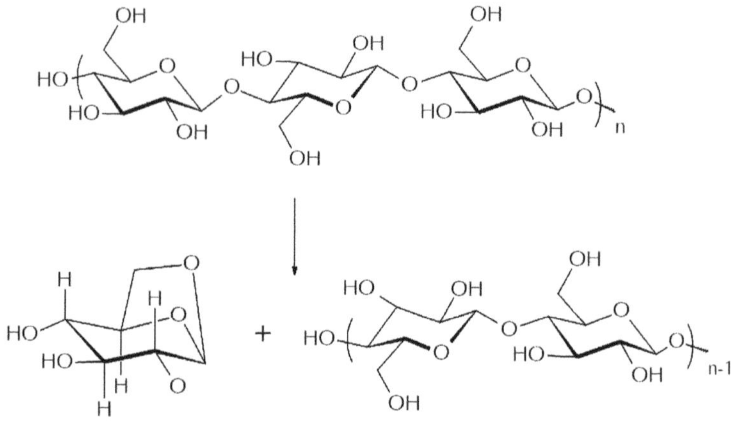

SCHEME 2   Formation of levoglucosan *via* unzipping of the cellulose chain.

The third reaction is the decomposition of the dehydrated product to yield char and volatile products. In addition, the levoglucosan can further decompose to yield volatile products and tars. Moreover, levoglucosan may also repolymerization. Below 270°C (Figure 6, curve 4), the dehydration reaction and the unzipping reaction proceed at comparable rates, and the basic skeletal structure of the cellulose is retained. During cellulose pyrolysis it is possible the inter-chain ether formation. This process would generate a three-dimensional polymer more stable than cellulose because the new ether bonds are more stable than the acetal-ether bonds. The increased thermal stability may allow further dehydration and possible dehydrogenation with less chain depolymerization and may be a step toward the char formation in cellulose pyrolysis. At higher temperatures (400°C) the cross-linked dehydrated cellulose and the repolymerized levoglucosan form polynuclear aromatic structures and carbonaceous char.

The reaction kinetics for thermal decomposition of cellulose in inert atmosphere has been reported in the literature as far back as 1956. Stamm [34] was among the first to report activation energy (109 kJ mol^{-1}). Hirata et al. [35] reported two processes occurring during the thermal decomposition of cellulose: an initial reaction and a propagation reaction with activation energies of 165 and 112 kJ mol^{-1}, respectively.

Building on the multistep mechanism Shafizadeh and co-workers [36] developed a three step kinetic model in which an initiation step forms "active cellulose," which subsequently decomposes by two competitive first-order reactions, one yielding volatiles and the other forming char and gas. More recently, Mamleev et al. [37-39] proposed a two step kinetic model to explain all observable phenomena related to the pyrolysis of cellulose, describing mass loss by two competing pathways of cellulose degradation, transglycosylation with E_{tar} 190–200 kJ mol^{-1} and elimination with E_{gas} 250 kJ mol^{-1}.

Compared to cellulose for the ethyl cellulose the transglycosidation reaction and formation of levoglucosan is unlikely because of ethylated (substitutes) fragments [40]. However, many pyrolysis products of ethyl cellulose are the corresponding ethylated compounds of the products generated in cellulose pyrolysis [41]. At the same time, the reverse aldolization and the elimination seem to occur in the same manner as for cellulose [41].

Recent analysis of the thermal degradation behavior of polymeric compositions (PE, PP, and PS) with the presence of cellulose additives showed a slight increase in the degradation temperature of such polymers [42, 43]. **Jakab et al [43] studied the effect of cellulose derivatives (wood, cellulose, and lignin) on the thermal degradation of PS, PP, and PE.** Thermogravimetry–mass spectrometry and pyrolysis–gas chromatography–mass spectrometry revealed that cellulose derivatives had a similar influence on PS and PE thermal decomposition under slow and fast heating conditions, respectively. Analysis of the influence of cellulose derivatives on PP, PS, and PE thermal degradation showed that the presence of char-forming cellulosic materials produced a slight increase in the degradation temperature associated with a change in the degradation mechanism of polymers. According to [43] charcoal promotes the hydrogenation of the unsaturated products and the hydrogenated products evolve at higher temperature. It has been noted that cellulose has a small effect on the thermogravimetric curves of PE [43].

27.7.2 KINETIC ANALYSIS OF CELLULOSE–LDPE AND ETHYL CELLULOSE–LDPE BLEND THERMAL DEGRADATION

Kinetic studies of material degradation have been carried out for many years using numerous techniques to analyze the data. Most often, TGA is the experimental method of choice and the only technique to be explored here. The TGA involves placing a sample of polymer on a microbalance within a furnace and monitoring the weight of the sample during some temperature program. It is commonly accepted that the degradation of materials follows the base Equation (1) [44].

$$dc/dt = -F(t, T c_o\, c_f) \tag{1}$$

where t—time, T—temperature, c_o—initial concentration of the reactant, and c_f—concentration of the final product. The right-hand part of the equation $F(t,T,c_o,c_f)$ can be represented by the two separable functions $k(T)$ and $f(c_o,c_f)$:

$$F(t,T,c_o,c_f) = k(T(t(\cdot f(c_o,c_f)) \tag{2}$$

Arrhenius Equation (3) will be assumed to be valid for the following:

$$k(T) = A\cdot exp(-E/RT) \tag{3}$$

Therefore,

$$dc/dt = -A\cdot exp(-E/RT)\cdot f(c_o,c_f) \tag{4}$$

A series of reaction types (classic homogeneous reactions and typical solid state reactions) are listed in Table 8 [44].

TABLE 8   Reaction types and corresponding reaction equations, $dc/dt = -A \cdot exp(-E/RT) \cdot f(c_o,c_f)$

Name	$f(c_o,c_f)$	Reaction type
F_1	C	first-order reaction
F_2	c^2	second-order reaction
F_n	c^n	nth-order reaction
R_2	$2 \times c^{1/2}$	two-dimensional phase boundary reaction
R_3	$3 \times c^{2/3}$	three-dimensional phase boundary reaction
D_1	$0.5/(1-c)$	one-dimensional diffusion
D_2	$-1/ln(c)$	two-dimensional diffusion
D_3	$1.5 \times e^{1/3}(c^{-1/3}-1)$	three-dimensional diffusion (Jander's type)
D_4	$1.5/(c^{-1/3}-1)$	three-dimensional diffusion (Ginstling-Brounstein type)
B_1	$c_o \times c_f$	simple Prout-Tompkins equation
B_{na}	$c_o^n \times c_f^a$	expanded Prout-Tompkins equation (na)
C_{1-X}	$c \times (1 + K_{cat} \times X)$	first-order reaction with autocatalysis through the reactants, X. $X = c_f$
C_{n-X}	$c^n \times (1 + K_{cat} \times X)$	nth-order reaction with autocatalysis through the reactants, X
A_2	$2 \times c \times (-ln(c))^{1/2}$	two-dimensional nucleation
A_3	$3 \times c \times (-ln(c))^{2/3}$	three-dimensional nucleation
A_n	$N \times c \times (-ln(c))^{(n-1)/n}$	n-dimensional nucleation/nucleus growth according to Avrami/Erofeev

The analytical output of measurements must fit with different temperature profiles by means of a common kinetic model.

Kinetic analysis of LDPE, cellulose, ethyl cellulose, and their blends at heating rates of 2.5, 5, and 10 K/min was carried out using Netzsch thermokinetics software according to research technique published elsewhere [45, 46]. Model-free methods of evaluations (Friedman-Analysis) [47] of all specimens were chosen as the starting points in kinetic analysis of sample thermal degradation for determining the activation energy in the development of the model.

Nonlinear fitting procedure established the two-stage model scheme of successive reactions for LDPE:

$$A \rightarrow F_n \rightarrow B \rightarrow F_n \rightarrow C \qquad (5)$$

For cellulose and ethyl cellulose the fits were attempted using nonlinear regression with triple-stage model of successive reactions, where three-dimensional diffusion (Jander's type), water loss, reaction type was used as the first step, whereas the nth-order (F_n) reaction type was used for the two last steps of the overall process of thermal degradation (Figure 8).

$$A \rightarrow D_3 \rightarrow D \rightarrow F_n \rightarrow E \rightarrow F_n \rightarrow F \qquad (6)$$

For cellulose–LDPE and ethyl cellulose–LDPE blends at the beginning stage and at the end of degradation, the degree of conversion depends on the heating rate (Figure 9), such dependence is strong evidence in favor of a branched reaction path. For this case, the approach has provided a four-stage model scheme comprising two pairs of competitive reactions of nth-order (F_n).

$$\begin{array}{ccccc} A & F_n & E & F_n & F \\ & & & & \\ F_n & B & F_n & C \\ & & & \\ 3 & & 4 \end{array} \qquad (7)$$

Here the first and second stages represent the thermal degradation of cellulose and ethyl cellulose additives, whereas third and fourth stages characterize the thermal degradation of LDPE matrix.

Calculated values of apparent kinetic parameters of the thermal degradation of cellulose, ethyl cellulose, LDPE, and their blends are listed in Table 9.

TABLE 9 Kinetic parameters of the thermal degradation of compositions based on LDPE, cellulose, and ethyl cellulose

Sample	E_1, kJ/mol	lg A_1, sec^{-1}	n_1	E_2, kJ/mol	lg A_2, sec^{-1}	n_2	E_3, kJ/mol	lg A_3, sec^{-1}	n_3	E_4, kJ/mol	lg A_4, sec^{-1}	n_4	Corr. co-eff.
LDPE	190	12.0	1.59	266	16.8	0.82	-	-	-	-	-	-	0.99965
Cellulose*	150	11.2	2.1	180	12.9	0.81	-	-	-	-	-	-	0.99991
Ethyl cellulose*	176	12.5	0.61	238	18.7	1.75	-	-	-	-	-	-	0.99982
Cellulose–LDPE 30:70 wt%	112	5.9	0.77	180	9.3	0.68	156	11.4	1.45	275	17.1	0.7	0.99962
Cellulose–LDPE 40:60 wt%	108	6.6	1.25	176	9.8	0.78	157	11.1	1.13	271	16.8	0.98	0.99978
Ethyl cellulose–LDPE 50:50 wt%	157	10.6	1.07	170	9.8	0.71	155	10.1	1.05	269	16.3	1.03	0.99831
Ethyl cellulose–LDPE 20:80 wt%	153	10.0	0.85	180	10.0	1.22	157	10.7	0.92	273	17.3	0.88	0.99782

* Kinetic parameters of water loss (D_3) for cellulose and ethyl cellulose: $E = 47$ and 54 kJ/mol, lg $A = 3.2$ and 1.8 sec^{-1}, respectively.

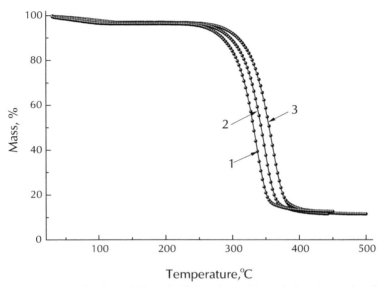

FIGURE 8 Nonlinear kinetic modeling of cellulose thermal degradation. Comparison between experimental TG data (dots) and model results (firm lines) at several heating rates—1–2.5, 2–5, and 3–10 K/min.

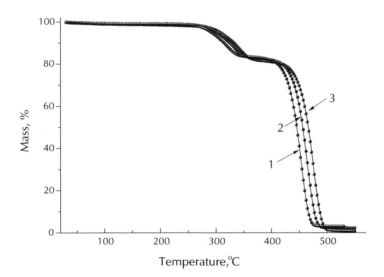

FIGURE 9 Nonlinear kinetic modeling of cellulose–LDPE (40:60 wt%). Comparison between experimental TG data (dots) and model results (firm lines) at several heating rates—1–2.5, 2–5, and 3–10 K/min.

Assuming a character of radical chain mechanism (Scheme 1), following the LDPE thermal degradation, the calculated values of the apparent energy of activation and pre-exponential factor correspond with early published data from isothermal analysis and dynamic TGA [28, 46]. At the same time, the values of activation energy found at the third stage of thermal degradation for cellulose–LDPE and ethyl cellulose–LDPE (7) are lower than those for LDPE (Table 9). This difference might be attributed to an influence of the degraded cellulosic derivatives (possessing radical character) on the initiation reaction of LDPE thermal degradation. However, the analysis of the activation energy at the fourth stage of cellulose–LDPE and ethyl cellulose–LDPE thermal degradation (7) indicates slight increase of E_4 (Table 9). This fact can be explained by means of transition of the thermal degradation process towards diffusion-limiting zone. In addition, it is possible to assume that polyethylene macro radicals R_p (Scheme 1) undergo termination reactions with lignocellulosic chair.

The thermal degradation of cellulose and ethyl cellulose as additives to LDPE blends show a decrease of apparent activation energies as compared with pristine cellulose and ethyl cellulose (Table 9). Notably, the value of activation energy of cellulose at the first stage of thermal degradation (7), decreases from 150 to 112 kJ/mol (cellulose–LDPE (30:70 wt%)) and 108 kJ/mol (cellulose–LDPE (40:60 wt%)) in contrast to increase of onsets and T'_{max} (Table 7). Pre-exponential factors also reduce from $10^{11.2}$ sec^{-1} to $10^{9.3}$ sec^{-1} and $10^{9.8}$ sec^{-1}, respectively (Table 9). This fact may be an evidence of the diffusion limitations caused by the LDPE melt on the volatilization of the cellulose and ethyl cellulose thermal degradation products.

The value of activation energy of ethyl cellulose at the first stage (7) decreases from 176 to 157 kJ/mol (ethyl cellulose–LDPE (50:50 wt%)) and 153 kJ/mol (ethyl cellulose–LDPE (20:80 wt%)). Pre-exponential factors reduce from $10^{12.5}$ sec^{-1} to $10^{9.8}$ sec^{-1} and 10^{10} sec^{-1}, respectively (Table 9).

The same trend can be observed for second stage (7) of thermal degradation of cellulose and ethyl cellulose additives in LDPE blends (Table 9).

The obtained results testify the observed changes of the blend morphology caused by amorphization, which directly influence the values of activation energies and pre-exponential factors (Table 9).

All these findings can be interpreted by the influence of the diffusion limitations on the interphase boundary of the degraded LDPE matrix and the volatilization of the cellulose and ethyl cellulose thermal degradation products.

27.7.3 THERMAL STABILITY OF TERNARY BLEND BASED ON TWO POLYSACCHARIDES

For comparison the curves of thermal degradation of LDPE, cellulose, their binary composition, and also the curve of decomposition of cellulose–LDPE–PEO$_{Lm}$ ternary blend are presented in Figure 10. As is seen from the presented data, the onset of the blend degradation occurs at temperatures higher than those for the individual cellulose, but the addition of LDPE leads to a decrease in the composition degradation temperature.

As an example, TG curve of cellulose–LDPE–chitin blend is given in Figure 10, curve 5. The comparison of the thermal stabilities of the ternary blends based on two synthetic polymers with the blends based on two polysaccharides showed that the introduction of the other polysaccharide also leads to decrease of the material thermal stability.

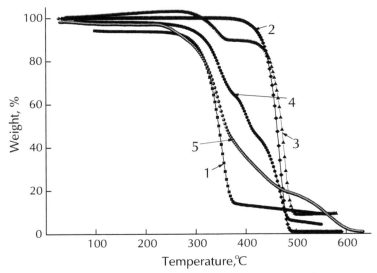

FIGURE 10 The TG curves of cellulose (1), LDPE (2), cellulose–LDPE (30:70 wt%) (3), cellulose–LDPE–PEO$_{Lm}$ (40:40:20 wt%) (4), and cellulose–LDPE–chitin (30:40:30 wt%) (5) blends obtained in an argon atmosphere.

27.8   INVESTIGATION OF THE BLEND BIODEGRADABILITY

The tests on the fungus resistance were performed with films obtained from cellulose–LDPE blend at component ratio of 40:60 wt%. On examination of the films under microscope only separate sprouted spores and a weakly developed mycelium are observed. The intensity of mold fungus growth measured according to five-balls scale was estimated at 1, so the investigated material was considered as funginert that is this blend contains nutrients in the amount providing an insignificant fungus growth.

It was occurred that the introduction of PEO into the system promotes the fungus growth. Hence, while for the films obtained from the cellulose–LDPE (40:60 wt%) blend the maximal intensity of fungus growth corresponds to 1, the addition of 20% of PEO$_{Lm}$ increases significantly the fungus amount and they can be clearly seen by the naked eye. In this case the estimation corresponds to 5 (the film composition – cellulose–LDPE–PEO$_{Lm}$ (40:40:20 wt%)) (Figure 11).

Changes occurred in the films from binary blends and compositions based on two polysaccharides after holding in soil at 30°C during five months were investigated by determining their weight losses. The curves of weight loss for the samples placed in soil are given in Figure 12. As illustrated in this figure, the weight loss of the binary composition occurs mainly during first two months (Figure 12, curve 1). Meanwhile the weight loss in the case of the ternary compositions occurs more intensively (Figure 12, curves 2 and 3); the microcracks and spots can be clearly seen, besides, the investigated samples become notably more fragile.

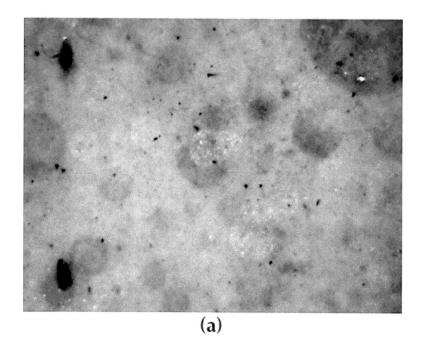

(a)

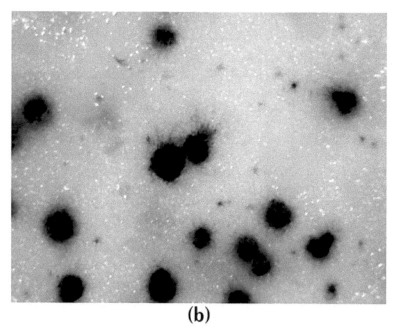

(b)

FIGURE 11 Surface micrographs of films from cellulose–LDPE (40:60 wt%) (a) and cellulose–LDPE–PEO$_{Lm}$ (40:40:20 wt%) and (b) blends infected with fungus spores.

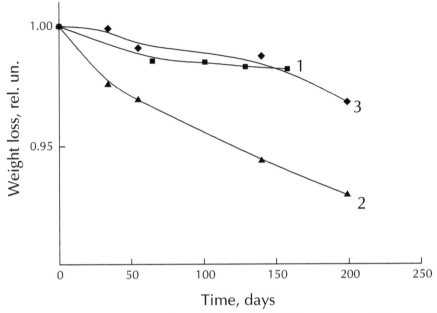

FIGURE 12 Weight loss of films from cellulose–LDPE (40:60 wt%) (1), cellulose–LDPE–chitin (2), and cellulose–LDPE–chitosan (30:40:30 wt%) (3) blends.

27.9 CONCLUSIONS

Powder blends of LDPE with cellulose and ethyl cellulose were prepared under conditions of shear deformations in a rotor disperser at different initial-component ratios. It was shown that during mixing of these polymers the homogeneous distribution of the components in the blend is observed. The mechanical properties and biodegradability of these mixtures generally depend on the composition of blend. Both addition of PEO and other polysaccharide (chitin or chitosan) to cellulose–LDPE compositions lead to an increase of the material biodegradability with insignificant deterioration of the mechanical parameters. The data of X-ray diffraction analysis testify the change of the structure of cellulose and ethyl cellulose under the action of shear deformation and, as a result, the rearrangement of the main chains and amorphization of the polymers take place. In this case the partial amorphization of the crystalline structure of LDPE is observed. Thus, the destruction proceeds in the whole volume of sample and the rate constants are determined by the mutual diffusion of the reagents. Since the rate of diffusion is much less than the chemical reaction rate, so it is one of the limiting stages of the thermal destruction process.

Considered schemes of the equations are only some variants of application of the simple approximation of the diffusion laws and simple dependence of rate on the component concentration. Validity of these reactions for relatively exact descriptions of the process is obvious in such cases, when as the reaction proceeds the estimation of the diffusion is more complicated.

KEYWORDS

- Biodegradable polymer blends
- Kinetics
- Mixing
- Polyethylene
- Polysaccharides
- Thermal degradation

ACKNOWLEDGMENTS

This chapter was supported by the Ministry of Education and Science of the Russian Federation (contract no. 8563 dated September 13, 2012). We are grateful to A. Ya. Gorenberg for analysis of the powder and film samples performed by SEM. The equipment of CKP MIPT and REC "Nanotechnology" of MIPT (SEM JEOL JSM-7001F) was used in this work.

REFERENCES

1. *Polymer blends*. D. R. Paul and S. Newman (Eds.) Academic Press. London, New York, and San Francisco (1978).
2. Bastioli, C. *Handbook of biodegradable polymers*. Rapra Technology Ltd, Shrewsbury (2005).
3. Nishio, Y. and Manley, R. St. *J. Polym. Eng. Sci.* **30**(2), 71–82 (1990).
4. Masson, J. F. and Manley, R. St. J. *Macromolecules.* **25**(2), 589–592 1992.
5. Masson, J. F. and Manley, R. St. J. *Macromolecules.* **24**(22), 5914–5921 (1991).
6. Çifci, C. and Kaya, A. *Desalination.* **253**, (1–3), 175–179 (2010).
7. Kaczmarek, H. and Ołdak, D. *Polym. Degrad. Stab.* **91**(10), 2282–2291 (2006).
8. Kim, T. J. and Lee, Y. M. *Im S-S. Polym. Compos.* **1**(3), 273–282 (1997).
9. Hendenberg, P. and Gatenholm, P. *J. Appl. Polym. Sci.* **60**(13), 2377–2385 (1996).
10. Karlsson, J. O., Blachot, J. F., Peguy, A., and Gatenholm, P. *Polym. Compos.* **17**(2), 300–304 (1996).
11. Pasquini, D., de Morais Teixeira, E., da Silva Curvelo, A. A., Belgacem, M. N., and Dufresne, A. *Compos. Sci. Technol.* **68**(1), 193–201 (2008).
12. Soulestin, J., Quiévy, N., Sclavons, M., and Devaux, J. *Polym. Eng. Sci.* **47**(4), 467–476 (2007).
13. Serratoni, M., Newton, M., Booth, S., and Clarke, A. *Eur. J. Pharm. Biopharm.* **65**(1), 94–98 (2007).
14. Larsson, M., Hjärtstam, J., Berndtsson, J., Stading, M., and Larsson, A. *Eur. J. Pharm. Biopharm.* **76**(3), 428–432.
15. Tan, J., Li, Y., Liu, R., Kang, H., Wang, D., Ma, L., Liu, W., Wu, M., Huang, Y. *Carbohydr. Polym.* **81**(2), 213–218 (2010).
16. Bai, Y., Jiang, C., Wang, Q., and Wang, T. *Carbohydr. Polym.* **96**(2), 522–527 (2013).
17. Girija, B. G., Sailaja, R. R. N., Sharmistha, B., and Deepthi, M. V. *J. Appl. Polym. Sci.* **116**(2), 1044–1056 (2010).

18. Sogal, K. and Ismaeil, H. *Int. J. Pharm.* **408**(1–2), 1–8 (2011).
19. Crowley, M. M., Schroeder, B., Fredersdorf, A., Obara, S., Talarico, M., Kucera, S., Mc-Ginity, M. J. *Int. J. Pharm.* **269**(2), 509–522 (2004).
20. Prut, E. V. *Polym. Sci., Ser. A.* **36**(4), 601–607 (1994).
21. Prut, E. V. and Zelenetskii, A. N. *Russ. Chem. Rev.* **70**(1), 65–79 (2001).
22. Bamford, C. H. and Tipper, C. F. H. *Comprehensive Chemical Kinetics in Degradation of Polymers.* C. H. Bamford and C. F. H. Tipper (Eds.), Elsevier Scientific publ. Amsterdam-Oxford-New York, **14**, 33–42 (1975).
23. Paabo, M. and Levin, B. C. *Fire Mater.* **11**, 55–70 (1987).
24. Lattimer, R. P. *J. Anal. Appl. Pyrolysis.* **31**, 203–226 (1995).
25. Kuroki, T., Sawaguchi, T., Niikuni, S., and Ikemura, T. *Macromolecules.* **15**, 1460–1462 (1982).
26. Kiran, E., Gillham, J. K. *J. Appl. Polym. Sci.* **4**, 931–947 (1976).
27. Poutsma, M. L. *Macromolecules.* **36**, 8931–8957 (2003).
28. Hornung, U., Hornung, A., and Bockhorn, H. *Chem. Eng. Tech.* **21**, 332–337 (1998).
29. Bockhorn, H., Hornung, A., Hornung, U., and Schawaller, D. *J. Anal.Appl. Pyrol.* **48**, 93–109 (1999).
30. Bockhorn, H., Hornung, A., and Hornung, U. *J. Anal. Appl. Pyrolysis.* **50**, 77–101 (1999).
31. Albertson, A. C. and Karlsson, S. *Chemistry and biochemistry of polymer biodegradation in Chemistry and Technology of Biodegradable Polymers.* G. J. L. Griffin (Ed.), Blackie Academic and Professional, Glasgow, pp. 7–17 (1994).
32. Rogovina, S. Z., Alexanyan, Ch. V., and Prut, E. V. *J. Appl. Polym. Sci.* **121**(3), 1850–1859 (2011).
33. Beyler, C. L. and Hirschler, M. M. *Thermal Decomposition of Polymers in SFPE Handbook of Fire Protection Engineering* (3rd ed.). Ph. J. DiNenno (Ed.), National Fire Protection Association, Quincy, Massachusetts, pp. 1–110 (2002).
34. Stamm, A. *J. Ind. Eng. Chem.* **48**, 413–417 (1956).
35. Hirata, T. *Ringyo Shikenjo Kenkyu Hokoku.* **263**, 17–33 (1974).
36. Bradbury, A. G. W., Sakai, Y., and Shafizadeh, F. *J. Appl. Polym. Sci.* **23**, 3271–3280 (1979).
37. Mamleev, V. Bourbigot, S., Le Bras, M., and Yvon, J. *J. Anal. Appl. Pyrolysis.* **84**, 1–17 (2009)
38. Mamleev, V., Bourbigot, S., and Yvon, J. *J. Anal. Appl. Pyrolysis.* **80**, 141–150 (2007).
39. Mamleev, V., Bourbigot, S., and Yvonl, J. *J. Anal. Appl. Pyrolysis.* **80**, 51–165 (2007).
40. **Boon, J. J., Pastorova, I., Botto, R. E, and Arisz, P. W.** *Biomass and Bioenergy.* **7**, 25–32 (1994).
41. Moldoveanu, S. C. *Analytical Pyrolysis of Polymeric Carbohydrates* in *Analytical Pyrolysis of Natural Organic Polymers.* S. C. Moldoveanu (Ed.), Elsevier Science, Amsterdam, pp. 217–273 (1998).
42. Pielichowski, K. and Njuguna, J. *Natural Polymers, in Thermal Degradation of Polymeric Materials.* K. Pielichowski and J. Njuguna (Ed.) Rapra Technology Ltd., Shawbury, Shropshire, pp. 133–138 (2005).
43. Jakab, E., Varhegyi, G.,and Faix, O. *J. Anal. Appl. Pyrolysis.* **56**, 273–285 (2000).
44. Opfermann, J. *J. Therm. Anal. Calorim.* **60**, 641–658 (2000).
45. Lomakin, S. M, Dubnikova, I. L, Berezina, S. M., and Zaikov G. E. *Polym. Int.* V. **54**, 999–1006 (2005).
46. Lomakin, S. M., Novokshonova, L. A., Brevnov, P. N., and Shchegolikhin, A. N. *J. Mater. Sci.* **43**, 1340–1353 (2008).
47. Friedman, H. L. *J. Polym. Sci.* **C6**, 183–195 (1965)

CHAPTER 28

STRUCTURE AND RELAXATION PROPERTIES OF ELASTOMERS

N. M. LIVANOVA and S. G. KARPOVA

CONTENTS

28.1 INTRODUCTION

The paramagnetic probe method with the use of free radicals of different dimensions (2,2,6,6-tetramethyl-1-piperidinyloxy (TEMPO) and 4-benzoate-2,2,6,6-tetramethyl-1-piperidinyloxy (BZONO)) has been employed to study the effect of the isomeric composition of butadiene units in polybutadienes (PB) and butadiene–acrylonitrile copolymers on the number and dimensions of ordered structures. The nature of density fluctuations and defective regions, that is, the regions in which the radicals are sorbed, has been ascertained. It has been shown that the ordered regions are composed of stereoregular chain fragments, while defective regions are enriched with butadiene isomers different from those present in prevailing amounts.

In [1], the micro and supramolecular structure of ethylene–propylene–diene ternary copolymers was studied by infrared (IR) and electron paramagnetic resonance (EPR) spectroscopy. The paramagnetic probe measurements showed that the temperature dependences of correlation time τ_c plotted in Arrhenius coordinate for microblock copolymers with high degrees of isotacticity of propylene units demonstrate transitions relevant to the breakdown of supramolecular structures composed of not only ethylene sequences but also propylene sequences [1]. Later on, the existence of these transitions was confirmed by the thermomechanical analysis of the copolymers [2]. Under the assumption that the radical solubility and distribution over rotation frequencies are the same in amorphous regions of homo and copolymers, the correlation time of a probe in ethylene–propylene–diene copolymers was calculated with a high accuracy on the basis of the additive contribution of each component corresponding to its weight fraction after deduction of the amounts of ordered structures. The latter were determined as the contents of fractions insoluble in toluene at room temperature. Thus, the paramagnetic probe method is a delicate instrument for investigating the molecular and supramolecular organization of elastomers.

Here, this method was applied for studying the molecular mobility in butadiene elastomers with different stereoisomerisms of units and in butadiene–acrylonitrile copolymers (acrylonitrile–butadiene rubbers (NBRs)). In most cases, they were mostly composed of isomeric butadiene units of one kind (*cis*-1,4- in PBs or *trans*-1,4- in NBRs) (Table 1). The temperature dependence of the rotational mobility of free-radical probes with different dimensions (TEMPO and BZONO) in these elastomers was investigated [3, 4].

TABLE 1 Isomeric composition of butadiene units in PBs of different brands and butadiene–acrylonitrile copolymers

Elastomer	Isomer content*, %		
	trans-1,4	1,2	cis-1,4
cis-1,4-*PB*	0.89	0.6	98.5
SKD	3.8	2.5	93.7

TABLE 1   *(Continued)*

SKDL	47	8.8	44.2
BNKS-18	82.0	8.2	9,8
BNKS -28	76.4	14.4	9.2
BNKS -40	93.0	4.4	2.6

* In the butadiene part of the polymer

28.2   EXPERIMENTAL

The objects of investigation were PB, such as *cis*-PB, SKD PB, and SKDL PB (a rubber synthesized in the presence of a lithium catalyst), with different stereoregularities of macromolecules and butadiene–acrylonitrile copolymers of the BNKS-18, BNKS-28, and BNKS-40 brands (acrylonitrile contents of 18, 28, and 40 wt%, respectively). The isomeric composition of butadiene units, that is, the contents of *trans*-1,4, *cis*-1,4, and 1,2-isomers, was determined by IR spectroscopy from bands at 967, 730, and 911 cm^{-1}, respectively, [5] through the use of film samples cast from solutions in CCl$_4$ or chloroform. The calculations were performed with the extinction ratios reported in [6]; the results are listed in Table 1.

The EPR spectroscopy was used to study the rotational mobility, which is characterized by correlation time τ_c of TEMPO and BZONO radicals [3, 4], in a temperature range of 20–90°C with an EPR-V radiospectrometer. The elastomers were saturated with TEMPO from the gas phase at room temperature for three days. Then, samples were stored at room temperature for a few days. During storage, the radical was desorbed from large defective regions of the samples (pores, cavities, and so on), in which polymers have almost no effect on the rotational mobility of the radical. The BZONO was incorporated into PB from solutions in acetone with a concentration of 10^{-3} mol/l with subsequent evaporation of the solvents at room temperature. The EPR spectra of the TEMPO probe in PBs and BNKS-18 show the classical triplet pattern. The spectra of both radicals measured in BNKS-28 and BNKS-40 at room temperature attest to a heterogeneous structure, that is, to the existence of fast and slow components. In the case of BZONO, the pattern of the spectra is typical of anisotropic rotation characterized by different frequencies of rotation about different molecular axes. The τ_c values determined at 24.5°C are listed in Table 2. The error in the measurements of τ_c is 5%.

TABLE 2   Arrhenius parameters of probe rotation in polymers and τ_c values at 24.5°C

Elastomer	$\tau_c \times 10^+$ ^{10}, sec (24°C)	E_1	E_2	$\tau_o^{(1)}$, sec	$\tau_o^{(2)}$, sec	Tempera-ture of the transition onset, K
		\multicolumn{2}{c}{kJ/mol}			
TEMPO						
cis-1.4-PB	0.89	33.5	56	1.1×10^{-16}	$2.5 \times \cdot 10^{-20}$	323
SKD	0.5	24.4	54.1	2.7×10^{-15}	2.8×10^{-20}	314
SKDL	0.9	17.2	48.5	8.5×10^{-14}	4.4×10^{-19}	308
BNKS -18	7.4	33.8	54.6	9.3×10^{-16}	1.5×10^{-19}	311
BNKS -28	17.8	14.4	65.2	5.6×10^{-12}	1.2×10^{-20}	306
BNKS -40	23.4	20.1	76.4	7.8×10^{-13}	3.8×10^{-22}	317
BZONO						
cis-1.4-PB	4.0	44.3	33,0	5.8×10^{-18}	4.5×10^{-16}	313
SKD	5.0	44.0	35.8	8.0×10^{-18}	2.0×10^{-16}	309
SKDL	7.9	46.5	43.3	5.0×10^{-18}	1.7×10^{-17}	308

Note: E_i and $\tau_o^{(i)}$ are the effective activation energies and preexponential factors at temperatures below ($i = 1$) and above ($i = 2$) the temperature of the transition onset.

28.3   DISCUSSION AND RESULTS

The effective activation energy E_{ef} was calculated for the rotation of the radicals (Table 2) from the temperature dependences of correlation time τ_c of TEMPO and BZONO in PBs and TEMPO in butadiene–acrylonitrile copolymers (Figure 1). The calculations were performed for different parts of the curves plotted in Arrhenius coordinates. These parts corresponded to different relaxation processes relevant to supramolecular structures. During approximation of the linear parts of the curves in Figure 1, correlation coefficients R^2 were 0.82–0.97.

(a)

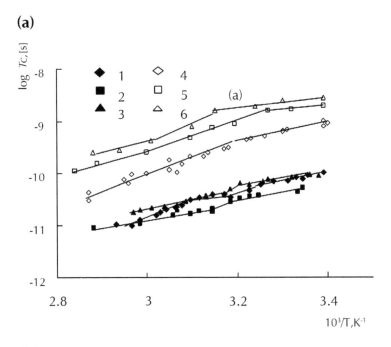

(b)

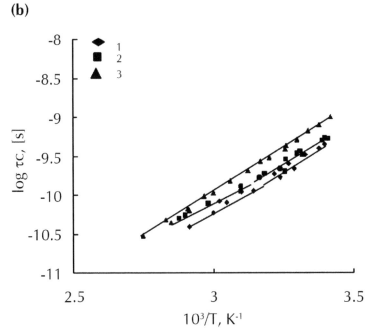

FIGURE 1 Temperature dependences of correlation times of the rotational mobility plotted for (a) TEMPO and (b) BZONO radicals in Arrhenius coordinates—(1) *cis*-PB, (2) SKD, (3) SKDL, (4) BNKS-18, (5) BNKS-28, and (6) BNKS-40.

For stereoregular PB (SKD rubbers) and NBRs, in the region of the high-elasticity plateau, relaxation spectrometry data indicate as many as three λ-relaxation processes relevant to the molecular mobility of structural elements that are larger than segments; that is, microblocks whose activation energy is equal to that of the viscous flow of the elastomers [7]. Hence, segments play the role of kinetic units of slow relaxation λ processes. The mechanism governing the molecular mobility of the microblocks consists in their formation and decomposition through the fluctuation attachment and detachment of segments as individual kinetic units [7]. Moreover, the thermomechanical analysis, the energy parameters of dissolution and diffusion of low-molecular-mass substances, and IR spectroscopy data suggest the existence of fluctuation structures in elastomers [8-12].

The presence of ordered microscopic regions and their size distribution is, as is (and in [13]), governed by the microtacticity (the geometry similarity of chain fragments) and temperature. The molecular dynamics of polymer chain fragments located in a structural defect depends on the mobility of regular chain fragments comprising ordered structures. The transitions are explained by the unfreezing of the molecular mobility in fluctuation structures composed of units occurring in the prevailing configuration and by their rearrangements [7-13].

Like any other low-molecular-mass substance, radicals are sorbed in the polymer regions with subsequent deteriorations of short-range order in the arrangement of chains [14]. Such regions most probably appear at the joints of different isomers of butadiene units, because there is no geometric similarity between them [13, 15]. When samples are stored for a few days at room temperature, TEMPO remains in the regions in which its desorption is hindered, that is, in regions with rather dense macromolecular packing but a free volume sufficient for accommodation of the radical. Defects in the chain stereoregularity may give rise to the appearance of such a free volume. The radical cannot penetrate regions with ordered structure and hindered molecular mobility.

The stereoregular PBs containing more than 99% *cis*-1,4-units form crystals with a maximum melting temperature of 12.5°C. When the content of *cis*-1,4-isomer is 98%, the melting temperature of the polymers is 0°C; at room temperature, their ordered structures are preserved [16]. At a *cis*-1,4-unit content of 90%, the maximum crystallinity of SKD is 55% (–37°C), while at 81%, the crystallinity is insignificant. It was shown that, during photochemical isomerization, crystallization occurs when the content of *trans*-1,4-units is higher than 75% [17]. However, when the content of *trans*-1,4-isomers is lower than 20%, *cis*-1,4-structures undergo low-temperature crystallization [18].

Experimental melting temperatures T_m depend on the contents of units occurring in different configurations and on the duration and temperature of crystallization and the rate of melting [16]. At a *cis*-1,4-unit content of 96.4%, the differential thermal analysis (DTA) method demonstrated the appearance of two overlapping endothermic peaks during rapid cooling and slow heating [19]. The DTA and dilatometry data depend on the frequency and randomness of the repetition of units with a structure different from the *cis*-1,4-configuration along a polymer chain [16]. Thus, the capability of ordering or crystallization is predetermined by the stereoregularity of elastomer chains. This conclusion is confirmed below by the EPR data.

For *cis*-PB and SKD and SKDL with different isomeric butadiene unit ratios, the effective activation energy E_1 of TEMPO radical rotation at temperatures below the transition linearly diminishes with an increase in the content of isomers present in small amounts, that is, *trans*-1,4- and 1,2-units (Figure 2). The total amount of *trans*-1,4- and 1,2- units in *cis*-PB and SKD is small (less than 10%) (Table 1), therefore, it is assumed that they are randomly distributed over polymer chains. According to [20], SKDL, which is synthesized in the presence of *n*-BuLi as a catalyst, contains blocks of cis and trans isomers connected by vinyl units. Hence, only *cis*-1,4-unit–1,2-unit–*trans*-1,4-unit joints are present in it. *cis*-1,4 and *trans*-1,4 isomers form fluctuation structures [11]. Thus, the regions with the lowest order are enriched with 1,2-units located between *cis*-1,4- and *trans*-1,4-isomers. Hence, the E_1 value of SKDL falls on the linear dependence of E_1 on the isomeric composition of butadiene units (Figure 2) if the content of 1,2-units (8.8%) (Table 1) is plotted on the abscissa axis.

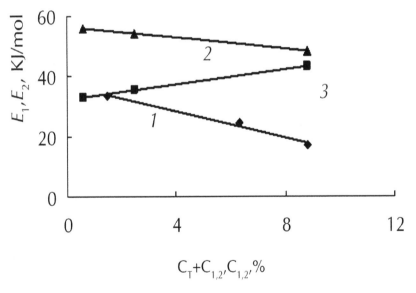

FIGURE 2 Effective activation energies (1) E_1 and (2, 3) E_2 of the rotational mobility of radicals as functions of the content of (1) *trans*-1,4- and 1,2-units ($C_T + C_{1,2}$) and (2, 3) 1,2-conformers ($c_{1,2}$)—(1, 2) TEMPO and (3) BZONO.

The effective activation energy of probe rotation decreases in proportion to reduction in the stereoregularity of chains and the dimensions of ordered structures (Tables (1-2), Figure 2). This fact agrees with the data reported in [16], where a PB sample synthesized in the presence of an ALR$_3$/TiCl$_4$-type catalyst was separated into fractions with different contents of *cis*-1,4-units by selective extraction.

In the case of *cis*-PB containing 98.5% *cis*-1,4-units and a content of *trans*-1,4- and 1,2-isomers of ~15%, the activation energy of the probe rotation in the initial region of the $\tau_c(T)$ curve (20–50°C) is 33.5 kJ/mol (Table 2). This result aggress with the data from [7], where the activation energy of the low-frequency relaxation processes, which are relevant to the mobility of microblocks of supramolecular structures and λ processes, for monodisperse PBs was found to be 34 kJ/mol.

In SKD containing 93.7% *cis*-1,4-units, structures composed of sequences of these units are found; however, because of a higher content of *trans*-1,4- and 1,2-units than that in *cis*-PB and, accordingly, a lower chain stereoregularity, the amount and dimensions of these structures are smaller than those in *cis*-PB. Therefore, the temperature of transition is lower (Table 2). The effective activation energy of TEMPO rotation in SKD (22.5 kJ/mol) is close to the activation energy of viscous flow determined for PB with a *cis*-1,4-unit content of 92% (23.5 kJ/mol) and the activation energy of sulfur diffusion in SKD (23.1 kJ/mol) [7, 10, 21]. Thus, the E_{ef} value of SKD is equal to the activation energy of the viscous flow of PB with a close isomeric composition. In SKD, the radical is sorbed by regions with defective chain microtacticity that are composed of the same sequences of isomers as those in *cis*-1,4-PB. In SKDL containing almost equal amounts of *cis*-1,4- and *trans*-1,4-isomers, the effective activation energy of radical rotation is 17.5 kJ/mol at temperatures below the transition temperature (35°C). The obtained transition temperature agrees with [11].

Hence, if the regular parts of chains are composed of ordered structures, polymer chains located in unordered regions (the regions of radical sorption) are enriched with butadiene isomers that are present in smaller amounts.

Similarly to E_1, the transition temperature linearly decreases with an increase in the content of *trans*-1,4- and 1,2-isomers in PBs (Figure 3). This fact lends support to the conclusion that, at low temperatures, the temperature dependence of the radical mobility is governed by the dimensions of stereoregular fragments of chains and related ordered structures.

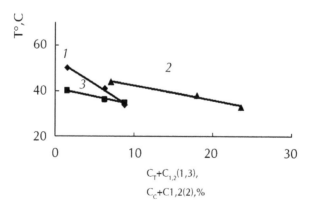

FIGURE 3 Transition temperatures of (*1, 3*) PBs and (*2*) NBRs versus the contents of (*1, 3*) *trans*-1,4- and 1,2-units ($c_T + c_{1,2}$), (*2*) *cis*-1,4- and 1,2-butadiene units ($c_c + c_{1,2}$)—(*1, 2*) TEMPO and (*3*) BZONO.

The effective activation energy of TEMPO radical rotation, E_2, corrresponding to the region above the transition temperature is higher than E_1 and slightly declines according to a linear relationship with an increase in the 1,2-isomer content (Figure 2). Hence, the transition is associated with the breakdown of primary structures and separation of chains related to an increase in segmental mobility; as a result, only vinyl units containing side double bonds influence the radical rotation.

In butadiene–acrylonitrile copolymers with different contents of polar units and isomer ratios of the butadiene comonomer (Tables (1–2)), the effective activation energies of TEMPO probe rotation that correspond to the initial part of Arrhenius dependence $\tau_c(T)$ likewise linearly decrease with an increase in the amount of all defects in the regularity of the butadiene component ($\sum$AN-units + cis-1,4-units + 1,2-units) (Figure 4). Energy E_2 corresponding to the region above the transition temperature linearly increases with the content of acrylonitrile units in the copolymers, that is, with the concentration of polar units exhibiting a high cohesion energy, because of a strong dipole–dipole interaction of nitrile groups, which noticeably increase chain rigidity. Thus, the transition improves the segmental mobility so much that chains are separated and the effect of butadiene unit isomers becomes unnoticeable against the background of the strong interaction between the polar groups, which strongly reduce chain flexibility.

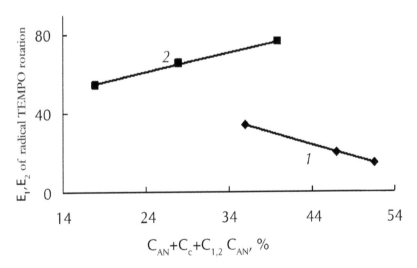

FIGURE 4 Effective activation energies (1) E_1 and (2) E_2 of radical TEMPO rotation versus total contents of (1) acrylonitrile + cis-1,4- + 1,2-units ($c_{AN} + c_c + c_{1,2}$), (2) c_{AN}.

Above the transition temperature, the mobility of TEMPO in PBs and NBRs is influenced by side vinyl or nitrile groups. In both cases, the larger the amount of defects in the stereostructure of the butadiene part of a chain, the lower the effective energy of radical rotation.

Figure 5 shows the dependence of the intensity ratio between the band at 1053 cm^{-1} attributed to the *trans*-1,4-units contained in crystallites and the band at 967 cm^{-1} that characterizes the total content of *trans*-1,4-isomers [5] on the total content of 1,2- and *cis*-1,4 isomers. As their content increases, the order of *trans*-1,4-units diminishes. BNKS-28 and BNKS-40 have the least and the most ordered structures, respectively. As was shown in [16], at a *trans*-1,4-unit content above 75%, the crystallinity is observed at room temperature. The total amorphization attests to a random distribution of *trans*-1,4-units. Hence, at a high content of acrylonitrile units, BNKS-40 contains the greatest amounts of structures formed by both strong dipole–dipole bonds between nitrile groups and *trans*-1,4-butadiene sequences [7, 12, 22].

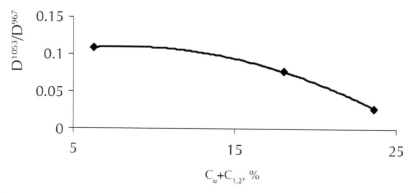

FIGURE 5   Optical density ratio for bands at 1053 and 967 cm^{-1} characterizing the fraction of ordered structures in NBRs versus total content of *cis*-1,4- and 1,2-units ($c_c + c_{1,2}$).

The BZONO probe, which is characterized by a high molecular mass, an asymmetric shape, and an anisotropic rotation, was incorporated into the elastomers from solutions; therefore, the sample structure may be different from the initial one. TEMPO molecules occupy defects that exist in a polymer. As regards BZONO, it seems to partly forms defects in the course of solvent evaporation. This fact may explain different patterns of the Arrhenius $\tau_c(T)$ dependences plotted for TEMPO and BZONO radicals (Figure 1).

When BZONO is used, relaxation transitions are distinctly observed for *cis*-PB and SKD (Figure 1, Table 2). Linear $\tau_c(T)$ dependences or small differences between E_{ef} values in the first and second regions suggest the absence of ordered structures or their insignificant effect on the radical mobility (their absence or small dimensions in SKDL).

In contrast to TEMPO, the BZONO radical occurs in the environment of long disordered defective butadiene fragments of chains. This circumstance seems to influence the E_{ef} value.

Equal E_1 values, which are observed for BZONO in PBs (Figure 2) irrespective of the stereoregularity of elastomer chains, may be interpreted as a consequence of the probe location in large defects possessing the same structure, while close temperatures of the relaxation transitions (Figure 3), which are lower than those for TEMPO, may be considered to be due to a smaller size of ordered structures. The E_2 value of BZONO slightly increases with an increase in the content of 1,2-isomers of butadiene units (similarly to TEMPO) (Figure 2). This finding may be explained by certain hindrances to the rotational mobility of the large probe from the side units of chains.

The results lead us to draw the following conclusions. A small TEMPO radical provides more complete information on the structure of polymers and the nature of defective regions in them. The incorporation of a probe from solutions is accompanied by changes in the initial structural organization of elastomers. For some systems, this phenomenon may be of interest. Radicals are sorbed by regions of a polymer matrix that contain butadiene units. Defective regions comprise chain fragments with alternating butadiene units of different isomeric structures. Their mobility is influenced by regular chain fragments composing ordered structures, a fact that additionally confirms the existence of the latter.

KEYWORDS

- **Benzoate-2,2,6,6-tetramethyl-1-piperidinyloxy**
- **Differential thermal analysis**
- **Electron paramagnetic resonance**
- **Polybutadienes**
- **2,2,6,6-Tetramethyl-1-piperidinyloxy**

REFERENCES

1. Livanova, N. M., Karpova, S. G., and Popov, A. A. Polymer Science, Ser., **A45**, 238–243 (2003) [*Vysokomol. Soedin., Ser.* **A 45**, 417–423 (2003)].
2. Livanova, N. M., Evreinov, Yu. V., Popov, A. A., and Shershnev, V. A. Polymer Science, Ser., **A45**, 530–536 (2003). [Vysokomol. Soedin., Ser. A 45, 903-912 (2003)].
3. Vasserman, A. M. and Kovarskii, A. L. Spin Labels and Probes in Physical Chemistry of Polymers (in Russian), Nauka, Moscow (1986).
4. Vasserman, A. M., Buchachenko, A. L., Kovarskii, A. L., and Neiman, M. B. Vysokomol. Soedin., Ser., **A10**, 1930–1936 (1968).
5. Dechant, J., Danz, R., Kimmer, W., and Schmolke, R. Ultrarotspektroskopische Untersuchungen an Polymeren (Akademie, Berlin, 1972; Khimiya, Moscow, 1976).
6. Kozlova, N. V., Sukhov, F. F., and Bazov, V. P. Zavod. Lab. **31**, 968–970 (1965).
7. Bartenev, G. M. Structure and Relaxation Properties of Elastomers (in Russian), Khimiya, Moscow, (1979).
8. Kargin, V. A., Berestneva, Z. Ya., and Kalashnikova, V. G. *Usp. Khim.*, **36**, 203–216 (1967).
9. Lebedev, V. P. *Usp. Khim.*,s **47**, 127–151 (1978).
10. Bartenew, G. M. *Plast. Kautsch.*, **17**, 727–730 (1970).
11. Sokolova, L. V., Chesnokova, O. A., Nikolaeva, O. A., and Shershnev, V. A. *Vysokomol. Soedin., Ser.*, **A27**, 352–361 (1985).

12. Grishin, V. S., Tutorskii, I. A., and Yurovskaya, I. S. *Vysokomol. Soedin., Ser.*, **A20**, 1967–1973 (1978).
13. Kuleznev, V. N. *Polymer Science, Ser.*, **B35**, 1156 (1993) [Vysokomol. Soedin., Ser. B 35, 1391-1402 (1993)].
14. Shlyapnikov, Yu. A., Kiryushkin, S. G., and Mar'in, A. P. *Antioxidation Stabilization of Polymers*(in Russian). Khimiya, Moscow, (1986).
15. Kulichikhin, V. G., Vinogradov, G. V., Malkin, A. Ya., et al. *Vysokomol. Soedin., Ser.*, **B10**, 739–742 (1968).
16. *The Stereo Rubbers*, (Ed.) W. M. Saltman. Wiley, New York, **2**, (1979).
17. Yagfarov, M. Sh. *Vysokomol. Soedin., Ser.*, **A10**, 1264–1269 (1968).
18. Collins, E. A. and Chandler, L. A. *Rubber Chem. Technol.*, **39**, 193 (1966).
19. Golub, M. *J. Polym. Sci.*, **25**, 373–377 (1957).
20. Mochel, V. D. *J. Polym. Sci.* Part A-1, **10**, 1009–1018 (1972).
21. Grishin, V. S., Tutorskii, I. A., and Potapov, E. E. *Vysokomol. Soedin., Ser.*, **A16**, 130–135 (1974).
22. Sokolova, L. V., Konovalova, O. A., and Shershnev, V. A. *Kolloidn. Zh.*, **44**, 716–721 (1982).

CHAPTER 29

A STUDY ON CARBON NANOTUBES STRUCTURE IN POLYMER NANOCOMPOSITES

Z. M. ZHIRIKOVA, V. Z. ALOEV, G. V. KOZLOV, and G. E. ZAIKOV

CONTENTS

29.1 INTRODUCTION

At present it is considered that carbon nanotubes (CNT) are one of the most perspective nanofillers for polymer nanocomposites [1]. The high anisotropy degree (their
length to diameter large ratio) and low transverse stiffness are CNTs specific features.
These factors define CNTs ring-like structures formation at manufacture and their introduction in polymer matrix. Such structures radius depends to a considerable extent
on CNTs length and diameter. Thus, the strong dependence of nanofiller structure on
its geometry is CNTs application specific feature. Therefore the present work purpose
is to study the dependence of nanocomposites butadiene-styrene rubber/carbon na-
norubes (BSR/CNT) properties on nanofiller structure, received by chemical vapor
deposition (CVD) method with two catalysts usage.

The CNT structure in polymer nanocomposites was studied. It has been shown that
this nanofiller feature is its "rolling up" in ring-like structures. This factor plays a crucial role in determination of nanocomposites structural and mechanical characteristics.

29.2 EXPERIMENTAL

The nanocomposites BSR/CNT with CNT content of 0.3 mass% have been used as
the study object. The CNT have been received in the Institute of Applied Mechanics of Russian Academy of Sciences by the vapors catalytic CVD, based on carbon
– containing gas thermochemical deposition on nonmetallic catalyst surface. Two
catalysts – Fe/Al_2O_3 (CNT-Fe) and Co/Al_2O_3 (CNT-Co) – have been used for the
studied CNT. The received nanotubes have diameter of 20 nm and length of order
of 2 mcm.

The nanofiller structure was studied on force-atomic microscope Nano-DST (Pacific Nanotechnology, USA) by a semi-contact method in the force modulation regime. The received CNT size and polydispersity analysis was made with the aid of the
analytical disk centrifuge (CPS Instrument, Inc., USA), allowing to determine with
high precision the size and distribution by sizes in range from 2 nm up to 50 nm. The
nanocomposites BSR/CNT elasticity modulus was determined by nanoindentation
method on apparatus Nano-Test 600 (Great Britain).

29.3 DISCUSSION AND RESULTS

In Figure 1 the electron microphotographs of CNT coils are adduced, which demonstrate ring-like structures formation for this nanofiller. In Figure 2 the indicated
structures distribution by sizes was shown, from which it follows, that for CNT-Fe
narrow enough monodisperse distribution with maximum at 280 nm is observed
and for CNT-Co – polydisperse distribution with maximums at ~ 50 nm and 210
nm.

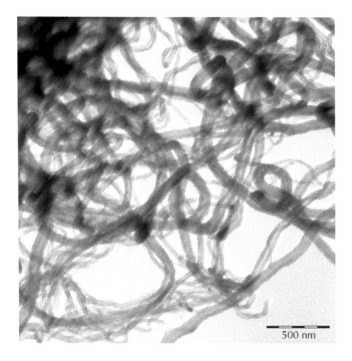

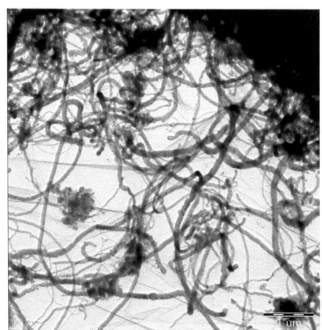

FIGURE 1 Electron micrographs of CNT structure, received on transmission electron microscope.

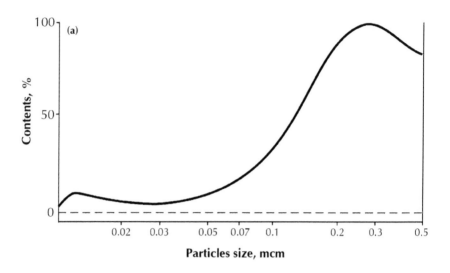

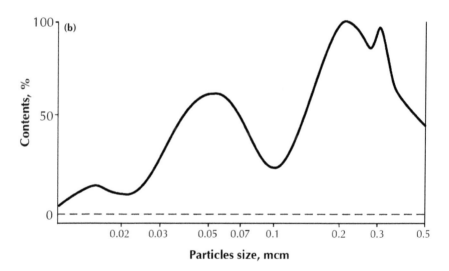

FIGURE 2　The particles distridution by sizes for CNT-Fe (1) and CNT-Co (2).

Further let us carry out the analytical estimation of CNT formed ring-like structures radius R_{CNT}. The first method uses the following formula, obtained within the frameworks of percolation theory [2]:

$$\varphi_n = \frac{\pi L_{CNT} r_{CNT}^2}{\left(2R_{CNT}\right)^3} \tag{1}$$

Where ϕ_n is CNT volume content, L_{CNT} and r_{CNT} are CNT length and radius, respectively.

The value ϕ_n was determined according to the well-known equation [3]:

$$\phi_n = \frac{W_n}{\rho_n} \qquad (2)$$

Where W_n and ρ_n are mass content and density of nanofiller, respectively.

In its turn, the value ρ_n was calculated as follows [4]:

$$\rho_n = 0.188 D_{CNT}^{1/3} \qquad (3)$$

Where D_{CNT} is CNT diameter.

The second method is based on the following empirical formula application [5]:

$$R_{CNT} = \left(\frac{D_{CNT}}{D_{CNT}^{st}} \right)^2 \left(0.64 + 4.5 \times 10^{-3} \phi_n^{-1} \right), mcm, \qquad (4)$$

Where D_{CNT}^{st} is a standard nanotube diameter, accepted in chapter [5] equal to CNT of the mark "Taunite" diameter (45 nm).

The values R_{CNT}, calculated according to the Equation (1) and Equation (4), are adduced in Table 1, from which their good correspondence (the discrepancy is equal to ~ 15%) follows. Besides, they correspond well enough to Figure 2 data.

TABLE 1   The structural and mechanical characteristics of nanocomposites BSR/CNT

Catalyst	E_n, MPa	E_n/E_m	E_n/E_m, the Equation (5)	R_{CNT}, nm, the Equation (1)	R_{CNT}, nm, the Equation (4)	b, the Equation (6)
Fe/Al$_2$O$_3$	4.9	1.485	1.488	236	278	8.42
Co/Al$_2$O$_3$	3.1	~ 1.0	1.002	236	278	0.27

Footnote: The value E_m for BSR is equal to 3.3 MPa.

In Table 1 the values of elasticity modulus E_n for the studied nanocomposites and E_m for the initial BSR are also adduced. As one can see, if for the nanocomposite BSR/CNT-Fe the very high (with accounting of the condition $W_n = 0.3$ mass %) reinforcement degree $E_n/E_m = 1.485$ was obtained, then for the nanocomposite BSR/CNT-Co reinforcement is practically absent (with accounting for experiment error) $E_n \approx E_m$. Let us consider the reasons of such essential distinction.

As it is known [4], the reinforcement degree for nanocomposites polymer/CNT can be calculated as follows:

$$\frac{E_n}{E_m} = 1 + 11 (c\phi_n b)^{1.7} \qquad (5)$$

Where c is proportionality coefficient between nanofiller ϕ_n and interfacial regions ϕ_{if} relative fractions, b is the parameter, characterizing interfacial adhesion polymer matrix-nanofiller level.

The parameter b in the nanocomposites polymer/CNT case depends on nanofiller geometry as follows [5]:

$$b = 80 \left(\frac{R_{CNT}^2}{L_{CNT} D_{CNT}^2} \right) \qquad (6)$$

Calculated according to the Equation (6) values b (for BSR/CNT-Fe and BSR/CNT-Co R_{CNT} magnitudes were accepted equal to 280 nm and 50 nm, respectively) are adduced in Table 1. As one can see, R_{CNT} decreasing for the second from the indicated nanocomposites results to b reduction in more than 30 times.

The coefficient c value in the Equation (5) can be calculated as follows [4]. First the interfacial layer thickness l_{if} is determined according to the equation [6]:

$$l_{if} = a \left(\frac{r_{CNT}}{a} \right)^{2(d - d_{surf})/d} \qquad (7)$$

Where a is lower linear scale of polymer matrix fractal behavior, accepted equal to statistical segment length l_{st}, d is dimension of Euclidean space, in which fractal is considered (it is obvious that in our case $d = 3$), d_{surf} is CNT surface dimension, which for the studied CNT was determined experimentally and equal to 2.89.

The indicated dimension d_{surf} has a very large absolute magnitude ($2 \leq d_{surf} < 3$ [4]), that supposes corresponding roughness of CNT surface, which BSR macromolecule, simulated by rigid statistical segments sequence [7], cannot be "reproduced". Therefore in practice the effective value d_{surf} (d_{surf}^{ef} for $d_{surf} > 2.5$) is used, which is equal to [7]:

$$d_{surf}^{ef} = 5 - d_{surf} \cdot \qquad (8)$$

And at last, the statistical segment length l_{st} is estimated according to the equation [4]:

$$l_{st} = l_0 C_\infty \tag{9}$$

Where l_0 is the length of the main chain skeletal bond, C_∞ is characteristic ratio. For BSR $l_0 = 0.154$ nm, $C_\infty = 12.8$ [8]. Further, simulating an interfacial layer as cylindrical one with external radius $r_{CNT} + l_{if}$ and internal radius r_{CNT}, let us obtain from geometrical considerations the formula for φ_{if} calculation [6]:

$$\phi_{if} = \phi_n \left[\left(\frac{r_{CNT} + l_{if}}{r_{CNT}} \right)^3 - 1 \right] \tag{10}$$

According to which the value c is equal to 3.47.

The reinforcement degree E_n/E_m calculation results according to the Equation (5) are adduced in Table 1. As one can see, these results are very close to the indicated parameter experimental estimations. From the Equation (5) it follows unequivocally, that the values E_n/E_m distinction for nanocomposites BSR/CNT-Fe and BSR/CNT-Co is defined by the interfacial adhesion level difference only, characterized by the parameter b, since the values c and φ_n are the same for the indicated nanocomposites. In its turn, from the Equation (6) it follows so unequivocally, that the parameters b distinction for the indicated nanocomposites is defined by R_{CNT} difference only, since the values L_{CNT} and D_{CNT} for them are the same. Thus, the fulfilled analysis supposes CNT geometry crucial role in nanocomposites polymer/CNT mechanical properties determination.

Let us note, that the usage of the average value R_{CNT} for nanocomposites BSR/CNT-Co according to Figure 2 data in the Equation (6), which is equal to 130 nm, will not change the conclusions made above. In this case $E_n/E_m = 1.036$, that is again close to the obtained experimentally practical reinforcement absence for the indicated nanocomposite.

29.4   CONCLUSION

Thus, the obtained in the present work results have shown that the nanotubes geometry, characterized by their length, diameter, and ring-like structures radius, is nanocomposites polymer/CNT specific feature. This factor plays a crucial role in the interfacial adhesion polymer matrix – nanofiller level determination and, as consequence, in polymer nanocomposites, filled with CNT, mechanical properties formation.

KEYWORDS

- **Carbon nanotubes**
- **Interfacial adhesion**
- **Nanocomposite**
- **Reinforcement degree**
- **Ring-like structures**

REFERENCES

1. Yanovskii, Yu. G. *Nanomechanics and Strength of Composite Materials*. Publishers of IPRIM RAN Moscow, p. 179 (2008).
2. Bridge, B. *J. Mater. Sci. Lett.*, **8**, № 2, 102–103 (1989).
3. Sheng, N., Boyce, M. C., Parks, D. M., Rutledge, G. C., Abes, J. I., and Cohen, R. E. *Polymer*, **45**, № 2, 487–506 (2004).
4. Mikitaev, A. K., Kozlov, G. V., and Zaikov, G. E. *Polymer Nanocomposites: Variety of Structural Forms and Applications*. Nova Science Publishers Inc., New York, p. 319 (2008).
5. Zhirikova, Z. M., Kozlov, G. V., and Aloev, V. Z. *Mater of VII Intern. Sci.-Pract. Conf. "New Polymer Composite Materials"*. Nal'chik, KBSU, pp. 158–164 (2011).
6. Kozlov, G. V., Burya, A. I., and Lipatov, Yu. S. *Mekhanika Kompozitnykh Materialov*, **42**, № 6, 797-802 (2006).
7. Van Damme, H., Levitz, P., Bergaya, F., Alcover, J. F., Gatineau, L., and Fripiat, J. J. *J. Chem. Phys.*, **85**, № 1, 616–625 (1986).
8. Yanovskii, Yu. G., Kozlov, G. V., and Karnet, Yu. N. *Mekhanika Kompozitsionnykh Materialov i Konstruktsii*, **17**, № 2, 203–208 (2011).

CHAPTER 30

KEY ELEMENTS ON SYNTHESIS, STRUCTURE, PHYSICOCHEMICAL PROPERTIES, AND APPLICATION OF POLYACETYLENE

V. A. BABKIN and G. E. ZAIKOV

CONTENTS

30.1 INTRODUCTION

Features of synthesis, structure, properties, and use of polyacetylenes are considered in this monography. The catalytic polymerization of acetylene using different catalysts are shown. The plasmachemical synthesis of carbines is considered. The results of studying the structure of polyacetylenes by electron spectroscopy are presented. The results of the research of the surface morphology of polyacetylene are presented. Agency of receiving methods of polyacetylene on its properties is shown. It should be noted that the first chapter of present monography (synthesis, structure, physicochemical properties, and application of polyacetylene) prepared for publication by Professor Rakhimov A. I. and Associate Professor Titova E. S. (Volgograd State Technical University). The remaining chapters prepared for publication by Professor Ponomarev O.A. (Bashkirskii State University), Professor Babkin V.A. (Volgograd State Architect-build University, Sebrykov Departament), Associate Professor Titova E. S. and Professor Zaikov G. E. (Moscow, Institute of Biochemical Physics, Russian Academy of Sciences).

30.2 SYNTHESIS AND STRUCTURE OF POLYACETYLENE

The chemical element in periodic system of D. I. Mendeleev carbon possesses a variety of unique properties. It is the reason of that, as carbon and its compounds, and materials on its basis serve as objects of basic researches and are applied in the most various areas.

Scientists thought, that two forms exists of crystal carbon only–diamond and graphite (opinion beginning of 60th years of 20 century). These forms are widespread in the nature and they are known to mankind with the most ancient times.

The question on an opportunity of existence of forms of carbon with sp-hybridization of atoms was repeatedly considered theoretically. In 1885 German chemist A. Bayer tried to synthesize chained carbon from derivatives of acetylene by a step method. However Bayer's attempt to receive polyin has appeared unsuccessful. He received the hydrocarbon consisting from four molecules of acetylene, associated in a chain, and appeared extremely unstable.

A. M. Sladkov, V. V. Korshak, V. I. Kasatochkin, and Yu. P. Kudryavcev [1] observed loss of a black sediment of polyin compound of carbon having the linear form at transmission acetylene in a water-ammonia solution of salt Cu (II) (oxidizing dehydropolycondensation of acetylene led obviously to polyacetylenides of copper). This powder blew up at heating in a dry condition, and in damp at a detonation. The process of oxidative dehydropolycondensation of acetylene can be written down in a following kind schematically [1] at $x + y + z = n$:

$$n \ H-C\equiv C-H$$

$$\Big\downarrow Cu^{2+}$$

$$H(-C\equiv C-)_x Cu \ + \ H(-C\equiv C-)_y H \ + \ Cu(-C\equiv C-)_z H$$

$$\Big\downarrow FeCl_3$$

$$H(-C\equiv C-)^-_n H$$

At surplus of ions Cu^{2+} the mix of various polyins and polyacetylenides of copper various molecular weights are formed. Additional oxidation of products received at this stage (with help $FeCl_3$ or $K_3[Fe(CN)_6]$) leads to formation polyins with the double molecular weight. The last do not blow up any more at heating and impact, but contain a plenty of copper. Possibly, trailer atoms of copper stabilize polyins by dint of to complexation.

The content by carbon was 90% of clean polyin (cleaning cleared from copper and impurity of other components of the reactionary medium). Only multihours heating of samples of polyin at 1000°C in vacuum has allowed receiving analytically pure samples of α-carbyne. Similar processing results not only in purification, but also to partial crystallization of polyacetylene.

Under A. M. Sladkov's offer such polyacetylene have named carbyne (from Latin *carboneum* (carbon) with the termination in, accepted in organic chemistry for a designation of acetylene bond).

By acknowledgement of polyin structures in chains is formation of oxalic acid after ozonation hydrolysis of carbine [2, 3]:

$$(-C=C-)_n \ \xrightarrow{O_3} \ \left(\begin{array}{c} -C\equiv C- \\ | \quad | \\ O \ O \\ V \\ O \end{array} \right)_n \ \xrightarrow{H_2O} \ nHOOC - COOH.$$

New linear polymer with cumulene bonds was received [2, 3]. It has named polycumulene. The proof of such structure became that fact, that at ozonation of polycumulene is received only carbon dioxide:

$$(=C=C)_n \ \xrightarrow{O_3} \ 2nCO_2$$

Cumulene modification of carbyne (β-carbyne) has been received on specially developed by Sladkov two-stage method [3]. At the first stage spent polycondensation of suboxide of carbon (C_3O_2) with dimagnesium dibromine acetylene as Grignard reaction with formation polymeric glycol:

$$nO=C=C=C=O+ nBrMgC=CMgBr \longrightarrow \left(-C\equiv C-\underset{\underset{OH}{|}}{C}=C=\underset{\underset{OH}{|}}{C}-\right)_n$$

At the second stage this polymeric glycol reduced by stannous chloride hydrochloric acid:

$$\left(-C\equiv C-\underset{\underset{OH}{|}}{C}=C=\underset{\underset{OH}{|}}{C}-\right)_n \xrightarrow[- (HCl + SnO_2)]{+ SnCl2} (=C=C=C=C=C=)_n.$$

High-molecular cumulene represents an insoluble dark-brown powder with the developed specific surface (200–300 m²/g) and density 2.25 g/sm³. At multihours heating at 1000°C and the depressed pressure polycumulene partially crystallizes. Two types of monocrystals have been found out in received after such annealing a product by means of transmission electronic microscopy. Crystals corresponded to α- and β-modifications of carbyne.

One of the most convenient and accessible methods of reception carbyne or its fragments–reaction of dehydrohalogenation of the some polymers content of halogens (GP). The feature of this method is formation of the carbon chain at polymerization corresponding monomers. The problem at synthesis carbyne consists only in that at full eliminating of halogen hydride with formation of linear carbon chain. Exhaustive dehydrohalogenation is possible, if the next atoms of carbon have equal quantities of atoms of halogen and hydrogen. Therefore convenient GP for reception of carbyne were various polyvinyliden halogenides (bromides, chlorides, and fluorides), poly (1,2-dibromoethylene), poly (1,1,2 or 1,2,3-trichlorobutadiene), for example:

$$(-CH_2-CHal_2-)_n \xrightarrow[-nHHal]{+B^-} (-CH=CHal-)_n \xrightarrow[-nHHal]{+B^-} (=C=C=)_n.$$

The reaction of dehydrohalogenation typically carries out at presence of solutions of alkalis (B^-) in ethanol with addition of polar solvents. At use of tetrahydrofuran synthesis goes at a room temperature. This method allows avoiding course of collateral reactions. The amorphous phase only cumulene modification of carbyne is received as a result. Then, crystal of β-carbyne is synthesized from amorphous carbine by solid-phase crystalization.

The next method is dehydrogenation of polyacetylene. At interaction of polyacetylene with metallic potassium at 800°C and pressure 4 GPa led to dehydrogenation and formation of potassium hydride, the carbon matrix containing potassium. After removal potassium from products (acid processing) precipitate out brown plate crystals of β-carbyne in hexagonal forms by diameter ~ 1 mm and thickness up to 1 micrometer.

Carbyne also can be received by various methods of chemical sedimentation from a gas phase.

30.2.1 PLASMOCHEMICAL SYNTHESIS OF CARBYNE

At thermal decomposition of hydrocarbons (acetylene, propane, heptane, and benzol), carbone tetrachloride, carbon bisulfide, and acetone in a stream nitrogen plasma is received the disperse carbon powders containing carbyne. The monocrystals of white color and (white or brown) polycrystals remain after selective oxidation of aromatic hydrocarbon. It is positioned, that formation of carbyne does not depend by nature initial organic compound. The moderate temperature of plasma (~3200K) and small concentration of reagents promote process.

30.2.2 LASER SUBLIMATION OF CARBON

Carbyne has been received at sedimentation on a substrate of steams of negative ions of the carbon after laser evaporation of graphite in 1971. The silvery-white layer was received on a substrate. This layer, according to data X-ray and diffraction researches, consists from amorphous and crystal particles of carbyne with the average size of crystallites> 10^{-5} sm.

30.2.3 ARC CRACKING OF CARBON

The evaporation in electric arc spectrally pure coals with enough slow polymerization and crystallization of a carbon steam on a surface of a cold substrate yields to product in which prevail carbyne forms of carbon.

30.2.4 ION-STIMULATED PRECIPITATION OF CARBYNE

At ion-stimulated condensation of carbon on a lining simultaneously or alternately the stream of carbon and a stream of ions of an inert gas moves. The stream of carbon is received by thermal or ionic evaporation of graphite. This method allows to receive carbyne films with a different degree of orderliness (from amorphous up to monocrystalline layers), carbynes of the set updating, and also a film of other forms of carbon. Annealing of films of amorphous carbon with various near order leads to crystallization of various allotropic forms carbon, including carbyne.

Sladkov has drawn following conclusions on the basis of results of experiments on synthesis carbyne by methods of chemical sedimentation from gas phase:

- White sediments of carbyne are received, possibly, in the softest conditions of condensation of carbon—high enough vacuum, small intensity of a stream, and low energy of flying atoms or groups of atoms, small speed of sedimentation,
- Chains, apparently, grow perpendicularly to a lining, not being crosslinked among themselves,

- Probably, being an environment monovalent heteroatoms stabilise chains, do not allow them to be crosslinked.

Reception of carbyne from carbon graphite materials leads by heating of cores from pyrolytic graphite at temperature 2700–3200K in argon medium. This leads to occurrence on the ends of cores a silvery white strike (already through 15–20 s). This strike consists of crystals carbyne that is confirmed by data of method electron diffraction.

In 1958 Natta with employees are polymerized acetylene on catalyst system $Al(C_2H_5))_3 — Ti(OC_3H_7)_4$ [7, 8].

The subsequent researches [9-11] led to reception of films stereoregular polyacetylene. The catalyst system $Al(Et)_3–Ti(OBu)_4$ provides reception of films of polyacetylene predominantly (up to 98%).

The films of polyacetylene are formed on a surface of the catalyst or practically too any lining moistened by a solution of the catalyst (it is preferable in toluene), in an atmosphere of the cleared acetylene [11]. The temperature and pressure of acetylene control growth of films [12, 13]. Homogeneous catalyst system before use typically maintain at a room temperature. Thus reactions of maturing of the catalyst occur [14]:

$$Ti(OBu)_4 + AlEt_3 \rightarrow EtTi(OBu)_3 \rightarrow AlEt_2(OBu)$$

$$2EtTi(OBu)_3 \rightarrow 2Ti\,(OBu)_3 + CH_4 + C_2H_6$$

$$Ti(OBu)_3 + AlEt_3 \rightarrow EtTi(OBu)_2 + Al(Et)_2OBu$$

$$EtTi(OBu)_2 + AlEt_3 \rightarrow EtTi(OBu)_2 \cdot Al\,(Et)_3$$

Ageing of the catalyst in the beginning raises its activity. However eventually the yield of polyacetylene falls because of the further reduction of the titan:

$$EtTi(OBu)_3 + Al\,(Et)_3 \rightarrow Ti(OBu)_4 + Al(Et)_2(OBu) + C_2H_4 + C_2H_6$$

The jelly-like product of red color is formed if synthesis led at low concentration of the catalyst. This product consists from confused fibrils in the size up to 800 Å. The foam material with density from 0.04 to 0.4 g/sm^3 is possible to receive from the dilute gels by sublimation of solvent at temperature below temperature of its freezing [15].

Research of speed dependence for formation of films of polyacetylene from catalyst concentration and pressure of acetylene has allowed to find an optimum parity of components for catalyst Al/Ti = 4. Increase of this parity up to 10 leads to increase in the sizes of fibrils [16]. Falling of speed of reaction in the end of process speaks deterioration of diffusion a monomer through a layer of the film formed on a surface of the catalyst [17]. During synthesis the film is formed simultaneously on walls of a flask and on a surface a catalytic solution. Gel collects in a cortex. The powder of polyacetylene settles at the bottom of a reactor. The molecular weight (M_n) a powder below, than gel, also is depressed with growth of concentration of the catalyst up to 400–500 [18]. The molecular weight of jellous polyacetylene slightly decreases with growth of concentration of the catalyst and grows from 2×10^4 up to 3.6×10^4 at rise in temperature from 78°C up to 10°C. The molecular weight of polyacetylene in a film is twice less, than in gel [19].

The greater sensitivity of the catalyst to impurity does not allow estimating unequivocally influence of various factors on M_p of polyacetylene. The low-molecular products with $Mp \sim -1200$ are formed at carrying out of synthesis in the medium of hydrogen [20]. The concentration of acetylene renders significant influence on M_p—at increase of its pressure up to 760 mm Hg increases M_p up to 1,20,000 [14, 21]. Apparently, the heterogeneity of a substratum arising because of imperfection of techniques for synthesis is the reason for some irreproducibility properties of the received polymers. It is supposed, that synthesis is carried out on "surface" of catalytic clusters. Research by electron paramagnetic resonance (EPR) method has allowed distinguishing in the catalyst up to 4 types of complexes [22]. Polyacetylene *cis*-transoid structures are formed as a result.

The formation of *trans*-structure at heats speaks thermal isomerization. The alternative opportunity *trans*-disclosing of triple bond in a catalytic complex for a transitive condition is forbidden spatially [22, 23]. The structure of a complex and kinetics of polymerization are considered in works in more detail [14, 24].

Parshakov A. S. with co-authors [25] have offered a new method of synthesis organo-inorganic composites nanoclusters transitive metals in an organic matrix by reactions of compounds of transitive metals of the maximum degrees of oxidation with monomers which at the first stage represent itself as a reducer. Formed thus clusters metals of the lowest degrees of oxidation are used for catalysis of polymerization a monomer with formation of an organic matrix.

Thus it is positioned, that at interaction $MoCl_5$ with acetylene in not polar mediums there is allocation hydrochloric acid (HCl), downturn of a degree of oxidation of molybdenum and formation metalloorganic nanoclusters. Two distances Mo-Mo are found out in these nanoclusters by method extended x-ray absorption fine structure (EXAFS) spectroscopy. In coordination sphere Mo there are two nonequivalent atom of chlorine and atom of carbon. On the basis of results matrix-assisted laser desorption/ionization time of flight (MALDI-TOF) mass-spectrometry the conclusion is made, that cluster of molybdenum has 12 or the 13 nuclear metal skeleton and its structure can be expressed by formulas $[Mo_{12}Cl_{24}(C_{20}H_{21})]^-$ or $[Mo_{13}Cl_{24}(C_{13}H_8)]^-$.

The greater sensitivity of the catalyst impurity does not allow to estimate unequivocally influence of various factors on M_n of polyacetylene. The low-molecular products with $M_n \sim 1200$ are formed at carrying out of synthesis in the environment of hydrogen [20]. Concentration of acetylene renders significant influence on M_n—M_n increases up to 1,20,000 at increase of its pressure up to 760mm Hg [14, 21]. Apparently, the heterogeneity of a substratum arising because of imperfection of techniques of synthesis is the reason for some nonrepeatability properties of the received polymers. It is supposed, that synthesis is carried out on "surface" of catalytic clusters. Research by method EPR has allowed to distinguish in the catalyst up to 4 types of complexes [22].

A Polyacetylene *cis*-transoid structures is as a result formed. The formation a trance-structure at heats speaks thermal isomerization, previous crystallization of circuits. The alternative opportunity–a trance-disclosing of triple bond in a catalytic complex for a transitive condition is forbidden spatially [22, 23]; in more detail the structure of a complex and a kinetics of polymerization are considered in works [14, 24].

Parshakov A.S. with co-authors [25] has offered a new method of synthesis of organo-inorganic composites nanoclusters of transitive metals in an organic matrix by reactions of connections of transition metals of the maximum degrees of oxidation with monomers. Monomers represent itself as a reducer at the first stage. Formed thus clusters of metals of the lowest degrees of oxidation are used for a catalysis of polymerization a monomer with formation of an organic matrix.

Thus it is positioned, that at interaction $MoCl_5$ with acetylene in not polar mediums there is allocation HCl, downturn of a degree of oxidation of molybdenum and formation metalloorganic nanoclusters. Two distances Mo-Mo are found out in these clusters by a method of EXAFS-spectroscopy. Two nonequivalent atoms of chlorine and atom of carbon are available in coordination sphere Mo. The conclusion is made on the basis of results MALDI-TOF of mass-spectrometry, that cluster molybdenum has 12 or the 13 nuclear metal skeleton and its structure can be expressed by formulas $[Mo_{12}Cl_{24}(C_{20}H_{21})]^-$ - or $[Mo_{13}Cl_{24}(C_{13}H_8)]^-$.

The set of results of Infrared, Raman, MASS, NMR ^{13}C and RFES has led to a conclusion, that the organic part of a composite represents polyacetylene a *trans*-structure. Polymeric chains lace, and alongside with the interfaced double bonds, are present linear fragments of twinned double $-HC=C=CH-$ and triple $-C≡C-$ bonds.

Reactions $NbCl_5$ with acetylene also is applied to synthesis of the organo-inorganic composites containing in an organic matrix clusters of transitive elements not only VI, but other groups of periodic system.

Solutions $MoCl_5$ sated at a room temperature in benzene or toluene used for reception of organo-inorganic composites with enough high concentration of metal. Acetylene passed through these solutions during 4–6 hr. Acetylene preliminary refined and drained from water and possible impurity. The color of a solution is varied from dark-yellow-green up to black in process of transmission acetylene. The solution heated up, turned to gel of black color, and after a while the temperature dropped up to room. The reaction was accompanied by formation HCl. The completeness of interaction pentachloride with acetylene judged on the termination of its allocation.

The sediment, similar to gel, settled upon termination of transmission acetylene. It filtered off in an atmosphere of argon, washed out dry solvent, and dried up under vacuum.

The received substances are fine dispersed powders of black color, insoluble in water, and in usual organic solvents.

The solutions after branch of a deposit represented pure solvent according to NMR. Formed compounds of molybdenum and products of oligomerization acetylene precipitated completely. The structure of products differed under the maintenance of carbon depending on speed and time of transmission acetylene a little. Thus relation C:H was conserved close to unit, and Cl:Mo close to two. The structure of products of reaction differed slightly in benzene and toluene (Table 1) and was close to $MoCl_{1.9 \pm 0.1}(C_{30 \pm 1}H_{30 \pm 1})$.

TABLE 1 Data of the element analysis of products for reaction $MoCl_5$ with acetylene in benzene and toluene

Solution	Percentage and weight%							
	C		H		Cl		Mo	
	Findings	Calcu-lated	Findings	Calcu-lated	Findings	Calcu-lated	Findings	Calcu-lated
Benzene	65,94	64,67	5,38	5,38	11,70	12,73	16,98	17,22
Toluene	66,23		5,24		11,87		16,66	

The presence on diffraction patterns the evolved products of a wide maximum at small corners allowed to assume X-ray amorphous or nanocrystalline a structure of the received substances. By a method of scanning electronic microscopy (SEM) it was revealed, that substances have low crystallinity and nonfibrillary morphology (Figure 1(a)).

a   ⊢—9μm—⊣

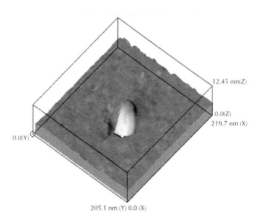

FIGURE 1   Photomicrographes $MoCl_{1.9 \pm 0.1}(C_{30 \pm 1}H_{30 \pm 1})$ according to SEM (a) and ASM (b).

30.3   PHYSICOCHEMICAL PROPERTIES AND APPLICATION OF POLYACETYLENE

By results of atomic-power microscopy (ASM) particle size can be estimated within the limits of $10 \div 15$ nm (Figure 1(b)). By means of translucent electronic microscopy has been positioned, that the minimal size of morphological element $MoCl_{1.9 \pm 0.1}(C_{30 \pm 1}H_{30 \pm 1})$ makes $1 \div 2$nm.

The substances are steady and do not fly in high vacuum and an inert atmosphere up to 300°C. Formation of structures $[Mo_{12}Cl_{24}(C_{20}H_{21})]^-$ and $[Mo_{13}Cl_{24}(C_{13}H_8)]$ is supposed also on the basis of mass-spectral of researches.

The spectrum EPR of composite $MoCl_{1.9 \pm 0.1}$ $(C_{30 \pm 1}H_{30 \pm 1})$ at 300 K (Figure 2(a)) consists of two isotropic lines. The intensive line $g = 1.935$ are carried to unpaired electrons of atoms of molybdenum. The observable size of the g-factor is approximately equal to values for some compounds of trivalent molybdenum. For example, in $K_3[InCl_6]\cdot 2H_2O$, where the ion of molybdenum Mo $(+ 3)$ isomorphically substitutes In $(+ 3)$, and value of the g-factor makes 1.93 ± 0.06.

The line of insignificant intensity with $g = 2.003$, close to the g-factor free electron $- 2.0023$, has been carried to unpaired electrons atoms of carbon of a polyacethylene matrix. Intensity of electrons signals for atoms molybdenum essentially above, than for electrons of carbon atoms of a matrix. It is possible to conclude signal strength, that the basic contribution to paramagnetic properties of a composite brings unpaired electrons of atoms of molybdenum.

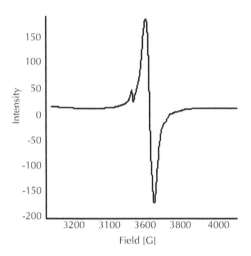

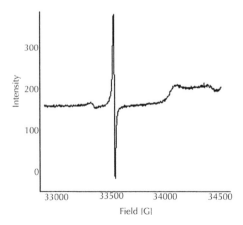

FIGURE 2 Spectrum EPR of composite $MoCl_{1.9 \pm 0.1}$ $(C_{30 \pm 1}H_{30 \pm 1})$ at 300 K, removed in a continuous mode in X-range (a) and W-range (b).

The spectrum EPR which has been removed in a continuous mode at 30 K, (Figure 2(b)) has a little changed at transition from X-to a high-frequency W-range. Observable three wide lines unpaired electrons atoms of molybdenum have been carried to three axial components with $g_1 = 1.9528$, $g_2 = 1.9696$ and $g_3 = 2.0156$ accordingly. Unpaired electrons atoms of carbon of a polyacethylene matrix the narrow signal $g = 2.0033$ answers. In a pulse mode of shooting of spectra EPR at 30 K (Figure 3) this line decomposes on two signals with $g_1 = 2,033$ and $g_2 = 2,035$. Presence of two signals EPR testifies to existence in a polyacethylene matrix of two types of the paramagnetic centers of the various nature which can be carried to distinction in their geometrical environment or to localized and delocalized unpaired electrons atoms of carbon polyacethylene chain.

The measurement of temperature dependence of a magnetic susceptibility X_g in the field of temperatures $77 \div 300$ K has shown, that at decrease in temperature from room up to 108 K the size of a magnetic susceptibility of samples is within the limits of sensitivity of the device or practically is absent. The sample started to display a magnetic susceptibility below this temperature. The susceptibility sharply increased at the further decrease in temperature.

The magnetic susceptibility a trance-polyacetylene submits to Curie law and is very small on absolute size. Comparison of a temperature course X_g^x a composite and pure allows to conclude a trance-polyacetylene, that the basic contribution to a magnetic susceptibility of the investigated samples bring unpaired electrons atoms of molybdenum in cluster. Sharp increase of a magnetic susceptibility below 108 K can be connected with reduction of exchange interactions between atoms of metal.

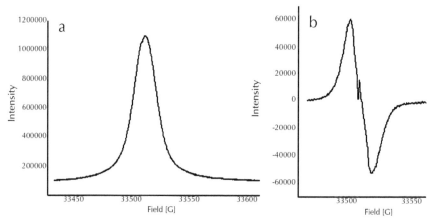

FIGURE 3 Spectrum absorption EPR of composite $MoCl_{1.9 \pm 0.1}$ $(C_{30 \pm 1}H_{30 \pm 1})$, removed in a pulse mode in a W-range at 30 K (a) and its first derivative (b).

The size of electroconductivity compressed samples. $MoCl_{1.9 \pm 0.1}(C_{30 \pm 1}H_{30 \pm 1})$, measured at a direct current at a room temperature $(1.3 \div 3.3) \cdot 10^{-7}$ $Ohm^{-1} \cdot cm^{-1}$ is in a range of values for a trance-polyacetylene and characterizes a composite as weak dielectric or the semiconductor. The positioned size of conductivity of samples at an alternating current $\sigma = (3.1 \div 4.7) \cdot 10^{-3}$ Ohm$^{-1} \cdot$cm^{-1} can answer presence of ionic (proton) conductivity that can be connected with presence of mobile atoms of hydrogen at structure of polymer.

Research of composition, structure, and properties of products of interaction NbCl$_5$ with acetylene in a benzene solution also are first-hand close and differ a little with the maintenance of carbon (Table 2).

TABLE 2 Data of the element analysis of products of interaction NbCl5 with acetylene in a solution and at direct interaction

| | The weights content, % | | | | | | |
| | C | | H | | Cl | | Nb | |
	Findings	Calculated	Findings	Calculated	Findings	Calculated	Findings	Calculated
In a solution	45.00	45.20	3.60	3.76	20.50	22.20	26.50	28.80
At direct interaction	41.60	45.20	4.02	3.76	22.81	22.20	27.78	28.80

To substances formula $NbCl_{2\pm0.1}(C_{12\pm1}H_{12\pm1})$ can be attributed on the basis of the received data. Interaction can be described by the equation:

$$NbCl_{5(solv/solid)} + nC_2H_2 \rightarrow NbCl_{2\pm0.1}(C_{12\pm1}H_{12\pm1})\downarrow + (n\text{-}12)C_6H_6 + 3HCl\uparrow + Q$$

The Wide line was observed on diffraction pattern. $NbCl_{2\pm0.1}(C_{12\pm1}H_{12\pm1})$ at $2\theta =$ 23–24°C. It allowed to assume a nanocrystalline structure of the received products.

Studying of morphology of surface $NbCl_{2\pm0.1}(C_{12\pm1}H_{12\pm1})$, received by direct interaction, method SEM has shown, that particles have predominantly the spherical form, and their sizes make less than 100 nm (Figure 4).

9μm

FIGURE 4   Microphoto of particles $NbCl_{2\pm0.1}(C_{12\pm1}H_{12\pm1})$, received by method SEM.

The Globular form of particles and their small size testify to the big size of their specific surface. It will be coordinated with high catalytic activity $NbCl_{2\pm0.1}(C_{12\pm1}H_{12\pm1})$.

The fibrils are formed in many cases as a result of synthesis [24]. The morphology of polyacetylene films practically does not depend on conditions of synthesis. The diameter of fibrils can change depending on these conditions and typically makes 200–800 Å [14, 24]. At cultivation of films on substrates the size fibrils decreases. The same effect is observed at reception of polyacetylene in the medium of other polymers. Time of endurance (ageing) of the catalyst especially strongly influences of the size of fibrils. The size of fibrils increases with increase in time of ageing. Detailed research of growth fibrils on thin films a method of translucent electronic microscopy has allowed to find out microfibrillar branchings on the basic fibril (the size 30–50 Å) and thickenings in places of its gearing, and also presence of rings on the ends of fibrils.

The essential changes in morphology of a film at isomerization of polyacetylene are not observed. The film consists from any way located fibrils. Fibrils sometimes are going in larger formations [23]. Formation of a film is consequence of interaction fibrils among themselves due to adhesive forces.

The big practical interest is represented catalytic system AlR_3-$Ti(oBu)_4$ [48]. Poly-acetylene is received on it at $-60°C$, possesses fibrous structure and can be manu-factured usual, accepted in technology of polymers by methods. The particle size in-creases from 100 up to 500 Å at use of the mixed catalyst and the increase in density of films are observed. A filtration of suspension it is possible to receive films of any sizes. On the various substrates possessing good adhesive properties, it is possible to receive films dispersion of gel. Polymer easily doping AsF_5, $FeCl_3$, I_2 and others electron acceptors. Preliminary tests have shown some advantages of such materials at their use in accumulators [48]. Gels polyacetylene with the similar properties, received on others catalytic systems, represent the big practical interest [49]. Data on techno-logical receptions of manufacturing of polyacetylene films from gels with diameter of particles 0.01–1.00 mm are in the patent literature [50]. Films are received at presence of the mixed catalyst at an interval of temperatures from $-100°C$ up to $-48°C$. These results testify to an opportunity of transition to enough simple and cheap technology of continuous process of reception of polyacetylene films. Rather accessible catalysts are WCl_6 and $MoCl_6$. Acetylene polymerizes at $20°C$ and pressure up to 14 atm at their presence. However the received polymers contain carbonyl groups because of presence of oxygen and possess low molecular weight [51]. The complex systems including in addition to WCl_6 or $MoCl_6$ tetraphenyltin are more perspective for recep-tion of film materials. The Film of doped predominantly (90%) a trance-structure with fibrous morphology is formed on a surface of a concentrated solution of the vanadic catalyst. Diameter of fibrils of the polymer received on catalyst $MoCl_5 - Ph_4Sn$, can vary within the limits of 300 Å–10 000 Å, in case of catalyst $WCl_6 - Ph_4Sn$ it reaches 1.2×10^5 Å [30].

Research of a kinetics of polymerization of acetylene on catalysts $Ti(OBu)_5$–$Al(Et)_3$, WCl_6-PhtSn, $MoCl_5$-Ph_4Sn, $Ti(CH_2CeH_5)$, has shown, that speed of process falls in the specified number [31]. Films, doped by various acceptors $[CH(SbF_5)_{0.7}]_n$, $[CH(CF_3SO_3H)_{0.8}]_m$ had conductivity 10–20 $Ohm^{-1}\cdot cm^{-1}$ at $20°C$.

New original method of reception of films doped in a trance-form is polymeriza-tion of a 7,8-bis(trifluoromethyl) tricyclo[4,2,2,0]deca-3,7,9-trien (BTFM) with dis-closing a cycle.

The polymerization occurs on catalytic system $WCl_6 - Sn(CH_3)_4$, precipitated on surfaces of a reactor. The film prepolymer as a result of heating in vacuum at 100–150° detaches trifluoromethylbenzene. The silvery film polyacetylene is formed. The den-sity of polymer reaches 1.1 g/cm^3 and comes nearer to flotation density polyacety-lene, received in other ways. Received this method of amorphous polyacetylene has completely a trance-configuration and does not possess fibrous structure. The rests of 1,2-bis(trifluoromethylbenzene) are present at polymer according to Infrared-spec-troscopy. Crystal films of polyacetylene with monoclinic system and $\beta = 91.5°$ are re-ceived at long heating of prepolymer on networks of an electronic microscope at 100° in vacuum. Improvement of a method has allowed to receive films and completely oriented crystal polyacetylene. In the further ways of synthesis of polymers from oth-ers monomers have been developed.

Naphthalene evolves at heat treatment of prepolymer in the first case. Anthracene evolves in the second case. Purification polymer from residual impurities occurs when the temperature of heat treatment rise. According to spectral researches, the absorption caused by presence of sp-hybrid carbon is not observed in films. Absorption in the field of 1480 cm^{-1} caused by presence of C = C bonds in Spectra KP of considered polymers, is shifted compared to the absorption observed in polyacetylene, received by other methods (1460 cm^{-1}).

It is believed that it is connected with decrease of size of interface blocks. The obtained films are difficult doping in a gas phase due to its high density. Conductivity of initial films reaches 10–200 (Ohm·cm^{-1}) when doped with bromine or iodine in a solution [51].

Solutions of complex compound cyclopentadienyl-dititana in hexane and sodium cyclopentadienyl complex have high catalytic activity. Films with a metallic luster can be obtained by slowly removing the solvent from the formed gel polyacetylene in vacuum. It is assumed that the active complex has a tetrahedral structure. The polymerization mechanism is similar to the mechanism of olefin polymerization on catalyst Ziegler - Natta. Obtained at 80°C *cis*-polyacetylene films after doping had a conductivity of 240 (Ohm·cm^{-1}).

The classical methods of ionic and radical polymerization do not allow to receive high-molecular polymers with system of the conjugated bonds because of isomerization the active centers [2] connected with a polyene chain. Affinity to electron and potential of ionization considerably varies with increase in effective conjugated. One of the methods, allowing to lead a cation process of polymerization, formation of a complex with the growing polyconjugated chains during synthesis. Practical realization of such process probably at presence of the big surplus of a strong acceptor of electrons in the reaction medium. In this case the electronic density of polyene chain falls. The probability of electron transfer from a chain on the active center decreases accordingly.

The polyacetylene films were able to synthesize on an internal surface of a reactor in an interval of temperatures from − 78 to − 198°C at addition of acetylene to arsenic pentafluoride. The strips of absorption *cis*-polyacetylene are identified in the field of 740 cm^{-1} and doped complexes in the field of 900 and 1370 cm^{-1}.

Similar in composition films were prepared by polymerization vinylacetylene in the gas phase in the presence SbF_5. However, to achieve a metallic state has failed. Soluble polyenes were obtained in solution AsF_3 in the by cationic polymerization, including soluble agents. Polymerization was carried out at the freezing temperature of acetylene. From the resulting solution were cast films with low conductivity, characteristic of weakly doped polyacetylene with 103 molecular weight. Spectral studies confirmed the presence of the polymer obtained in the *cis*-structure. Practical interest are insoluble polymers with a conductivity of 10^{-3} (Ohm-cm^{-1}), obtained at − 78°C polymerization of acetylene in the presence AsF_5, NaAsF, SiF, AsF_3, BF_3, SbF_5, and PF_6.

The effective co-catalyst of cationic polymerization of acetylene and its derivatives are compounds of bivalent mercury and its organic derivatives. As a result, the reactions of oxide or mercury salts with proton and aprotic acids into saturated hydrocarbon formed heterogeneous complexes. They are effectively polymerized acetylene vinylacetylene, phenylacetylene, and propargyl alcohol. Polyacetylene predominantly *trans*-structure with a crystallinity of 70% was obtained in the form of films during the polymerization of acetylene on the catalyst surface. According to X-ray studies, the main reflection corresponds to the interplanar distance d = 3.22 Å. Homogeneous catalysts obtained in the presence of aromatic ligands. Active complex of the catalyst with acetylene is stable at low temperatures. The alkylation of solvent and its interpolymerization with acetylene is in the presence of aromatic solvents (toluene and benzene). In the infrared spectrum of the films revealed absorption strias corresponding to the aromatic cycle. The patterns of polymerization of acetylene monomers, the effect of temperature and composition of the catalyst on the structure and properties of the resulting polymers were studied. Morphology of the films showed the absence of fibrils.

Almost all of the methods of synthesis polyacetylene with high molecular weight lead to the formation of insoluble polymers. Their insolubility due to the high intermolecular interaction and form a network structures. For the polymers obtained by Ziegler catalyst systems, the crosslink density of the NMR data of 3–5%. This is confirmed by ozonolysis. The quantum chemical calculations confirm that the isomerization process intermolecular bonds are formed.

The most accessible and promising method for synthesis of the polyacetylene of linear structure is the polymerization of acetylene in the presence of metals VIII group, in the presence with reducing agents (catalyst Luttlnger). Polyacetylene with high crystallinity was obtained by polymerization of acetylene on the catalyst system $Co(NO_3)_2$–$NaBH_4$ with component ratio 1:2. Both components are injected into the substrate containing monomer, to prevent the death of a catalyst. Raising the temperature and the concentration of sodium borane leads to partial reduction of the polymer. The activity of nickel complex can be significantly improved if the polymerization leads in the presence of $NaBH_2$. Crystalline polymers, obtained at low temperatures, do not contain fibrils. Crystallites have dimensions of 70 Å.

A typical reflex observed at 23.75°C (d = 3.74 Å) confirms the *trans*-structure of the polymer. Palladium complexes are ineffective in obtaining high molecular weight polymers. Catalyst Luttlnger enables one to obtain linear polymers of *cis*-structure, characterized by high crystallinity. The yield of polyacetylene is 25–30 g/catalyst.

The chlorination of the freshly prepared polymers at low temperatures allows to obtain soluble chlorpolymers about 104 molecular weight [27]. Although the authors argue that the low-temperature chlorination, in contrast to hydrogenation, there is no polymer degradation, data suggest that an appropriate choice of temperature and solvent derived chlorinated polymers have a molecular weight up to $2.5 \cdot \times 10^5$. Destruction more visible at chlorination on light and on elevated temperatures. The proof of the linear structure of polyacetylene is the fact that soluble iodinated polymers are obtained by iodination polyacetylene suspension in ethanol.

Systematic studies of methods for the synthesis of polyacetylene allowed to develop a simple and convenient method of obtaining the films on various substrates wetted by an ethereal solution of the catalyst. The disadvantages of these films, as well as films produced by Shirakava [24, 25] are difficult to clean them of residual catalyst and the dependence of properties on the film thickness. Much more manufacturable methods for obtaining films of polyacetylene spray pre-cleaned from residues of the catalyst in a stream of polyacetylene gels inert gas or a splash of homogenized gels [49]. The properties of such films depend on the conditions of their formation. Free film thickness of 2–3 mm are filtered under pressure in an inert atmosphere containing a homogenized suspension of polyacetylene 5–10 g/L. The suspension formed in organic media at low temperatures in the presence of the catalyst $Co(NO_3)_2$–$NaBH_4$. The films obtained by spraying a stream of inert gas, homogeneous, have good adhesion to substrates made of metal, polyurethane, polyester polyethylenetereftalate, polyimide, and so on.

Suspension of polyacetylene changes its properties with time significantly. Crosslinking and aggregation of fibrils observed in an inert atmosphere at temperatures above –20°C. This leads to a decrease in the rate of oxidation and chlorination. The morphology of the films changes particularly striking during the aging of the suspension in the presence of moisture and oxygen—increasing the diameter of the fibrils, decreasing their length, breaks and knots are formed. The suspension does not change its properties in two weeks.

The preparation of soluble polymers with a system of conjugated double bonds, and high molecular weight is practically difficult because of strong intermolecular interactions. Sufficiently high molecular weight polyacetylene were obtained in the form of fine-dispersed particles during the synthesis on a Luttinger catalyst in the presence surfactants—copolymer of styrene and polyethylene with polyethylene oxide. Acetylene was added to the catalyst and the copolymer solution in a mixture of cyclohexane tetrahydrofuran at $-60°C$ and heated to $-30°C$. Stable colloidal solutions with spherical particles in size from 40 to 2000 Å formed. The density of selected films is 1.15 g/cm³. Colloidal solutions of polyacetylene can be obtained in the presence of other polymers that prevent aggregation of the forming molecules of polyacetylene.

Polyacetylene obtained in the presence of Group VIII metals, in combination with $NaBH_4$, has almost the same morphology, as a polymer synthesized by Shirakawa [25]. The dimensions of fibrils lay in the range 300–800 Å, and depend on the concentration of the catalyst, the synthesis temperature and medium [27].

The thermogravimetric curves for the polyacetylene, there are two exothermic peaks at 145°C and 325°C [16]. The first of these corresponds to an irreversible *cis-trans* isomerization. Migration of hydrogen occurs at 325°C, open chain and crosslinking without the formation of polyacetylene volatile products. The color of the polymer becomes brown. A large number of defects appear. In the infrared spectrum there are absorption bands characteristic of the CH_2, CH_3, $-C=C-$ and $-C_2H_5$–groups [16].

Structuring polymer occurs in the temperature range 280–380°C. But 72% of initial weight of polyacetylene losses at 720°C. The main products of the decomposition of polyacetylene are benzene, hydrogen, and lower hydrocarbons [15]. The crystallinity of polyacetylene reduced when heated in air to 90° after several hours. The brown amorphous substance, similar cuprene, obtained after 70 hr.

Catalytic hydrogenation of polyacetylene leads to the formation of cross-linked product [29]. Non-crosslinked and soluble products are obtained in the case of hydrogenation of polyacetylene doped with alkali metals [30, 31]. Polyacetylenes are involved in redox reactions that occur in processing strong oxidizing and reducing agents (iodine, bromine, AsF_5, and Na-naphthalene in order to significantly increase the electrical conductivity [32].

The practical use of polyacetylene is complicated by its easy oxidation by air oxygen [33]. Oxidation is easily exposed to the polymer obtained by polymerization of acetylene [34, 35]. The *cis*-or *trans*-$(CH)_x$ in air or oxygen for about an hour exposed to the irreversible oxidative degradation [36, 37]. The limiting value of weight gain due to absorption of polyacetylene (absorption) of oxygen from air oxidation at room temperature is 35% [33]. The resulting product is characterized by the formula $[(C_2H_2)O_{0.9}]_n$. The ease of oxidation depends on the morphology of the polymer and changes in the series of crystal < amorphous component < the surface of the fibrils [35]. The absorption of oxygen begins at the surface of fibrils, and then penetrates. Polyacetylene globular morphology is more stable to the effects of O_2 than polymer fibrillar structures [38].

Polyacetylene obtained by polymerization of acetylene in Ziegler-Natta catalysts, after doping Cl_2, Br_2, I_2, AsF_5 becomes a semiconductor in the form of flexible, silvery films "organic metals" [39]. Doping with iodine increases the amorphous samples δ 6 × 10^{-5}, and crystal to 7 × 10^2 $Ohm^{-1}·cm^{-1}$ [40]. The highest electrical conductivity of the polyacetylene compared with those obtained by other methods, the authors [41] explain the presence of catalyst residues. In their view, the concentration of the structure of sp^3 - hybridized carbon atoms is relatively little effect on the conductivity as compared with the influence of catalyst residues. Doping with iodine films of polyacetylene obtained by metathesis polymerizing cyclooctatetraene leads to an increase in their electrical conductivity 10^{-8} to 50–350 $Ohm^{-1}·cm^{-1}$ [42], and have received polymerization of benzvalene with ring opening from 10^{-8}–10^{-5} to 10^{-4}–10^{-1} $Ohm^{-1}·cm^{-1}$ [43].

Conductivity increases when pressure is applied to polyacetylene, obtained by polymerization of acetylene [44] and interphase dehydrochlorination of poly vinyl chloride (PVC) [45]. Anomalously large (up to ten orders of magnitude) an abrupt increase in conductivity when the load is found for iodine-doped crystalline polivinilena - conversion product of PVC [46].

Magnetic properties of polyacetylene significantly depend on the configuration of chains [47]. In the EPR spectrum of the polymerization of polyacetylene singlet line with g-factor of 2.003 [48] and a line width (ΔH) of 7 to 9.5 Oe for the *cis*-isomer [49] and from 0.28 to 5 Oe for the *trans*-isomer [50] observed. According to other reports [47], the *cis*-isomer, syn-synthesized by polymerization of acetylene at 195K, the EPR signal with g-factor = 2.0025 is not observed. This signal appears when the temperature of polymerization increases, when the *trans*-isomer in the form of short chains mainly at the ends of the molecules is 5–10 wt% [51]. The morphology of polyacetylene also has an effect on the paramagnetic properties. The concentration of PMC in the amorphous polyacetylene is ~ 1018 spin/g, and in the crystal 1019 spin/g [51].

KEYWORDS

- Dehydrohalogenation
- Hydrocarbons
- Monocrystals
- Polyacetylenes
- Scanning electronic microscopy

REFERENCES

1. Sladkov, A. M., Kasatochkin, V. I., Korshak, V. V., and Kudryavcev, Yu. P. Diploma on discovery. *Bulletin of inventions*, (107), 6 (1992).
2. Korshak, V. V., Kasatochkin, V. I., Sladkov, A. M., Kudryavcev, Yu. P., and Usunbaev, K. About synthesis and properties of polyacetylene. Lecture Academy of Sciences the USSR, **136**(6), 1342 (1991).
3. Sladkov, A. M. Carbyne– the third allotropic form of carbon. *M. Science*, p. 152 (2003).
4. Heimann, R. B. and Evsyukov, S. E. Allotropy of carbon. *Nature*, (8), p. 66 (2003).
5. Sladkov, A. M. and Kudryavcev, Yu. P. Diamond, graphite, carbyne – allotropic forms of carbon. *Nature*, (5), p. 37 (1969).
6. Kudryavcev, Yu. P., Evsyukov, S. E., Guseva, M. B., Babaev V. G., and Xvostov, V. V. Carbyne– the third allotropic form of carbon. Proceedings of the Academy of Sciences. *Series Chemical*, (3), p. 450 (1993).
7. Natta, G., Pino, P., and Mazzanti, G. Patent. Hal. 530753 Italy//C. A. **52**, 15128 (1958).
8. Natta, G., Mazzanti, G., Corradini, P.//АШ Accad. Naz. Lincei, Cl. Sci. Fis. Mat. Nat. Rend. **25**, p. 2 (1998).
9. Watson, W. H., Memodic, W. C, and Lands, L. G. 3. *Polym. Sci.*, **55**, 137 (1961).
10. Shirakawa, H. and Ikeda, S.//*Polym. J.*, **2**, 231 (1971).
11. Ito, T., Shirakawa, H., and Ikeda, S. *J. Polym. Sci. Polym. Chem.* **12**, 11 (1974).
12. Tripathy S. K, Rubner, M., and Emma, T. et all/Ibid. **44**, C3–37 (1993).
13. Wegner, G. *Macromol. Chem.*, **4**, 155 (1981).
14. Schen, M. A., Karasz, F. E., and Chien, L C. *J. Polym. Sci.: Polym. Chem.* **21**, 2787 (1983).
15. Wnek, G. E., Chien, J. C, and Karasz, F. E. *et al./JJ. Polym. Sci. Polym. Lett.* **17**, 779.
16. Aldissi, M. Synthetic Metals. **9**, 131 (1984).
17. Schue, F. and Aldissi, Af. //Colloq. Int. Nouv. Orient. Compos. Passifs. Mater. Technol. Mises Ocure. Paris. p. 225 (1982).
18. Chien, M. A, Karasz, F. E., and Chien, J. C. *Macromol. Chem. Rapid Communs.*, **5**, 217 (1984).
19. Chien, J. C. *J. Poli. Sci. Polym. Lett.*, **21**, 93 (1983).
20. Saxman, A. M., Liepins, R., and Aldissi, M. *Progr. Polym. Sci.*, **11**, 57 (1985).
21. Chien, J. C, Karasz, F. E., Schen, M. A., and Hirsch, T. 4.//*Macromolecules*. **16**, 1694 (1983).
22. Chien J. C, Karasz, F. E., MacDiarmid, A. G., and Heeger, A. /.//*J. Polym. Sci. Polym. Lett.*, **18**, 45 (1980).
23. Chien, J. C. *Polymer News*, **6**, 53 (2012).
24. Dandreaux, G. F., Galuin, M. E., and Wnek G. E. *J. Phys.* **44**, C3–135 (1983).

25. Parshakov, A. S., Ilin, E. G., Parshakov, A. S., Buryak, A. K., Kochubei, D. I., Drobot, D. V., and Nefedov, V. I. Interaction pentachloride molybdenum with acetylene - a new method for the synthesis of nanoscale composite materials, Moscow. Abstract of a thesis IONCh RAS. 2010.; DAN, **427**(5), 641–645 (2009).

26. Matnishyan, A. A. *Advances of chemistry*, **57**(4), 656–683 (1988).

27. Natta, G., Pino, P., and Mazzanti, G. Patent. Hal. 530753 Italy//C. A. **52**, 15128 (1958).

28. Natta, G., Mazzanti, G., and Corradini, P. //AIII Accad. Naz. Lincei, Cl. Sci. Fis. Mat. Nat. Rend. **25**, 2 (1998).

29. Chasko, B., Chien, J. C. W., Karasz, F. E., Mc Diarmid, A. G., and Heeger, A. J. *Bull. Am. Phys. Soc.*, (24), pp. 480–483 (1979).

30. Shirakawa, H., Sato, M., Hamono, A., Kawakami, S., Soga, K., and Ikeda, S. *Macromolec.*, **13**(2), 457–459 (1980).

31. Soga, K., Kawakami, S., Shirakawa, H., and Ikeda, S. *Makromol. Chem., Rapid. Commun.*, **1**(10) 643–646 (1980).

32. Lopurev, V. A., Myachina, G. F., Shevaleevskiy, O. I., and Hidekel, M. L. *High-molecular compounds. A*,**30**(10), 2019–2037 (1988).

33. Kobryanskiy, V. M., Zurabyan, N. J., Skachkova, V. K., and Matnishyan, A. A. *High-molecular compounds.B*, **27**(7), 503–505 (1985).

34. Berlyn, A. A., Geyderih, M. A., and Davudov, B. E. *Chem. of polyconjugate systems. M.: Chem.*, p. 272 (1972).

35. Yang, X. Z. and Chien, J. C. W. *J. Polym. Sci.: Polym. Chem.*, **23**(3), 859–878 (1985).

36. MacDiarmid, A. G., Chiang, J. C., and Halpern, M. et al. *Amer. Chem. Soc. Polym. Prepr.*, **25**(2), 248–249 (1984).

37. Gibson, H. and Pochan, J. *Macromolecules*, **15**(2), 242–247 (1982).

38. Kobryanskii, V. M. *Mater. Sci.*, **27**(1) 21–24 (1991).

39. Deits, W., Cukor, P. Rubner, M., and Jopson, H. *Electron. Mater*, **10**(4), 683–702 (1981).

40. Heeger, A. J., MacDiarmid, A. G., and Moran, M. J. *Amer. Chem. Soc.Polym. Prepr.*, **19**(2), 862 (1978).

41. Arbuckle, G. A., Buechelev, N. M., and Valentine, K. G. *Chem. Mater*, **6**(5), 569–572 (1994).

42. Korshak, J. V., Korschak, V. V., Kanischka, G., and Hocke H. *Makromol. Chem. Rapid Commun.*, **6**(10) 685–692 (1985).

43. Swager, T. M. and Grubbs, R. H. *Synth. Met.*, **28**(3), D57–D62, (1989) 51.

44. Matsushita, A., Akagi, K., Liang, T. S., and Shirakawa, H. *Synth. Met.*, **101**(1–3), 447–448 (1999).

45. Salimgareeva, V. N., Prochuhan, Yu. A., Sannikova, N. S. and others. *High-molecular compounds*, **41**(4), 667–672 (1999).

46. Leplyanin, G. V., Kolosnicin, V. S., Gavrilova, A. A and others. *Electro chemistry*, **25**(10), 1411–1412 (1989).

47. Zhuravleva, T. S. *Advances of chemistry*, **56**(1), 128–147 (1987).

48. Goldberg, I. B., Crowe, H. R., Newman, P. R., Heeger, A. J., and MacDiarmid, A. G. *J. Chem. Phys.*, **70**(3)1132–1136 (1979).

49. Bernier, P.,Rolland, M., Linaya, C., Disi, M., Sledz, I., Fabre, I. M., Schue, F., and Giral, L. *Polym. J.*, **13**(3), 201–207 (1981).

50. Holczer, K., Boucher, J. R., Defreux, F., and Nechtschein, M. *Chem. Scirpta.*, **17**(1–5), 169–170 (1981).

51. Krinichnui, V. I. *Advances of chemistry*, **65**(1), 84 (1996).

CHAPTER 31

IMMUNE SYSTEM: COMPONENTS AND DISORDERS

ANAMIKA SINGH and RAJEEV SINGH

CONTENTS

31.1 INTRODUCTION

The immune system is a biological structure within organism and it helps to protect the organisms against disease. Immune system can detect a wide range of agents like viruses, worms, spores, fungi, bacteria, and other foreign elements. The immune systems identify and destroy the foreign pathogen through different defense mechanisms [1]. If there will be any change in the immune system it causes abnormality in defense mechanism and it leads to disease like inflammatory, cancer or autoimmune disease [2, 3].

Immune system is an adaptive defense system in vertebrates (Figure 1). It provides protection against the infectious agents [1].

There are two components of an effective immune system [4, 5]:

1. **Recognition:** Which provides specific recognition of the foreign pathogen while discriminating body's own cells and proteins.
2. **Response:** Elimination of invading organism.

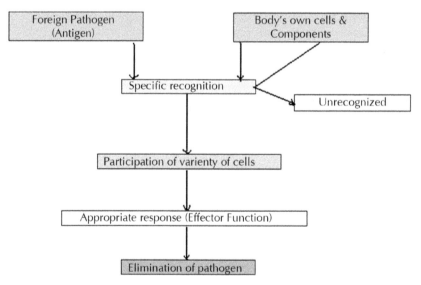

FIGURE 1 Components of immune system.

31.2 COMPONENTS OF IMMUNE SYSTEM

The major components of the immune system include (Figure 2):

31.2.1 LYMPH NODES

A lymph node is Small, oval-shaped structures that produce and store cells, that fight infection and disease. The lymph nodes are distributed throughout the body. Lymph nodes are actually protector of B, T, and other immunity cells. Main role of Lymph nodes is to catch and filter the foreign pathogens by which the whole immune system will work. Lymph nodes contain a fluid which helps to carries the cells in different part of the body. When body suffers with any infection the lymph nodes swells and become enlarged. When swollen, inflamed or enlarged, lymph nodes can be hard, firm or tender [6-8]

31.2.2 SPLEEN

It is largest lymphatic organ in the body and it contains white blood cells that help to fight against infection. It controls the amount of blood in the body as well as removes old and damaged blood cells from the body [9, 10]. Spleen synthesizes antibodies and it also stores half of the body monocytes [11]. These monocytes help in tissue healing [11-13].

31.2.3 BONE MARROW

Bone marrow presents in the interior of bones and are flexible in nature. Red blood cells are produced in the heads of long bones. Bone marrow helps in body systemic circulation [14]. Bone marrow is a major support for immune system as produces lymphocytes and also helps in lymphatic system [15]. Bone marrow transplantation is a very successful method to control the severe disease like cancer, in spite of that bone marrow cell can be transformed into functional neural cells [16, 17]. The yellow tissue in the center of the bones produces white blood cells.

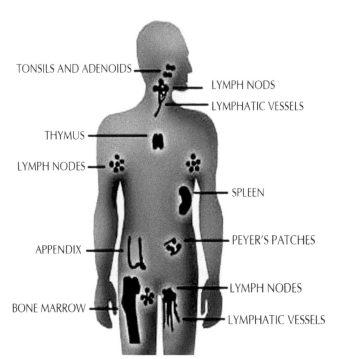

FIGURE 2   Body showing organs involved in immunity.

31.2.4 LYMPHOCYTES

Lymphocytes are a type of white blood cells (WBC) and are a most important part of vertebrate immune system [18]. Three major types of lymphocytes are present in the body and theses are T cells, B cells and natural killer (NK) cells. These WBC play a large role in defending the body against disease [19-21].

31.2.5 THYMUS

It is a small organ help to mature the T cells. This small organ is where T-cells mature [22]. The thymus "educates" T-lymphocytes (T cells), which are critical cells of the adaptive immune system. T cells are very specific and are able attack at a specific antigen [23, 24].

31.2.6 LEUKOCYTES

The WBC or leukocytes help to fight against infections. Five different types of WBCs are produced a specialized part of bone marrow called hematopoietic stem cell [25]. The average life span of a WBC is three to four days and later they were replaced by the new one. Leucocytes are present throughout the body, blood and lymphatic system [26]. The change in the number of leukocytes in the blood is often an indicator of disease. In general leucocytes are approximately 1% of the total blood volume in a healthy adult [27].

31.3 DISEASES OF THE IMMUNE SYSTEM

When the immune system is lacking one or more of its components, the result is an immunodeficiency disorder. These can be inherent, acquired through infections or produced as an inadvertent side effect of drug such as those used to treat cancer or transplant patients.

An immune disorder is a dysfunction of the immune system. These disorders characterized in several different ways:
- By the affected immune system components,
- Over expressive and under expressive immune system,
- Congenital or acquired conditions.

International Union of Immunological Societies has been characterizing more 150 primary immunodeficiency diseases (PIDs) [28]. However, the numbers of acquired immunodeficiencies diseases are more than PIDs [29]. It is possible that most of the people may have at least one primary immuno deficiency immunodeficiency [30]. As immune system is redundant theses PIDs are not detected. Autoimmune diseases, inflammatory diseases and cancer all occurs due to abnormalities in immune system. Fever and fatigue are the most common signs, which immune system is not functioning properly. While symptoms of immune diseases vary, that the immune system is not functioning properly.
- Allergy
- Autoimmune disease
- Immune complex Disease

31.3.1 ALLERGY

An allergy occurs due to hypersensitivity of the immune system [31] .Allergic reactions takes place when the immune system of a person work against normal harmless agents or harmless environment factor, these agents are known as allergens. Allergic reactions are very fast and instant. Commonly an allergy is also known as immediate hypersensitivity. Specific white cells also known as mast cells are mainly involved in allergic reactions. In spite of that an antibody called Immunoglobulin E is also taking part in allergic reactions.

In general allergic reactions occur by fever, red eye, itching, running nose eczema or asthama attack. Allergy may occur because of adverse environmental conditions or by contaminated food or water and so on [32]. There are so many diagnostic tests, which are able to prove a variety of allergic conditions in a person. In these tests an allergen should be placed on skin of a person and examine for the reactions, like swelling and blood response to the allergen [33].

Few allergic diseases are Asthama, Atopic, eczema, Anaphylaxis, Food allergies, and so on.

31.3.2 AUTOIMMUNE DISEASES

It arises when the body's own substances and tissues work against the body (autoimmunity). These immune responses are organ specific like autoimmune thyroiditis or restricted to specific tissues like Goodpasture's disease where basement membrane of both lungs and kidneys are affected. The best way to treat an autoimmune disease Witebsky and colleagues in 1957 gives certain postulates for autoimmune diseases. [34] for a disease to be regarded as an autoimmune disease it needs to answer to *Witebsky's postulates* (first formulated by Ernst Witebsky and colleagues in 1957 and modified in 1994) [35].

In autoimmune diseases the immune system may begin to produce an antibody that instead of fighting against infections, attack the body's own tissues. Examples of autoimmune diseases include:

MULTIPLE SCLEROSIS (MS)

The effected part is nerve cells. Symptoms include pain, blindness, weakness, poor coordination, and muscle spasms [36].

TYPE 1 DIABETES MELLITUS

Immune system produces antibodies that attack and destroy the insulin-producing cells of the pancreas [34].

RHEUMATOID ARTHRITIS

In this case immune system produces antibodies that attack the lining of joints, causing inflammation, swelling, and pain. If this arthritis is untreated it causes permanent joint damage [34]. If untreated, rheumatoid arthritis causes gradually causes permanent joint damage. Oral and inject able medicines re able to control the rheumatoid arthritis.

31.3.3 IMMUNE DEFICIENCY DISEASES

It is a state of the body in which the immune system fails its ability to fight against the infections and diseases. Most of the cases immunodeficiency diseases are acquired (secondary) or from birth (primary) [37]. The infected persons are said to be immuno-compromised as the person may be particularly vulnerable to opportunistic infections, in addition to normal infections.

Immune deficiency disease may occur because of:

GRANULOCYTE DEFICIENCY

It may be due to decreased numbers of granulocytes (granulocytopenia) or, due to absence of granulocytes (agranulocytosis).

ASPLENIA

It takes place due to nonfunctional spleen.

HUMORAL IMMUNE DEFICIENCY

It may be because of decrease of one or more types of antibodies known as hypogam-maglobulinemia or due to lack of all or most antibody production called agamma-globulinemia [37].

T-CELL DEFICIENCY

This type of deficiency is acquired immune deficiency syndrome and is secondary in nature [38].

COMPLEMENT DEFICIENCY

This type of deficiency is where the function of the complement system is deficient Examples of immune deficiency diseases include:

HUMAN IMMUNODEFICIENCY VIRUS/ACQUIRED IMMUNE DEFICIENCY SYNDROME (HIV/AIDS)

In this case a virus is able to suppress and destroy immune system cells that normally fight infections. Due to this disease the number of immune cell decreases the person will become more prone for a disease [39].

SEVERE COMBINED IMMUNE DEFICIENCY (SCID)

It is a genetic disorder and it causes severe impairment in multiple areas of the immune system.

COMMON VARIABLE IMMUNE DEFICIENCY (CVID)

It is a genetic defect, and the immune system produces only few selective immune cells. So body misses the whole immune cells. Sit may be treated to inject the missing antibody.

DRUG-INDUCED IMMUNE DEFICIENCY

It occurs as the side effect of the immunosuppressive drugs. People taking immune-suppressing drugs for long periods and suffering from this disease [40].

31.4   CONCLUSION

Although the immune system is working for the body defense mechanism, but whenever there is any change or disturbance in immune system the whole system suffers a lot. There are so many disease caused due to altered immune system however the immune system is helping us to cure and protect from the infections.

KEYWORDS

- **Bone marrow**
- **Common variable immune deficiency (CVID)**
- **Natural killer (NK) cells**
- **Primary immunodeficiency diseases (PIDs)**
- **Severe combined immune deficiency (SCID)**
- **White blood cells (WBC)**

REFERENCES

1. Beck, Gregory, and Habicht, Gail S. Immunity and the Invertebrates (PDF). *Scientific American*, **275**(5), 60–66 (November 1996) doi: 10.1038/scientificamerican1196-60.
2. Lisa, M. Coussens and Zena Werb, Inflammatory Cells and Cancer. *Journal of Experimental Medicine*, **193**(6), pages F23-26 (March 19, 2001).
3. O'Byrne, K. J. and Dalgleish, A. G. , Chronic Immune Activation and Inflammation as the Cause of Malignancy. *British Journal of Cancer*, **85**(4), 473–483 (August 2001).
4. Kumar, H, Kawai, T, and Akira, S. Pathogen recognition in the innate immune response. *Biochem J.*, **420**(1), 1–16 (Apr 28, 2009).
5. Kumagai, Y, Takeuchi, O, Akira, S. Pathogen recognition by innate receptors. *J Infect Chemother.*, **14**(2):86-92 (30 Apr 2008) doi: 10.1007/s10156-008-0596-1. Epub Review. http://children.webmd.com/tc/swollen-glands-and-other-lumps-under-the-skin-topic-overview
6. Warwick, R. and Peter, L. W. (1858) Angiology (Chapter 6). Gray's anatomy. illustrated (35th ed.). E. M. Richard Moore (Ed), Longman, London, 588–785 (1973).
7. Kaldjian, E. P., Gretz J. E., Anderson, A. O., Shi, Y. and Shaw, S. Spatial and molecular organization of lymph node T cell cortex: a labyrinthine cavity bounded by an epithelium-like monolayer of fibroblastic reticular cells anchored to basement membrane-like extracellular matrix. *International Immunology (Oxford Journals)*, 13(10), 1243–1253 (October, 2001).
8. Spleen, *Internet Encyclopedia of Science*
9. Mebius, R. E., Kraal, G. Structure and function of the spleen. *Nat Rev Immunol.* **5**(8), 606–16 (2005) PMID 16056254.
10. Swirski, F. K. , Nahrendorf, M, Etzrodt, M, Wildgruber, M, Cortez-Retamozo, V, Panizzi, P, Figueiredo, J. L., Kohler, R. H., Chudnovskiy, A, Waterman, P, Aikawa, E, Mempel, T. R., Libby, P, Weissleder, R., and Pittet, M. J. Identification of Splenic Reservoir Monocytes and Their Deployment to Inflammatory Sites. *Science*, **325**, 612–616 (2009).
11. Jia, T and Pamer, E. G. Dispensable But Not Irrelevant. *Science*, **325**, 549–550 (2009).
12. Angier, N. Finally, the Spleen Gets Some Respect. The New York Times, (August 3, 2009).
13. Vunjak-Novakovic, G., Tandon, N. Godier, A., Maidhof, R., Marsano,A., Martens T. P., and Radisic, M. Challenges in Cardiac Tissue Engineering. *Tissue Engineering: Part B*, **16**(2) (2010).
14. The Lymphatic System. Allonhealth.com, Retrieved (December 5, 2011).

15. Antibody Transforms Stem Cells Directly Into Brain Cells. *Science Daily*, 22 April, 2013, Retrieved (April 24, 2013).
16. Research Supports Promise of Cell Therapy for Bowel Disease. *Wake Forest Baptist Medical Center*, 28 February 2013, Retrieved (March 5, 2013).
17. The process of B-cell maturation was elucidated in birds and the B most likely means bursa-derived referring to the bursa of Fabricius. B Cell. Merriam-Webster Dictionary. *Encyclopædia Britannica*, Retrieved (October 28, 2011).
18. Janeway, C., Travers, P., Walport, M., and Shlomchik, M. Immunobiology, (5th ed.). New York and London: Garland Science (2001).
19. Abbas, A. K. and Lichtman, A. H. Cellular and Molecular Immunology (5th ed.). Saunders, Philadelphia (2003).
20. Berrington, J. E., Barge, D., Fenton, A. C., Cant, A. J., and Spickett, G. P. Lymphocyte subsets in term and significantly preterm UK infants in the first year of life analysed by single platform flow cytometry. *Clin Exp Immunol*, 140 (2), 289–292 (2005).
21. Miller, J. F. The discovery of thymus function and of thymus-derived lymphocytes. *Immunol. Rev.*, 185, 7–14 (2002).
22. Miller, J. F. Events that led to the discovery of T-cell development and function--a personal recollection. *Tissue Antigens,* 63(6), 509–17 (2004). http://www.thymusfunctions.com
23. LaFleur-Brooks, M. *Exploring Medical Language: A Student-Directed Approach* (7th Ed.). Mosby Elsevier, St. Louis, Missouri, USA, p. 398 (2008).
24. Maton, D., Hopkins, J., McLaughlin, Ch. W., Johnson, S., Warner, M. Q., LaHart, D., Wright, J. D., and Kulkarni, D. V. Human Biology and Health. Englewood Cliffs, New Jersey, Prentice Hall, USA (1997).
25. Alberts, B. Leukocyte functions and percentage breakdown. *Molecular Biology of the Cell.*, NCBI Bookshelf. Retrieved (April 14, 2007).
26. Geha, R. S., Notarangelo, L. D., Casanova, J. L., et al. Primary immunodeficiency diseases: an update from the International Union of Immunological Societies Primary Immunodeficiency Diseases Classification Committee. *J. Allergy Clin. Immunol*, 120 (4), 776–94 (October 2007)
27. Kumar, A, Teuber, S. S., and Gershwin, M. E. Current perspectives on primary immunodeficiency diseases. *Clin. Dev. Immunol.* 13 (2–4), 223–59 (2006).
28. Casanova, J. L., Abel, L. Primary immunodeficiencies: a field in its infancy. *Science,* 317 (5838), 617–9 (August 2007).
29. Dorland's Medical Dictionary. Allergy. (1890)
30. Kay, A. B. Overview of allergy and allergic diseases: with a view to the future. *Br. Med. Bull.*, 56(4), 843–64, 33 (2000).
31. Cox, L., Williams, B., Sicherer, S., Oppenheimer, J., Sher, L., Hamilton, R., and Golden, D. Pearls and pitfalls of allergy diagnostic testing: report from the American College of Allergy, Asthma and Immunology/American Academy of Allergy, Asthma and Immunology Specific IgE Test Task Force. .American College of Allergy, Asthma and Immunology Test Task, Force; American Academy of Allergy, Asthma and Immunology Specific IgE Test Task, Force. Annals of allergy, asthma & immunology: official publication of the American College of Allergy, *Asthma & Immunology* 101(6), 580–92 (December 2008).
32. Witebsky, E., Rose, N. R., Terplan, K., Paine, J. R., and Egan, R. W. Chronic thyroiditis and autoimmunization. *J. Am. Med. Assoc.* 164(13), 1439–47 (1957).
33. Rose, N. R. and Bona, C. Defining criteria for autoimmune diseases (Witebsky's postulates revisited). *Immunol Today*, 14 (9), 426–30. http://www.nejm.org/doi/full/10.1056/NEJMoa1110740".

34. *Immunodeficiency*. Dr. Saul Greenberg (Ed.), University of Toronto. Last updated on (February 5, 2009).
35. Schwartz, R. A (MD, MPH). Medscape,T-cell Disorders. Chief Harumi Jyonouchi (Ed.), Updated on (May 16, 2011).
36. Basic Immunology: Functions and Disorders of the Immune System, 3rd Ed. (2011).
37. Ammatikos, A., Tsokos, G. Immunodeficiency and autoimmunity: lessons from systemic lupus erythematosus. *Trends Mol Med*, 18 (2), 101–108 (2012).
http://aids.gov/hiv-aids-basics/just-diagnosed-with-hiv-aids/hiv-in-your-body/immune-system-101/

CHAPTER 32

PROGRESS IN PORE STRUCTURE ANALYSIS OF POROUS MEMBRANES

BNTOLHODA HADAVI MOGHADAM and MAHDI HASANZADEH

CONTENTS

32.1   INTRODUCTION

Nanoporous membranes are an important class of nanomaterials that can be used in many applications, especially in micro and nanofiltration. Electrospun nanofibrous membranes have gained increasing attention due to the high porosity, large surface area per mass ratio along with small pore sizes, flexibility, and fine fiber diameter, and their production and application in development of filter media. Image analysis is a direct and accurate technique that can be used for characterization of porous media. This technique, due to its convenience in detecting individual pores in a porous media, has some advantages for pore measurement. The three-dimensional reconstruction of porous media, from the information obtained from a two-dimensional analysis of photomicrographs, is a relatively new research area. In the present chapter, we have reviewed the recent progress in pore structure analysis of porous membranes with emphasis in image analysis technique. Pore characterization techniques, properties, and characteristics of nanoporous structures are also discussed in this chapter.

Nanofibrous membranes have received increasing attention in recent years as an important class of nanoporous materials. Although, there are various techniques to produce polymer nanofiber mats, electrospinning is considered as one of the most efficient ways to obtain nonwoven nanofiber mats with pore sizes ranging from tens of nanometers to tens of micrometers [1-3].

In recent years, significant progress has been done in the understanding and modeling of pore-scale processes and phenomena. By using increased computational power, realistic pore-scale modeling in recognition of tomographic and structure of porous membranes can be obtained [4-6]. Information about the pore structure of membranes is often obtained by several methods including mercury intrusion porosimetry [7-9], liquid extrusion porosimetry [9-12], flow porosimetry [9-15], and image analysis of thin section images [16-23]. Image analysis is a useful technique that is gaining attention due to its convenience in detecting individual pores in the membrane image.

The three-dimensional reconstruction of porous media, such as nanofiber mats, from the information obtained from a two-dimensional micrograph has attracted considerable interest for many applications. A successful reconstruction procedure leads to significant improvement in predicting the macroscopic properties of porous media [23-44].

This short review intends to introduce recent progress in pore structure analysis of porous membranes, with emphasis on electrospun polymer nanofiber mats. This chapter presents polymer membranes and their types. It is also deals with the nanofibrous membranes as one of the most important porous media. It presents the porosity of membranes and the techniques used to evaluate the pore characteristics of porous membranes. Finally, it surveys the most characteristic and important recent examples, which image analysis techniques was used for characterization of porous media, especially three-dimensional reconstruction of porous structure. The report ends with a conclusion.

32.2 POLYMER MEMBRANES

Membrane technologies are already serving as a useful tool for industrial processes, such as health sector, food industry, sustainable water treatment, and energy conversion and storage. Membrane is a selective barrier between two phases and defined as a very thin layer or cluster of layers that allows one or more selective components to permeate through readily when mixtures of different kinds of components are driven to its surface, thereby producing a purified product. The ability to control the diffusion rate of a chemical species through the membrane is key property of membranes [45-54]. Several membrane processes have been proposed based on the barrier structure, including microfiltration [54, 55], ultrafiltration [56, 57], nanofiltration [54, 58], and reverse osmosis [58]. The membrane processes can be categorized based on the barrier structure by different driving forces as shown in Figure 1.

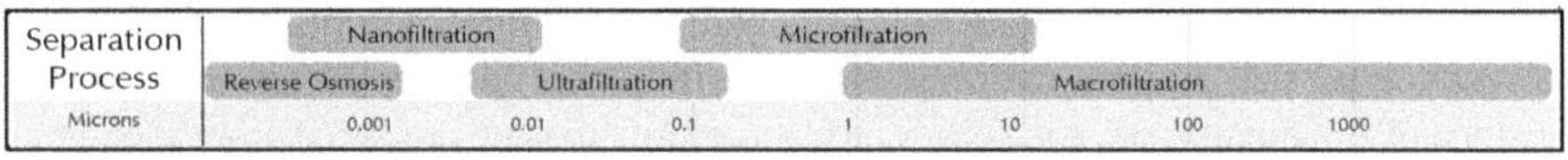

Separation Process	Nanofiltration			Microfiltration			
	Reverse Osmosis	Ultrafiltration			Macrofiltration		
Microns	0.001	0.01	0.1	1	10	100	1000

FIGURE 1 Size range of particles in various membrane separation processes.

Pore size distribution, specific surface area, outer surface, and cross section morphology are some of the most important characteristics of membranes.

32.2.1 TYPES OF MEMBRANES

According to the morphology, membranes can be classified into two main types:
- Isotropic
- Anisotropic

Isotropic membranes include microporous, nonporous, and dense membranes and are made of single layer with uniform structure through the depth of the membrane. On the other hand, anisotropic membranes consist of more than one layer supported by a porous substrate.

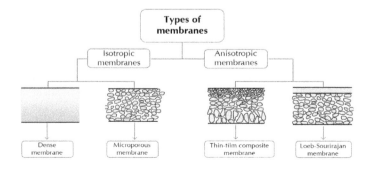

FIGURE 2 Classification of membranes.

ISOTROPIC MEMBRANES

Most of the available membranes are porous or consist of a dense top layer on a porous structure. Isotropic dense membranes can be prepared:
- By melt extrusion of polymer
- By solution casting (solvent evaporation) [50-52]

Isotropic microporous membranes have a rigid, highly voided structure, and interconnected pores. These membranes have higher fluxes than isotropic dense membranes. Microporous membranes are prepared by some methods, such as track-etching, expanded-film, and template leaching [50, 59].

32.2.2 ANISOTROPIC MEMBRANES

Anisotropic membranes are consists of a very thin (0.1 to 1 pm) selective skin layer on a highly permeable microporous substrate and highly membrane fluxes in which the porosity, pore size, or even membrane composition change from the top to the bottom surface of the membrane. An asymmetric structure are now produced from a wide variety of polymers which are currently applied in pressure driven membrane processes, such as reverse osmosis, ultrafiltration, or gas separation.

Anisotropic membranes may be prepared by various techniques, including phase inversion [50, 59], interfacial polymerization [50], solution coating [50], plasma deposition [50], and electrospinning in the laboratory or on a small industrial scale [60]. Nanofibrous membranes, as simple and interesting anisotropic membranes, are described here.

32.3 NANOFIBROUS MEMBRANE

Nanofibrous membranes have high specific surface area, high porosity, small pore size, and flexibility to conform to a wide variety of sizes and shapes. Therefore, they have been suggested as excellent candidate for many applications, especially in micro and nanofiltration. Nanofibrous membrane can be processed by a number of techniques such as drawing [61], template synthesis [62], phase separation [63], self-assembly [64], and electrospinning [65]. Among them electrospinning has an advantage with its comparative low cost and relatively high production rate.

With regard to the low mechanical properties of nanofibrous membrane, many attempts have been made to improve mechanical properties of nanofibers. In this regard, electrospinning process is the best method to fabricate the mixed membrane of microfiber and nanofibers simultaneously. The good physical and mechanical properties of microfibers (as a substrate) are favorable for enhancing the mechanical performance of nanofibrous membrane.

32.3.1 ELECTROSPINNING PROCESS

Within the past several years, electrospinning process has garnered increasing attention, due to its capability and feasibility to generate large quantities of nanofibers with well-defined structures. Figure 3 shows a schematic illustration of electrospinning setup. In this process, a strong electric field is applied between polymer solution contained in a syringe with a capillary tip and grounded collector. When the electric field overcomes the surface tension force, the charged polymer solution forms a liquid jet and travels towards collection plate. As the jet travels through the air, the solvent evaporates and dry fibers deposits on the surface of a collector.

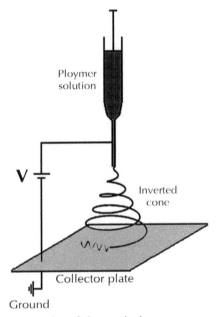

FIGURE 3   Schematic representation of electrospinning process.

The morphology and the structure of the electrospun nanofibrous membrane are dependent upon many parameters which are mainly divided into three categories:
- Solution properties (the concentration, liquid viscosity, surface tension, and dielectric properties of the polymer solution)
- Processing parameters (applied voltage, volume flow rate, tip to collector distance, and the strength of the applied electric field)
- Ambient conditions (temperature, atmospheric pressure and humidity) [66-80].

32.3.2   POROUS NANOFIBERS

There are various techniques for the fabrication of highly porous nanofiber, including fiber bonding, solvent casting, particle leaching, phase separation, emulsion freeze-drying, gas foaming, and 3D printing. These porous membranes can also be produced by a combination of electrospinning and phase inversion techniques. Nanofibrous membrane produced by this approach can generate additional space and surface area within as-spun fibrous scaffolds. Due to the fine fiber size and large expected surface area, electrospun nanofibrous membranes have a desirable property for filter media, catalyst immobilization substrates, absorbent media, and encapsulated active ingredients, such as activated carbon and various biocides.

Studied showed that the formation of pores on nanofibers during electrospinning process affected by many parameters such as humidity, type of polymer, solvent vapor pressure, electrospinning conditions, and so on. Although no generally agreed set of definitions exists, porous materials can be classified in terms of their pore sizes into various categories including capillaries (>200 nm), macropores (50–200 nm), mesopores (2–50 nm), and micropores (0.5–2 nm). According to the literature, the mechanism that forms porous surface on polymer casting film is applicable to the phenomenon on electrospun nanofibers [81-88]. The rapid solvent evaporation and subsequent condensation of moisture into water particles result in the formation of nano or micropores on the fiber surface. When the environment humidity increases, the pore size becomes larger. However, this result was observed only when the solution used a highly volatile organic solvent, such as chloroform, tetrahydrofuran and acetone.

32.3.3 POTENTIAL APPLICATIONS

The research and development of electrospun nanofibrous membrane has evinced more interest and attention in recent years due to the heightened awareness of its potential applications in various fields. The electrospun nanofibrous membrane, due to their high specific surface area, high porosity, flexibility, and small pore size, have been suggested as excellent candidate for many applications including filtration, multifunctional membranes, reinforcements in light weight composites, biomedical agents, tissue engineering scaffolds, wound dressings, full cell, and protective clothing [72, 91, 92].

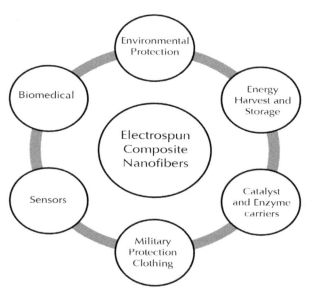

FIGURE 4   Potential applications of electrospun nanofibrous membranes.

Membranes have been widely used for a great variety of applications such as bioseparations, sterile filtration, bioreactors, and so on. Each application depends on the membrane material and structure. For microfiltration and ultrafiltration process, the efficiency was determined by the selectivity and permeability of the membrane.

32.4 MEMBRANE POROSITY

As it has been mentioned earlier, the most important characteristics of membranes are:
- Thickness
- Pore Diameter
- Solvent permeability
- Porosity

Moreover, the filtration performance of membranes is also strongly related to their pore structure parameters, that is, percent open area, and pore size distribution. Hence, the porosity and pore structure characteristics play a more important role in membrane process and applications [93-97].

The porosity, ε_V, is defined as the percentage of the volume of the voids, V_v, to the total volume (voids plus constituent material), V_t, and is given by

$$\varepsilon_V = \frac{V_v}{V_t} \times 100 \tag{1}$$

Similarly, the percent open area, ε_A, that is defined as the percentage of the open area, A_0, to the total area, A_t, is given by

$$\varepsilon_A = \frac{A_0}{A_t} \times 100 \tag{2}$$

Usually, porosity is determined for membranes with a three-dimensional structure (example, relatively thick nonwoven fabrics). Nevertheless, for two-dimensional structures such as woven fabrics and relatively thin nonwovens, it is often assumed that porosity and percent open area are equal [98].

32.4.1 PORE CHARACTERIZATION TECHNIQUES

Development and application of effective procedures for membrane characterization are one of indispensable components of membrane research. Pore structure, as the main characteristics of porous membrane, has significant influence on the performance of membranes. In general, there are three types of pores in membrane:
1. The closed pore that are not accessible.
2. The blind pore that terminate within the material.
3. The through pore that permit fluid flow through the material and determine the barrier characteristics and permeability of the membrane (see Figure 5).

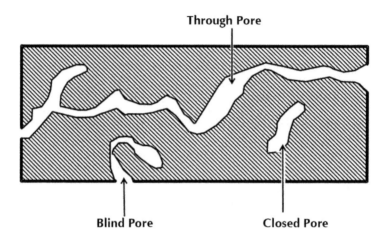

FIGURE 5   Schematic representation of closed, blind and through pores.

Pore structure characteristics, as one of the main tools for evaluating the performance of any porous membrane, can be performed using microscopic and macroscopic approaches. Microscopic techniques usually consist of high-resolution light microscopy, electron microscopy, and X-ray diffraction. The major disadvantage of this technique is that it cannot determine flow properties and also is time consuming and expensive. Various techniques may be used to evaluate the pore characteristics of porous membranes through macroscopic approach, including mercury intrusion porosimetry, liquid extrusion porosimetry, and liquid extrusion flow porometry. These techniques are also used for pore structure characterization of nanofibrous membranes, but the low stiffness and high pressure sensitivity of nanofiber mats limit application of these techniques [99-108].

MERCURY INTRUSION POROSIMETRY

Mercury intrusion porosimetry is a well-known method which is often used to evaluate pore characteristics of porous membranes. This technique provides statistical information about various aspects of porous media such as pore size distribution, or the volume distribution.

Due to the fact that mercury, as a non-wetting liquid, does not intrude into pore spaces (except under application of sufficient pressure), a relationship between the size of pores and the pressure applied can be found. In this technique, a porous membrane is completely surrounded by mercury and pressure is applied to force the mercury into pores. As mercury pressure increases the large pores are filled with mercury first and the pore sizes are calculated as the mercury pressure increases. At higher pressures, mercury intrudes into the fine pores and, when the pressure reaches a maximum, total pore volume and porosity are calculated [98].

According to Jena and Gupta [9-11], the relationship of pressure and pore size is determined by the Laplace equation:

$$D = -\frac{4\gamma\cos\theta}{p} \tag{1}$$

Where D is pore diameter, γ is surface tension of mercury, θ is contact angle of mercury and p is pressure on mercury for intrusion into the pore.

In addition to the pore size and its distribution, the total pore volume and the total pore area can also be determined by mercury intrusion porosimetry method. On the other hand, this method gives no information about the number of pores and is generally applicable to porous membrane with pore sizes ranging from 0.0018 μm to 400 μm. Moreover, mercury intrusion porosimetry does not account for closed pores because mercury does not intrude into them. Due to the application of high pressures, sample collapse and compression is possible, hence it is not suitable for fragile compressible materials such as nanofiber sheets. Other concerns include the fact that it is assumed that the pores are cylindrical, which is not the case in reality [98].

LIQUID EXTRUSION POROSIMETRY

In liquid extrusion porosimetry, a wetting liquid spontaneously intrudes into the pores and then is extrude from pores by a non-reacting gas. The differential pressure p is related to pore diameter D by

$$D = -\frac{4\gamma\cos\theta}{p} \tag{2}$$

Where γ is surface tension of wetting liquid, and θ is contact angle of wetting liquid. The volume of extruded liquid and the differential pressure is measured by this technique.

In this method, a membrane is placed under the sample such that the largest pore of the membrane is smaller than the smallest pore of interest in the sample. First, the pores of the sample and the membrane are filled with a wetting liquid and then the pressure on gas is increased to displace the liquid from pores of the sample. Because the gas pressure is inadequate to empty the pores of the membrane, the liquid filled pores of the membrane allow the extruded liquid from the pores of the sample to flow out while preventing the gas to escape [9, 10].

Through pore volume and diameter were determined by measuring the volume of the liquid flowing out of the membrane and differential pressure, respectively. It should be noted that liquid extrusion porosimetry measures only the volume and diameters of through pores (blind pore are not measured), whereas mercury intrusion porosimetry measures all pore diameter [11, 12].

FLOW POROSIMETRY (BUBBLE POINT METHOD)

Flow porosimetry or bubble point method is based on the principle that a porous membrane will allow a fluid to pass only when the pressure applied exceeds the capillary attraction of the fluid in the largest pore. In this test method, the pore of the membrane is filled with a liquid and continues air flow is used to remove liquid from the pores. At a critical pressure, the first bubble will come through the largest pore in the wetted specimen. As the pressure increases, the smaller pores are emptied of liquid and gas flow increases. Once the flow rate and the applied pressure are known, particle size distribution, the number of pores, and porosity can be derived. In flow porosimetry, the membrane with pore sizes in the range of 0.013–500 µm can be measured [98].

It is important to note that flow porosimetry measures only the throat diameter of each through pore and cannot measure the blind pore. This technique is based on the assumption that the pores are cylindrical, which is not the case in reality [15].

IMAGE ANALYSIS

Image analysis technique, due to its convenience in detecting individual pores in a nonwoven image, has some advantages for pore measurement. Image analysis technique has been used to measure the pore characteristics of electrospun nanofiber webs [98]. To measure the pore characteristics of electrospun nanofibrous membranes using image analysis, images (or micrograph) of the nanofiber webs, which are usually obtained by scanning electron microscopy (SEM), transmission electron microscopy (TEM) or atomic force microscopy (AFM), are required. This is highly relevant in that a picture to be used for image analysis must be of high-quality and taken under appropriate magnifications [98]. The major advantage of porosity characterization by image analysis technique is that the cross sections provide detailed information about the spatial and size distribution of pores as well as their shape.

In this technique, initial segmentation of the micrographs is required to produce binary images. The typical way of producing a binary image from a grayscale image is by 'global thresholding' in which a single constant threshold is applied to segment the image. All pixels up to and equal to the threshold belong to the object and the remaining belong to the background. Global thresholding is very sensitive to inhomogeneities in the gray-level distributions of object and background pixels. In order to eliminate the effect of inhomogeneities in global thresholding, local thresholding scheme could be used. Firstly, the image is divided into sub-images where the inhomogeneities are negligible. Then the optimal thresholds are found for each sub-image. It can be found that this process is equivalent to segmenting the image with locally varying thresholds [98]. Figure 6 shows global thresholding and local thresholding of electrospun nanofibrous mat. It is obvious that global thresholding resulted in some broken fiber segments. However, this problem was solved by using local thresholding. It should be mentioned that this process is extremely sensitive to noise contained in the image. So, a procedure to clean the noise and enhance the contrast of the image is necessary before the segmentation [19, 98].

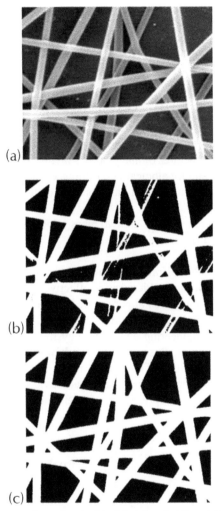

FIGURE 6   (a) SEM image of a real web, (b) global thresholding, (c) local thresholding.

32.4.2 *APPROPRIATE TECHNIQUE FOR NANOFIBROUS MEMBRANES*

As mentioned above, through pore volume of nanofibrous membranes can be measured by mercury intrusion porosimetry and liquid extrusion porosimetry. Due to the high pressure that is applied to the nanofibrous membranes in mercury intrusion porosimetry method, the pores can get enlarged, which leads to overestimation of porosity values. Blind pores in the nanofiber mat are negligible. Therefore, porosity of the nanofibrous membranes can be obtained from the measured pore volume and bulk density of the material. Liquid extrusion technique can give liquid permeability and surface area of through pores, which could not be measured by mercury intrusion porosimetry. It is important to note that for many applications such as filtration, pore throat diameters of nanofiber mats are required in addition to pore volume. While mercury intrusion and liquid extrusion porosimetry cannot measure pore throat diameter, flow porosimetry can measures pore throat diameters without distorting pore structure. Therefore, flow porosimetry is more suited for pore characterization of nanofibrous membranes [18, 109, 110].

32.5 SUMMARY

Establishing the quantitative relationships between the microstructure of porous media and their properties are an important goal, with a broad relevance to many scientific sectors and engineering applications. Since variations in pore shape and pore space connectivity are intrinsic features of many porous media, a pore structure model must involve both geometric and topological descriptions of their complex microstructure. Nowadays modeling and simulation of nanoporous membrane is of special interest to many researchers.

According to the literature, there are two types of pore-scale modeling:
- Lattice Boltzmann (LB) model
- Pore network model

The LB models capable of simulating flow and transport in the actual pore space. Pore network model has been considered as an effective tools used to investigate macroscopic properties from fundamental pore-scale behavior of processes and phenomena based on geometric volume averaging. This model has been used in chemical engineering, petroleum engineering and hydrology fields to study a wide range of single and multiphase flow processes. Pore network model utilizes an idealization of the complex pore space geometry of the porous media. For this purpose the pore space is represented by pore elements having simple geometric shapes such as pore-bodies and pore-throats that have been represented by spheres and cylinders, respectively [31, 111-113].

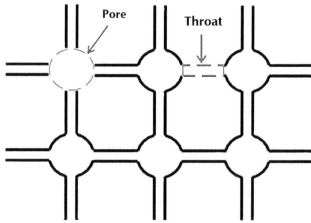

FIGURE 7   Schematic of a pore network illustrating location of pores and throats.

32.5.1 2D IMAGE ANALYSIS OF POROUS MEDIA

Definition of pore network structure, such as pore-body locations, pore-body size distributions, pore-throat size distributions, connectivity, and the spatial correlation between pore-bodies, is an important step towards analysis of porous media. Many valuable attempts have been made for characterization of porous media based on image analysis techniques. For example, Masselin et al. [20] employed image analysis to determine the parameters, such as porosity, pore density, mean pore radius, pore size distribution, and thickness of five asymmetric ultrafiltration membranes. The results obtained from image analysis for the pore size were found to be in good agreement with rejection data.

Ekneligoda et al. [114] used image analysis technique to extract the area and perimeter of each pore from SEM images of two sandstones, Berea and Fontainebleau. The compressibility of each pore was calculated using boundary elements, and estimated from a perimeter-area scaling law. After the macroscopic bulk modulus of the rock was estimated by the area-weighted mean pore compressibility and the differential effective medium theory, the predicted results were compared with experimental values of the bulk modulus. The resulting predictions are close to the experimental values of the bulk modulus.

Bazylak et al. [115] discuss the application of pore network modeling to investigate transport phenomena in a porous media, such as the full cell gas diffusion layer (GDL). They employed transparent experimental microfluidic chips containing these pore networks to compliment the development of a GDL representative pore network model. They described the procedure to design isotropically and diagonally biased networks. They found that the implementation of directed water transport in GDLs has the potential to improve liquid water management and improve the lifetime and durability of the proton exchange membrane fuel cell.

A variety of image analysis techniques was used by Lange et al. [19] to characterize the pore structure of cement-based materials, including plain cement paste, pastes with silica fume, and mortars. These techniques include sizing, two-point correlation, and fractal analyses. Backscattered electron images of polished sections were used to observe the pore structure of cement pastes and mortars. They measured pore size distribution of specimen by using image analysis techniques and compared with mercury intrusion porosimetry derived pore size distribution curves. They found that the image-based pore size distribution was able to better describe the large porosity than the mercury intrusion porosimetry.

In the study on pore structure of electrospun nanofibrous membranes by Ziabari et al. [98], a novel image analysis-based method was developed for measuring pore characteristics of electrospun nanofiber webs. Their model was direct, very fast, and presents valuable and comprehensive information regarding pore structure parameters of the webs. In this method, SEM images of nanofiber webs were converted to binary images and used as an input. First, voids connected to the image border are identified and cleared by using morphological reconstruction where the mask image is the input image and marker image is zero everywhere except along the border. Total area, which is the number of pixels in the image, is measured. Then the pores are labeled and each considered as an object. Here the number of pores may be obtained. In the next step, the number of pixels of each object as the area of that object is measured. Having the area of pores, the porosity may be calculated.

They also investigated the effects of web density, fiber diameter and its variation on pore characteristics of the webs by using some simulated images and found that web density and fiber diameter significantly influence the pore characteristics, whereas the effect of fiber diameter variations was insignificant. Furthermore, it seemed that the changes in number of pores were independent of variation of fiber diameter and that this could be attributed to the arrangement of the fibers.

In another study, Ghasemi-Mobarakeh et al. [116] demonstrated the possibility of porosity measurement of various surface layers of nanofibers mat using image analysis. They found that porosity of various surface layers is related to the number of layers of nanofibers mat. This method is not dependent on the magnification and histogram of images. Other methods such as mercury intrusion porosimetry, indirect method, and also calculation of porosity by density measurement cannot be used for porosity measurement of various surface layers and measure the total porosity of nanofibers mat. These methods show high porosity values (higher than 80%) for the nanofibers mat, while the porosity measurements based on thickness and apparent density of nanofibers mat demonstrated the porosity of between 60% and 70% [117]. This value was calculated using the following equation:

$$\varepsilon_V = 1 - \frac{\rho_a}{\rho_b} \times 100 \qquad (1)$$

$$\rho_a = \frac{m}{T \times A} \qquad (2)$$

Where ρ_a and ρ_b are apparent density and bulk density of nanofiber mat, m is nanofiber mat mass, A is nanofiber mat area, and T is thickness of nanofiber mat [117].

32.5.2   3D IMAGE ANALYSIS OF POROUS MEDIA

Several instrumental characterization techniques have been suggested to obtain 3D volume images of pore space, such as X-ray computed micro tomography and magnetic resonance computed micro tomography. However, these techniques may be limited by their resolution. So, the 3D stochastic reconstruction of porous media from statistical information (produced by analysis of 2D photomicrographs) has been suggested. Although pore network models can be two or three-dimensional, 2D image analysis, due to their restricted information about the whole microstructure, was unable to predict morphological characteristics of porous membrane. Therefore 3D reconstruction of porous structure will lead to significant improvement in predicting the pore characteristics. Recently research work has focused on the 3D image analysis of porous membranes.

Wiederkehr et al. [118] in their study of three-dimensional reconstruction of pore network utilized an image morphing technique to construct a three-dimensional multiphase model of the coating from a number of such cross section images. They show that the technique can be successfully applied to light microscopy images to reconstruct 3D pore networks. The reconstructed volume was converted into a tetrahedron-based mesh representation suited for the use in finite element applications using a marching cubes approach. Comparison of the results for three-dimensional data and two-dimensional cross-section data suggested that the 3D-simulation should be more realistic due to the more exacter representation of the real microstructure.

Delerue et al. [119] utilized skeletization method to obtain a reconstructed image of the spatialized pore sizes distribution that is a map of pore sizes, in soil or any porous media. The Voronoi diagram, as an important step towards the calculation of pore size distribution both in 2D and 3D media was employed to determine the pore space skeleton. Each voxel has been assigned a local pore size and a reconstructed image of a spatialized local pore size distribution was created. The reconstructed image not only provides a means for calculating the global volume versus size pore distribution, but also performs fluid invasion simulation which take into account the connectivity of and constrictions in the pore network. In this case, mercury intrusion in a 3D soil image was simulated.

Al-Raoush et al. [31] employed a series of algorithms, based on the three-dimensional skeletonization of the pore space in the form of nodes connected to paths, to extract pore network structure from high-resolution three-dimensional synchrotron microtomography images of unconsolidated porous media systems. They used dilation algorithms to generate inscribed spheres on the nodes and paths of the medial axis to represent pore-bodies and pore-throats of the network, respectively. The authors have also determined the pore network structure, which is three-dimensional spatial distribution (x, y, and z-coordinates) of pore-bodies and pore-throats, pore-body and pore-throat sizes, and the connectivity, as well as the porosity, specific surface area, and representative elementary volume analysis on the porosity. They show that X-ray microtomography is an effective tool to non-destructively extract the structure of porous media. They concluded that spatial correlation between pore-bodies in the network is important and controls many processes and phenomena in single and multiphase flow and transport problems. Furthermore, the impact of resolution on the properties of the network structure was also investigated and the results showed that it has a significant impact and can be controlled by two factors:

1. The grain size/resolution ratio
2. The uniformity of the system.

In another study, Liang et al. [5] proposed a truncated Gaussian method based on Fourier transform to generate 3D pore structure from 2D images of the sample. The major advantage of this method is that the Gaussian field is directly generated from its autocorrelation function and also the use of a linear filter transform is avoided. Moreover, it is not required to solve a set of nonlinear equations associated with this transform. They show that the porosity and autocorrelation function of the reconstructed porous media, which are measured from a 2D binarized image of a thin section of the sample, agree with measured values. By truncating the Gaussian distribution, 3D porous media can be generated. The results for a Berea sandstone sample showed that the mean pore size distribution, taken as the result of averaging between several serial cross-sections of the reconstructed 3D representation, is in good agreement with the original thresholded 2D image. It is believed that by 3D reconstruction of porous media, the macroscopic properties of porous structure such as permeability, capillary pressure, and relative permeability curves can be determined.

Diógenes et al. [120] reported the reconstruction of porous bodies from 2D photomicrographic images by using simulated annealing techniques. They proposed the following methods to reconstruct a well-connected pore space:

- Pixel-based Simulated Annealing (PSA)
- Objective-based Simulated Annealing (OSA)

The difference between the present methods and other research studies, which tried to reconstruct porous media using pixel-movement based simulated techniques, is that this method is based in moving the microstructure grains (spheres) instead of the pixels. They applied both methods to reconstruct reservoir rocks microstructures, and compared the 2D and 3D results with microstructures reconstructed by truncated Gaussian methods. They found that PSA method is not able to reconstruct connected porous media in 3D, while the OSA reconstructed microstructures with good pore space connectivity. The OSA method also tended to have better permeability determination results than the other methods. These results indicated that the OSA method can reconstruct better microstructures than the present methods.

In another study, a 3D theoretical model of random fibrous materials was employed by Faessel et al. [121]. They used X-ray tomography to find 3D information on real networks. Statistical distributions of fibers morphology properties (observed at microscopic scale) and topological characteristics of networks (derived from mesoscopic observation), is built using mathematical morphology tools. The 3D model of network is assembled to simulate fibrous networks. They used a number of parameter describing a fiber, such as length, thickness, and parameters of position, orientation and curvature, which derived from the morphological properties of the real network.

32.6 CONCLUSION

In recent years, great efforts have been devoted to nanoporous membranes. As a conclusion, much progress has been made in the preparation and characterization of porous media. Among several porous membranes, electrospun nanofibrous membranes, due to the high porosity, large surface area-to-volume ratios, small pores, and fine fiber diameter, have gained increasing attention. Useful techniques for evaluation of the pore characteristics of porous membranes are reviewed. Image analysis techniques have been suggested as a useful method for characterization of porous media due to its convenience in detecting individual pores. It is believed that the three-dimensional reconstruction of porous media, from the information obtained from a two-dimensional analysis of photomicrographs, will bring a promising future to nanoporous membranes.

KEYWORDS

- Image analysis
- Nanofibrous membrane
- Porous media
- Three-dimensional analysis

REFERENCES

1. Rutledge, G. C., Li, Y., and Fridrikh, S. *Electrostatic Spinning and Properties of Ultrafine Fibers*, National Textile Center Annual Report, November (2001) M01-D22.
2. Yao, C., Li, X., and Song, T. Electrospinning and Crosslinking of Zein Nanofiber Mats, *Journal of Applied Polymer Science*, **103**, 380–385 (2007).
3. Kim, G. and Kim, W. Highly Porous 3D Nanofiber Scaffold Using an Electrospinning Technique, *Journal of Biomedical Materials Research Part B: Applied Biomaterials* (2006).
4. Silina, D.and Patzekb, T. Pore space morphology analysis using maximal inscribed spheres, *Physica A*.
5. Liang, Z. R., Fernandes, C. P., Magnani, F. S., and Philippi, P.C. A reconstruction technique for three-dimensional porous media using image analysis and Fourier transforms, *Journal of Petroleum Science and Engineering*, **21**, 273–283 (1998).
6. Kim, K. J., Fanen, A. G., Ben Aimb, R., Liub, M. G., Jonsson, G., TessaroC, I. C., Broekd, A. P., and Bargeman, D. A comparative study of techniques used for porous membrane characterization: pore characterization, *Journal of Membrane Science*, **81**, 35–46 (1994).
7. Dullien, F. A. L. and Dhawan, G. K. Characterization of Pore Structure by a Combination of Quantitative Photomicrography and Mercury Porosimetry, *Journal of Colloid and Interface Science,* **47**(2) (May 1974).
8. Liabastre, A. A. and Orr, C. An Evaluation of Pore Structure by Mercury Penetration, *Journal of Colloid and Interface Science*, **64**(1) (March 15, 1978).
9. Jena, A. and Gupta, K. Pore Volume of Nano fiber Nonwovens, Porous Materials Inc., Ithaca, NY 14850, USA. (2005).
10. Jena, A. and Gupta, K. Characterization of Pore Structure of Filtration Media, Porous Materials, Inc, 83 Brown Road, Ithaca, NY 14850.
11. Jena, A. and Gupta, K. Characterization of Pore Structure of Fuel Cell Components Containing Hydrophobic and Hydrophilic Pores, Porous Materials, Inc., 20 Dutch Mill Road, Ithaca, NY 14850, USA.
12. KC, K., CY, F., and T, M. Membrane Characterization, *Water and Wastewater Treatment Technologies*.
13. Calvo, J. I., Herna Ndez, A., Pra Danos, P., Martibnez, L., and Bowen, W. R. Pore Size Distributions in Microporous Membranes, *Journal of Colloid and Interface Science* **176**, 467–478 (1995).
14. Jena, A. and Gupta, K. Porosity Characterization of Microporous Small Ceramic Components, Porous Materials, Inc., 20 Dutch Mill Road Ithaca, NY 14850.
15. Nassehi, V., Das, D. B., Shigidi, I. M. T. A., and Wakeman, R. J. Numerical Analyses of Bubble Point Tests used for Membrane Characterisation: Model Development and Experimental Validation, Department of Chemical Engineering, Loughborough University, Loughborough, Leicestershire LE11 3TU, UK.

16. Gribble, C. M., Matthews, G. P., Laudone, G. M., Turner, A., Ridgway, C. J., Schoelkopf, J., and Gane P. A. C. Porometry, porosimetry, image analysis and void network modelling in the study of the pore-level properties of filters, *Chemical Engineering Science* (2011).

17. Deshpande, S., Kulkarni, A., Sampath, S., and Herman, H. Application of image analysis for characterization of porosity in thermal spray coatings and correlation with small angle neutron scattering, *Surface & Coatings Technology,* **187**, 6–16 (2004).

18. Tomba, E., Facco, P., Roso, M., Modesti, M., Bezzo, F., and Barolo, M. Artificial Vision System for the Automatic Measurement of Interfiber Pore Characteristics and Fiber Diameter Distribution in Nanofiber Assemblies, *Ind. Eng. Chem. Res*, **49** (2010).

19. Lange, D. A. Image Analysis Techniques For Characterization of Pore Structure of Cement-Based Materials, *Cement and Concrete Research*, **24**(5), 841–853 (1994).

20. Masselin, I., Durand-Bourlier, L., Laine, J. M., Sizaret, P. Y., Chasseray, X., and Lemordant, D. Membrane characterization using microscopic image analysis, *Journal of Membrane Science,* **186**, 85–96 (2001).

21. Garboczi, E. J., Bentz, D. P., and Martys, N. S. Digital Images and Computer Modeling, *Experimental Methods in the Physical Sciences*, **35**, Methods in the Physics of Porous Media. Chapter 1, Academic press, San Diego, CA, 1–41 (1999).

22. Mickel, W., Munster, S., Jawerth, L. M., Vader, D. A., Weitz, D. A., Sheppard, A. P., Mecke, K., Fabry, B., and Schroder-Turk, G. E. Robust Pore Size Analysis of Filamentous Networks from Three-Dimensional Confocal Microscopy, *Biophysical Journal*, **95**, 6072–6080 (December 2008).

23. Roysam, B., Lin, G., Amri Abdul-Karim, M., Al-Kofahi, O., Al-Kofahi, K., Shain, W., Szarowski, D. H., and Turner, J. N. Automated Three-Dimensional Image Analysis Methods for Confocal Microscopy, *Handbook of Biological Confocal Microscopy*, 3rd edition, Springer, New York (2006).

24. Quiblier, J. A. A New Three-Dimensional Modeling Technique for Studying Porous Media, *Journal of Colloid and Interface Science*, **98**(1) (March 1984).

25. Santos, L. O .E., Philippi, P. C., Damiani, M. C., and Fernandes, C. P. Using three-dimensional reconstructed microstructures for predicting intrinsic permeability of reservoir rocks based on a Boolean lattice gas method, *Journal of Petroleum Science and Engineering*, **35** 109–124 (2002).

26. Sambaer, W., Zatloukal, M., and Kimmer, D. 3D modeling of filtration process via polyurethane nanofiber based nonwoven filters prepared by electrospinning process, *Chemical Engineering Science,* **66**, 613–623 (2011).

27. Shin, C. H., Seo, J. M., and Bae, J. S. Modification of a hollow fiber membrane and its three-dimensional analysis of surface pores and internal structure for a water reclamation system, *Journal of Industrial and Engineering Chemistry,* **15**, 784–790 (2009).

28. Ye, G., van Breugel, K., and Fraaij, A. L. A. Three-dimensional microstructure analysis of numerically simulated cementitious materials, *Cement and Concrete Research,* **33**, 215–222 (2003).

29. Holzer, L., Münch, B., Rizzi, M., Wepf, R., Marschall, P., and Graule, T. 3D-microstructure analysis of hydrated bentonite with cryo-stabilized pore water, *Applied Clay Science,* **47**, 330–342 (2010).

30. Liang, Z., Ioannidis, M. A., and Chatzis, I. Geometric and Topological Analysis of Three-Dimensional Porous Media: Pore Space Partitioning Based on Morphological Skeletonization, *Journal of Colloid and Interface Science,* **221**, 13–24 (2000).

31. Al-Raoush, R. I. and Willson, C. S. Extraction of physically realistic pore network properties from three-dimensional synchrotron X-ray microtomography images of unconsolidated porous media systems, *Journal of Hydrology,* **300**, 44–64 (2005).

32. Fenwick, D. H. and Blunt, M. J., Three-dimensional modeling of three phase imbibition and drainage, *Advances in Wafer Resources,* **21**(2), D-143 (1998).
33. Santos, L. O. E., Philippi, P. C., Damiani, M. C., and Fernandes, C. P. Using three-dimensional reconstructed microstructures for predicting intrinsic permeability of reservoir rocks based on a Boolean lattice gas method, *Journal of Petroleum Science and Engineering,* **35**, 109–124 (2002).
34 Mendoza, F., Verboven, P., Mebatsion, H. K., Kerckhofs, G., Wevers, M., and Nicolaı, B. Three-dimensional pore space quantification of apple tissue using X-ray computed microtomography, *Planta,* **226**, 559–570 (2007).
35. Bakke, S. and Øren, P. 3D Pore-Scale Modelling of Sandstones and Flow Simulations in the Pore Networks, *SPE Journal,* **2** (June 1997).
36. Yee Ho, A. Y., Gao, H., Cheong Lam, Y., and Rodrıguez, I. Controlled Fabrication of Multitiered Three-Dimensional Nanostructures in Porous Alumina, *Adv. Funct. Mater,* **18**, 2057–2063 (2008).
37. Sakamoto, Y., Kim, T. W., Ryoo, R., and Terasaki, O. Three-Dimensional Structure of Large-Pore Mesoporous Cubic Ia3d Silica with Complementary Pores and Its Carbon Replica by Electron Crystallography, *Angew. Chem,* **116**, 5343–5346 (2004).
38. Al-Raoush, R. I. Extraction of Physically-Realistic Pore Network Properties from Three-Dimensional Synchrotron Microtomography Images of Unconsolidated Porous Media, *PHD Thesis, Department of Civil and Environmental Engineering* (2002).
39. Boissonnat, J. D. Geometric Structures for Three-Dimensional Shape Representation, *ACM Transactions on Graphics,* **3**(4) (1984)., P
40. Holzer, L., Indutnyi, F., Gasser, Ph., Münch, B., and Wegmann, M., Three-dimensional analysis of porous BaTiO$_3$ ceramics using FIB nanotomography, *Journal of Microscopy,* **216**(1), 84–95 (2004).
41. Pothuaud, L., Porion, P., Lespessailles, E., Benhamou, C. L., and Levitz, P. A new method for three-dimensional skeleton graph analysis of porous media: application to trabecular bone microarchitecture, *Journal of Microscopy,* **199** (Pt 2), 149-161 (2000).
42. Yeong, C. L. Y. and Torquato, S. Reconstructing random media. II. Three-dimensional media from two-dimensional cuts, *Physical Review E,* **58**(1) (1998).
43. Biswal, B., Manwart, C., and Hilfer, R. Three-dimensional local porosity analysis of porous media, *Physica A* **255**, 221–24 (1998).
44. Desbois, G., Urai, J. L., Kukla, P. A., Konstanty, J., and Baerle, C. High-resolution 3D fabric and porosity model in a tight gas sandstone reservoir: A new approach to investigate microstructures from mm- to nm-scale combining argon beam cross-sectioning and SEM imaging, *Journal of Petroleum Science and Engineering,* **78**, 243–257 (2011).
45. Ulbricht, M. Advanced functional polymer membranes, *Polymer,* **47**, 2217–2262 (2006).
46. Amendt, M. A. Nanoporous Thermosetting Membranes using Reactive Block Polymer Templates, PHD Thesis, University Of Minnesota (2010).
47. Szewczykowski, P. Nano-porous Materials from Diblock Copolymers and its Membrane Application, PHD Thesis, University of Denmark (2009).
48. Gullinkala, T. Evaluation of Poly (Ethylene Glycol) Grafting as a Tool for Improving Membrane Performance, PHD Thesis, University of Toledo (2010).
49. Naveed, S. and Bhatti, I. Membrane Technology and Its Suitability for Treatment of Textile Waste Water in Pakistan, *Journal of Research (Science),* **17**(3), 155-164 (2006).
50. Baker, R. W. Membrane Technology and Applications, John Wiley & Sons, England (2004).
51. Roychowdhury, A. Fabrication of Perforated Polymer Membranes Using Imprinting Technology, MSC Thesis, Louisiana State University (2007).

52. Catherina, K., W. F. Membrane Formation by Phase Inversion in Multicomponent Polymer Systems, PHD Thesis, University of Twente (1998).

53. Buckley-Smith, M. K. The Use of Solubility Parameters to Select Membrane Materials for Pervaporation of Organic Mixtures, PHD Thesis, University of Waikato (2006).

54. Nunes, S. P. and Peinemann, K. V. Membrane Technology in the Chemical Industry, WILEY-VCH, Germany (2001).

55. Li, W. Fouling Models for Optimizing Asymmetry of Microfiltration Membranes, PHD Thesis, University of Cincinnati (2009).

56. Childress, A. E., Le-Clech, P., Daugherty, J. L., Chen, Caifeng, and Leslie, Greg L. Mechanical analysis of hollow fiber membrane integrity in water reuse applications, *Desalination*, **180**, 5–14 (2005).

57. Li, L., Szewczykowski, P., Clausen, L. D., Hansen, K. M., Jonsson, G. E., and Ndoni, S. Ultrafiltration by Gyroid Nanoporous Polymer Membranes, *Journal of Membrane Science*, **384**, 126–135 (2011).

58. Chaoyiba. Design of Advanced Reverse Osmosis and Nanofiltration Membranes for Water Purification, PHD Thesis, University of Illinois at Urbana-Champaign, (2010).

59. Yen, C. Synthesis and Surface Modification of Nanoporous Poly(ε-caprolactone) Membrane for Biomedical Applications, PHD Thesis, Ohio State University (2010).

60. Zon, X., Kim, K., Fang, D., Ran, S., Hsiao, B. S., and Chu, B. Structure and process relationship of electrospun bioabsorbable nanofiber membranes, *Polymer*, **43**, 4403–4412 (2002).

61. Ondarçuhu, T. and Joachim, C. Drawing a Single Nanofibre Over Hundreds of Microns, *Europhysics Letters*, **42**, 215–220 (1998).

62. Feng, L., Li, S., Li, Y. Li, H., Zhang, L., Zhai, J., Song, Y., Liu, B., Jiang, L., and Zhu, D. Super-Hydrophobic Surfaces: From Natural to Artificial, *Advanced Materials*. **14**, 1221–1223 (2002).

63. Ma, P. X. and Zhang, R., Synthetic Nano-Scale Fibrous Extracellular Matrix, *Journal of Biomedical Materials Research.*, **46**, 60–72 (1999).

64. Liu, G., Ding, J., Qiao, L., Guo, A., Dymov, B. P., Gleeson, J. T., Hashimoto, T. K., and Saijo. Polystyrene-Block-Poly(2-Cinnamoyl ethyl Methacrylate) Nanofibers Preparation, Characterization, and Liquid Crystalline Properties, *Chemistry-A European Journal*. **5**, 2740–2749 (1999).

65. Doshi, J. and Reneker, D. H., Electrospinning Process and Applications of Electrospun Fibers, *Journal of electrostatics.*, **35**, 151–160 (1995).

66. Reneker, D. H. and Yarin, A. L., Electrospinning jets and polymer nanofibers, *Polymer* **49** 2387-2425 (2008).

67. Yordem, O. S., Papila, M., and Menceloglu, Y. Z., Effects of electrospinning parameters on polyacrylonitrile nanofiber diameter: An investigation by response surface methodology, *Materials and Design*, **29**, 34–44 (2008).

68. Gibson, P. and Schreuder-Gibson, H. Patterned electrospun polymer fiber structures, *e-Polymers*, paper no. T002 (2003).

69. Gibson, P. W., Schreuder-Gibson, H. L., and Rivin, D. Electrospun Fiber Mats: Transport Properties, *AIChE Journal*, **45**(1) (1999).

70. Theron, A., Zussman, E., and Yarin, A. L. Electrostatic field-assisted alignment of electrospun nanofibers, *Nanotechnology*, **12**, 384–390 (2001).

71. Teo, W.E., Inai, R., and Ramakrishna, S., Technological advances in electrospinning of nanofibers, Science And Technology Of Advancedmaterials, *Sci. Technol. Adv. Mater.* **12**, (013002), 19 (2011).

72. Subbiah, T., Bhat, G. S., Tock, R. W., Parameswaran, S., and Ramkumar, S. S., Electrospinning of Nanofibers, *Journal of Applied Polymer Science*, **96**, 557–569 (2005).

73. Zong, X., Kim, K., Fang, D., Ran, S., Hsiao, B. S., and Chu, B. Structure and process relationship of electrospun bioabsorbable nanofiber membranes, *Polymer*, **43**, 4403–4412 (2002).

74. Tan, S., Huang, X., and Wu, B., Some fascinating phenomena in electrospinning processes and applications of electrospun nanofibers, *Polym Int,* **56**, 1330–1339 (2007).

75. Burger, C., Hsiao, B. S., and Chu, B. Nanofibrous materials And Their Applications, *Annu. Rev. Mater. Res.,***36,** 333–368 (2006).

76. Zhang, C., Li, Y., Wang, W., Zhan, N., Xiao, N., Wang, S., Li, Y., and Yang, Q. A novel two-nozzle electrospinning process for preparing microfiber reinforced pH-sensitive nano-membrane with enhanced mechanical property, *European Polymer Journal* **47**, 2228–2233 (2011).

77. Choi, J. Nanofiber Network Composite Membranes for Proton Exchange Membrane Fuel Cells, PHD thesis, Department of Chemical Engineering CASE WESTERN RESERVE UNIVERSITY (2010).

78. Reneker, D. H., Yarinb, A. L., Zussman, E., and Xu, H. Electrospinning of Nanofibers from Polymer Solutions and Melts, *Advances In Applied Mechanics*, **41** (2007).

79. Rutledge, G. C. and Shin, M. Y. A Fundamental Investigation of the Formation and Properties of Electrospun Fibers, *National Textile Center Annual Report*, M98-D01 (2001).

80. Zander, N. E. Hierarchically Structured Electrospun Fibers, *Polymers.*, **5**, 19–44 (2013).

81. Bhardwaj, N. and Kundu, S. C., Electrospinning: A fascinating fiber fabrication technique, *Biotechnology Advances*, **28**, 325–347 (2010).

82. Wang, N., Burugapalli, K., Song, W., Halls, J., Moussy, F., Ray, A., and Zheng, Y. Electrospun fibro-porous polyurethane coatings for implantable glucose Biosensors, *Biomaterials*, **34**, 888–901 (2013).

83. Jung, H. R., Ju, D. H., Lee, W. J., Zhang, X., and Kotek, R. Electrospun hydrophilic fumed silica/polyacrylonitrile nanofiber-based composite electrolyte membranes, *Electrochimica Acta*, **54**, 3630–3637 (2009).

84. Gong, Z., Ji, G., Zheng, M., Chang, X., Dai, W., Pan, L., Shi, Y., and Zheng, Y. Structural Characterization of Mesoporous Silica Nanofibers Synthesized Within Porous Alumina Membranes, *Nanoscale Res Lett*, **4**, 1257–1262 (2009).

85. Wang, Y., Zheng, M., Lu, H., Feng, S., Ji, G., and Cao, J. Template Synthesis of Carbon Nanofibers Containing Linear Mesocage Arrays, *Nanoscale Res Lett*, **5**, 913–916 (2010).

86. Yin, G. B., Analysis of Electrospun Nylon 6 Nanofibrous Membrane as Filters, *Journal of Fiber Bioengineering and Informatics*, **3**(3) 2010.

87. Lee, J. B., Jeong, S. I., Bae, M. S., Yang, D. H., Heo, D. N., Kim, C. H., Alsberg, E., and Kwon, I. K. Highly Porous Electrospun Nanofibers Enhanced by Ultrasonication for Improved Cellular Infiltration, *Tissue Engineering: Part A,* **17**(21–22) (2011).

88. AA, T., Q. J, L. F, Z. B, Preparation and application of amino functionalized mesoporous nanofiber membrane via electrospinning for adsorption of Cr3+ from aqueous solution, *J Environ Sci (China)*, **24**, 610 (2012).

89. Kim, G. H. and Kim, W. D. Highly Porous 3D Nanofiber Scaffold Using an Electrospinning Technique, *Journal of Biomedical Materials Research Part B: Applied Biomaterials* (2006).

90. Zhang, Y. Z., Feng, Y., MHuang, Z., Ramakrishna, S., and Lim, C.T. Fabrication of porous electrospun nanofibers, *Nanotechnology*, **17**, 901–908 (2006).

91. Ramakrishna, S., Fujihara, K., Teo, W. E., Lim, T. C., and Ma, Z. An Introduction to Electrospinning and Nanofibers, World Scientific Publishing Co., Singapore, (2005).

92. Burger, C., Hsiao, B. S., and Chu, B. Nanofibrousmaterials and Their Applications, *Annu. Rev. Mater. Res*, **36**, 333–368 (2006).

93. Berkalp, O. B. Air Permeability & Porosity in Spun-laced Fabrics, *Fibres & Textiles in Eastern Europe*, **14**(3), 57 (2006).

94. Choat, B., Jansen, S., Zwieniecki, M. A., Smets, E., and Holbrook, N. M. Changes in pit membrane porosity due to deflection and stretching: the role of vestured pits, *Journal of Experimental Botany*, **55**(402), 1569–1575 (2004).

95. Alrawi, A. T., and Mohammed, S. J. DetermenationThe Porosity of CdS Thin Film by SeedFilling Algorithm, *International Journal on Soft Computing (IJSC)*, **3**(3) (2012).

96. Krajewska, B. and Olech, A. Pore structure of gel chitosan membranes. I. Solute diffusion measurements, *Polymer Gels and Networks*, **4**, 33–43 (1996).

97. Esselburn, J. D. Porosity and Permeability in Ternary Sediment Mixtures, MSC Thesis, Wright State University, (2009).

98. Ziabari, M., Mottaghitalab, V., and Haghi, A. K. Evaluation of electrospun nanofiber pore structure parameters, *Korean J. Chem. Eng.*, **25**(4), 923–932 (2008).

99. Shrestha, A. Characterization of Porous Membranes via Porometry, *MSC* Thesis, University of Colorado (2012).

100. Borkar, N. Characterization of microporous membrane filters using Scattering techniques, MSC Thesis, B.S. University of Cincinnati (2010).

101. Cuperus, F. P., and Smolders, C. A. Characterization of UF Membranes, *Advances in Colloid and Interface Science*, **34**, 135–173 (1991).

102. Mart'ınez, L., Florido-D'ıaz, F. J., Hernández, A., and Prádanos, P. Characterisation of three hydrophobic porous membranes used in membrane distillation Modelling and evaluation of their water vapour permeabilities, *Journal of Membrane Science*, **203**, 15–27 (2002).

103. Bloxson, J. M. Characterization of the Porosity Distribution Within the Clinton Formation, Ashtabula County, Ohio by Geophysical Core and Well Logging, MSC Thesis, Kent State University (2012).

104. Cao, G. Z., Meijerink J., Brinkman, H. W., and Burggraa, A. J. Permporometry study on the size distribution of active pores in porous ceramic membranes, *Journal of Membrane Science*, **83**, 221–235 (1993).

105. Cuperus, F. P., Bargeman, D., and Smolders, C. A. Permporometry. The determination of the size distribution of active pores in UF membranes, *Journal of Membrane Sczence*, **71**, 57-67 (1992).

106. Fernando, J. A. and Chuung, D. D. L. Pore Structure and Permeability of an Alumina Fiber Filter Membrane for Hot Gas Filtration, *Journal of Porous Materials*, **9**, 211–219 (2002).

107. Shobana, K. H., Kumar M. S., Radha, K. S., and Mohan, D. Preparation and characterization of pvdf/ps blendultrafiltration membranes, *Scholarly Journal of Engineering Research*, **1**(3), 37–44 (2012).

108. Cañas, A., Ariza, M. J., and Benavente, J. Characterization of active and porous sublayers of a composite reverse osmosis membrane by impedance spectroscopy, streaming and membrane potentials, salt diffusion and X-ray photoelectron spectroscopy measurements, *Journal of Membrane Science*, **183**, 135–146 (2001).

109. Frey, M. W. and Li, L. Electrospinning and Porosity Measurements of Nylon-6/Poly (ethylene oxide) Blended Nonwovens, *Journal of Engineered Fibers and Fabrics*, **2**(1) (2007).

110. ˇSirc, J., Hobzov, R., Kostina, N., Munzarov, M., Jukl´ı˘ckov, M., Lhotka, M., Kubinov, S., Zaj´ıcov, A., and Mich´alek, J. Morphological Characterization of Nanofibers: Methods and Application in Practice, *Journal of Nanomaterials*, Volume, Article ID 327369, p. 14 (2012).

111. Venkatarangan, A. B. Geometric and Statistical Analysis of Porous Media, PHD Thesis, University of New York (2000).

112. Zhou, B. Simulation of Polymeric Membrane Formation in 2D and 3D, PHD Thesis, Massachusetts Institute of Technology (2006).

113. Manwart, C., Aaltosalmi, U., Koponen, A., Hilfer1, R., and Timonen, J. Lattice-Boltzmann and finite-difference simulations for the permeability for three-dimensional porous media, *Phys.Rev.E* (2002).

114. Ekneligoda, T. C. and Zimmerman, R. W. Estimating the Elastic Moduli of Sandstones Using Two-Dimensional Pore Space Images, Royal Institute of Technology, Stockholm, Sweden.

115. Bazylak, A., Berejnov, V., Sinton, D., and Djilali, N. Pore network modelling for fuel cell diffusion media, Department Dept. of Mechanical Engineering and Institute for Integrated Energy Systems, University of Victoria, Victoria, British Columbia, Canada.

116. Ghasemi-Mobarakeh, L., Semnani, D., and Morshed, M. A Novel Method for Porosity Measurement of Various Surface Layers of Nanofibers Mat Using Image Analysis for Tissue Engineering Applications, *Journal of Applied Polymer Science*, **106**, 2536–2542 (2007).

117. He, W., Ma, Z., Yong, T., Teo, W. E., and Ramakrishna, S. Fabrication of collagen-coated biodegradable polymer nanofiber mesh and its potential for endothelial cells growth, *Biomaterials*, **26**, 7606–7615 (2005).

118. Wiederkehr, T., Klusemann, B., Gies, D., Müller, H., and Svendsen, B., An image morphing method for 3D reconstruction and FE-analysis of pore networks in thermal spray coatings, *Computational Materials Science*, **47**, 881–889, (2010).

119. Delerue, J. F., Perrie, E., Yu, Z. Y. and Velde, B. New Algorithms in 3D Image Analysis and their Application to the Measurement of a Spatialized Pore Size Distribution in Soils, *Phys. Chem. Earth (A)*, **24**(7), 639–644 (1999).

120. Diógenes, A. N., dos Santos, L. O. E., Fernandes, C. P., Moreira, A. C., and Apolloni, C. R. Porous Media Microstructure Reconstruction Using Pixel-Based And Object-Based Simulated Annealing – Comparison With Other Reconstruction Methods, *Engenharia Térmica (Thermal Engineering)*, **8**(2), 35–41 (2009).

121. Faessel, M., Delisee, C., Bos, F., and Castera, P. 3D Modelling of random cellulosic fibrous networks based on X-ray tomography and image analysis, *Composites Science and Technology* **65**, 1931–1940 (2005).

INDEX

Milton Keynes UK
Ingram Content Group UK Ltd.
UKHW022052141024
449569UK00031B/1603